MOLECULAR BIOLOGY
INTELLIGENCE
UNIT

Nature's Versatile Engine: Insect Flight Muscle Inside and Out

Jim O. Vigoreaux, Ph.D.
Department of Biology
and
Department of Molecular Physiology and Biophysics
Cell and Molecular Biology Program
University of Vermont
Burlington, Vermont, U.S.A.

LANDES BIOSCIENCE / EUREKAH.COM
GEORGETOWN, TEXAS
U.S.A.

SPRINGER SCIENCE+BUSINESS MEDIA
NEW YORK, NEW YORK
U.S.A.

Nature's Versatile Engine: Insect Flight Muscle Inside and Out

Molecular Biology Intelligence Unit

Landes Bioscience / Eurekah.com
Springer Science+Business Media, Inc.

ISBN: 0-387-25798-5 Printed on acid-free paper.

Springer Science+Business Media, Inc., 233 Spring Street, New York, New York 10013, U.S.A.
http://www.springeronline.com

Please address all inquiries to the Publishers:
Landes Bioscience / Eurekah.com, 810 South Church Street, Georgetown, Texas 78626, U.S.A.
Phone: 512/ 863 7762; FAX: 512/ 863 0081
http://www.eurekah.com
http://www.landesbioscience.com

9 8 7 6 5 4 3 2 1

Library of Congress Cataloging-in-Publication Data

Vigoreaux, Jim O.
Nature's versatile engine : insect flight muscle inside and out / Jim O. Vigoreaux.
p. ; cm. -- (Molecular biology intelligence unit)
ISBN 0-387-25798-5
1. Insects--Flight--Physiological aspects. 2. Wings--Muscles--Physiology. I. Title. II. Series: Medical intelligence unit (Unnumbered : 2003)
[DNLM: 1. Insects--physiology. 2. Muscle Contraction--physiology. 3. Flight, Animal--physiology. 4. Muscle Proteins. 5. Muscle, Skeletal--anatomy & histology. QL 496.7 V691n 2006]
QL496.7V54 2006
573.7'98157--dc22

2005024666

To Johana and Francis

May your quest into Nature's wonders be exciting, rewarding, and ever-lasting.

CONTENTS

EDITOR

Jim O. Vigoreaux
Department of Biology
and Department of Molecular Physiology and Biophysics
Cell and Molecular Biology Program
University of Vermont
Burlington, Vermont, U.S.A.
Email: jvigorea@uvm.edu
Chapters 7, 15, 21

CONTRIBUTORS

Gereon Ader
Institut fur Zoophysiologie
Westfalische Wilhelms-Universität Munster
Munster, Germany
Chapter 8

Juan Jose Arredondo
Departamento de Bioquimica
Facultad de Medicina
Universidad Autonoma de Madrid
Madrid, Spain
Chapter 6

Agnes Ayme-Southgate
Department of Biology
College of Charleston
Charleston, South Carolina, U.S.A.
Email: southgatea@cofc.edu
Chapter 13

Byron Barton
Department of Biology
University of Vermont
Burlington, Vermont, U.S.A.
Email: bbarton@zoo.uvm.edu
Chapter 7

Gernot Beinbrech
Institut fur Zoophysiologie
Westfalische Wilhelms-Universität Munster
Munster, Germany
Email: beinbre@uni-muenster.de
Chapter 8

Sanford I. Bernstein
Department of Biology
San Diego State University
San Diego, California, U.S.A.
Email: sbernst@sunstroke.sdsu.edu
Chapter 5

Belinda Bullard
European Molecular Biology Laboratory
Heidelberg, Germany
Email: belinda.bullard@embl-heidelberg.de
Chapters 11, 14

Margarita Cervera
Departamento de Bioquimica
Facultad de Medicina
Universidad Autonoma de Madrid
Madrid, Spain
Email: margarita.cervera@uam.es
Chapter 6

Richard M. Cripps
Department of Biology
University of New Mexico
Albuquerque, New Mexico, U.S.A.
Email: rcripps@unm.edu
Chapter 1

Raquel Marco Ferreres
Departamento de Bioquimica
Facultad de Medicina
Universidad Autonoma de Madrid
Madrid, Spain
Chapter 6

Alberto Ferrús
Instituto Cajal, CSIC
Madrid, Spain
Email: aferrus@cajal.csic.es
Chapter 10

Michael A. Geeves
School of Biosciences
University of Kent
Canterbury, U.K.
Chapter 19

Bernd Heinrich
Department of Biology
University of Vermont
Burlington, Vermont, U.S.A.
Email: bheinric@zoo.uvm.edu
Foreword

Joshua A. Henkin
Department of Biology
University of Vermont
Burlington, Vermont U.S.A.
Email: jhenkin@zoo.uvm.edu
Chapter 21

Thomas C. Irving
Department of Biological, Chemical and Physical Sciences
Illinois Institute of Technology
Chicago, Illinois, U.S.A.
Email: irving@biocat1.iit.edu
Chapter 16

Robert K. Josephson
School of Biological Sciences
University of California at Irvine
Irvine, California, U.S.A.
Email: rkjoseph@uci.edu
Chapter 3

Mark C. Leake
Clarendon Laboratory
University of Oxford
Oxford, U.K.
Chapter 14

Fritz-Olaf Lehmann
BioFuture Research Group
University of Ulm
Department of Neurobiology
Ulm, Germany
Email: fritz.lehmann@biologie.uni-ulm.de
Chapter 18

Kevin R. Leonard
Structural Biology Programme
European Molecular Biology Laboratory
Heidelberg, Germany
Email: leonard@embl.de
Chapters 11, 14

James H. Marden
Department of Biology
Pennsylvania State University
University Park, Pennsylvania, U.S.A.
Email: jhm10@psu.edu
Chapter 17

David Maughan
Department of Molecular Physiology and Biophysics
University of Vermont College of Medicine
Burlington, Vermont, U.S.A.
Email: maughan@physiology.med.uvm.edu
Chapter 20

Becky M. Miller
Department of Biology and Molecular Biology Institute
San Diego State University
San Diego, California, U.S.A.
Chapter 5

Jeffrey R. Moore
Department of Molecular Physiology and Biophysics
University of Vermont College of Medicine
Burlington, Vermont, U.S.A.
Email: jmoore@physiology.med.uvm.edu
Chapter 4

Mary C. Reedy
Department of Cell Biology
Duke University Medical Center
Durham, North Carolina, U.S.A.
Email: mary.reedy@cellbio.duke.edu
Chapter 2

Judith D. Saide
Department of Physiology
Boston University School of Medicine
Boston, Massachusetts, U.S.A.
Email: jsaide@bu.edu
Chapter 12

Richard Southgate
Department of Biology
College of Charleston
Charleston, South Carolina, U.S.A.
Chapter 13

John C. Sparrow
Department of Biology (Area 10)
University of York
York, U.K.
Email: jcs1@york.ac.uk
Chapters 9, 19

Douglas Swank
Department of Molecular Physiology and Biophysics
University of Vermont
Burlington, Vermont, U.S.A.
Chapter 20

Vivek Vishnudas
Department of Biology
University of Vermont
Burlington, Vermont, U.S.A.
Email: vvishnud@zoo.uvm.edu
Chapter 15

FOREWORD

A Naturalist's View of Insect Flight Muscle

Bernd Heinrich

By almost any measure, insects as a group are an astounding evolutionary achievement. Through their diversity and their adaptation to a great range of life-styles, forms, and physical and biological environments, they offer unparalleled insights into the constraints, selective pressures and the range of possible solutions to almost any conceivable contingency that animals have faced on earth.

Insects evolved to fly at least 50 million years before any other animal. As in other animals, that capacity resulted from coopting and development of structures previously used for other functions. Pterosaurs, birds and mammals elaborated their forelimbs that were used originally for terrestrial locomotion, that had in turn evolved from fins. The insects' precursors for wings are less clear and several mutually-compatible possibilities have been proposed, but in my opinion the one hypothesis most consistent with the available fossil evidence, and natural history and structures of current forms, is that they evolved from aquatic gills of a mayfly-like ancestor.[1] Gills of aquatic larvae still serve a dual function of moving the medium for respiration, and for locomotion—propelling the body through the medium.

As in all flight, it is not the original precursors that are as much in doubt as the intervening steps and selective pressures. However, one thing is not in doubt, namely that muscle is a precursor for flight, and flight requires unique specialization of muscle for high power output.

We associate animal movement with life, and essentially all animal movement is accomplished with that most remarkable of all tissues, muscle. I therefore give tribute to muscle, as such; because it is hard to imagine a tissue of greater importance to almost all aspects of an animal's life. Indeed, the invention of the actin-myosin complex probably helped set the evolution of animal life on a path distinct from that of plants.

DNA has arguably been the most important molecule that has elevated our understanding of Life from a philosophical to a scientific concept, so that we can now see a virus, and maybe even a bacterium, as a very complex chemical reaction. Plants, with their innovation of chlorophyll for capturing the sun's energy that drives most of life, are still almost inanimate in the popular consciousness. Plants are ultimately an incredibly complicated biochemical mechanism of capturing, converting, and using the energy of the sun. But in the general consciousness, the almost philosophical concept of liveness is reserved for the capacity of movement that is independent of the immediate constraints of gravity and other physical forces. Of course since the introduction of Huxley's sliding filament model of muscle contraction we now have a mechanistic explanation of one of the most fundamental properties of Life, in terms of a molecular structure. The evolutionary invention of the elegant mechanism of muscle contraction is, I submit, as significant, intellectually exciting, and fundamental as those of DNA and chloroplasts, that opened the possibility for Life; and for plant life specifically.

There are other mechanisms of movement, but they had limited applicability and never "took off." Movement by growth, changes in turgor pressure, and flagellar beating, apparently didn't have much potential for either speed or for innovation of larger life forms on land. But when a fiber evolved that consisted of two filaments, one set sliding over the other so that they could be part of an enzymatic catalysis powered by the already-existing ATP mechanism, then the age of animals began. Movement begat movement in arms races to reach evermore or different plant resources, and these arms races generated still others as herbivores and then also carnivores became prey.

As muscles became better designed for faster contraction, to promote faster locomotion, especially that required to conquer the air, they needed evermore support systems for their greater energy demands. Muscles themselves became adapted for use in different organ systems to facilitate the movement of nutrients through the gut, to facilitate nutrient distribution through the body to other muscles and other organs, and to help in gas exchange. Finally, as one set of muscles increased the need for evolving other muscles to supply them in a positive feed-back loop, the huge increases of metabolic rate despite ever-greater body size became possible.

Our understanding of vertebrate comparative physiology of muscle is largely restricted to smooth (involuntary), cardiac and skeletal muscle, and in skeletal muscle to those specialized to produce rapid bursts of power versus those that compromise power for more endurance. Synchronous insect flight muscle is anatomically most analogous to vertebrate skeletal (striated) muscle, whereas asynchronous (fibrillar) flight muscle[2] is much like vertebrate cardiac muscle. Mutations in both flies and humans that affect the stretch activation response impair flight in flies and cause cardiomyopathies in humans. Insights gleaned from insect flight muscle give us a unique window into that type of muscle because the demands to which that muscle has been put, and hence the comparative insights as well as the experimental possibilities, are a goldmine of opportunity that has been opened over the last six-plus decades. Insect flight muscles allow us to see the limits of what muscles can accomplish and to study the underlying constraints and parameters for specific muscle performances. Numerous variables exist that can affect which wing muscles are used and how they are used.

In most insects both sets of wings either operate together as a unit (Lepidoptera, Cicadidae, Hymenoptera), one of the two sets of wings is not or only marginally used for power generation (Coleoptera), or it is missing, being coopted as halteres for gyroscopic sensors in flight control (Diptera).

One of the major variables for specific flight muscle performance includes mechanical hook-ups. The most obvious of these are probably the direct versus indirect attachment to the wings that differentiate between whether a muscle acts on the wing directly or whether the wing is indirectly depressed or elevated due to changes in shape of the thoracic box to harness resonant mechanical properties of the flight apparatus. In all insects, the muscles used for controlling the trajectory of the wings are direct muscles whereas in the vast majority of insects the power of the up-stroke and down-stroke of the wings results from two antagonistic sets of muscles that, unlike in all vertebrates, pull not on the wings themselves but on the thoracic box. The Odonates are an exception. In them the power-producing wing muscles pull directly on the wings. Additionally, the two sets of wings can operate independently of one another for flight control. The apparent advantage of the almost universal adoption of the more common indirect muscle-to-wing hook-ups is not clear, but it presumably relates to/with mechanical energetic advantage such as the "click mechanism"[3] that allows a large wing-beat amplitude with as little as a 1 percent muscle shortening, hence permitting greater wing-beat frequency when associated with stretch-activated muscle fibers.

The wings are attached to the thoracic box presumably to most efficiently harness the power output of the muscles, but a compromise may be required to permit their control for flight maneuvers. However, the size of the wings, in turn, affects wing-beat frequency. In general, large wings have lower wing-beat frequency and demand a lower overall rate of energy expenditure for flight.

Ultimately the variables that affect flight performance depend on the muscles themselves. Muscle dimensions that are selected for could potentially be their mass and their length. Predictably we can assume that muscle mass relative to body mass has been minimized by evolution, and a high muscle mass/body mass ratio would be predicted for high power output such as that required for rapid acceleration for predator avoidance,[4] and for aerial combat of males in sexual selection.[5] Different muscle length would probably not have been of much relevance once the indirect wing attachment had evolved, although it could possibly still apply in dragonflies, if there is evolutionary pressure to alter wing-beat amplitude.

A great number of variables that affect flight muscle performance are found at both the cellular and molecular levels. The muscle cell can be activated to contract by stretching or by a direct one-to-one neural innervation to initiate the contraction cycle. These two types of muscles, the asynchronous and synchronous, each have their own properties of time course of contraction, passive stiffness, force production, and these properties are related to development and extent of the sarcoplasmic reticulum and myofibrillar diameter, among others. The asynchronous mode of muscle action is a unique insect adaptation (in Diptera, Hymenoptera, Heteroptera and Coleoptera) for attaining high-frequency wingbeats with concomitant high power output[2] that are also related to molecular differences of protein isoforms: different myosins result in different shortening velocities, others affect passive stiffness.

In large-bodied flying insects the same flight muscles are commonly used for at least two and sometimes several functions in the same species or individual. Each function has its own requirements for neurological control, power, and speed of contraction. In the katydid *Neoconocephalus robustus* for example, the synchronous thoracic muscles contract at near 20-30 Hz during flight in both males and females, but the males contract those same muscles at close to 200 Hz during stridulation.[6] Sexual selection in these insects has resulted in unique muscle morphology to permit such different function. Different function of the same muscles is routine in insects. Most flying insects above a modest mass of about 50-100 mg use their flight muscles alternately for preflight warm-up, flight, and still other behaviors. Foraging bumblebees for example regularly use their flight muscles by contracting them into a near-tetanus during warm-up when there are no wing-vibrations,[7] but they also vibrate their whole bodies through flight muscle oscillations to shake pollen loose from flower anthers, in a behavior called "buzz pollination". Similar buzz-vibrations serve as warning signals to predators, and honeybees communicate within the hive by a variety of sounds that are all produced with the thoracic muscles vibrating the thorax and wings.

The flight muscle contractions of all insects are highly temperature dependent, even though the body temperature of most insects is relatively labile. The flight muscles are adapted to operate at the temperatures that they are most likely to experience during activity.[8] Given vast differences in body size and consequent cooling rates and diverse ecological conditions, the suitable temperature range may be large. For example, the small winter-flying geometrid moth *Operophtera bruceata* can fly with a muscle temperature near 0°C[9] whereas some thick-bodied tropical sphinx moths can't begin to fly until their flight muscles are heated up to near 35°C[10] and some moths, depending on wing-loading, may operate their flight motor at near 44°C.[11] The biochemical and/or morphological mechanisms that may account for the ability to work at high rates at very different temperatures are not clear, but presumably there are both biochemical and structural compromises involving sarcoplasmic reticula, mitochondrial numbers and membrane permeability changes. Curiously, in several species of moths that have different flight muscle temperature optima, the mitochondrial enzymes do not have significantly different optima,[9] suggesting that their amounts may vary instead. However, wing morphology, which proximally affects muscle temperature[12] and energy expenditure in flight, may also be altered evolutionarily to compensate for inability to generate sufficient power.

A related problem of flight muscle in large-bodied insects is that a high muscle temperature is not only a prerequisite for activity, it is also a consequence of the activity. Thus, in order to fly large insects need to first exercise their muscles to be able to then operate them at levels suitable to support flight. Some cuculiinid winter moths that require a muscle temperature over 30°C to fly are already able to exercise them at the phenomenally low temperature of 0°C for shivering, whereas honeybees that normally fly at similar muscle temperature but can warm up in the warmth of the hive are unable to shiver until they experience much higher muscle temperatures.[13] An insect, such as a bumblebee, that can also shiver already at near 0°C has at least half of its muscle volume committed to mitochondria and tracheal tubes for gas exchange.[14]

Presumably, an overcommitment to heat production compromises power. All insect flight muscles are used first and foremost for power production for flight. However, some vertebrates have maximized heat production at the expense of power by evolving muscle cells without contractile fibers to serve as "brown fat" for heat production alone. Lastly, in both vertebrates and some insects, muscle can also be used as an energy source in emergencies. We catabolize muscle during starvation, and during extreme exercise such as migratory birds experience, muscle mass declines as it is used as flight fuel. The regular use of muscle degeneration and regeneration in energy economy is, again, perhaps most extreme in some insects (Hemiptera such as *Gerris* and *Oncopeltus* and Coleoptera such as *Leptinotarsa* and *Ips*)[15] that degenerate their flight muscles when flight activity stops and reproductive activity begins, and then regenerate them when flight again becomes necessary.

Numerous puzzles of muscle design remain to be elucidated, and closer examination of the natural history of insects will uncover them. The flight muscles of insects provide an extraordinary opportunity to elucidate the design constraints and possibilities, molecular mechanisms, and genetics, of one of the most remarkable, important, and unique tissue in the animal world.

References

1. Kukalova-Peck J. Origin and evolution of insect wings and their relation to metamorphosis, as documented by the fossil record. J Morphol 1978; 156:53-126.
2. Josephson RK, Malamud JG, Stokes DR. Asynchronous muscle: A primer. J Exp Biol 2000; 203(Pt 18):2713-2722.
3. Boettiger EG, Furshpan E. The mechanism of flight movements in Diptera. Biol Bull (Woods Hole) 1952; 102:200-211.

4. Chai P, Srygley RB. Predation and the flight morphology and the temperature of neotropical butterflies. Am Natur 1989; 135:748-765.
5. Marden JH. Bodybuilding dragonflies: Costs and benefits of maximizing flight muscle. Physiol Zool 1989; 62:505-521.
6. Josephson RK, Halverson RC. High frequency muscles used in sound production by a katydid. I. Organization of a motor system. Biol Bull (Woods Hole) 1971; 141:411-433.
7. Esch H, Goller F, Heinrich B. How do bees shiver? Naturwissenschaften 1991; 78:325-328.
8. Heinrich B. Why have some animals evolved to regulate a high body temperature? Am Natur 1977; 111:623-640.
9. Heinrich B, Mommsen TP. Flight of winter moths near 0°C. Science 1985; 228:177-179.
10. Heinrich B. The Hot-blooded Insects: Strategies and Mechanisms of Thermoregulation. Cambridge: Harvard University Press, 1993.
11. Heinrich B, Casey TM. Metabolic rate and endothermy in sphinx moths. J Comp Physiol 1973; 82:195-206.
12. Heinrich B, Bartholomew GA. A field study of flight temperatures in moths in relation to body weight and wing loading. J Exp Biol 1973; 58:123-135.
13. Esch H. The effect of temperature on flight muscle potentials in honeybees and cuculiinid winter moths. J Exp Biol 1988; 135:109-117.
14. Heinrich B. Bumblebee Economics. Cambridge: Harvard University Press, 1979.
15. Bhakthan NMG, Borden JH, Nair KK. Fine structure of degenerating and regenerating flight muscles in a bark beetle, Ips confusus. J Cell Science 1970; 6:807-819.

PREFACE

"God in His wisdom made the fly
And then forgot to tell us why."
—Ogden Nash 'The Fly' (1942)

As a beginning graduate student I was once scorned by a biochemistry professor for asking why a certain mechanism worked the way he had just described. "Why? Only God knows *why*, as a scientist I am only concerned about *how* it works" was part of his stern response that still resonates loud and clear. As documented throughout the pages of this book, many a scientist has asked the question 'how do insects fly'? Ever since Aristotle scientists and philosophers alike have been fascinated with flying insects: the speed of a mosquito, the agility of a housefly, the grace of a damselfly, and the fluttering of a butterfly are just a handful of examples that have sparked our curiosity. The immense diversity of flying insects (and there are more species of flying insects than any other type of animal known) is a testament that flight is a desirable and advantageous trait. The subject of this book, insect flight muscle, has made this diversity possible.

It will become evident to the reader of this book that 'it takes a village' to address the question of how insects fly, or even the less mechanistically complex question of how the flight muscle works. From field observations, to wet bench experimentation with laboratory-made trinkets, to the application of biotechnology and state-of-the-art imaging and bioengineering techniques, up to the 'big science' of high energy physics, scientists have stopped at nothing in their pursuit of answers to this most fascinating of questions. It is not surprising then that scientists working on insect flight muscle speak different dialects, for they have been trained in disciplines that otherwise overlap little or not at all. Insect flight muscle thus provides a venue for the evolutionary ecologist, the structural biologist, the geneticist, the biophysicist, and the physiologist, among others, to come together and tear down the 'disciplinary' barriers that often impede the progress of science. In this era of 'omics' and 'systems biology' insect flight provides a valuable paradigm for integrative studies. Systems biology is largely based on theoretical, computational, and modeling methods in combination with large scale, high throughput approaches such as DNA microarrays that display global gene expression profiles. The increasing emphasis on proteomic approaches, especially those that examine protein-protein interactions on a large scale, have been a major push for the conception of systems biology as a discipline. Systems biology is best developed in model systems in which a strong foundation of interdisciplinary and complementary approaches is already in place.

We are at the crucial point in the study of insect flight where we can begin to connect the dots from the behavior of molecules to the behavior of the animal. The goal of this book is to provide a sampling of the ongoing research on insect flight muscle, across a wide spectrum of biological disciplines, in a language that is appealing to the specialist and the novice alike.

This book is organized in three sections. Section I provides an overview of the historical contributions of experimental insect flight systems. *Drosophila*, via its genetic amenability (Chapter 1), and *Lethocerus*, via its highly ordered flight muscle (Chapter 2), are the two key players in the story that unfolds in later chapters. Chapter 3 expands this repertoire to include the flight muscles of beetles and locust and describes how a comparative analysis of these flight muscles has increased our understanding of the evolutionary history of insects. Section II, entitled 'Components of the Myofibril', examines one by one the major filament systems and structures of the contractile organelle and the properties of the proteins that constitute them. Chapters 5 through 8 discuss thick filaments, Chapters 9, 10 and 11 discuss thin filaments, Chapter 12 discusses the Z band, and Chapters 13 and 14 discuss the connecting filaments. Section III, entitled 'Towards a Systems Level Analysis of Muscle', is divided in two subsections: (i) 'From the Outside In' (Chapters 15, 16, 17 and 18) examines how muscle function is studied in the intact animal and how results are interpreted in an ecological context (ii) 'From the Inside Out' (Chapters 19, 20 and 21) describe molecular type assays whose results can be extrapolated and interpreted at a systems wide level.

J.O.V.
Burlington, Vermont
2005

Section I
Experimental Insect Flight Systems: Historical Contributions

Chapter 1

The Contributions of Genetics to the Study of Insect Flight Muscle Function

Richard M. Cripps

Abstract

The utility of *Drosophila* as a model genetic organism has had a profound impact upon our understanding of muscle assembly and function. This has arisen from the large number of mutant alleles that have been isolated and characterized using a variety of screens, and also reflects an highly efficient method for generating transgenic animals. Combining these two methodologies has permitted genetic rescue experiments using both wild-type and engineered alleles, to probe precisely the functions of proteins or their domains in myofibril assembly and contractility. Genetical approaches using suppressor screens are now being used to more precisely determine protein function in the myofibril. In this review I shall summarize how these approaches have been informative in understanding muscle formation and function, and discuss how *Drosophila* genetics will serve the muscle field in the future.

Introduction

The use of genetics to dissect complex biological processes has a noted history. While mutations affecting muscle function have been described in a variety of organisms from invertebrates to man, systematic screens for muscle mutants have been attempted in only a few, predominantly the fruit fly *Drosophila melanogaster* and the nematode *Caenorhabditis elegans*.

Drosophila muscle mutants have been isolated based mostly upon a simple flight-testing assay in which flightless mutants are collected. Consequently, many of these mutations affect the function of the indirect flight muscles (IFMs). Since a number of myofibrillar proteins have IFM-specific isoforms (see below), mutations affecting the IFMs can be isolated which in the laboratory do not cause major effects upon the viability or fertility of mutant stocks. The IFMs function to power flight and have individual fibers approaching 1mm in length. These muscles are therefore amenable to biochemical, physiological, and ultrastructural analyses which can proceed alongside the genetic studies. Such multi-pronged approaches have been successful in unraveling IFM biology.

By contrast, *C. elegans* mutants isolated based upon poor locomotion have been found to affect predominantly the innervation and function of the body wall muscles.[1] While these screens have elegantly permitted insight into myofibril assembly (see for examples refs. 2, 3), the small size of the muscles and their divergent ultrastructure when compared to IFM or vertebrate skeletal muscle, has resulted in relatively less effort in this species towards understanding muscle formation and function.

This chapter will concentrate upon the genetic screens used to isolate mutations affecting muscle in *Drosophila*, and how the mutants isolated in these screens continue to provide important information into how muscles function.

Nature's Versatile Engine: Insect Flight Muscle Inside and Out, edited by Jim Vigoreaux.

Isolation of Mutations Affecting Flight Muscle Function

Drosophila genetics in the 1970s was characterized by a flurry of screens aimed at identifying the genetic components of behavior, probably inspired by the efforts of Benzer.[4] The first screen designed to isolate mutations affecting flight behavior in *Drosophila* was described by Benzer himself[4,5] who screened the offspring of mutagenised flies using a clear plastic cylinder coated on the inside surface with oil. Flies introduced into the top of the cylinder either flew laterally and were immobilized in the oil, or flightless individuals would fall through the cylinder to a collection vessel at the bottom. Similar apparatuses were later used by several researchers.[6] Sheppard used a different apparatus where mutagenized flies were encouraged to fly from the lower chamber of a plastic box into an higher chamber, and those that were unable to fly were retained in the lower chamber.[7,8] Flightless individuals from either strategy were subsequently crossed to maintain the flightless mutation and subjected to genetic analysis.

Most of the early mutagenesis schemes aimed to isolate flightless mutations linked to the X chromosome, since both dominant and recessive alleles could be isolated in F1 males, rather than the more time-consuming F2 screens required for the autosomes. While some of the X-linked mutations had dominant effects upon flight, many of the mutations isolated were recessive. Even now, X-linked flightless mutants are still over-represented in the literature since no systematic screens for autosomal recessive flightless mutants have been carried out.

A recent screen for second chromosome recessive mutations affecting flight muscle development has been described.[9] These authors analyzed ethyl methane sulfonate-induced recessive viable mutants for defects in flight muscle organization, via direct observation under polarized light. Under this illumination, the birefringent muscles can be easily observed and any defects recorded. A strong advantage to this approach was that it did not require a flightless phenotype to be present, and indeed was highly successful in identifying several new loci required for normal muscle development.

As flight behavior results from the combined effects of a number of different tissues in the fly, it is not surprising that some of the flightless mutants isolated in these early screens arose due to defects outside of the musculature. This was elegantly determined by early gynandromorph fate mapping techniques (see refs. 4, 10, 11); for example the mutation *vertical wing* was thought to affect the development or function of the tracheae rather than of the muscles.[10] Since such fate mapping required the affected loci to be X-linked and involved challenging analysis, this difficult approach was not widely used. Instead, ultrastructural and biochemical assays became popular to attempt to understand the molecular bases of flightless behaviors, by looking for defects in the musculature either in phase contrast images of isolated IFM myofibrils, by electron microscopy, or by two-dimensional SDS-PAGE analysis of dissected IFMs.[6,10,12]

At that time, SDS-PAGE analysis had become an highly powerful analytical tool due to both the resolution of the technique and the efforts of Mogami et al.[13] The latter authors developed the technology to reproducibly identify *Drosophila* thoracic proteins based upon their pI and mass in 2-D gel electrophoresis, and assigned numbers to each of 186 detectable proteins. Many of these proteins were designated as IFM-specific or mostly IFM-specific based upon 2-D gel analyses of dissected tissues. Their technique was not amenable to the characterization of particularly large polypeptides nor those with extreme isoelectric points, since they would not separate efficiently in these gels. Nevertheless, a number of subsequent authors were successful in identifying the proteins represented by each spot number as major myofibrillar structural proteins, and on this basis 2-D gel analysis has been considered an highly successful tool (reviewed in refs. 14, 15). The challenge remained to isolate and identify mutations in the genes encoding each polypeptide.

A strategy to more directly isolate mutations affecting myofibrillar protein genes rather than all mutations affecting flight behavior was proposed and carried out by Mogami and Hotta.[12] These authors reasoned that since the power requirement for flight is extremely high and since protein levels are often dependent upon gene dosage, mutations in major myofibrillar protein genes might show dominant flightless phenotypes resulting from a reduction in total power output. This prompted the isolation of autosomal dominant flightless mutants which could be achieved using F1 screens. While Mogami and Hotta[12] concentrated upon dominant flightless mutations that also showed disruptions in the accumulation of the major myofibrillar proteins as visualized by SDS-PAGE, a similar screen for dominant flightless mutants carried out by Cripps et al[16] did not require obvious protein accumulation defects to be present.

These screens identified five autosomal loci which when mutated showed haploinsufficiency for flight, in addition to the X-linked haploinsufficient loci previously discovered. Many of the autosomal loci have proven to correspond to muscle protein genes: the single *Myosin heavy-chain* gene (*Mhc*);[17] the *Tropomyosin 2* gene (*Tm2,* also referred to as *TmI*);[18] and the largely IFM-specific actin gene *Act88F.*[19] Two additional autosomal loci, *Ifm(2)11*[12] and *l(3)Laker*[16] are thought to encode muscle proteins, however the affected genes have yet to be identified.

The screens of Mogami and Hotta[12] and Cripps et al[16] clearly did not isolate mutations affecting all myofibrillar protein genes expressed in the IFMs. It has since been determined that recessive lethal mutations isolated in the autosomal locus encoding MLC2 show dominant effects upon flight ability.[20] Lethal mutations isolated in the gene encoding Titin/Kettin have also been reported to cause dominant flightlessness,[21] although not all alleles appear to show this phenotype.[22] It is assumed that the dominant flightless phenotypes of mutants in the latter two genes are sufficiently mild to have prevented their detection in screens for flightless flies.

There are also a number of muscle protein gene loci which do not mutate to cause dominant flightlessness, including *Projectin, Paramyosin, Tropomyosin 1* and *Flightin* (see refs. 23-29). For several of these, unique screens have been carried out to isolate mutants.[20,26,29] Taken together, these studies underline the diversity of screens available for gene discovery in *Drosophila*, as well as emphasize the potential need for distinct screens to isolate mutations in specific genes.

There is currently no satisfactory explanation of why some myofibrillar protein loci show stronger haploinsufficiency for flight than others. Perhaps those showing more mild effects as heterozygotes are expressed at relatively lower levels in the flight muscles, thus their reduction is less critical to muscle function; alternatively haploinsufficiency for some loci may have greater effects upon myofibril ultrastructure than other loci, and this is likely to be a critical parameter for muscle function as discussed below.

Isolation of Myofibrillar Protein Genes

The onset of gene cloning technology around the time of the early screens also allowed the isolation of genes encoding muscle proteins irrespective of the existence of mutant alleles for that gene. While much of the power of *Drosophila* as a model system arises from the ability to isolate mutants and subsequently to use those mutations to identify the gene(s) affected, none of the major myofibrillar protein genes were initially cloned based upon their mutant phenotypes. Rather, *Drosophila* muscle protein genes were predominantly identified by either low-stringency library screens using DNA probes from other species, by screening *Drosophila* cDNA expression libraries with antibodies generated to muscle proteins in other species, or by identification of putative muscle genes using hybrid selected translation. This has introduced a clear bias into the muscle protein genes isolated to date, and the full power of *Drosophila* genetics has yet to be brought to bear upon the identification of novel myofibrillar protein genes.

On the other hand, there was an immediate and striking correlation observed between the chromosomal locations of the isolated genes and the clusters of mutants that were isolated by Hotta and Benzer,[4] Deak et al,[10,30] Homyk and Sheppard,[8] Mogami and Hotta[12] and Cripps et al.[16] Many of the newly-isolated genes therefore could be correlated with a mutant phenotype, allowing a rapid and detailed understanding of the role of the encoded protein in muscle formation and function.

Table 1 lists the known myofibrillar protein genes in *Drosophila*, summarizes how the genes were cloned and how mutants in them were isolated, and describes briefly our current understanding of the genetic analysis of these genes coupled with some details of the proteins

Table 1. Genes encoding* Drosophila *myofibrillar proteins; their isolation and characterization; features of the encoded proteins

Gene Name	Cytological Location	Gene Isolation	Mutant Isolation	Heterozygote Phenotype	Homozygote Phenotype	Protein Size (kD)	Spot Number#	IFM-Specific Isoform ?	Refs.
Titin/Kettin/ sallimus	62C	Cross-species antibody; genetic mapping	Modifiers of gene interactions; abnormal neuro-muscular junctions	Variable	Embryonic lethal	1 000, 700,540	Not known	Not known	21,22, 59-61
bent/Projectin	102CD	Antibody; cross-species PCR	Existing spontaneous	Normal flight	Embryonic lethal	1 001	Not known	Yes, by alternative splicing	23,24,62
Myosin heavy chain	36B	Cross-species DNA probe	Dominant flightless mutants	Flightless	Embryonic lethal@	220	2	Yes, by alternative splicing	12,17, 35,63-67
Z(210)/Zetalin	–	Not known	None known	Not known	Not known	210	Not known	Expressed only in IFM and Large cells of TDT	68
Alpha-actinin	2C	Cross-species antibody	X-linked flightless mutants; screen for lethal complement-ation groups	Mostly flighted; one flightless	Early larval lethal	97	26,27	No, but an adult isoform exists by alternative splicing	8,68-73
Paramyosin	66D	Cross-species antibody	P-element excision screen for lethal complementation groups	Normal flight	Embryonic lethal	90	19,20,21	No, but isoform diversity generated by alternate promoters	25,26, 74,75
Tropomyosin1 /TnH	88E	Chromosomal walking; mRNA enriched in muscle	Screen for embryonic lethals	Normal flight	Embryonic lethal	70-80	33,34, 128	Two IFM-specific heavy isoforms (33 &34) generated by alternative splicing	27,28, 76-78
Troponin T	12A	Cross-species DNA probe	X-linked flightless mutants	Flightless	Viable, flightless*	47	Not known	No, but an adult isoform exists by alternative splicin	30,79-81

continued on next page

Table 1. Continued

Gene Name	Cytological Location	Gene Isolation	Mutant Isolation	Heterozygote Phenotype	Homozygote Phenotype	Protein Size (kD)	Spot Number#	IFM-Specific Isoform ?	Refs.
sanpodo/ tropomodulin	99F7-8	Transposon tagging	Neuronal defects	Not known	Embryonic lethal	45	Not known	Yes, by alternate promoter usage; a more ubiquitous isoform also present in IFMs	82-84
Actin88F	88F	Cross-species DNA probe	Dominant flightless mutants	Flightless	Flightless	43	100,101 74 (arthrin)	Yes; *Act88F* expression is predominantly IFM-specific	12,85-90
Tropomyosin2	88E	Chromosomal walking; mRNA enriched in muscle	Dominant flightless mutants	Flightless	Embryonic lethal	35	127	Yes, by alternative splicing	12,18,27
Troponin I	16F	Chromosomal walking; mRNA enriched in muscle	X-linked flightless mutants	Flightless	Viable, flightless*	35	Not known	Yes, by alternative splicing	4,10,33, 49,91,92
Glutathione-S-transferase 2	53F	Mass spectrometry of purified protein	Screen for P-element induced lethals	Not known	Lethal	32	137	No, but enriched in IFMs	93,94
Myosin light chain-2	99DE	mRNA enriched in muscle	Screen for recessive lethals	Flightless	Embryonic lethal	30	138,148, 149	Yes, by alternative post-translational modification	20,95-97
Flightin	76DE	Antibody to fly protein	Screen for protein nulls	Slight flight defect	Flightless	20	158,159, 160	Yes, *Flightin* expression is IFM-specific	29,98,99
Myosin light chain-alkali	98B	mRNA enriched in muscle	No mutants	Not known	Not known	18	185	Yes, by alternative splicing	100-102
Troponin C 41F	41F	Genome analysis	No mutants	Not known	Not known	17	Not known	Predominant IFM isoform; also other tissues at lower levels	103,104
Troponin C 41C	41C	Degenerate PCR	No mutants	Not known	Not known	17	Not known	Minor IFM isoform; also other tissues	103-105

Notes: # from reference 13; @ some alleles that affect IFM-specific exons are homozygous viable and flightless; *although mutant alleles of both TnT and TnI have been characterized, they are each dominant flightless and viable as heterozygotes. There are no lethal alleles of these genes. However, since the mutants were isolated in F1 screens of hemizygous adult males, the screens selected for viable alleles in these genes. Given the broad expression patterns of these genes it is likely that true null alleles would be recessive lethal.

encoded. It is interesting to note that the vast majority of myofibrillar proteins have IFM-specific isoforms which are thought to contribute to the unique structural and mechanical properties of this particular muscle. Furthermore, these unique isoforms are generated in a number of distinct manners: as members of multigene families; by alternative transcript splicing; by alternative promoter usage; and by differential post-translational modification. The extensive isoform diversity explains why many of the dominant flightless mutations isolated were homozygous viable despite affecting major myofibrillar proteins expressed in most muscle types. More importantly, this fact aided in the ascension of the *Drosophila* IFMs as a premiere model in which to analyze muscle function, since these fibers could be readily manipulated genetically with minimal effects upon viability in the laboratory.

Use of Genetics to Study Myofibril Assembly

Armed with a number of mutations affecting muscle and a knowledge in several cases of the proteins encoded by the affected genes, it became possible to ask directed questions concerning the roles of individual myofibrillar components in muscle formation and function.

An interesting observation initially made by Mogami and Hotta[12] was that many of the mutants upon which they concentrated showed a failure to accumulate a number of myofibrillar proteins. Since only single genes were affected in each mutant studied, it was hypothesized that proteins which normally associate with each other in the myofibril are interdependent in their accumulation. Thus, if Myosin heavy chain is missing from the myofibril, all of the myosin-associated proteins might be turned over due to the absence of a stable association. This relationship was clarified by Chun and Falkenthal[31] who showed that in the absence of myofibrillar Myosin heavy chain, myosin light chains were synthesized at wild-type levels but were rapidly turned over; some additional myofibrillar proteins behaved in a similar manner and were therefore also thought to be myosin-associated. In the same mutants, actin and tropomyosin steady-state levels were unaffected. Other studies demonstrated that mutants lacking Actin88F showed strong reductions in the levels of thin filament proteins including Tropomyosin and Troponin-H,[16,32] although effects upon thick filament-associated proteins were also sometimes observed.[32]

Taken together these findings formed a large body of evidence to indicate that the stability of thick- or thin-filament associated proteins within the myofibril were critically dependent upon the integrity of the filaments themselves. While this result might have been predicted based upon established methods to biochemically purify thick or thin filament proteins, this was the first in vivo demonstration of the fact. Furthermore, these in vivo studies also demonstrated that the formation of myofibrils with normal organization and dimensions depends not only upon the polypeptide capable of polymerization to form filaments (actin or myosin heavy-chain) but is also critically dependent upon proteins associated with core polypeptides, including Myosin light-chains,[20] Tropomyosin,[18] Flightin,[29] and Troponin-I.[33]

The stability of actin and its associated proteins in the absence of myofibrillar myosin also indicated that the major filament systems formed independently of one another: in the absence of myofibrillar actin, thin filaments were absent yet electron microscopy revealed thick filaments with apparently normal morphology and associated M-lines.[19,32] Conversely, in the absence of myofibrillar myosin heavy chain, skeins of thin filaments were observed associated with electron-dense material reminiscent of Z-discs.[31,34,35] These conclusion arose from the combined findings of a number of researchers, and were distilled by Beall et al.[36]

While actin is dispensable for myosin accumulation and vice versa, research from a number of quarters indicated that changes in gene dosage could have a profound effect upon myofibril organization. Thus haploinsufficiency for *Mhc* or *Mlc2* caused flightless phenotypes and concomitant fractures in the structure of the IFM myofibrils,[36-38] and haploinsufficiency for *Act88F* or *Tropomyosin 2* caused flightless phenotypes with myofibrils showing excesses of thick filaments at their peripheries.[18,27,36,39] Furthermore, increased gene dosage of *Mhc* caused an overaccumulation of MHC protein and a flightless phenotype, with myofibrils showing defects similar to those observed for reductions in actin gene dosage.[40] By contrast, increasing the gene dosage of *Act88F* in an otherwise wild-type background dids not significantly affect muscle function.[41] While it is interesting to note the unique effects of reducing or increasing actin or myosin gene dosages, the key conclusion here is that changes in the dosage of major myofibrillar protein genes frequently have a profound effect upon myofibril structure. Given the precise interactions that have to take place between thick and thin filaments during the cross bridge cycle, these studies suggested that haploinsufficiency for flight might arise as much from defective myofibrillar structure as from a general reduction in power output as originally proposed by Mogami and Hotta.[12]

This hypothesis was further tested by Beall et al,[36] who studied myofibril structure and flight ability in mutants doubly heterozygous for null alleles of *Act88F* and *Mhc*. Such mutants showed a more wild-type myofibril structure compared to single mutant heterozygotes and a partial rescue of flight ability, indicating that comparable dosages of actin and myosin heavy chain genes are critical to normal myofibrillogenesis. Therefore both myofibril structure and total muscle size are likely to be important for normal flying ability.

Transgenic Approaches to Muscle Assembly and Function

Clearly the isolation of null mutants for numerous muscle protein genes has been informative in determining critical parameters for filament formation and myofibril organization. Combining these null alleles with transgenic technology has allowed the reintroduction of cloned muscle protein genes into the genome. The first example of such an approach, the partial rescue of the *raised* allele of *Act88F* by transgenic *Act88F*$^+$, demonstrated for the first time that a muscle disease can be rescued by transgenic methods.[32] Since then, transgenic approaches have permitted the investigator to probe more subtle aspects of muscle assembly and function using both wild-type and mutant muscle protein genes engineered in vitro. While such approaches will be described elsewhere in this volume in regard to specific myofibrillar protein genes, some examples can be given.

Transgenic approaches have been used to understand the functional significance of multiple protein isoforms. Fyrberg et al[42] generated alleles of *Act88F* in which amino acid substitutions were encoded at sites differing between Actin88F and Actin42A (a cytoplasmic actin). These alleles were introduced into *Act88F* null flies and flight ability and myofibril structure were assayed. While individual amino acid substitutions within Actin88F had mild effects upon muscle performance, multiple substitutions resulted in a loss of normal function. These results, and results from other chimeric genes generated in the same work, clearly indicated that actin isoforms were nonequivalent.

Similarly, Wells et al[43] introduced into an *Mhc*-null background a transgene encoding solely an embryonic isoform of *Mhc*. These lines, expressing a single myosin isoform in all muscles of the body, were viable but showed poor locomotion and severe degeneration of the flight muscle myofibrils. This result indicated that the embryonic MHC isoform is incapable of supporting the functions of all muscles in the body.

The studies of Fyrberg et al[42] and Wells et al[43] underlined the important contributions made by specific muscle isoforms to the performance of individual muscles, and this is now a generally accepted finding. By contrast Miller et al[44] demonstrated that either of two tropomyosin isoforms encoded by the *Tm2* gene was sufficient for normal function in all muscles, suggesting that isoform-specific functions might in some cases be more subtle.

Some effort has also been expended towards understanding the roles of protein domains in myofibril assembly. Cripps et al[45] used the *Act88F* promoter to express a truncated form of the *Mhc* gene predicted to encode a molecule comprising the myosin rod and lacking the globular head region. By crossing transgenic lines of this construct into a mutant background in which no endogenous MHC was produced in the IFMs, the behavior of the truncated myosin could be analyzed in a normal cellular environment. It was found that the headless myosin was capable of assembling into thick filaments and myofibrils, however myofibril size and dimensions were abnormal. These results indicated that the myosin head was dispensable for filament formation but that the myosin head probably interacted with other myofibrillar components to fashion a normal myofibril.

Finally, transgenic approaches have been used to precisely probe the functions of specific residues or domains in the function of particular protein isoforms, as will be described in detail in individual subsequent chapters.

Use of Genetic Interactions to Understand Muscle Function

The functional myofibril is a complex machine comprising a number of distinct polypeptides working together to generate regulated force. The interdependence of these components suggests that mutations affecting one protein are likely to indirectly affect additional proteins. By extension, it should be possible to relieve or enhance the effects of mutations at one locus by second-site mutations at different loci. Such genetic suppression approaches have been particularly useful in dissecting gene function in *C. elegans* muscle (see for example ref. 46), and offer fruitful opportunities for gene discovery and functional analysis in *Drosophila* (reviewed in refs. 47, 48). This concept is vindicated by the observations of Mogami and Hotta[12] in showing that mutations in a single myofibrillar protein gene affects the accumulation of multiple others.

In fact, Mogami and Hotta[12] were the first to determine if mutations affecting flight showed intergenic interactions. The authors generated double-heterozygous mutants and scored for abnormalities in resting wing position, an indirect assay of IFM function. The authors showed that adults heterozygous for mutations in both *Ifm(2)11* and *Mhc* showed an enhancement of the abnormal wing position phenotype. However since the severity of this effect between autosomal alleles was influenced by the sex of the adults tested, it has been suggested that conclusions drawn based upon wing position abnormality might be subject to genetic background effects and present difficulties in interpretation.

A more broad analysis of genetic interaction between existing loci affecting muscle was performed by Homyk and Emerson.[49] These authors looked for interactions among a large number of X-linked and autosomal flightless mutations, and scored the results based upon an enhancement of wing position abnormality, flightlessness, or lethality. The most striking finding was the observation of synthetic lethality when the viable X-linked *hdp*^{2} allele of the *TnI* gene was hemizygous in combination with heterozygosity for several *Mhc* alleles. This was an unusual result for two reasons: firstly, it was interesting to note that the combination of two unlinked viable alleles resulted in lethality; secondly, since *Mhc* and *TnI* mutations interacted genetically it suggested a functional connection between TnI and MHC proteins that had not been previously suspected based upon the known interactions between thick and thin filaments. The latter conclusion should be considered with caution however, since it is also possible that long-range effects occur between muscle proteins that do not directly interact with each other.

The only series of studies aimed at dissecting at the molecular level the bases of genetic interactions in *Drosophila* muscle have arisen from the isolation of suppressors of the TnI mutation *hdp*^{2} (reviewed by ref. 48). Beall and Fyrberg[33] showed that the *hdp*^{2} allele arises from a missense mutation in a constitutive exon of the TnI gene, changing alanine 116 to valine. This mutation causes hypercontraction of the IFMs which results in a highly penetrant wings-up phenotype. Beall and Fyrberg[33] showed that *hdp*^{2} myofibrillar degeneration could be suppressed if the animals also were homozygous for an IFM-specific null allele of *Mhc*. This finding suggested strongly that inappropriate myosin-actin interactions were the cause of hypercontraction.

Possible mechanisms for how the amino acid substitution in *hdp*^{2} might impact actomyosin regulation have been proposed, either by making the thin filament more sensitive to Ca^{2+}, or by attenuating muscle relaxation.[50] In either case, Prado et al[51] initiated a genetic screen to identify suppressors of *hdp*^{2} wing position abnormalities, to characterize the molecular basis for this hypercontraction in greater detail. In this instance, new mutations were isolated based upon a suppression of the *hdp*^{2} wing position phenotype, rather than testing for interactions with existing alleles of muscle protein genes.

The first such suppressor characterized mapped within the *TnI* gene itself and is therefore an intragenic suppressor.[51] This arose from a second-site mutation resulting in substitution of leucine 188 with phenylalanine adjacent to the actin-binding domain of TnI, and suggested a role for this region of the protein specific for the IFMs.

A second suppressor of *hdp*^{2} mapped to the *Tropomyosin 2* gene, and dominantly suppressed the wings-up and flightless phenotypes of *hdp*^{2} heterozygotes or homozygotes.[50] Sequencing revealed that the suppressor mutation altered serine 185 to phenylalanine, a location close to the TnT binding domain of tropomyosin. Mechanistically, this mutation is likely to add a large side chain pointing out from the axis of tropomyosin and the steric effects of this change must impact troponin-based regulation.

A final class of *hdp*^{2} suppressors mapped to the *Mhc* locus on the second chromosome, and mutations in four alleles were clustered in the myosin head.[52] Since it had already been shown that IFM-null alleles of *Mhc* could suppress the hypercontraction effects[33] the isolation of *Mhc* alleles in the suppressor screens was perhaps unsurprising. *Mhc* suppressors contrasted with the *Tm2* suppressor in that the *Mhc* alleles did not rescue flight ability, and indeed caused a dominant flightless phenotype even in the absence of the *hdp*^{2} allele. Thus one mechanistic explanation for suppression was that the *Mhc* mutations result in "dead heads". Mutant heterozygotes for these *Mhc* alleles likely have a reduced power output in the flight muscles, and this effect could account for both an inability of the muscles to pull apart during development in the *hdp*^{2} mutant background, as well as insufficient power for flight in an otherwise wild-type background. Such an argument was supported by the findings of Nongthomba et al,[53] who demonstrated that *hdp*^{2} hypercontraction could also be suppressed by replacement of some full-length MHC by headless myosin molecules. Another argument supporting the conclusion that these suppressors reduce force production is that they are nonspecific—they suppress hypercontraction phenotypes resulting from mutations in a number of different muscle structural genes.

An alternative explanation for the suppression of *hdp*^{2} by selected *Mhc* alleles is that there is a direct functional interaction between MHC and TnI in the *Drosophila* IFMs, and indeed there is some support for this hypothesis from structural studies,[54] which is discussed at length by Kronert et al.[52] In either case, these experiments present strong evidence that suppressor screen can provide important mechanistic insight into muscle formation and function, and it is likely that more extensive use of this approach in the future will result in increased mechanistic understandings as well as the discovery of new genes critical to muscle formation and function.

Concluding Remarks

This review has focused upon historical aspects of *Drosophila* muscle genetics and it is clear that genetics offers the flight muscle scientist opportunities on a number of fronts: the identification of genes critical to flight muscle function; the ability to determine the requirement for a particular protein in myofibril formation or function; the ability to probe the roles of functional domains by mutant isolation and transgenic rescue experiments; and the ability to interrogate protein-protein interactions in vivo via genetic interaction screens. Many of these studies also depend upon advances in other fields of muscle cell analysis and a great strength of the *Drosophila* system results from the variety of different approaches available from the muscle community to the investigator. On these bases alone, a strong foundation for future approaches has been laid down in which genetics will be carefully integrated with other levels of systems analysis, as will be evident throughout the rest of this book.

This review also reveals that each of these areas have still to be fully impacted by genetical studies: for example no systematic screens have attempted to identify all of the loci which can mutate to give rise to recessive flightless phenotypes. In fact, most of the major IFM myofibrillar proteins that have been isolated and characterized in *Drosophila* are direct homologues of vertebrate muscle proteins. This results at least partially from the fact that many of the *Drosophila* muscle protein genes were isolated either using probes from other species, or based upon the biochemical properties of myofibrillar proteins in vertebrates. Although it is clear that the major myofibrillar components are highly conserved across diverse Phyla, there is a bias here in that proteins accumulating at lower levels have not been identified. Thus while *Drosophila* presents a useful system to identify muscle protein genes in the absence of biochemical information, this has yet to be effected via genetic screen.

Furthermore, proteins unique to *Drosophila* muscle are likely still to be discovered. This is supported by the observations of Mogami et al[13] where several numbered proteins thought to be IFM-specific have yet to be identified. Although mass spectrometry combined with a knowledge of the *Drosophila* genome is an effective tool for such an analysis, its use in the *Drosophila* muscle field has been limited (see ref. 55 and Chapter by Henkin and Vigoreaux in this volume).

Transgenic approaches are also providing critical insight into the roles of muscle-specific isoforms as well as the functions of specific residues in the cross-bridge cycle, and this is clearly a demonstrated strength of the fly system. However, the functions of some unique features of myofibrillar proteins could be effectively addressed by expressing engineered myofibrillar protein gene alleles in transgenic animals. For example, a modified *Tm1* cDNA could be expressed in which the long C-terminal extensions of Tropomyosin 1 found in the TnH molecules are deleted, allowing the investigator to specifically probe the function of that protein domain. In addition, *Act88F* alleles could be engineered which are incapable of ubiquitination to the arthrin isoform, addressing the role of this unique protein in muscle biology.[56]

Finally the study of the indirect flight muscle, while valuable in its own right, also serves to provide important basic information for vertebrate muscle biology and pathology. As with the indirect flight muscles, vertebrate cardiac muscle is stretch-activated (see the Chapter by Moore), and understanding the mechanistic basis for contraction in one muscle type is likely to be informative for other muscle types. In addition, some myopathies are proposed to occur from haploinsufficiency for structural protein genes (see for example refs. 57, 58) thus an understanding of the mechanisms of haploinsufficiency in flies are likely to be broadly useful.

Clearly the future is bright for genetics to play a major role in an integrated approach to the study of insect flight muscle function, and its relevance to the muscle biology field as a whole.

Acknowledgements

I am grateful to Jim Vigoreaux for critical review of the manuscript, and Jennifer Brower for assistance with manuscript preparation. Research in the Cripps Laboratory is supported by grants from the National Institutes of Health (GM61738) and the Muscular Dystrophy Association.

References

1. Anderson P. Molecular genetics of nematode muscle. Ann Rev Genet 1989; 23:507-525.
2. Epstein HF. Genetic analysis of myosin assembly in Caenorhabditis elegans. Mol Neurobiol 1990; 4:1-25.
3. Hoppe PE, Waterston RH. A region of the myosin rod important for interaction with Paramyosin in Caenorhabditis elegans striated muscle. Genetics 2000; 156:631-643.
4. Hotta Y, Benzer S. Mapping of behaviour in Drosophila mosaics. Nature 1972; 240:527-535.
5. Benzer S. Genetic dissection of behavior. Sci Amer 1973; 229:24-37.
6. Koana T, Hotta Y. Isolation and characterization of flightless mutants in Drosophila melanogaster. J Embryol Exp Morphol 1978; 45:123-143.
7. Sheppard DE. A selective procedure for the separation of flightless adults from normal flies. Drosophila Inform Serv 1974; 51:150.
8. Homyk Jr T, Sheppard DE. Behavioral mutants of Drosophila melanogaster. I. Isolation and mapping of mutations which decrease flight ability. Genetics 1977; 87:95-104.
9. Nongthomba U, Ramachandra NB. A direct screen identifies new flight muscle mutants on the Drosophila second chromosome. Genetics 1999; 153:261-274.
10. Deak II. Mutations of Drosophila melanogaster that affect muscles. J Embryol Exp Morphol 1977; 40:35-63.
11. Homyk Jr T. Behavioral mutants of Drosophila melanogaster. II. Behavioral analysis and focus mapping. Genetics 1977; 87:105-128.

12. Mogami K, Hotta Y. Isolation of Drosophila flightless mutants which affect myofibrillar proteins of indirect flight muscle. Mol Gen Genet 1981; 183:409-417.
13. Mogami K, Fujita SC, Hotta Y. Identification of Drosophila indirect flight muscle myofibrillar proteins by means of two-dimensional electrophoresis. J Biochem 1982; 91:643-650.
14. Bernstein SI, O'Donnell PT, Cripps RM. Molecular genetic analysis of muscle development, structure, and function in Drosophila. Int Rev Cytol 1993; 143:63-151.
15. Vigoreaux JO. Genetics of the Drosophila flight muscle myofibril: A window into the biology of complex systems. BioEssays 2001; 23:1047-1063.
16. Cripps RM, Ball E, Stark M et al. Recovery of dominant, autosomal flightless mutants of Drosophila melanogaster and identification of a new gene required for normal muscle structure and function. Genetics 1994; 137:151-164.
17. Bernstein SI, Mogami K, Donady J et al. Drosophila muscle myosin heavy chain encoded by a single gene in a cluster of muscle mutations. Nature 1983; 302:393-397.
18. Karlik CC, Fyrberg EA. An insertion within a variably spliced Drosophila tropomyosin gene blocks accumulation of only one encoded isoform. Cell 1985; 41:57-66.
19. Karlik CC, Coutu MD, Fyrberg EA. A nonsense mutation within the Act88F actin gene disrupts myofibril formation in Drosophila indirect flight muscles. Cell 1984; 38:711-719.
20. Warmke JW, Kreuz AJ, Falkenthal S. Colocalization to chromosome bands 99E1-3 of the Drosophila melanogaster Myosin light chain-2 gene and a haploinsufficient locus that affects flight behavior. Genetics 1989; 122:139-151.
21. Hakeda S, Endo S, Saigo K. Requirements of Kettin, a giant muscle protein highly conserved in overall structure in evolution, for normal muscle function, viability, and flight activity of Drosophila. J Cell Biol 2000; 148:101-114.
22. Kolmerer B, Clayton J, Benes V et al. Sequence and expression of the kettin gene in Drosophila melanogaster and Caenorhabditis elegans. J Mol Biol 2000; 296:435-448.
23. Ayme-Southgate A, Vigoreaux J, Benian G et al. Drosophila has a twitchin/titin-related gene that appears to encode projectin. Proc Natl Acad Sci USA 1991; 88:7973-7977.
24. Fyrberg CC, Labeit S, Bullard B et al. Drosophila projectin: Relatedness to titin and twitchin and correlation with lethal(4)102Cda and bent-Dominant mutations. Proc R Soc Lond B 1992; 249:33-40.
25. Becker KD, O'Donnell PT, Heitz JM et al. Analysis of Drosophila paramyosin: Identification of a novel isoform which is restricted to a subset of adult muscles. J Cell Biol 1992; 116:669-681.
26. Liu H, Mardahl-Dumesnil M, Sweeney ST et al. Drosophila paramyosin is important for myoblast fusion and essential for myofibril formation. J Cell Biol 2003; 160:899-908.
27. Kreuz AJ, Simcox A, Maughan D. Alterations in flight muscle ultrastructure and function in Drosophila tropomyosin mutants. J Cell Biol 1996; 135:673-687.
28. Tetzlaff MT, Jackle H, Pankratz MJ. Lack of Drosophila cytoskeletal tropomyosin affects head morphogenesis and the accumulation of oskar mRNA required for germ cell formation. EMBO J 1996; 15:1247-1254.
29. Reedy MC, Bullard B, Vigoreaux JO. Flightin is essential for thick filament assembly and sarcomere stability in Drosophila flight muscles. J Cell Biol 2000; 151:1483-1499.
30. Deak II, Bellamy PR, Bienz M et al. Mutations affecting the indirect flight muscles of Drosophila melanogaster. J Embryol Exp Morphol 1992; 69:61-81.
31. Chun M, Falkenthal S. Ifm(2)2 is a Myosin heavy chain allele that disrupts myofibrillar assembly only in the indirect flight muscle of Drosophila melanogaster. J Cell Biol 1988; 107:2613-2621.
32. Mahaffey JW, Coutu MD, Fyrberg EA et al. The flightless Drosophila mutant raised has two distinct genetic lesions affecting accumulation of myofibrillar proteins in flight muscles. Cell 1985; 40:101-110.
33. Beall CJ, Fyrberg EA. Muscle abnormalities in Drosophila melanogaster heldup mutants are caused by missing or aberrant troponin-I isoforms. J Cell Biol 1991; 114:941-951.
34. O'Donnell PT, Collier VL, Mogami K et al. Ultrastructural and molecular analyses of homozygous-viable Drosophila melanogaster muscle mutants indicate there is a complex pattern of myosin heavy-chain isoform distribution. Genes Dev 1989; 3:1233-1246.
35. Collier VL, Kronert WA, O'Donnell PT et al. Alternative myosin hinge regions are utilized in a tissue-specific fashion that correlates with muscle contraction speed. Genes Dev 1990; 4:885-895.
36. Beall CJ, Sepanski MA, Fyrberg EA. Genetic dissection of Drosophila myofibril formation: Effects of actin and myosin heavy chain null alleles. Genes Dev 1989; 3:131-140.
37. O'Donnell PT, Bernstein SI. Molecular and ultrastructural defects in a Drosophila myosin heavy chain mutant: Differential effects on muscle function produced by similar thick filament abnormalities. J Cell Biol 1988; 107:2601-2612.

38. Warmke J, Yamakawa M, Molloy J et al. Myosin light chain-2 mutation affects flight, wing beat frequency, and indirect flight muscle contraction kinetics in Drosophila. J Cell Biol 1992; 119:1523-1539.
39. Drummond DR, Peckham M, Sparrow JC et al. Alteration in crossbridge kinetics caused by mutations in actin. Nature 1990; 348:440-442.
40. Cripps RM, Becker KD, Mardahl M et al. Transformation of Drosophila melanogaster with the wild-type myosin heavy-chain gene: Rescue of mutant phenotypes and analysis of defects caused by overexpression. J Cell Biol 1994; 126:689-699.
41. Sparrow JC, Drummond DR, Hennessey ES et al. Drosophila actin mutants and the study of myofibrillar assembly and function. In: el Haj A, ed. Molecular Biology of Muscle. Soc Exp Biol Symp 1992; 46:111-129.
42. Fyrberg EA, Fyrberg CC, Biggs JR et al. Functional nonequivalence of Drosophila actin isoforms. Biochem Genet 1998; 36:271-287.
43. Wells L, Edwards KA, Bernstein SI. Myosin heavy chain isoforms regulate muscle function but not muscle assembly. EMBO J 1996; 15:4454-4459.
44. Miller RC, Schaaf R, Maughan DW et al. A nonflight muscle isoform of Drosophila tropomyosin rescues an indirect flight muscle tropomyosin mutant. J Muscle Res Cell Motil 1993; 14:85-98.
45. Cripps RM, Suggs JA, Bernstein SI. Assembly of thick filaments and myofibrils occurs in the absence of the myosin head. EMBO J 1999; 18:1793-1804.
46. Moerman DG, Plurad S, Waterston RH. Mutation in the unc-54 myosin heavy chain gene of Caenorhabditis elegans that alter contractility but not muscle structure. Cell 1982; 29:773-781.
47. Epstein HF, Bernstein SI. Genetic approaches to understanding muscle development. Dev Biol 1992; 154:231-244.
48. Ferrús A, Acebes A, Marin MC et al. A genetic approach to detect muscle protein interactions in vivo. Trends Cardiovasc Med 2000; 10:293-298.
49. Homyk Jr T, Emerson Jr CP. Functional interactions between unlinked muscle genes within haploinsufficient regions of the Drosophila genome. Genetics 1988; 119:105-121.
50. Naimi B, Harrison A, Cummins M et al. A tropomyosin-2 mutation suppresses a troponin I myopathy in Drosophila. Mol Biol Cell 2001; 12:1529-1539.
51. Prado A, Canal I, Barbas JA et al. Functional recovery of troponin I in a Drosophila heldup mutant after a second site mutation. Mol Biol Cell 1995; 6:1433-1441.
52. Kronert WA, Acebes A, Ferrús A et al. Specific myosin heavy chain mutations suppress troponin I defects in Drosophila muscles. J Cell Biol 1999; 144:989-1000.
53. Nongthomba U, Cummins M, Clark S et al. Suppression of muscle hypercontraction by mutations in the myosin heavy chain gene of Drosophila melanogaster. Genetics 2003; 164:209-222.
54. Reedy MC. Reedy MK, Leonard KR et al. Gold/Fab immuno electron microscopy localization of troponin H and troponin T in Lethocerus flight muscle. J Mol Biol 1994; 239:53-67.
55. Ashman K, Houthaeve T, Clayton J et al. The application of robotics and mass spectrometry to the characterization of the Drosophila melanogaster indirect flight muscle proteome. Lett Peptide Science 1997; 4:57-65.
56. Schmitz S, Schankin CJ, Prinz H et al. Molecular evolutionary convergence of the flight muscle protein arthrin in Diptera and Hemiptera. Mol Biol Evol 2003; 20:2019-2033.
57. Yu B, French JA, Carrier L et al. Molecular pathology of familial hypertrophic cardiomyopathy caused by mutations in the cardiac myosin binding protein C gene. J Med Genet 1998; 35:205-210.
58. Thierfelder L, Watkins H, MacRae C et al. Alpha-tropomyosin and cardiac troponin T mutations cause familial hypertrophic cardiomyopathy: A disease of the sarcomere. Cell 1994; 77:701-712.
59. Adams MD, Celniker SE, Holt RA et al. The genome of Drosophila melanogaster. Science 2000; 287:2185-2195.
60. Kennison JA, Tamkun JW. Dosage-dependent modifiers of Polycomb and Antennapedia mutations in Drosophila. Proc Natl Acad Sci USA 1988; 85:8136-8140.
61. Zhang Y, Featherstone D, Davis W et al. Drosophila D-titin is required for myoblast fusion and skeletal muscle striation. J Cell Sci 2000; 113:3103-3115.
62. Ayme-Southgate A, Southgate R, Saide J et al. Both synchronous and asynchronous muscle isoforms of projectin (the Drosophila bent locus product) contain functional kinase domains. J Cell Biol 1995; 128:393-403.
63. George EL, Ober MB, Emerson Jr CP. Functional domains of the Drosophila melanogaster muscle myosin heavy-chain gene are encoded by alternatively spliced exons. Mol Cell Biol 1989; 9:2957-2974.
64. Rozek CE, Davidson N. Differential processing of RNA transcribed from the single-copy Drosophila Myosin heavy-chain gene produces four messenger RNAs that encode two polypeptides. Proc Natl Acad Sci USA 1986; 83:2128-2132.

65. Kazzaz JA, Rozek CE. Tissue-specific expression of the alternatively processed Drosophila myosin heavy-chain messenger RNAs. Dev Biol 1989; 133:550-561.
66. Hastings GA, Emerson Jr CP. Myosin functional domains encoded by alternative exons are expressed in specific thoracic muscles of Drosophila. J Cell Biol 1991; 114:263-276.
67. Kronert WA, Edwards KA, Roche ES et al. Muscle-specific accumulation of Drosophila myosin heavy chains: A splicing mutation in an alternative exon results in an isoform substitution. EMBO J 1991; 10:2479-2488.
68. Saide JD, Chin-Bow S, Hogan-Sheldon J et al. Characterization of components of Z-bands in the fibrillar flight muscle of Drosophila melanogaster. J Cell Biol 1989; 109:2157-2167.
69. Fyrberg E, Kelly M, Ball E et al. Molecular genetics of Drosophila alpha-actinin: Mutant alleles disrupt Z disc integrity and muscle insertions. J Cell Biol 1990; 110:1999-2011.
70. Vigoreaux JO, Saide JD, Pardue ML. Structurally different Drosophila striated muscles utilize distinct variants of Z-band-associated proteins. J Muscle Res Cell Motil 1991; 12:340-354.
71. Roulier EM, Fyrberg C, Fyrberg E. Perturbations of Drosophila alpha-actinin cause muscle paralysis, weakness, and atrophy but do not confer obvious nonmuscle phenotypes. J Cell Biol 1992; 116:911-922.
72. Homyk Jr T, Szidonya J, Suzuki DT. Behavioral mutants of Drosophila melanogaster III. Isolation and mapping of mutations by direct visual observation of behavioral phenotypes. Mol Gen Genet 1980; 177:553-565.
73. Perrimon N, Engstrom L, Mahowald AP. Developmental genetics of the 2C-D region of the Drosophila X chromosome. Genetics 1985; 111:23-41.
74. Vinós J, Maroto M, Garesse R et al. Drosophila melanogaster paramyosin: Developmental pattern, mapping and properties deduced from its complete coding sequence. Mol Gen Genet 1992; 231:385-394.
75. Arrendondo JJ, Ferreres RM, Maroto M et al. Control of Drosophila paramyosin/ miniparamyosin gene expression: Differential regulatory mechanisms for muscle-specific transcription. J Biol Chem 2001; 276:8278-8287.
76. Karlik CC, Mahaffey JW, Coutu MD et al. Organization of contractile protein genes within the 88F subdivision of the D. melanogaster third chromosome. Cell 1984; 37:469-481.
77. Hanke PD, Storti RV. The Drosophila melanogaster tropomyosin II gene produces multiple proteins by use of alternative tissue-specific promoter and alternative splicing. Mol Cell Biol 1988; 8:3591-3602.
78. Bautch VL, Storti RV, Mischke D et al. Organization and expression of Drosophila tropomyosin genes. J Mol Biol 1982; 162:231-250.
79. Fyrberg E, Fyrberg CC, Beall C et al. Drosophila melanogaster troponin-T mutations engender three distinct syndromes of myofibrillar abnormalities. J Mol Biol 1990; 216:657-675.
80. Benoist P, Mas JA, Marco R et al. Differential muscle-type expression of the Drosophila troponin T gene: A 3-base pair microexon is involved in visceral and adult hypodermic muscle specification. J Biol Chem 1998; 273:7538-7546.
81. Bullard B, Leonard K, Larkins A et al. Troponin of asynchronous flight muscle. J Mol Biol 1988; 204:621-637.
82. Mardahl-Dumesnil M, Fowler VM. Thin filaments elongate from their pointed ends during myofibril assembly in Drosophila indirect flight muscle. J Cell Biol 2001; 155:1043-1053.
83. Salzberg A, Develyn D, Schulze KL et al. Mutations affecting the pattern of the PNS in Drosophila reveal novel aspects of neuronal development. Neuron 1994; 13:269-287.
84. Dye CA, Lee JK, Atkinson RC et al. The Drosophila sanpodo gene controls sibling cell fate and encodes a tropomodulin homolog, and actin/tropomyosin-associated protein. Development 1998; 125:1845-1856.
85. Fyrberg EA, Kindle KL, Davidson N. The actin genes of Drosophila: A dispersed multigene family. Cell 1980; 19:365-378.
86. Ball E, Karlik CC, Beall CJ et al. Arthrin, a myofibrillar protein of insect flight muscle, is an actin-ubiquitin conjugate. Cell 1987; 51:221-228.
87. Tobin SL, Zulauf E, Sánchez F et al. Multiple actin-related sequences in the Drosophila melanogaster genome. Cell 1980; 19:121-131.
88. Nongthomba U, Pasalodos-Sanchez S, Clark S et al. Expression and function of the Drosophila Act88F actin isoform is not restricted to the indirect flight muscles. J Muscle Res Cell Motil 2001; 22:111-119.
89. Hiromi Y, Hotta Y. Actin gene mutations in Drosophila; heat shock activation in the indirect flight muscles. EMBO J 1985; 4:1681-1687.
90. Fyrberg EA, Mahaffey JW, Bond BJ et al. Transcripts of the six Drosophila actin genes accumulate in a stage- and tissue-specific manner. Cell 1983; 33:115-123.

91. Barbas JA, Galceran J, Krahjentgens I et al. Troponin-I is encoded in the haplolethal region of the Shaker gene-complex of Drosophila. Genes Dev 1991; 5:132-140.
92. Barbas JA, Galceran J, Torroja L et al. Abnormal muscle development in the heldup[3] mutant of Drosophila melanogaster is caused by a splicing defect affecting selected troponin I isoforms. Mol Cell Biol 1993; 13:1433-1439.
93. Beall C, Fyrberg C, Song S et al. Isolation of a Drosophila gene encoding glutathione-S-transferase. Biochem Genet 1992; 30:515-527.
94. Clayton JD, Cripps RM, Sparrow JC et al. Interaction of troponin-H and glutathione S-transferase-2 in the indirect flight muscles of Drosophila melanogaster. J Muscle Res Cell Motil 1998; 19:117-127.
95. Toffenetti J, Mischke D, Pardue ML. Isolation and characterization of the gene for myosin light chain 2 of Drosophila melanogaster. J Cell Biol 1987; 104:19-28.
96. Parker VP, Falkenthal S, Davidson N. Characterization of the myosin light-chain-2 gene of Drosophila melanogaster. Mol Cell Biol 1985; 5:3058-3068.
97. Takano-Ohmura H, Hirose G, Mikawa T. Separation and identification of Drosophila myosin light chains. J Biochem 1983; 94:967-974.
98. Vigoreaux JO, Saide JD, Valgeirsdottir K et al. Flightin, a novel myofibrillar protein of Drosophila stretch-activated muscles. J Cell Biol 1993; 121:587-598.
99. Vigoreaux JO, Hernandez C, Moore J et al. A genetic deficiency that spans the flightin gene of Drosophila melanogaster affects the ultrastructure and function of the flight muscles. J Exp Biol 1998; 201:2033-2044.
100. Falkenthal S, Parker VP, Mattox WW et al. Drosophila melanogaster has only one myosin alkali light-chain gene which encodes a protein with considerable amino-acid sequence homology to chicken myosin alkali light-chains. Mol Cell Biol 1984; 4:956-965.
101. Falkenthal S, Parker VP, Davidson N. Developmental variations in the splicing pattern of transcripts from the Drosophila gene encoding myosin alkali light chain results in different carboxyl-terminal amino-acid sequences. Proc Natl Acad Sci USA 1985; 82:449-453.
102. Falkenthal S, Graham M, Wilkinson J. The indirect flight muscle of Drosophila accumulates a unique myosin alkali light chain isoform. Dev Biol 1987; 121:263-272.
103. Qiu F, Lakey A, Agianian B et al. Troponin C in different insect muscle types: Identification of an isoform in Lethocerus, Drosophila and Anopheles that is specific to asynchronous flight muscle in the adult insect. Biochem J 2003; 371:811-821.
104. Herranz R, Diaz-Castillo C, Nguyen TP et al. Expression patterns of the whole troponin C gene repertoire during Drosophila development. Gene Exp Patterns 2004; 4:183-190.
105. Fyrberg C, Parker H, Hutchison B et al. Drosophila melanogaster genes encoding three troponin-C isoforms and a calmodulin-related protein. Biochem Genet 1994; 32:119-135.

Chapter 2

3D Structure of Myosin Crossbridges in Insect Flight Muscle:

Toward Visualization of the Conformations during Myosin Motor Action

Mary C. Reedy

Abstract

Insect flight muscle (IFM) provides a model system that allows direct viewing of individual myosin head structures in situ that give rise to the average structures reported by X-ray patterns and by the mechanical behavior of the fibers. Coordinating X-ray diffraction, physiological monitoring and fast freezing with EM tomography, correspondence class averaging and atomic model building in IFM is providing 3D imaging of different myosin conformations in situ in relaxed, active and rigor states. Rigor has yielded the most detailed 3D structure, showing actin, myosin S2 and a distribution of variously flexed myosin lever arm in class averages. EM tomograms of fast frozen/freeze substituted isometric and stretch-activated contractions show that crossbridges in active contraction bind to actin target zones by only one head, in contrast to the most prominent class of rigor crossbridges that attach with both myosin heads to actin. In contrast to a ~5nm lever arm swing inferred during rigor induction, active myosin heads display a wide range of crossbridge angles, consistent with a power stroke greater than 10nm, that proceeds from a prestroke "up" configuration "down" to a rigor angle. However, measurements of isometrically active IFM crossbridges to determine their position, angle and frequency of attachment to actin indicate that the majority of crossbridges in isometric contraction are angled close to perpendicular to the filament axis (60% within 11°), results that are consistent with X-ray studies of vertebrate isometric contraction. X-ray modeling of ATP relaxed *Lethocerus* IFM shows that a myosin head conformation similar to "prestroke" crystal structures, is arrayed in the 14.5nm periodic "shelves" along thick filaments such that only one head of each molecule is well-positioned, as if poised to bind to actin upon activation, while the other head curves around the thick filament shaft.

Introduction

The indirect flight muscles (IFM) of certain insects, in particular that of the giant waterbug, *Lethocerus*, and the tiny fruit fly *Drosophila*, have presented a treasure trove of information about the structure and function of the actin/myosin molecular motor in situ. IFM presents a unique opportunity to view myosin heads in 3D as they generate tension in the intact muscle lattice. Actin and myosin show an almost crystalline order and lattice arrangement in IFM that makes it an excellent subject for 3D reconstruction. Furthermore, *Lethocerus* IFM is composed of very long (1 cm) parallel fibers. This, combined with the high degree of ordering, make it

Nature's Versatile Engine: Insect Flight Muscle Inside and Out, edited by Jim Vigoreaux.

ideal for X-ray diffraction studies that can be coordinated with 3D reconstructions from thin section electron microscopy. *Drosophila* IFM is similarly well-ordered but is less than 1 mm in length. *Drosophila's* small size has allowed X-ray diffraction of IFM in living flies[1] (see also chapter by Irving in this volume) and detailed study of IFM development during pupation.[2,3] The accessibility of *Drosophila* IFM for genetic manipulation of contractile proteins is unparalleled[4,5-9] and the structural effects of a number of mutants in contractile proteins have been described.[10-16] However, the most detailed structural studies of crossbridges have focused on *Lethocerus* IFM and these will be discussed in this chapter.

In 1965, Reedy, Holmes and Tregear[17] showed with coordinated X-ray diffraction and electron microscopy (EM) that in glycerinated *Lethocerus* IFM relaxed in MgATP and low calcium (pCa 6.5), myosin heads appeared to project at a ~90° angle from "shelves" every 14.5 nm along the thick filaments. When rigor was produced by washing ATP out of the muscle, the myosin heads formed ~45° angled crossbridges in a regular pattern of "chevrons" every 38.7 nm along the actin thin filaments (Fig. 1). Measurements of IFM fibers on a force transducer showed that induction of rigor produced substantial stiffness and tension, suggesting that rigor showed the myosin heads at the end of a power stroke. X-ray diffraction patterns indicated that crossbridge mass was concentrated at the 14.5 nm periodicity in ATP-relaxed, while in rigor, the large intensity increase on 38.7 nm layer line and intensity decrease on the 14.5 nm layer line were best modelled and matched by the angled crossbridges attached every 38.7 nm to the thin filament. These results supported the models of sliding filaments and independent force generators[18-21] and the swinging crossbridge hypothesis.[22] X-ray patterns recording the transition between relaxed and rigor IFM from recent studies[23] (Fig. 2) reflect a striking rearrangement of myosin heads and S2 by conversion of 4-fold myosin head origins to 2-fold origins, discussed below.

Advances in instrumentation and techniques have enabled experiments that have filled out and altered this early picture of the tilting power stroke of the myosin crossbridge. For almost three decades, various strategies, including chemical fixation and nucleotide analogs, were used to accumulate static arrays of IFM crossbridges in equilibrium states thought to mimic intermediate stages of a power stroke, seeking to visualize the hypothesized tilting of myosin heads from 90° to 45°.[24-26] Figure 3 compares crossbridge structures in electron micrographs and EM tomograms obtained from tilt series of longitudinal 25 nm "myac" layer sections of rigor and nucleotide-analog (AMPPNP) treated IFM. The rigor lead bridges are the crossbridge pair closer to the M-line in each 38.7 nm axial repeat, while the rear bridges are the pair closer to the Z-line. The lead and rear rigor crossbridges form what is called a "double chevron" motif. In all states observed thus far, in which crossbridges attach to actin, they bind in actin target zones, helically well-oriented segments of the thin filament. In rigor the target zones span about 4 actin protomers along each strand, while in the intermediate nucleotide states, with lower actin affinity, actin target zones appear to span only two or three protomers. In AMPPNP at room temperature, the nucleotide released the rear crossbridges and caused only a small change in axial angle of the lead crossbridges. When fibers were partially relaxed by adding ethylene glycol to AMPPNP, the long sought-after reversal of attached crossbridge angle was achieved: glycol-AMPPNP crossbridges bound to actin target zones with average angles around 90° to the long axis of the filaments.

Electron tomography, a nonaveraging method of 3D reconstruction,[27,28] succeeded in retaining and displaying the variety of crossbridge forms in these less uniformly ordered intermediate states.[25,29] At the same time, new methods of averaging were needed to improve signal to noise in the tomograms without "blurring" details or wiping out the variation in structure preserved in the tomograms.[30] The newer methods give an improved view of myosin head structure that supported efforts to fit crystallographic structures of myosin heads into the averaged crossbridge envelopes.

X-ray crystallography has provided the highest resolution views of the conformational changes that accompany alterations in the nucleotide state when myosin is not bound to actin.[31,32-36]

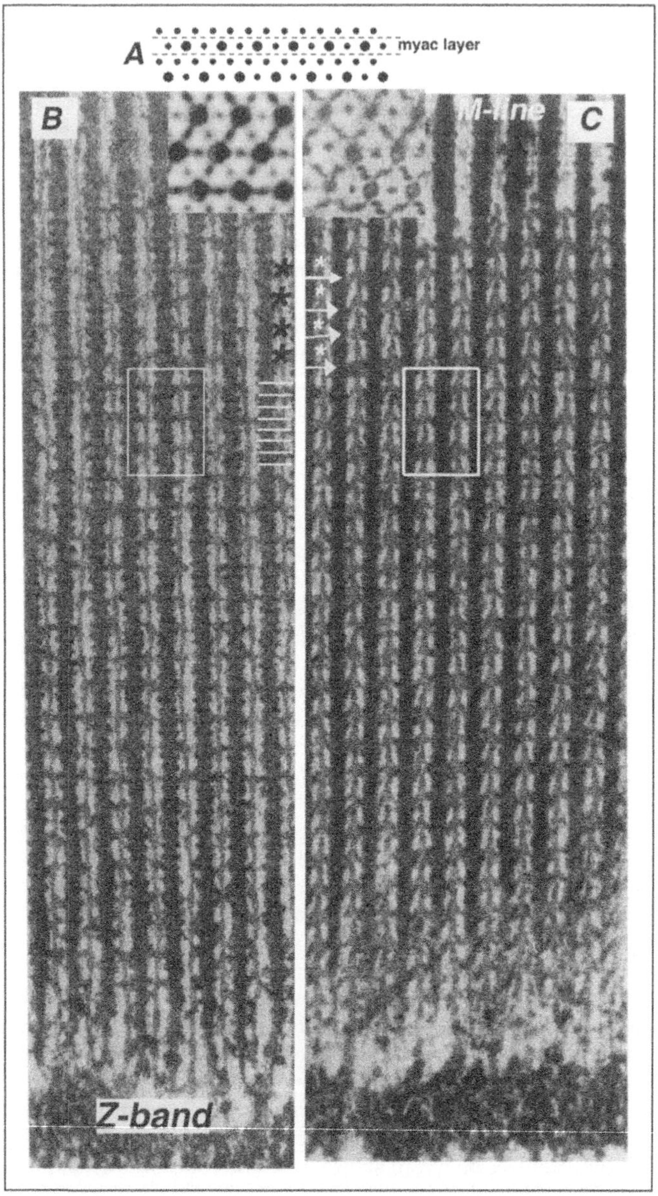

Figure 1. Electron micrographs of 25nm longitudinal sections of *Lethocerus* IFM comparing rigor and ATP-relaxed states at the same magnification. A) Diagram of a cross section view of the hexagonal thick and thin filament lattice of IFM with the actin filaments at the dyad positions between thick filaments. Dashed lines indicate the single filament layer of alternating myosin and actin filaments (myac layer) shown in the longitudinal sections throughout this paper. This arrangement of filaments allows myosin crossbridges to be seen from their origins on the thick filaments to their actin attachments, without inclusion of crossbridges from planes in the hexagonal lattice above or below the myac layer section. B) In ATP-relaxed IFM, the myosin heads form dense shelves every 14.5 nm along the thick filaments that project at a 90° angle toward the thin filaments. At many 38.7 nm intervals, the 90° projections appear to touch the thin filaments at the level of troponin. The figure legend is continued on the next page.

Figure 1, contiued. The black asterisks at the right highlight some of these bridging densities, which are in register across the sarcomere and which coincide with the head regions of troponin. The white lines on the right highlight the levels of eight 14.5 nm shelves, also in register across the sarcomere. The white box includes an 116nm axial repeat, consisting of three 38.7 nm target zones and eight 14.5 nm shelves. The inset shows a ~15 nm cross section view of the relaxed filament lattice that includes only a single "shelf" level on each filament. The thick filaments have a tetragonal outline because the relaxed myosin heads originate from four points around them. C) In rigor, the maximum number of myosin heads attach to actin at each half turn of the actin helix every 38.7 nm to form regular, angled chevrons that point toward the M-line. The alternating white lines and asterisks on the left highlight lead crossbridges (arrows) alternating with the head region of troponin (asterisks), also spaced every 38.7 nm. The white box outlines two thin/thick filament corridors over an 116nm axial repeat containing three 38.7 nm chevrons. The most prominent rigor crossbridges, the lead chevrons, bind in the actin target zone midway between the head regions of troponin, which appear as black dots on the thin filament between the chevrons. The inset shows a ~15 nm cross section view of rigor that includes only one level of bridges. Rigor bridges originate from two-fold positions across the thick filament to form the "flared X".

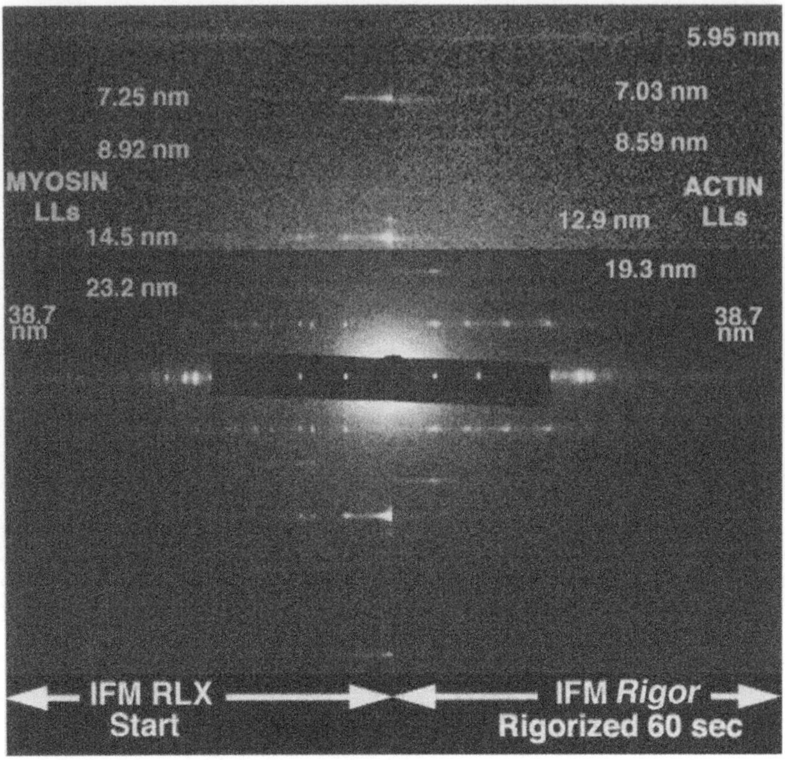

Figure 2. Time resolved synchrotron X-ray diffraction patterns of Lethocerus IFM recording the transition in 60 seconds from ATP-relaxed on the left side to rigor on the right. The 14.5 nm and 7.2 nm myosin-based reflections are very strong in relaxed IFM and weak in rigor, and the 38.7 nm and 19.3 reflections become stronger in rigor. As rigor develops, the weakening of the 14.5 nm and 7.2 nm relaxed reflections and the strengthening of the 38.7 nm and 19.3 nm layer lines reflects the striking rearrangement of myosin molecules as they leave the relaxed head array on the thick filament, attach to actin and go through a power stroke to form long-lasting, strong-binding rigor attachments to actin as ATP is exhausted. These synchrotron X-ray patterns were obtained at BIOCAT beamline at APS, Argonne.[69]

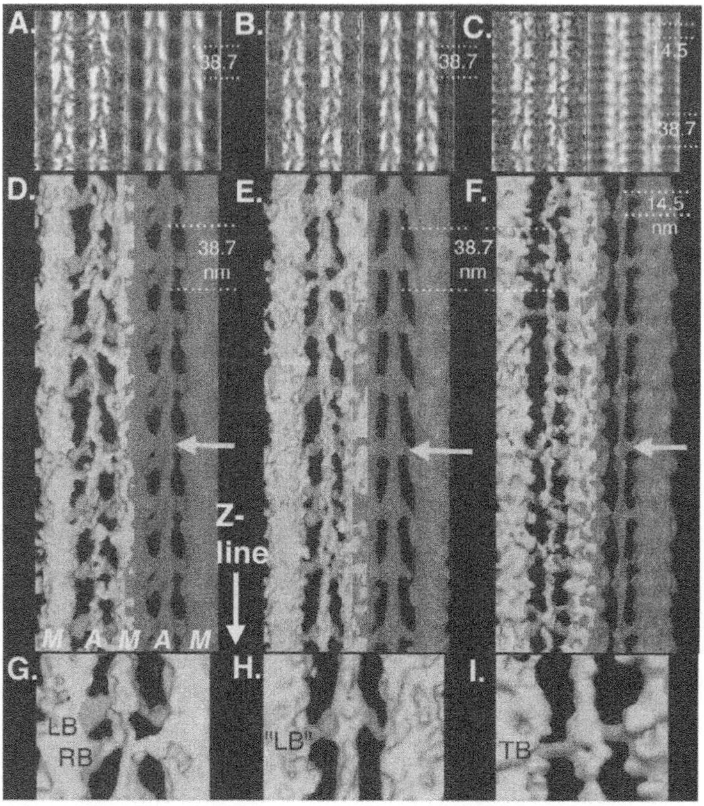

Figure 3. Comparison of *Lethocerus* IFM in chemically fixed, stable equilibrium states of rigor (A,D,G), AMPPNP (B,E,H) and GlycolAMPPNP (C,F,I). Electron micrographs (left side) and 2D filtered images (right side) are shown in (A,B,C). Single thin filament corridors from unaveraged tomograms of the respective states are shown on the left in yellow and their column average on the right in red in (D,E,F). Chartreuse arrows point to crossbridges at the center of the actin target zone (lead bridges (LB) in rigor and AMPPNP; equivalent target zone bridges (TB) in glycolAMPPNP). These bridges are also highlighted orange in the 38.7 nm repeats from the unaveraged tomogram in (G,H,I). RB is rear bridge. M is myosin filament; A is actin filament. Modified from Schmitz et al.[25,29] A color version of this figure is available online at http://www.Eurekah.com.

Crystallographic studies have shown that each myosin head, also known as subfragment 1 (S1), contains the first 843 N-terminal residues of the heavy chain and an essential light chain (ELC) and a regulatory light chain (RLC). Each S1 consists of a motor domain that contains the actin binding and ATPase sites and a light chain domain (LCD), a long alpha-helical segment of the myosin heavy chain encircled by one ELC and one RLC. Several models of the myosin working stroke[37,38] based on X-ray crystal structures propose that the LCD serves as a lever arm to amplify the small conformational changes in the motor domain driven by ATP-hydrolysis and product release. Figure 4A depicts the superimposition on one motor domain of the S1 crystal structures of the prestroke conformation[34] with bound nucleotide and the Rayment et al[32] nucleotide-free rigor conformation. The transition between the "up" and "down" positions of the lever arms suggested a 10.5 nm myosin power stroke. Different LCD positions consistent with the tilting lever arm hypothesis have been observed in acto-S1[39,40] and in actively contracting IFM.[41]

Rebuilding the crystal structures of acto-S1 to fit the envelopes of crossbridges in 3D reconstructions provides a link between the single average structure at high resolution obtained by X-ray crystallography of myosin crystals or X-ray diffraction of muscle fibers and lower resolution 3D tomograms that retain individual variations that contribute to the single average. Comparing in situ crossbridges to atomic structures of acto-S1 allows workers to infer the position of the motor and lever arm domains of S1 in the crossbridge envelopes. This allows different models of the myosin power stroke derived from X-ray diffraction, crystallography and fiber mechanics to be tested and visualized in muscle fibers.

The Rigor State

The most detailed 3D picture of myosin head conformations comes from rigor, in which the maximum number of myosin heads form stable attachments to actin. Even though myosin heads in rigor are strongly bound to actin at the end of the power stroke, the forms of myosin heads in rigor are expected to reflect a range of structures responsible for active force generation, but in a highly ordered, long-lasting state that facilitates structural analysis. Rigor has served, therefore, as a model for the development of several of the new methodologies. Chemically-fixed rigor IFM that showed excellent order, including preservation of the 5.9 and 5.1 nm actin layer lines in thin sections,[42] has been extensively utilized for developing averaging methods to apply to EM tomograms. The most recent method is 3D correspondence analysis (or multivariant statistical analysis).[30,43] Rather than averaging together all variable crossbridge structures throughout the tomogram or along a filament, each 38.7 nm repeat (a 3D crossbridge motif) is demarcated ("masked") in the tomogram. Correspondence analysis then identifies and groups motifs with self-similar 3D crossbridge structures and averages members of each group to form class averages.[44]

Figure 4B shows one class average from a tomogram of IFM in rigor that displays a double chevron. The envelope of the thick and thin filaments and crossbridges are shown in a transparent rendering that allows the rebuilt atomic models of acto-S1 fitted to the envelope to be seen. The lead chevrons, the M-ward crossbridge pair in each 38.7 nm motif, contain both heads of one myosin molecule. The S1 on the M-line side is the M-ward head; the S1 on the Z-line side is the Z-ward head. The Z-ward pair, the rear chevron, usually consists only of a single myosin head in each bridge.

The range of conformations of myosin heads under tension in the intact lattice has been explored by rebuilding the crystal structure of nucleotide-free S1[32] to fit rigor crossbridge envelopes in EM tomograms[44] (Fig. 4B). When S1 heads are bound to actin in vitro, the S1 heads are free from the common origin that constrains them in the intact molecule. The C-termini of S1s bound to successive actins are separated both axially and azimuthally (peek ahead to Fig. 5A,B). In contrast to the uniform structure of S1 in crystals or bound to actin in vitro, the two S1s in rigor crossbridges that contain both heads of one molecule, show different angles and shapes. The position of the motor domain of the S1 atomic model on actin matches rigor crossbridges, but the orientation of the LCD requires axial, azimuthal and twist adjustments to fit crossbridges.

When the motor domains of all refined S1 models rebuilt to fit chemically fixed rigor crossbridges[44] are superimposed onto a single actin and motor domain, the range of positions of the C-terminal heavy chain residue of the LCD of the rebuilt S1, K843, can be interpreted as reflecting an accumulation of all lever arm angles occurring during the final power stroke as the fiber enters rigor (Fig. 4C). In the rigor fittings, the positions of K843 on the myosin heavy chains define an arc that covers an axial range of 5-6 nm. The axial angles of the fitted lever arms of single-headed "rear" crossbridges also covered a 5-6 nm range. The range of LCD positions suggests a 5-6 nm lever arm stroke during rigor induction. The range of rigor lever arm angles is similar to the range observed in rebuilt S1 models inferred to be in late-stages of the power stroke in isometrically active IFM.[41]

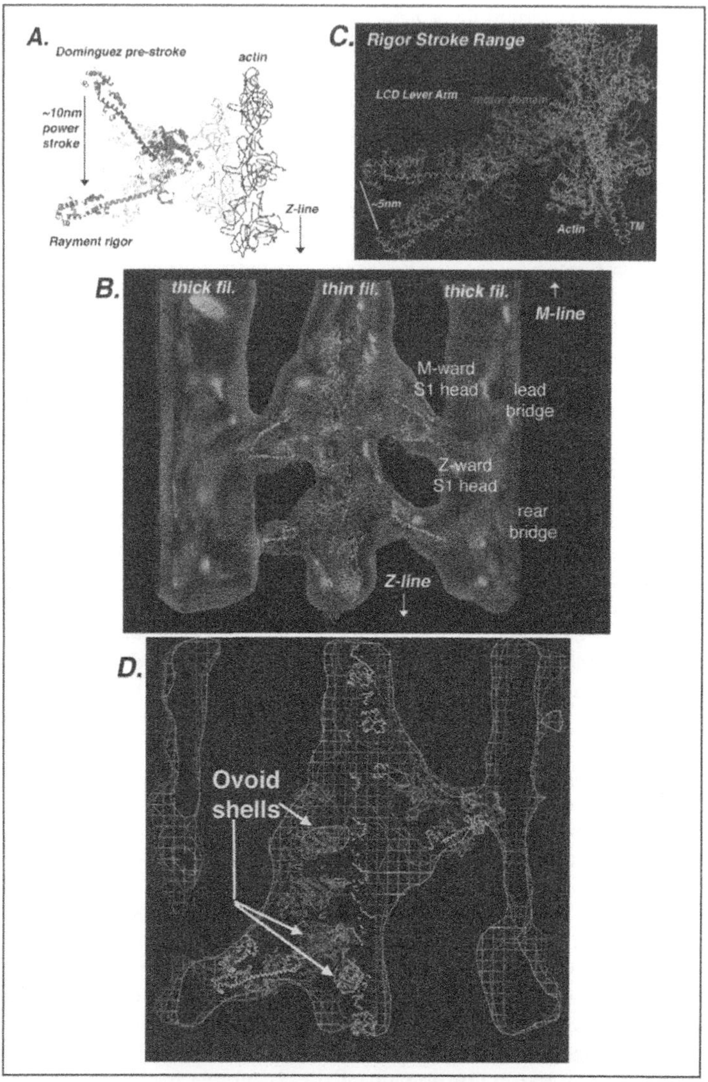

Figure 4. A) Superimposition of the crystal structures of the hypothesized prestroke and rigor S1 on the same motor domain docked on one actin protomer in the strong-binding rigor configuration. The atomic structure of chicken smooth myosin with ADP•AlF4 bound,[34] was proposed to mimic the prepowerstroke shape of the head on actin (i.e., the "A•M•ADP•Pi" state, although in the absence of actin). With the actin filament axis vertical and the Z-band towards the bottom, the crystal structure of rigor S1 has the lever arm angled down at about 45°.[37] A transition between these two forms would give a power stroke of ~10 nm. B) A class average from a tomogram of rigor IFM rendered in transparent envelope showing a double chevron and Rayment rigor S1s rebuilt to fit the crossbridge envelopes. Modified from Chen et al.[44] C) Superimposition of all S1s fitted to rigor crossbridges on one motor domain and actin shows a 5.8 nm range of LCD positions, implying a 5 nm lever arm swing during the transition into rigor. TM is tropomyosin. Actin strands are green/blue, the myosin motor domain is red, the LCD heavy chain is yellow, the converter domain is green, the ELC is purple, the RLC is brown. D) Slab through the central region of a white wireframe representation of a class average from a tomogram of stretched rigor IFM[45] showing the ovoid stain shells outlining the actin (green)/motor domain (red region). The thin slab includes different portions of the actin/motor domains at successive axial positions along the helical filament because successive actins are rotated ~25°. A color version of this figure is available online at http://www.Eurekah.com.

Correspondence class averages from tomograms of IFM rigor fibers fast frozen and freeze substituted after ramp stretches[45] showed actin filament substructure (Fig. 4D) for the first time in *averaged* images of IFM. This supports more precise alignment of the acto-S1 atomic models in the crossbridge envelopes (Fig. 5A-D). Original images and class averages also showed for the first time the position of a segment of S2 in some 38.7 nm crossbridge motifs (Fig. 5). Visualization of S2 allows us to estimate how close to bring the LCDs at the S1/S2 junction and how the "hook helices" must be directed in order to join the closely-spaced coiled-coil of S2. The atomic fittings of stretched rigor crossbridges (Fig. 5A,B) demonstrate that the LCD lever arm of the nucleotide-free S1crystal structure[32] lies outside the envelopes of rigor crossbridges. K843 at the end of the "hook" helix is much too far from K843 of the partner head to join the coiled-coil of S2 (Fig. 5A). The azimuthal position of Rayment S1[32] diverges even more from in situ crossbridges, requiring very large azimuthal shifts of the lever arm to fit rear crossbridges (Fig. 5B compared to 5D). This large azimuthal distortion, which is typical of rear bridges, reflects their position at the edge of the actin target zone and can account for their variable occupancy and single-headedness in rigor.

The rationale for stretching rigor fibers was to explore the range of flexibility of strongly bound myosin heads: Could rigor crossbridges be "backbent" toward the M-line by a stretch, and if so, where, and by how much, do they bend? The average M-ward shift of rigor bridges in vertebrate muscle reported by X-ray diffraction was very small, compared to the larger amount of average M-ward movement detected in isometrically active muscle.[46,47] Lui et al[45] found a variation in the amount of crossbridge response to stretch along the 116 nm long axial repeat of rigor IFM. In some class averages, crossbridge LCD lever arms were backbent M-ward by up to 4.5 nm (Fig. 4D), while the lever arm angles of other classes were the same as those from unstretched rigor class averages. In some class averages, apparent backbending in original EM images could be seen after atomic fitting to be largely due to the swing-out of a segment of S2 from the thick filament shaft. The exposure of S2 lends a very strong backbent appearance to the crossbridge in projection view, but does not actually involve extensive flexing of the lever arm toward the M-line (Fig. 5C). The *overall average* of M-ward shift of LCD lever arms from all class averages was only 1.4 nm, consistent with X-ray results from stretches of frog rigor muscle.[46-48] But the overall average does not reveal the variation in crossbridge response to a change in load or the distribution of the variation along the thin filament.

Correspondence analysis not only improves the resolution of the averages, it allows the distribution and position of particular crossbridge conformations to be located in the large array of crossbridges in the sarcomere. "Mapbacks" replace each 38.7 nm crossbridge repeat in the tomogram by its class average (Fig. 5E). In stretched rigor, the variation in lever arm backbending was regularly distributed along the 116nm long repeat (3 X 38.7 nm and 9 X 12.9 nm).[45,49,50] In rigor, the variability in crossbridge form along the thin filament inherent to the varying register of myosin head origins and actin targets has been smoothed over by movements of the lever arms and S2 to yield a seamless 38.7 nm repeat of chevrons. Mapbacks of class averages reveal the hidden strain and distortion exposed by the stretch in rigor. This variation in crossbridge strain along the filament is important in active contraction.

Active Contraction

The ultimate goal has been to elucidate the 3D structure of myosin crossbridges as they generate force during an active contraction. But for many years, 3D visualization of actively contracting crossbridges has been limited by the variability of active myosin head conformations due to the mismatch between bridge origins and actin targets and the millisecond time resolution required to trap the stretch-activated contractions characteristic of IFM.

Isometric, Calcium Activated Contraction

In order to by-pass the stringent time constraints required to capture stretch activated contractions, isometric, calcium-activated contractions of glycerinated *Lethocerus* fibers were developed for structural studies as the high static tension or HST state.[41] Stretch is unnecessary

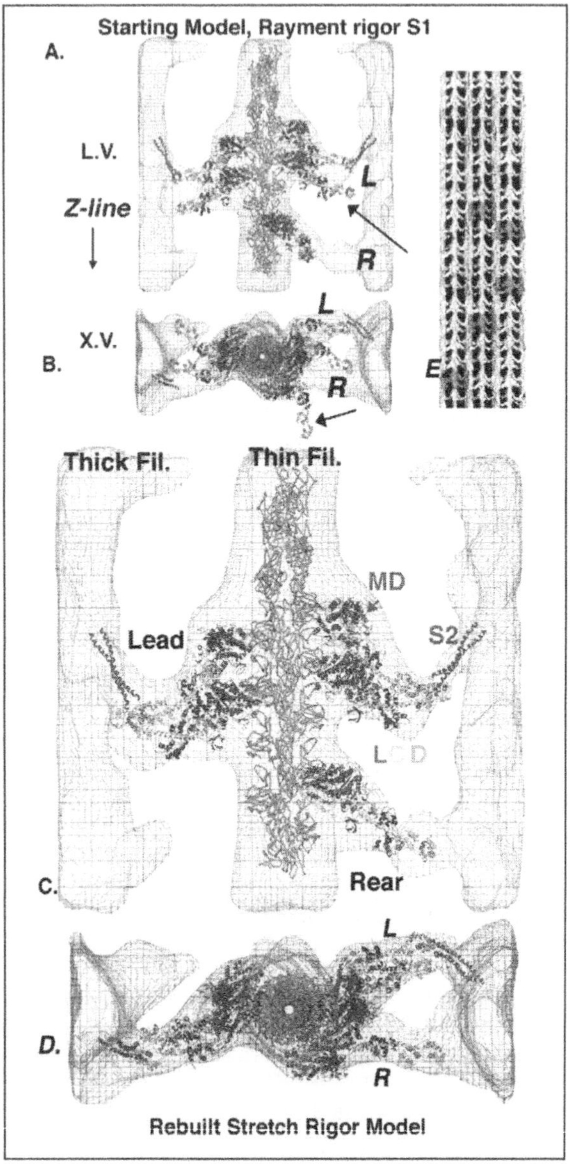

Figure 5. A-D) One class average from a tomogram of fast frozen, stretched rigor IFM comparing the unmodified (A,B) and rebuilt (C,D) atomic models of rigor S1[32] and the relation of the S1s to S2 (~50 residue segment, 7 heptads, magenta). A) Longitudinal view (L.V.) of unmodified rigor S1. The LCD lever arm often lies outside the bridge envelope (arrow) and clearly, the LCDs are too far apart to form a vertex at the junction with S2. Actin filament is green. B) Cross section view (X.V.) of Lead bridges (L) shown in panel (A). The unmodified S1 crystal structure lies far outside the bridge envelope particularly at the rear bridge (R). C) The crystal structure of Rayment S1 was rebuilt to fit the bridge envelope by pivoting about three residues, G710 between the motor (MD) and converter (green) domains, D780 between the converter and ELC (blue) and M806 between the ELC and RLC (cyan). D) Cross section view of (C). E) Small area from a "mapback" of a stretch rigor tomogram in surface rendering showing some examples (red squares) of the distribution of the class average shown in panels (A-D). Data from Liu et al.[45] A color version of this figure is available online at http://www.Eurekah.com.

for HST; in the presence of Mg•ATP, raising [Ca^{2+}] to >0.1 mM (pCa 4.5) produces strong isometric contraction that can last for many seconds. In vivo, prolonged isometric contraction by flight muscles is used for thermogenesis in the preflight warmup that brings IFM to ~40°C, as required for flight in most larger insects.[51]

Figure 6 illustrates a general overview of tomography and atomic model fitting of IFM of the isometric HST state.[41] EM tomograms of isometrically-activated IFM freeze-substituted after rapid freezing show well-ordered, single headed cross-bridges binding in actin target zones with a wide range of attachment angles, from prestroke to rigor-like end-stroke.[41] The variation in bridge angle appeared to correlate with the varying alignment of the actin target zones every 38.7 nm with the crossbridge origins every 14.5 nm. Actin targets and bridge origins go through a match/mismatch alignment every 116 nm (3 X 38.7 nm and 8 X 14.5 nm) along the thin filament in active IFM. Therefore, the signal to noise in the raw unaveraged tomogram was improved without averaging across the variable crossbridge array over the sarcomere, by averaging along each thick-thin column (column averaging) to reduce each thin filament to a single average 116 nm long repeat. The column averages showed that there was always at least one pair of crossbridges bound in a narrow actin target zone midway between the head regions of the troponin complex every 38.7 nm and one more crossbridge attached in the target zone in one or two of the three 38.7 nm repeats. The angles of isometric crossbridges varied from >100° (antirigor) in bridges that originated on the M-ward side of the target zone center, to rigor-like (~45°) in bridges originating on the Z-ward side of target zone center in each 38.7 nm repeat.

Rebuilding acto-S1 models to fit isometrically active crossbridge envelopes required various axial and azimuthal adjustments of the Rayment starting S1 crystal model[41] (Fig. 6). The active bridges near the rigor angle ("end stroke") needed only small azimuthal adjustments of the lever arm angle of the S1 model, while bridges at intermediate axial angles between ~60° to ~100° required larger adjustments of the LCD lever arm position. The "antirigor" angle (>100°) crossbridges could not be fitted with a lever arm adjustment of rigor S1. These putative "prestroke" bridges also required shifting the motor domain of rigor S1 M-ward of its position and very large M-ward tilting of the S1 lever arm to fit the envelope of the crossbridge. A crystal structure of S1 with bound nucleotide proposed to be in the "prestroke" conformation[52] was a closer match to the "antirigor"crossbridges. However, the motor domain of "prestroke" S1 was oriented on actin in the rigor position and did not precisely match the nonrigor position of the "anti-rigor" crossbridges.

The rebuilt acto-S1 atomic models suggested a two stage power stroke, in which the motor domain tilts on actin from a "weak" binding position to the strong rigor position, followed by tilting of the LCD lever arm from a prepower stroke position at high angle to a rigor position near 45°. Several models propose that following attachment to actin, tilting of the LCD can produce a ~10-11 nm working stroke, while the motor domain maintains a relatively constant rigor-like orientation.[37,38,53] The crossbridges in which the motor domain is not at the rigor position on actin and the lever arm is up at a high angle may represent weak contacts between myosin and actin at the beginning of a power stroke that realign on actin to reach a strong binding position. Or, they may be nonstereospecific contacts that do not evolve into force producing interactions.

The range of rebuilt S1 models in HST was ordered into a hypothetical sequence compatible with a continuously attached, progressive ~12 nm power stroke comprising an angular range of >100° to 45°.[41] However, the frequency and distribution of cross-bridge positions and angles within the full range were not determined.

Analysis of the tomograms of isometrically active IFM was extended to quantify the distribution and orientation of attached cross-bridges during isometric contractions.[54] The number and position of attached myosin heads were measured by tracing cross-bridges through the 3-D tomogram from their origins on 14.5 nm spaced shelves along the thick filament to their attachments in the actin target zones every 38.7 nm. Surprisingly, the frequency of crossbridge binding to actin plotted relative to the axial distance between the center of the actin target zone

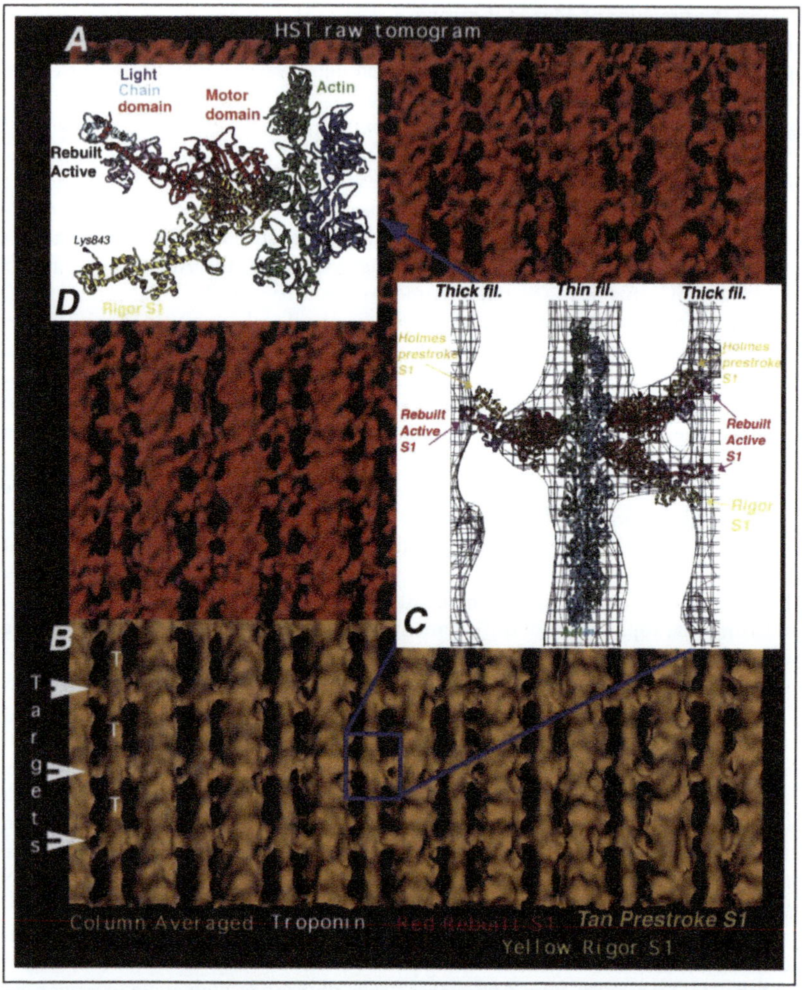

Figure 6. Summary of tomography and atomic model building of the calcium-activated, isometric high static tension (HST) state of Lethocerus IFM from Taylor et al.[41] A) Unaveraged (raw) tomogram (orange) of HST in surface rendering is very noisy. B) "Column averaging" (yellow) of each thin filament and crossbridges reduces each filament to three 38.7 nm averaged motifs (an 116 nm repeat) that reduces noise while retaining natural crossbridge variability. "Targets" indicate the actin target zones midway between troponins (T). The blue rectangle outlines the motif enlarged in (C). C) Wireframe rendering of an averaged 38.7 nm crossbridge motif (blue box) from the tomogram of the isometric HST state showing fitted active S1 in red. Unaltered Rayment rigor S1[32] in yellow is superimposed on the Z-ward bridge and unaltered Holmes prestroke S1 (tan)[52] is superimposed on the more M-ward bridges. D) One of the "prestroke" fitted S1 (red) and one of the fitted rigor-angled HST S1 are shown bound to the same actin to illustrate the range of observed myosin head angles in isometric contraction.

and the bridge origin showed a Gaussian distribution. The majority of isometric crossbridges are nearly perpendicular to the filaments (60% within 11°) with lower and equal numbers of bridges at prestroke and rigor angles. The results suggest that when the filaments cannot slide, individual cross-bridges generate tension with little change in axial translocation or angle.

Fast force transients and X-ray diffraction of vertebrate fibers also indicate the average bridge angle in isometric contraction is close to 90°. In isometrically active frog muscle, the behaviour of the 14.5 nm meridional X-ray reflection is best modelled with cross-bridges nearly perpendicular to the filaments.[47,46] X-ray interferometry indicates that in skeletal muscle at high load, the length of the myosin power stroke is short[55] and most tension-generating bridges are close to perpendicular to the filament axis.

Force generation with little or no lever-arm tilt suggests a flexing cantilever action of the lever arm, similar to the bending of a fishing pole when the hook is lodged in a very heavy load.[56,47] Such flexing is seen in class averages of stretched rigor (Fig. 5).

Stretch Activation

An important long-standing question about IFM has been whether stretch activation involves a fundamentally different mechanism than calcium activation. Recent studies of *Lethocerus* IFM have examined the relationship between Ca^{++}-activated isometric contraction and stretch activation.

Physiological experiments indicate that Ca^{++}-activation and stretch-activation are complementary mechanisms that trigger a common process of crossbridge attachment and force production.[57] In the absence of any stretch, at high calcium concentration (pCa4.5) *Lethocerus* IFM can reach nearly its maximum force (80 kN m^{-2}). A minimum amount of tension (5-10% of maximum) and crossbridge attachment must be activated by Ca^{++} in order to obtain stretch activation. The amount of increased force obtained after stretch is greatest when calcium-activated isometric force is ~20% of maximum.The amount of stretch-activated tension decreases as the starting Ca^{++}-activating tension increases to produce a nearly constant sum of isometric Ca^{++} activated and stretch-activated tension. Linari et al[57] propose that stretch distorts the requisite small number of crossbridges attached in response to a low level of calcium activation. The stretch induced distortion of initially attached crossbridges displaces tropomyosin (Tm) over a longer stretch of actin and opens many more sites for more crossbridge attachment, without requiring Ca^{++}/troponin induced movement of tropomyosin along the full length of the thin filament.

Other recent experiments have characterized two different isoforms of TnC in *Lethocerus* that suggest that the Tn/Tm regulatory system is adapted to allow stretch to trigger full crossbridge attachment and high force at [Ca^{++}] that are too low to fully activate the muscle.[58,59] Agianian et al[59] propose that the special troponin is somehow mechanically activated by stretch. EM images of ATP-relaxed *Lethocerus* IFM, like those in Figure 1, show that the most prominent bridging bars in relaxed IFM are aligned with the troponin densities every 38.7 nm. This suggests the possibility that these relaxed myosin/actin/troponin contacts are activated at low calcium concentration and distortion of these myosin heads by stretch mechanically activates the troponin/tropomyosin system along the entire thin filament, allowing rapid crossbridge attachment in actin target zones all along the thin filaments.

Our first glimpses of stretch activated crossbridge structure and arrangement have been enabled by advances in time-resolved fast freezing and synchrotron X-ray diffraction.Time-resolved X-ray diffraction of IFM during stretch activated contractions[60] provided evidence that myosin crossbridges attach preferentially to actin target zones midway between troponin head regions every 38.7 nm. Stretch-activation triggered by step stretches of *Lethocerus* IFM poised at the lowest threshold of calcium-activated force gave a large rise in active tension that peaked in 100-200 ms. The intensities of the lattice-sampled peaks of the pattern changed as active tension rose: the 14.5 and 7.2 nm meridionals fell, a first row line peak became visible on the 19.3 nm layer line, and the first row line peak on the 38.7 nm layer line fell. Tregear et al[60] concluded that stretch-activated tension under these conditions is produced by the binding of between 16-25% of the total number of myosin heads in IFM.

Electron micrographs of 25 nm longitudinal sections of stretch-activated IFM fibers at low calcium concentration, fast frozen at the plateau of tension (200 ms), and freeze substituted

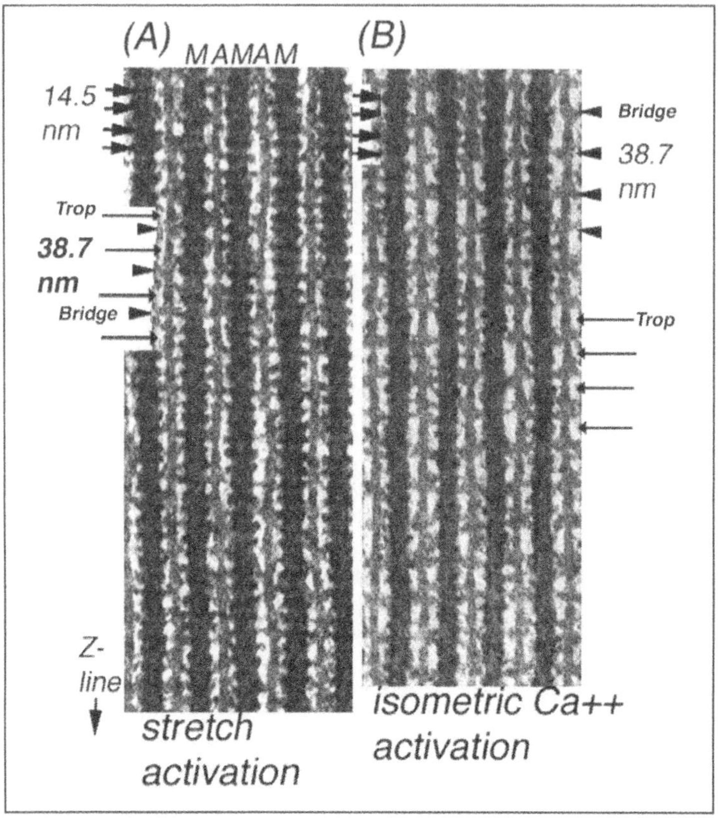

Figure 7. Electron micrographs of myac layers of stretch activated (A) and isometric HST (B) *Lethocerus* IFM. Note the prominence of the 14.5 nm shelves (short arrows) on the thick filaments at the plateau of stretch-activated tension compared to those in HST (short arrows in B).The regularity and number of crossbridges binding to actin is greater in HST, consistent with more heads moving out from the 14.5 nm shelves in HST. However, in both HST and stretch activation, bridges bind primarily in actin target zones every 38.7 nm (arrowheads), midway between troponin densities (trop, long arrows) also distributed every 38.7 nm. In (A), the arrows highlighting this alternating pattern of crossbridge (bridge) binding (arrowheads) with troponin density (long arrows, trop) are shown in sequence, while in (B) bridge and troponin arrows are shown separately. M is myosin filament; A is actin filament.

and epoxy embedded (Fig. 7A), show single myosin heads at varying angles attaching preferentially in actin target zones every 38.7 nm along thin filaments, in agreement with X-ray diffraction patterns. In contrast to rigor crossbridges, stretch activated bridges were single headed and very few were at or near the classical "45°" rigor bridge angle. Compared to isometrically active IFM (Fig. 7B), fewer crossbridges are bound to actin in stretch activation and the 14.5 nm shelves along the thick filament remain very prominent, indicating that many myosin heads are not attached to actin. However, stretch activated IFM exemplifies the difficulties in analyzing 3D structural details in crossbridge states that display a large variation in crossbridge forms and have a large population of myosin heads that are not attached to actin and remain related to the thick filament. Until very recently, there were no models for the disposition of myosin heads along the *Lethocerus* thick filament.

The ATP-Relaxed State

The conformation and arrangement of myosin heads in ATP relaxed muscle is important as the departure point for active contraction. In ATP-relaxed *Lethocerus* IFM, the shelves of density every 14.5 nm project from the thick filament toward actin at a 90° axial angle. The "right angle" appearance of the 14.5 nm crown shelves in electron micrographs and tomograms naturally suggested that relaxed myosin heads project toward actin at a 90° axial angle. However, X-ray modeling has recently revealed a surprising arrangement of the myosin heads in each 14.5 nm shelf.[61]

X-ray diffraction patterns from relaxed IFM have been modelled by testing which shape and arrangement of the 8 heads in a 14.5 nm shelf gave diffraction that best matched the native X-ray pattern. The different head shapes were obtained by pivoting the myosin head atomic coordinates around a hinge between the motor domain and the LCD lever arm. The myosin head shape that best matched the relaxed X-ray pattern (Fig. 8) resembled the crystal structure of S1 thought to be in a "prepower stroke" conformation.[34] In this configuration, the lever arm of the S1 is angled "up" relative to the motor domain, instead of angling "down" toward the Z-band, as in rigor, or projecting straight off the motor domain, as in Houdusse et al.[35] In the X-ray model, both heads of one myosin molecule assume the identical prestroke shape, but they adopt nonequivalent positions. One head projects out from the thick filament surface, emphasizing the projecting shelves of density seen in electron micrographs, while the second "inner" head curves circumferentially around the backbone to bring its ATP-binding cleft into contact with the LCD of the projecting head of an adjacent myosin molecule (Fig. 8). This differs from the contact between coheads within unphosphorylated smooth muscle heavy meromyosin (smHMM),[62] where the actin binding region of one head contacts the converter domain of the cohead. In IFM, the inner head tucked in behind the projecting head could act to maintain the position of the projecting head and stabilize the relaxed head arrangement. The LCD/ATP site contact in relaxed IFM could inhibit the ATPase of the inner head in the resting state and possibly also the activity of the projecting head.

The prestroke conformation of the myosin heads in ATP-relaxed *Lethocerus* IFM is consistent with data that the myosin head is cocked in an "up" conformation while detached from actin.[63] The nonrigor orientation of the actin binding cleft in the relaxed X-ray model (Fig. 8) suggests that a twisting movement of the motor domain or whole head may be needed to align the cleft for strong binding to actin. This realignment may be related to the twist of the lever arm observed during contraction in bi-functional probe experiments.[64,65] It is also consistent with the nonrigor orientation of the motor domain in highly angled "prestroke" bridges in isometrically contracting IFM.[41]

Concluding Remarks

The newer picture of the myosin power stroke suggests that the internal structure of the myosin head can drive large angle changes and 5-12 nm strokes, but, in situ, the length of the power stroke depends on the load. At high load, myosin molecules can generate force with very little change of angle, probably by a flexing of the lever arm. Rapid length and tension transients superimposed on isometrically active IFM are needed to explore the range of myosin lever arm positions associated with changes in load or length. Applying elastic network and normal mode analysis[66] to model building holds promise for defining the elastic deformation of crossbridges. Further improvement in correspondence analysis averaging is essential for atomic model building of crossbridges in tomograms of stretch-activated contractions. Mapbacks of correspondence class averages hold promise for defining the distribution of crossbridge conformations and strain in sarcomeres and should contribute insight into strain-dependent processes, such as ADP release,[67,68] that are increasingly recognized as central to myosin motor function.

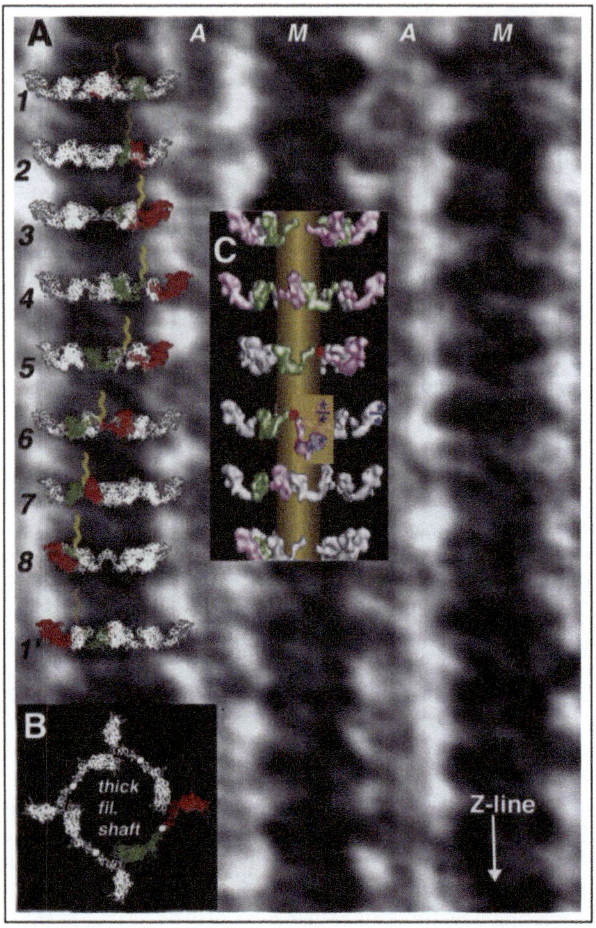

Figure 8. The myosin head conformation and arrangement that best matches the X-ray pattern of ATP-relaxed *Lethocerus* IFM. A) Thick filaments in a highly enlarged electron micrograph of a myac layer section of relaxed IFM are overlaid at approximately the same scale by crowns of S1 (numbered 1-8) whose prestroke conformation and arrangement fits the X-ray pattern.[61] Each crown is rotated 33.75° at successive 14.5 nm shelf levels. Eight 14.5 nm levels are equal to 116nm; the ninth level (1') rotation is superimposable on crown level (1). The yellow squiggly line indicates the direction of S2 along the thick filament shaft and is only illustrated for the single myosin molecule whose heads are shown in red and green, to emphasis the helical tracks formed by the red and green heads. All S1 are in the same "prestroke" conformation, but the projecting head (red) shows this more clearly than in inner head (green). B) "Cross section" view of one crown/shelf at the same rotation as level 4 in (A). The heads of one molecule are colored, the heads of the other three molecules at each level are white. The S2 joining coheads of one molecule are solid white circles. C) A surface rendering of the myosin heads in the best fit arrangement showing six 14.5 nm shelf levels, each rotated 33.75°, is shown over a gold thick filament shaft. In one myosin molecule at each level, the projecting head is pink and the inner head is green. The other three heads are white. At the same level as shelf six on the adjacent filament in (A), blue stars flank bars that indicate the orientation of the actin binding cleft of myosin. In the yellow box, the position of the cleft that matches the strong-binding rigor position is indicated by the cyan bar on the pink head. This projecting head, still in the prestroke conformation, has been shifted from its relaxed position (pale yellow, horizontal bar/cleft) by rotating around its junction with S2 (red dot) to orient its actin binding cleft in the strongly bound position on actin without changing the "prestroke" conformation of the head. The actin binding cleft of the projecting white head would need to be rotated ~60° by a twist of some part, or all, of the myosin head in order to orient in the strong binding position on actin.

Acknowledgements

The preparation of this paper was supported by NIH grant R01AR14317 to M.K. Reedy.

References

1. Irving TC, Maughan DW. In vivo X-ray diffraction of indirect flight muscle from Drosophila melanogaster. Biophys J 2000; 78(5):2511-2515.
2. Reedy MC, Beall C. Ultrastructure of developing flight muscle in Drosophila. I. Assembly of myofibrils. Dev Biol 1993; 160:443-465.
3. Reedy MC, Beall C. Ultrastructure of developing flight muscle in Drosophila. II. Formation of the myotendon junction. Dev Biol 1993; 160:466-479.
4. Emerson Jr CP, Bernstein SI. Molecular genetics of myosin. Annu Rev Biochem 1987; 56:695-726.
5. Bernstein SI, Milligan RA. Fine tuning a molecular motor: The location of alternative domains in the Drosophila myosin head. J Mol Biol 1997; 271:1-6.
6. Becker KD, Bernstein SI. Genetic and transgenic approaches to dissecting muscle development and contractility using the Drosophila model system. Trends Cardiovasc Med 1994; 4:243-250.
7. Fyrberg EA. Genetic and molecular analyses of Drosophila contractile protein genes. BioEssays 1986; 2:250-254.
8. Fyrberg EA, Mahaffey JW, Bond BJ et al. Transcripts of the six Drosophila actin genes accumulate in a stage- and tissue-specific manner. Cell 1983; 33:115-123.
9. Nongthomba U, Cummins M, Clark S et al. Suppression of muscle hypercontraction by mutations in the myosin heavy chain gene of Drosophila melanogaster. Genetics 2003; 164(1):209-222.
10. Reedy MC, Beall C, Fyrberg EA. Formation of reverse rigor chevrons by myosin heads. Nature 1989; 339:481-483.
11. Dickinson MH, Hyatt CJ, Lehmann FO et al. Phosphorylation-dependent power output of transgenic flies: An integrated study. Biophys J 1997; 73(6):3122-3134.
12. Sparrow J, Reedy MC, Ball E et al. Functional and ultrastructural effects of a missense mutation in the indirect flight muscle-specific actin gene of Drosophila melanogaster. J Mol Biol 1991; 222:963-982.
13. Fyrberg E, Kelly M, Ball E et al. Molecular genetics of Drosophila alpha-actinin: Mutant alleles disrupt Z disc integrity and muscle insertions. J Cell Biol 1990; 110:1999-2011.
14. Brault V, Sauder U, Reedy MC et al. Differential epitope tagging of actin in transformed Drosophila produces distinct effects on myofibril assembly and function in indirect flight muscles. Mol Biol Cell 1999; 10(1):135-149.
15. Cripps RM, Becker KD, Mardahl M et al. Transformation of Drosophila melanogaster with the wild-type myosin heavy-chain gene: Rescue of mutant phenotypes and analysis of defects caused by overexpression. J Cell Biol 1994; 126:689-699.
16. Reedy MC, Bullard B, Vigoreaux JO. Flightin is essential for thick filament assembly and sarcomere stability in Drosophila flight muscles. J Cell Biol 2000; 151(7):1483-1500.
17. Reedy MK, Holmes KC, Tregear RT. Induced changes in orientation of the cross-bridges of glycerinated insect flight muscle. Nature 1965; 207:1276-1280.
18. Huxley HE. The mechanism of muscular contraction. Sci Am 1965; 213:18-27.
19. Huxley HE, Hanson J. Changes in the cross-striations of muscle during contraction and stretch and their structural interpretation. Nature 1954; 173:973-976.
20. Huxley AF, Niedergerke R. Structural changes in muscle during contraction. Nature 1954; 173:971-973.
21. Huxley AF, Simmons RM. Proposed mechanism of force generation in striated muscle. Nature 1971; 233:533-538.
22. Huxley HE. The structural basis of muscular contraction. Proc R Soc Lond B: Biological 1971; 178:131-149.
23. Reedy MK, Squire JM, Baumann BAJ et al. X-ray fiber diffraction of the indirect flight muscle of Lethocerus indicus. Advanced Photon Source User Activity: Report 2000. Argonne: Argonne National Laboratory, 2000.
24. Schmitz H, Reedy MC, Reedy MK et al. Electron tomography of insect flight muscle in rigor and AMPPNP at 23°C. J Mol Biol 1996; 264:279-301.
25. Schmitz H, Reedy MC, Reedy MK et al. Tomographic three-dimensional reconstruction of insect flight muscle partially relaxed by AMPPNP and ethylene glycol. J Cell Biol 1997; 139:695-707.
26. Reedy MC. Visualizing myosin's power stroke in muscle contraction. J Cell Sci 2000; 113:3551-3562.
27. Taylor KA, Tang J, Cheng Y et al. The use of electron tomography for structural analysis of disordered protein arrays. J Struct Biol 1997; 120:372-386.

28. Frank J, ed. Electron Tomography: Three-dimensional imaging with the transmission electron microscope. New York: Plenum Press, 1992.
29. Schmitz H, Reedy MC, Reedy MK et al. Tomographic 3D-reconstruction of insect flight muscle in rigor and aqueous AMPPNP. J Muscle Res Cell Motil 1996; 17(1):119-119.
30. Winkler H, Taylor KA. Multivariate statistical analysis of three-dimensional cross- bridge motifs in insect flight muscle. Ultramicroscopy 1999; 77(3-4):141-152.
31. Himmel DM, Gourinath S, Reshetnikova L et al. Crystallographic findings on the internally uncoupled and near-rigor states of myosin: Further insights into the mechanics of the motor. Proc Natl Acad Sci USA 2002; 99(20):12645-12650.
32. Rayment I, Rypniewsky WR, Schmidt-Bäse K et al. Three-dimensional structure of myosin subfragment-1: A molecular motor. Science 1993; 261:50-58.
33. Fisher AJ, Smith CA, Thoden JB et al. X-ray structures of the myosin motor domain of Dictyostelium discoideum complexed with MgADP•BeF_x and MgADP•AlF_4^-. Biochemistry 1995; 34:8960-8972.
34. Dominguez R, Freyzon Y, Trybus KM et al. Crystal structure of a vertebrate smooth muscle myosin motor domain and its complex with the essential light chain: Visualization of the pre power stroke state. Cell 1998; 94:559-571.
35. Houdusse A, Kalabokis VN, Himmel D et al. Atomic structure of scallop myosin subfragment S1 complexed with MgADP: A novel conformation of the myosin head. Cell 1999; 97:459-470.
36. Gulick AM, Bauer CB, Thoden JB et al. X-ray structures of the Dictyostelium discoideum myosin motor domain with six nonnucleotide analogs. J Biol Chem 2000; 275(1):398-408.
37. Rayment I, Holden HM, Whittaker M et al. Structure of the actin-myosin complex and its implications for muscle contraction. Science 1993; 261:58-65.
38. Holmes KC. The swinging lever-arm hypothesis of muscle contraction. Curr Biol 1997; 7:R112-118.
39. Jontes JD, Wilson-Kubalek EM, Milligan RA. A 32° tail swing in brush border myosin I on ADP release. Nature 1995; 378:751-753.
40. Whittaker M, Wilson-Kubalek EM, Smith JE et al. A 35-Å movement of smooth muscle myosin on ADP release. Nature 1995; 378:748-751.
41. Taylor KA, Schmitz H, Reedy MC et al. Tomographic 3-D reconstruction of quick frozen, Ca2+-activated contracting insect flight muscle. Cell 1999; 99:421-431.
42. Reedy MK, Reedy MC. Rigor crossbridge structure in tilted single filament layers and flared-X formations from insect flight muscle. J Mol Biol 1985; 185:145-176.
43. Frank J, van Heel M. Correspondence analysis of aligned images of biological particles. J Mol Biol 1982; 161:134-137.
44. Chen LF, Winkler H, Reedy MK et al. Molecular modeling of averaged rigor crossbridges from tomograms of insect flight muscle. J Struct Biol 2002; 138:92-104.
45. Liu J, Reedy MC, Goldman YE et al. Electron tomography of fast frozen, stretched rigor fibers reveals elastic distortions in the myosin crossbridges. J Struct Biol 2004; 147(3):268-282.
46. Irving M, Piazzesi G, Lucii L et al. Conformation of the myosin motor during force generation in skeletal muscle. Nat Struct Biol 2000; 7(6):482-485.
47. Dobbie I, Linari M, Piazzesi G et al. Elastic bending and active tilting of myosin heads during muscle contraction [see comments]. Nature 1998; 396:383-387.
48. Reconditi M, Dobbie I, Irving M et al. Myosin head movements during isometric contraction studied by X-ray diffraction of single frog muscle fibres. Adv Exp Med Biol 1998; 453:265-270.
49. Crowther RA, Luther PK, Taylor KA. Computation of a three dimensional image of a periodic specimen from a single view of an oblique section. Electron Microsc Rev 1990; 3:29-42.
50. Schmitz H, Lucaveche C, Reedy MK et al. Oblique section 3-D reconstruction of relaxed insect flight muscle reveals the cross-bridge lattice in helical registration. Biophys J 1994; 67:1620-1633.
51. Heinrich B. The thermal warriors. Strategies of insect survival. Cambridge: Harvard University Press, 1996.
52. Holmes KC. Muscle proteins—their actions and interactions. Curr Opin Struct Biol 1996; 6:781-789.
53. Holmes KC. A molecular model for muscle contraction. Acta Crystallogr A 1998; 54:789-797.
54. Tregear RT, Reedy MC, Goldman YE et al. Cross-bridge number, position and angle in target zones of cryofixed isometrically active insect flight muscle. Biophys J 2004; 86(5):3009-3019.
55. Reconditi M, Linari M, Lucii L et al. The myosin motor in muscle generates a smaller and slower working stroke at higher load. Nature 2004; 428(6982):578-581.
56. Goldman YE. Wag the tail: Structural dynamics of actomyosin. Cell 1998; 93:1-4.
57. Linari M, Reedy MK, Reedy MC et al. Ca-activation and stretch-activation in insect flight muscle. Biophys J 2004; 87(2):1101-1111.

58. Qiu F, Lakey A, Agianian B et al. Troponin C in different insect muscle types: Identification of two isoforms in Lethocerus, Drosophila and Anopheles that are specific to asynchronous flight muscle in the adult insect. Biochem J 2003; 371(Pt 3):811-821.
59. Agianian B, Krzic U, Qiu F et al. A troponin switch that regulates muscle contraction by stretch instead of calcium. EMBO J 2004; 23:772-779.
60. Tregear RT, Edwards RJ, Irving TC et al. X-ray diffraction indicates that active crossbridges bind to actin target zones in insect flight muscle. Biophys J 1998; 74:1439-1451.
61. AL-Khayat HA, Hudson L, Reedy MK et al. Myosin head configuration in relaxed insect flight muscle: X-ray modelled resting crossbridges in a prepowerstroke state are poised for actin binding. Biophys J 2003; 85(2):1063-1079.
62. Wendt T, Taylor D, Trybus KM et al. Three-dimensional image reconstruction of dephosphorylated smooth muscle heavy meromyosin reveals asymmetry in the interaction between myosin heads and placement of subfragment 2. Proc Natl Acad Sci USA 2001; 98(8):4361-4366.
63. Geeves MA, Holmes KC. Structural mechanism of muscle contraction. Annu Rev Biochem 1999; 68:687-728.
64. Corrie JE, Brandmeier BD, Ferguson RE et al. Dynamic measurement of myosin light chain domain tilt and twist in muscle contraction. Nature 1999; 400(6743):425-430.
65. Hopkins SC, Sabido-David C, van der Heide UA et al. Orientation changes of the myosin light chain domain during filament sliding in active and rigor muscle. J Mol Biol 2002; 318(5):1275-1291.
66. Tama F, Brooks C. Normal mode based flexible fitting of high-resolution structure into low-resolution experimental data from cryo-EM. J Structural Biology 2004; in press.
67. Smith DA, Geeves MA. Strain-dependent cross-bridge cycle for muscle. Biophys J 1995; 69:524-537.
68. Cremo CR, Geeves MA. Interaction of actin and ADP with the head domain of smooth muscle myosin: Implications for strain-dependent ADP release in smooth muscle. Biochemistry 1998; 37:1969-1978.
69. Irving TC, Fischetti R, Rosenbaum G et al. Fiber diffraction using the BioCAT undulator beamline at the advanced photon source. Nuclear Instruments and Methods (A). 2000; 448:250-254.

Chapter 3

Comparative Physiology of Insect Flight Muscle

Robert K. Josephson

Abstract

Insect flight is powered by muscles that attach more-or-less directly to the wings (direct flight muscles) and muscles that bring about wing movement by distorting the insect's thorax (indirect flight muscles). Flight stability and steering are achieved by differential activation of power muscles and by the activity of control muscles that alter wing stroke amplitude and angle of attack. One evolutionary trend seen when comparing more advanced with less advanced fliers is a reduction in the number of power muscles and an increase in the number of control muscles. On the basis of the neural control of contraction, insect muscles may be divided into synchronous muscles and asynchronous muscles. In synchronous muscles there is neural input and evoked muscle action potentials associated with each contraction. Asynchronous muscles are turned on by neural input, but, when activated, they can contract in an oscillatory manner if attached to an appropriate, mechanically resonant load. The features of asynchronous muscles that allow oscillatory contraction are delayed stretch activation and delayed shortening deactivation. Because asynchronous muscles do not have to be turned on and off by neural input for each contraction, they are expected to be more efficient and more powerful than are synchronous muscles for high frequency operation.

Introduction

The earliest known fossil insects, found in Devonian strata formed about 390 million years ago, were wingless.[1] Insects radiated and flight systems appeared in the Carboniferous.[2] The structures from which the wings evolved are still uncertain (for discussion see Dudley).[3] The power required to move the wings in the earliest flying insects came from pre-existing thoracic musculature. Muscles attached to the lateral body wall became associated with cuticular plates on or attached to the primitive wings. Proximal leg muscles were co-opted to move wings as well as legs. Contractions of longitudinal and dorso-ventral muscles of the thoracic segments became coupled, through the resulting thoracic distortion, to up and down movements of the wings themselves. The Paleozoic radiation and subsequent evolution has resulted in about 30 orders of extant insects, most of which fly as adults. The flight systems of these insects include a wide range of wing morphologies and substantial variability in the organization and physiology of the muscles that power and control flight. It is the flight muscles that will concern us here. More information about insect wings, flight muscles, and other aspects of insect flight systems and flight performance, can be found in a recent, comprehensive and scholarly book by R. Dudley.[3]

Nature's Versatile Engine: Insect Flight Muscle Inside and Out, edited by Jim Vigoreaux.

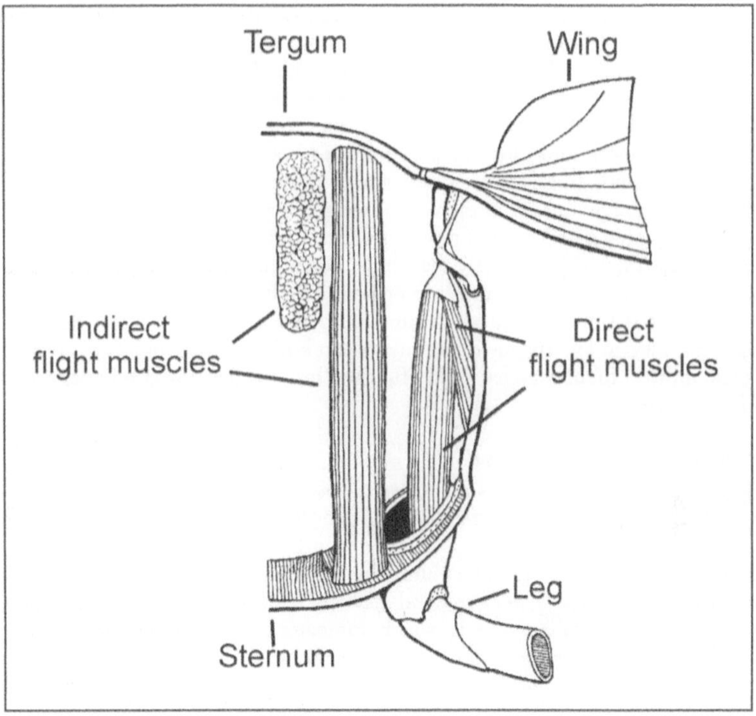

Figure 1. Direct and indirect flight muscles as seen in cross section of a half thorax (after Snodgrass[4]).

Direct and Indirect Flight Muscles

The muscles that power the oscillating wing movements of insect flight are customarily divided into direct flight muscles and indirect flight muscles (Fig. 1). Direct flight muscles insert on the base of a wing or on cuticular patches in the wing articulation that are in turn attached to the wings. The direct flight muscles lie ventral to the wings and their attachment to a wing is lateral to the wing's fulcrum. Consequently direct flight muscles are wing depressors, muscles whose contraction produces ventral movement of the wings. Indirect flight muscles induce wing movements by changing the position and shape of the tergum, the dorsal plate of the thorax. The suspension of the wings to the thorax is such that downward movement of the tergum, caused by contraction of the dorso-ventral thoracic muscles, causes wing elevation while upward bowing along the fore-and-aft axis, brought about by contraction of dorsal longitudinal muscles, causes wing depression.

Power muscles of flight can generally be distinguished from other muscles of the thorax, including control muscles of the wing, by their decidedly pinkish-brown color. Flight is a metabolically expensive way to move about. Estimates of muscle power output as work and heat during flight, based on oxygen consumption or carbon dioxide production, range from several hundred to over 2,000 W per kg muscle.[5,6] The metabolism of flight muscles is, as far as is known, nearly totally aerobic.[7] The high metabolic rates of the flight muscles are supported by a high concentration of mitochondria within the muscle fibers. Mitochondria make up 20-40% of the volume of flight muscle fibers, and only 10% or less of equivalent muscles from nonflying forms.[8,9] It is the high concentration of mitochondria, with their constituent cytochromes, that give flight muscle a distinct color.

The thoracic muscles of locusts (Order Orthoptera) and cockroaches (O. Blattaria) are more similar to those of primitively flightless silverfish (O. Thysanura) than are the muscles of bees (O. Hymenoptera) or flies (O. Diptera). Thus it may be surmised that the flight systems of cockroaches and locusts have diverged less from the condition in primitive fliers than have those of bees and flies. One apparent trend in the evolution of flight musculature seen in comparing less advanced and more advanced groups is a reduction in the number of power muscles and a corresponding increase in the size of those power muscles remaining. Flight is powered by 9-12 pairs of muscles in each of the wing-bearing segments of cockroaches, locusts and katydids,[10] by only 4 or 5 pairs of power muscles in the whole thorax of flies, and by but 2 pairs of large muscles, which fill nearly the whole thorax, in bees. Another trend is the increasing importance of indirect flight muscles in powering flight. Essentially all of the wing depressors in dragonflies (O. Odonata), which are considered to be relatively primitive fliers, are direct flight muscles; in locusts power production on both upstroke and downstroke is shared by direct and indirect flight muscles; and in flies and bees the few power muscles present are all indirect flight muscles.

Power and Control Muscles

In addition to the power muscles that drive the oscillations of the wings, the flight systems of insects contain control muscles; muscles whose contractions are involved in the continuous adjustment of wing stroke amplitude and angle of attack needed to produce stable, directed flight and for turning during flight. Certainly the best studied among the less advanced insect fliers are locusts of the genera *Schistocerca* and *Locusta*. In locusts there is a single control muscle for each wing, the pleuroaxillary muscle, which originates on the lateral body wall and attaches to a cuticular patch in the hinge region of each of the wings. The pleuroaxillary muscles are active during flight, and contraction of these muscles decreases wing pronation (= downward twisting of the leading edge of the wing) during the downstroke, and decreases wing supination (upward twisting of the leading edge) during the upstroke.[11] In addition to and probably more important than the flight control offered by the pleuroaxillary muscles is that provided by differential activation and changes in the timing of activation among the many power muscles of the thorax.[12,13] Of particular importance here are the basalar and subalar muscles; the former causing wing depression and pronation; the latter wing depression and supination. The basalar(s) and subalar muscles of a wing are thus synergists for wing depression but antagonists for altering the angle of attack of the wing. Control muscles become more important in advanced fliers in which there is a reduction in the total number of power muscles able to be individually activated, and in which the power muscles are largely indirect muscles that act by distorting the entire thorax rather than on individual wings. The honeybee, as mentioned, has two pairs of power muscles and five pairs of muscles that probably act as flight control muscles,[14] and in the dipteran *Drosophila*, with 4 pairs of power muscles, there are 18 pairs of small muscles in the thorax that contribute to the control of the one pair of wings.[15]

Bifunctional Muscles

Some of the power muscles of insect flight, both direct and indirect flight muscles, attach ventrally to proximal leg segments. Contraction of such muscles could cause limb movements, especially if the wings are stationary, or wing movements, especially if the leg is held in a fixed position. Thus these muscles are anatomically bi-functional.[16] There is some uncertainty about the extent to which these muscles are functionally as well as anatomically bi-functional. Based on electrical recordings from muscles and motorneurons, Wilson[16] and Ramirez and Pearson[17] concluded that the bifunctional muscles of locusts are active, but in different patterns, during both walking and flight. In contrast, Duch and Pflüger,[18] also working with locusts, found that two of the three bifunctional muscles that they examined were active during flight but not during walking, while one participated in both walking and flight but was turned on by different motorneurons during the two activities. It is actually a bit surprising that bifunctional muscles of locusts are sometimes used during walking. These muscles are large, produce twitches whose amplitude is a substantial fraction of the maximum force available from the muscle (ratio of twitch to tetanic tension = 60%), and have quite brief twitch contractions (approximately 40 ms, onset to 90% relaxation at 30°C [19]). The muscle kinetics are appropriate for flight, during which the operating frequency is about 20 Hz, but the muscles seem ill-designed for walking movements which require low power and fine control, and which occur at an operating frequency an order of magnitude or so lower than that of flight.

Synchronous and Asynchronous Muscles (or Fibrillar and Nonfibrillar Muscles)

Skeletal muscles in arthropods and vertebrates are made up of elongate, multinucleate cells termed muscle fibers. It has been known since the middle of the nineteenth century that flight muscle fibers of some insects differ histologically from the usual striated muscle fibers of arthropods and vertebrates in being composed mainly of large, easily dissociable fibrils (reviewed in Tiegs, ref. 10). The presence of large fibrils, which are up to 2 µm in diameter and which are readily seen with light microscopy, led to the flight muscles containing them being identified as fibrillar muscles. The muscle fibers are sometimes unusually large in fibrillar muscles. The fibers of ordinary striated muscles are typically 0.05 to 0.1 mm in diameter; in contrast those of the fibrillar muscles of diptera are up to 2 mm in diameter.[10]

About 50 years ago studies by Pringle,[20] and Roeder[21] and Boettiger[22,23] demonstrated that the flight muscles of insects could be divided into two classes on the basis of the neural control of contraction, synchronous muscles and asynchronous muscles. Synchronous muscles of insects are like striated muscles elsewhere in that each muscle contraction is initiated by impulses or bursts of impulses in one or more motorneurons to the muscle . The motorneuron impulses evoke depolarization of the innervated fibers of the muscle termed muscle action potentials. An action potential in a fiber triggers release of calcium from an internal store, the sarcoplasmic reticulum (SR), within the fiber. The released calcium diffuses into the myofibrils, binds to control sites on the contractile filaments, and in binding turns on contractile activity of the filaments. Contraction is terminated as the sarcoplasmic reticulum resequesters the released calcium, reducing its cytoplasmic concentration to a level below that needed for contractile activity. The muscles are called synchronous because with them there is a 1:1 relationship between muscle electrical activity and contraction (Fig. 2). Asynchronous muscles too are activated by muscle action potentials and triggered calcium release from the SR, but once activated asynchronous muscles can contract in an oscillatory manner if they are connected to an appropriate resonant load such as is offered, in life, by the wings and thorax of the insect. The oscillatory frequency of an active asynchronous muscle is typically substantially higher than the frequency of muscle action potentials that maintain activation of the muscle. It is because there is no congruence between action potentials and individual muscle contractions during oscillatory activity that the muscles are termed asynchronous.

The features of asynchronous muscle that allow oscillatory activity are delayed shortening deactivation and delayed stretch activation.[9] If an activated, asynchronous muscle is allowed to rapidly shorten, the shortening is followed by a transient decrease in the muscle's capacity to generate force (shortening deactivation). Stretch of an activated, asynchronous muscle results in a temporary increase in capacity for force generation (stretch activation). During oscillatory contraction at an appropriate frequency, the force generated by the muscle during shortening is greater, because the muscle has been activated by the preceding lengthening, than is the force during lengthening, which occurs while the muscle is partially deactivated following the preceding shortening. Because the force at any length is greater during the shortening half-cycle than during the lengthening half-cycle, more work is done by the muscle during shortening than is required to restretch the muscle after shortening, and there is net work output over a full shortening-lengthening cycle; work which is available to drive the wings in a flying insect

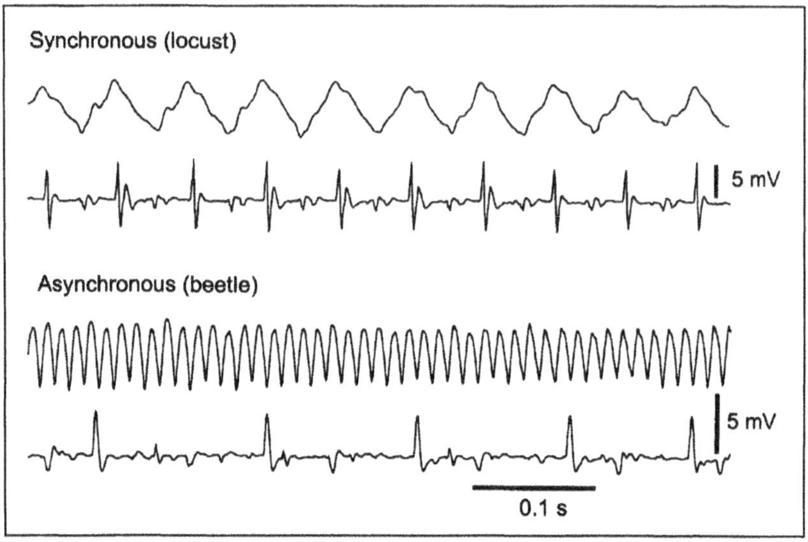

Figure 2. Wing thrust (top trace in each pair) and muscle action potentials during tethered flight. Animals were suspended from a force transducer, attached to the dorsal prothorax, which recorded thrust associated with wing beats during flight. Muscle action potentials were recorded with 100 μm silver wires implanted in wing muscles. The locust muscle was the metathoracic tergosternal muscle of *Schistocerca americana*; the beetle muscle was the metathoracic basalar muscle of *Cotinus mutabilis*. The wing stroke frequency for the locust was 16 Hz; that of the beetle 77 Hz.

(Fig. 3). This work output is, of course, not free, and there is increased ATP hydrolysis associated with the work output.[24,25]

As far as is known all fibrillar muscles of insects are asynchronous, and all asynchronous muscles are fibrillar. It is not obvious that asynchronous muscles need be fibrillar. The ability of asynchronous muscles to oscillate when activated and attached to a mechanically-resonant load is a feature of the contractile proteins and the way they are arranged in thick and thin filaments, and is seen even in isolated, glycerinated myofibrils.[26] There is no direct functional link requiring that for oscillatory behavior the thick and thin filaments of asynchronous muscles be arranged in unusually large fibrils which dissociate easily. It seems likely that the association between fibrillar structure and asynchronous operation is a result of co-evolution of functionally independent traits.

The large size and loose arrangement of the myofibrils are related to the rather slow rate at which asynchronous muscles are turned on and off by neural input. The initial events in muscle activation by neural input—neuromuscular transmission, generation of a muscle action potential, triggered calcium release—are quite fast, occurring in a few ms.[27] The rate limiting steps, those that determine how rapidly a muscle fiber can be turned on and off, are likely to be calcium diffusion into and out of the myofibrils, and, for the time course of relaxation, the rate at which released calcium can be taken up again by the sarcoplasmic reticulum. Muscles capable of producing short twitches have narrow myofibrils, which reduces diffusion distances and diffusion time, and well-developed sarcoplasmic reticulum, which increases the surface area of the SR available to transport calcium.[28,29] Muscles with long twitches have large fibrils and reduced SR. Although the operating frequency of asynchronous muscles may be quite high, often 100 Hz or more, isometric twitches recorded from them are rather slow, with durations of over 100 ms (onset to 50 or 90 % relaxation, 30°C [30-32]). A long period of activation following a muscle action potential is a useful feature of an asynchronous muscle, for it allows the muscle to be

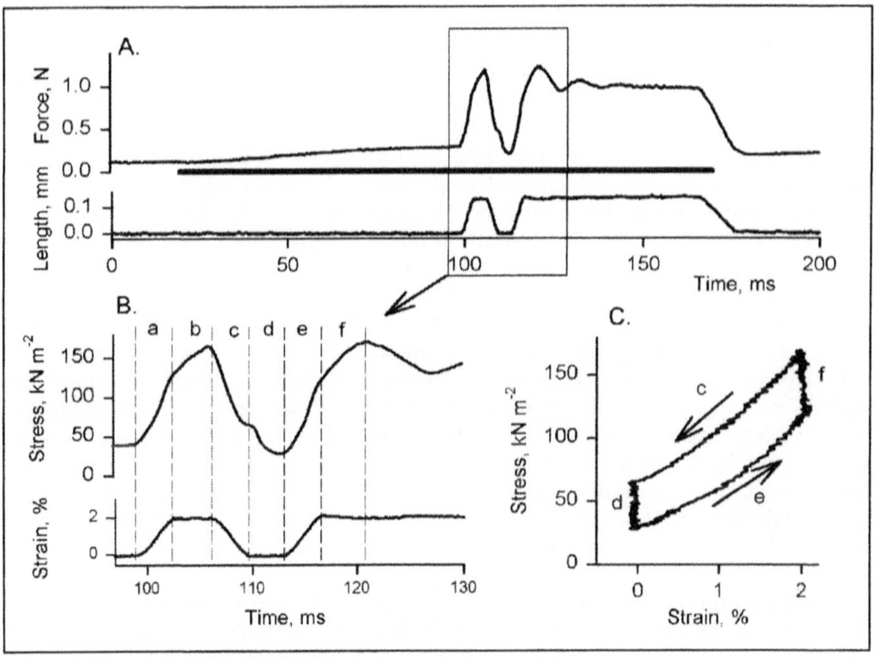

Figure 3. Stretch activation, shortening deactivation and work output from a beetle flight muscle. A) The muscle was stimulated with shocks at 100 Hz and, during the tetanic contraction, subjected to two stretch-hold-release cycles. The thick line beneath the force trace indicates the duration of the stimulation burst. The portion of (A) within the box is shown in (B) on an expanded time base and as strain and resulting stress. Note that the force rises during stretch (a), and continues to rise (stretch activation) during the interval at constant length (b) following the stretch. Similarly force declines during shortening (c), and continues to decline (shortening deactivation) for several ms following the shortening (d). C) The work loop formed by plotting stress against strain for the cycle defined by c-f. The area of the loop is the work output over the cycle. The loop is traversed counterclockwise, indicating that there is net work output.

maintained fully active with low frequency neural input. It is because the neural activation kinetics are slow that the frequency of muscle activation potentials can be much lower than that of oscillatory contractions. The slow activation kinetics are a consequence of the large size of the myofibrils in "fibrillar" muscle, and the scarcity of SR in these muscles. And it is because the SR is sparse and fibrils are not bound together by sheets of SR that ramify through the muscle, as is the case with fast, synchronous muscles, that the fibrils are easily dissociable.

Why Asynchronous Muscles?

Asynchronous muscle is an insect invention, not known to occur elsewhere. Flight is powered by asynchronous muscles in several of the most speciose insect orders, including beetles (O. Coleoptera), flies (O. Diptera), ants, bees and wasps (O. Hymenoptera), and true bugs (O. Hemiptera). Most insects that fly, and therefore most animals that fly, do so using asynchronous muscles. The distribution of asynchronous muscles among extant insect groups suggests that asynchronous muscle has evolved at least six times as flight muscle in different insect lines,[3] and independently again in cicadas in which the main sound producing-muscle of some but not all species is asynchronous.[30]

Asynchronous muscles are high-frequency muscles. The operating frequencies of asynchronous wing muscles during flight range from about 20 Hz in large, belostomid bugs[33] to 1000

Hz in small midges.[34] Not all high-frequency behaviors in insects are powered by asynchronous muscles. Wing beat frequencies in small moths, presumably using synchronous muscles, approach 100 Hz.[35] Synchronous muscles power sound production at contraction frequencies of 100-200 Hz in some long-horned grasshoppers,[36-38] 30 to 220 Hz in several different cicadas,[39] and an astonishing 550 Hz in one cicada species.[40] Synchronous muscles can operate at high frequencies, but it has been proposed that for high frequency operations asynchronous muscles are more efficient and more powerful than synchronous ones.[41,42]

Asynchronous muscles are likely to be more efficient than synchronous ones for high frequency activity because activation costs, those metabolic expenditures associated with calcium cycling, are expected to be much lower in asynchronous muscles than in synchronous ones. In synchronous muscles each contraction involves triggered calcium release from the SR followed by reuptake. The transport of calcium back into the SR is by an ATP-fueled calcium pump and consumes metabolic energy. In asynchronous muscles the frequency of muscle action potentials maintaining the muscle in an active state during oscillatory contraction is much lower, typically by a factor of 10 to 50, than the contraction frequency.[43] Thus the frequency of calcium release and re-uptake cycles, and the associated cost of these, is much lower for an asynchronous muscle than it would be for a synchronous one operating at a similar contraction frequency. Further, the extreme reduction in the volume of SR in asynchronous muscles suggests that the amount of calcium released per action potential in asynchronous muscles is likely to be substantially smaller than that in fast, synchronous muscles, further reducing the calcium cycling costs.

The activation cost for muscle contraction would be of little consequence if it was a trivial component in the overall energy budget sheet, but it is not. A rough guess as to the cost of activation can be obtained by estimating how much calcium must be released from the SR to activate the myofibrils, and how much energy must be spent to resequester this calcium and bring about relaxation. The relevant calcium is that which binds to control sites on troponin molecules within the myofibrils. The concentration of troponin in rabbit white muscle is about 0.07 mmol kg^{-1},[44] and I will assume that the concentration is similar in insect muscle, at least within the myofibrillar portion of insect muscle. A rabbit muscle fiber is composed almost entirely of myofibrils, while in insect flight muscle a large part of the fiber volume is occupied by mitochondria and SR. Therefore a correction is required to adjust for different myofibrillar densities in rabbit muscle and insect flight muscle. I will use, as my insect example, locust flight muscle, in which about 65% of the fiber volume is made up of myofibrils.[9] Adjusting for the nonfibrillar volume gives a troponin concentration in the whole fiber of 0.046 mmol (kg fiber)$^{-1}$. Each troponin molecule has two calcium binding sites involved in the regulation of contraction,[45,46] so the minimum amount of calcium needed to fully saturate the troponin is 0.092 mmol (kg fiber)$^{-1}$. Because of calcium binding to components in the fiber other than troponin, including ATP, the amount of calcium that must be released to ensure that the troponin is saturated is likely to be substantially greater than this minimal amount. Calcium is taken up into the SR by a pump whose stoichiometry is 2 calcium ions transported per ATP hydrolyzed.[47] To take up the amount of calcium required to saturate the troponin control sites requires the hydrolysis of 0.046 mmol ATP (kg fiber)$^{-1}$, which, allowing 50 kJ (mole ATP)$^{-1}$,[48] is equivalent to an energy expenditure for calcium pumping of 2.3 J (kg fiber)$^{-1}$ twitch^{-1}. The wing stroke frequency during flight for a locust is about 20 Hz, so the power required for cycling calcium is about 46 W kg^{-1}, which is more than half of the mechanical power output of the muscle during conditions mimicking flight.[49] If the flight frequency in the locust were 100 Hz, probably well beyond the capabilities of a locust but not an uncommon value for fliers with asynchronous muscles, the power involved in calcium cycling would rise to 230 W kg^{-1}, which is substantially greater than the maximum value for mechanical power output recorded from insect muscles. The procedure for estimating calcium cycling costs is not defensible in detail, but the conclusion that calcium cycling costs are substantial seems unavoidable (for more direct information on this see Schramm et al[50]).

It has been proposed that the power per unit muscle volume in high-frequency, synchronous muscles should be less than in asynchronous ones because of the large investment in structures required for calcium cycling and muscle activation in a high-frequency, synchronous muscle. A significant part of the volume of a high-frequency, synchronous muscle is occupied by SR, and is unavailable for myofibrils. Further, an additional mitochondrial volume above that required to support the mechanical power output is needed in a fast, synchronous muscle to provide the ATP needed for the high calcium cycling costs of high-frequency muscle activation. An extreme instance of SR and mitochondrial hypertrophy, at the expense of relative myofibril volume, is seen in the singing muscle of a cicada, mentioned above, which operates at more than 500 Hz. About one-third of the muscle fiber volume in this cicada is SR, another third mitochondria, and only 22% is made up of myofibrils.

To appropriately test the assertion that asynchronous muscles are more efficient and more powerful for high-frequency activity than are synchronous ones requires measurements of power and efficiency from examples of the two types of muscle, preferably operating at a similar high frequency. Unfortunately evidence is scarce. The efficiency of the conversion of fuel to mechanical energy has been measured directly for only two flight muscle; a synchronous muscle from a locust and an asynchronous muscle from a beetle.[6,51] The efficiency of the beetle muscle (14-16%) was indeed higher, by about two and one-half times, than that of the locust muscle (6.4%). The maximum mechanical power output measured with the work loop technique from synchronous flight muscles of katydids, locusts, sphinx moths and dragonflies ranged from 54 to 108 W kg^{-1} at 30°C.[19,49,52-54] The contraction frequencies in the different studies were generally chosen to match those during normal flight, and ranged from 25 to 37 Hz. The maximum power output from an asynchronous muscle of a beetle at 30°C was 73 W kg^{-1} at an optimum contraction frequency of 52 Hz.[32] Thus the power output of the asynchronous muscle was similar to that of the synchronous muscles at the same temperature but occurred at a higher contraction frequency. The thoracic temperature of the beetle during flight is about 35°C; at this temperature the maximum mechanical power output of the flight muscle is 144 W kg^{-1}.[51] In tettigoniids (katydids) the mesothoracic wing muscles of male animals are generally used both in flight and in sound production, and the contraction frequency can be quite high during the latter. The wing stroke frequency during singing by *Neoconocephalus triops* is about 100 Hz. At this frequency the maximum power output of the wing muscle was only 18 W kg^{-1},[38] providing a clear example of the limited power available from synchronous muscles during high frequency contraction.

Coda

One of the keys to the success of insects was the early evolution of flight in the group, and the opportunities for wide foraging and dispersion that flight allowed. Another factor in the success of insects is their size which, from an anthropocentric view, is rather small and which opens to them many niches not available to larger creatures. The largest flying insects are fist-sized beetles and plate-sized moths, but most adult insects are less than a cm long and many are tiny, the smallest fliers being tiny parasitic wasps whose body length, a bit over 0.2 mm, is about one-half the size of the period at the end of this sentence.[3] Being small poses some special problems for flight. For aerodynamic reasons there is an inverse relationship between the size of a winged flier and the wing beat frequency required for flight. For animals of similar shape but differing in linear dimension (L), the minimum wing beat frequency required to remain airborne is expected to be proportional to $(L)^{-1/2}$.[55] Asynchronous muscles give insects motors that, for high frequency operation, provide more power at less cost than is seemingly possible with ordinary muscle. The repeated invention among insects of asynchronous muscle has allowed insects to be both small and able to fly and has been a major contributor to their obvious success.

References

1. Labandeira CC, Beall BS, Hueber FM. Early insect diversification: Evidence from a lower Devonian bristletail from Québec. Science 1988; 242:913-916.
2. Wootton RJ. Paleozooic insects. Ann Rev Entomol 1981; 26:319-344.
3. Dudley R. The biomechanics of insect flight. Form, Function, Evolution. Princeton: Princeton Univ Press, 2000.
4. Snodgrass RE. Principles of Insect Morphology. New York: McGraw-Hill Book Co., 1935.
5. Casey TM, Ellington CP. Energetics of insect flight. In: Wieser W, Gnaiger E, eds. Energy Transformation in Cells and Organisms. Stuttgart: Georg Thieme, 1989:200-210.
6. Josephson RK, Stevenson RD. The efficiency of a flight muscle from the locust Schistocerca americana. J Physiol 1991; 442:413-429.
7. Beenakkers AMT. Insect flight metabolism. Insect Biochem 1984; 14:243-260.
8. Stokes DR, Malamud JG, Schreihofer DA. Gender specific developmental transformation of a cockroach bifunctional muscle. J Exp Zool 1994; 268:364-376.
9. Josephson RK, Malamud JG, Stokes DR. Asynchronous muscle: A primer. J Exp Biol 2000; 203:2713-2722.
10. Tiegs OW. The flight muscles of insects-their anatomy and histology; with some observations on the structure of striated muscle in general. Phil Trans Roy Soc Lond 1955; B238:221-347.
11. Wolf H. On the function of a locust flight steering muscle and its inhibitory innervation. J Exp Biol 1990; 150:55-80.
12. Baker PS. The role of forewing muscles in the control of direction in flying locusts. J Comp Physiol A 1979; 131:59-66.
13. Fischer H, Kutsch W. Timing of elevator muscle activity during climbing in free locust flight. J Exp Biol 1999; 202:3575-3586.
14. Snodgrass RE. Anatomy and Physiology of the Honeybee. New York: McGraw-Hill Book Co., 1925.
15. Dickinson MH, Tu MS. The function of dipteran flight muscle. Comp Biochem Physiol 1997; 116A:223-238.
16. Wilson DM. Bifunctional muscles in the thorax of grasshoppers. J Exp Biol 1962; 39:669-677.
17. Ramirez JM, Pearson KG. Generation of walking patterns for walking and flight in motoneurons supplying bifunctional muscles in the locust. J Neurobiol 1988; 19:257-282.
18. Duch C, Pflüger HJ. Motor patterns for horizontal and upside-down walking and vertical climbing in the locust. J Exp Biol 1995; 198:1963-1976.
19. Malamud JG, Mizisin AP, Josephson RK. The effects of octopamine on contraction kinetics and power output of a locust flight muscle. J Comp Physiol A 1988; 162:827-835.
20. Pringle JWS. The excitation and contraction of the flight muscles of insects. J Physiol 1949; 108:226-232.
21. Roeder KD. Movements of the thorax and potential changes in the thoracic muscles of insects during flight. Biol Bull 1951; 100:95-106.
22. Boettiger EG. The machinery of insect flight. In: Scheer BT, ed. Recent Advances in Invertebrate Physiology. Eugene: University of Oregon Press, 1957:117-142.
23. Boettiger EG. Insect flight muscles and their basic physiology. Ann Rev Entomol 1960; 5:1-16.
24. Steiger GJ, Rüegg JC. Energetics and "efficiency" in the isolated contractile machinery of an insect fibrillar muscle at various frequencies of oscillation. Pflügers Arch 1969; 307:1-21.
25. Pybus J, Tregear RT. The relationship of adenosine triphosphatase activity to tension and power output of insect flight muscle. J Physiol 1975; 247:71-89.
26. Jewell BR, Rüegg JC. Oscillatory contraction of insect fibrillar muscle after glycerol extraction. Proc Roy Soc London B 1966; 164:428-459.
27. Miledi R, Parker I, Zhu PH. Calcium transients evoked by action potentials in frog twitch muscle fibres. J Physiol 1982; 333:655-679.
28. Josephson RK, Young D. Fiber ultrastructure and contraction kinetics in insect fast muscle. Amer Zool 1987; 27:991-1000.
29. Schaeffer PJ, Conley KE, Lindstedt SL. Structural correlates of speed and endurance in skeletal muscle: The rattlesnake tailshaker muscle. J Exp Biol 1996; 199:351-358.
30. Josephson RK, Young D. Synchronous and asynchronous muscles in cicadas. J Exp Biol 1981; 91:219-237.
31. Josephson RK, Ellington CP. Power output from a flight muscle of the bumblebee Bombus terrestris. I. Some features of the dorso-ventral flight muscle. J Exp Biol 1997; 200:1215-1226.
32. Josephson RK, Malamud JG, Stokes DR. Power output by an asynchronous flight muscle from a beetle. J Exp Biol 2000; 203:2667-2689.

33. Barber SB, Pringle JWS. Functional aspects of flight in belostomatid bugs (Hemiptera). Proc Roy Soc London B 1966; 164:21-39.
34. Sotavalta O. Recordings of high wing-stroke and thoracic vibration frequency in some midges. Biol Bull 1953; 104:439-444.
35. Sotavalta O. The flight-tone (wing-stroke frequency) of insects. Acta Ent Fenn 1947; 4:1-117.
36. Josephson RK, Halverson RC. High frequency muscles used in sound production by a katydid. I. Organization of the motor system. Biol Bull 1971; 141:411-433.
37. Josephson RK. Contraction kinetics of the fast muscles used in singing by a katydid. J Exp Biol 1973; 59:781-801.
38. Josephson RK. Contraction dynamics of flight and stridulatory muscles of tettigoniid insects. J Exp Biol 1984; 108:77-96.
39. Young D, Josephson RK. Mechanisms of sound-production and muscle contraction kinetics in cicadas. J Comp Physiol A 1983; 152:183-195.
40. Josephson RK, Young D. A synchronous insect muscle with an operating frequency greater than 500 Hz. J Exp Biol 1985; 118:185-208.
41. Rome LC, Lindstedt SL. The quest for speed: Muscles built for high-frequency contractions. News Physiol Sci 1998; 13:261-268.
42. Syme DA, Josephson RK. How to build fast muscles: Synchronous and asynchronous designs. Integ and Comp Biol 2002; 42:762-770.
43. Kammer AE. Flying. In: Kerkut GA, Gilbert LI, eds. Comprehensive Insect Physiology, Biochemistry and Pharmacology. Vol. V: Nervous System: Structure and Motor Function. New York: Pergamon Press, 1984:491-552.
44. Ebashi S, Ohtsuki I. Control of muscle contraction. Quart Rev Biophysics 1969; 2:351-384.
45. Collins JH. Myosin light chains and troponin C: Structural and evolutionary relationships revealed by amino acid sequence comparisons. J Muscle Res Cell Motil 1991; 12:3-25.
46. Qiu F, Lakey A, Agianian B et al. Troponin C in different insect muscle types: Identification of two isoforms in Lethocerus, Drosophila and Anopheles that are specific to asynchronous flight muscles in the adult insect. Biochem J 2003; 371:811-821.
47. Inesi G. Mechanism of calcium transport. Ann Rev Physiol 1985; 47:573-601.
48. Woledge RC, Curtin NA, Homsher E. Energetic aspects of muscle contraction. London: Academic Press, 1985.
49. Mizisin AP, Josephson RK. Mechanical power output of locust flight muscle. J Comp Physiol A 1987; 160:413-419.
50. Schramm M, Klieber H-G, Daut J. The energy expenditure of actomyosin-ATPase, Ca2+-ATPase and Na+, K+-ATPase in guinea-pig ventricular muscle. J Physiol 1994; 481:647-662.
51. Josephson RK, Malamud JG, Stokes DR. The efficiency of an asynchronous flight muscle from a beetle. J Exp Biol 2001; 204:4125-4139.
52. Josephson RK. Mechanical power output from striated muscle during cyclic contraction. J Exp Biol 1984; 114:493-512.
53. Marden JH, Fitzhugh GH, Girgenrath M et al. Alternative splicing, muscle contraction and intraspecific variation: Associations between troponin T transcripts, Ca2+ sensitivity and the force and power output of dragonfly flight muscles during oscillatory contraction. J Exp Biol 2001; 204:3457-3470.
54. Stevenson RD, Josephson RK. Effects of operating frequency and temperature on mechanical power output from moth flight muscle. J Exp Biol 1990; 149:61-78.
55. Pennycuick CJ. Animal Flight. London: Edward Arnold, 1972.

Chapter 4

Stretch Activation:

Toward a Molecular Mechanism

Jeffrey R. Moore

Abstract

Insect flight is often powered by high wing beat frequencies. Surprisingly, the flight muscles of some insects are capable of driving high wing beats without extensive calcium cycling machinery. Rather than precisely timed signals from motor neurons driving each contraction, nervous stimulation is sporadic, which presumably serves to maintain a moderately elevated intracellular calcium concentration. In this calcium activated state the muscle will also produce a delayed increase in tension that is initiated by a stretch (stretch activation), produced when an antagonist muscle shortens. Although stretch activation is enhanced in insect flight and cardiac muscle, it is a general property of all muscles. Historically, the underlying mechanism of stretch activation has been studied using several model systems. Initial studies relied on mechanical and ultrastructural studies of giant water bug (*Lethocerus*) flight muscle and vertebrate cardiac muscle. More recently, studies of *Drosophila* flight muscles have allowed powerful genetic methods to be added to the researcher's arsenal. Using these systems, several mechanisms have been proposed to explain stretch activation: (i) matching of the thick and thin filament lattices, (ii) passive stress in the connecting filaments,[1] (iii) myosin regulatory light chain (RLC) phosphorylation,[2,3] and (iv) stretch sensitive calcium sensitivity.[1,4]

While popular, models proposing lattice matching have been challenged by more recent analysis of filament lattice geometries form several insect species.[5] Insect flight and cardiac muscle exhibit a high passive stiffness and are therefore very sensitive to applied stretch. Recent results obtained with insect flight and cardiac muscle preparations provide new insight into a possible molecular pathway that explains the effects of thick filament stress on crossbridge formation. Evidence suggests that stress is transmitted through connecting filaments that extend from the Z-band to the thick filament. We propose that thick filament stress relieves an inhibitory conformation of myosin, which has been observed by low angle X-ray diffraction of isolated *Lethocerus* fiber bundles.[6] Release of this inhibition by stretch/stress together with RLC phosphorylation increases the recruitment of force generating crossbridges and leads to stretch activation.

Stretch sensitive calcium sensitivity via the thin filament regulatory system is also an attractive hypothesis that has recently gained experimental support.[7] Agianian et al[7] showed that isometric tension and stretch activated tension are controlled by different isoforms of troponin C (TnC). Although these results can be explained by a stretch sensitive troponin complex,[7] the description does not provide a clear explanation for the large body of evidence suggesting thick filament stress and regulatory light chain phosphorylation are important for stretch activation. Any model for stretch activation must incorporate both thick and thin filament influences.

Nature's Versatile Engine: Insect Flight Muscle Inside and Out, edited by Jim Vigoreaux.

Here it is proposed that Ca^{++} binding to TnC activates isometric force generation at high calcium levels; however, IFM operate at low "permissive" calcium concentrations and that stretch and/or phosphorylation induced effects on myosin position are required for activation.

Introduction

The cytoskeleton is a network of cross-linked protein filaments which determines the mechanical properties of cells. It is analogous to the skeleton of a multicellular organism in that it provides mechanical strength and cell shape; however, it also provides the contractile forces for locomotion.

The cytoskeleton is designed to transmit mechanical signals derived from intracellular contractile forces or externally applied force perturbations. Striated muscles represent a cell type that is specialized for both the transmission and sensing of forces. Accordingly, they possess organelles (the myofibrils) that have a highly specialized cytoskeleton consisting of an ordered overlapping array of thick (myosin containing) and thin (actin containing) filaments. In striated muscle, thousands of myofibrils are arranged in a parallel array making the forces they generate additive and their ability to sense externally applied forces exquisite.

Striated muscles power locomotion for a wide range of animals. While all striated muscles produce force and motion via the cyclical interactions of molecular motors (myosin) with their substrate (actin), it is the specialized adaptations of a basic contractile mechanism that are used in various muscle types. For example, the fast contraction velocities observed in skeletal muscle are possible, at least partly, because of the fast ADP release rate of skeletal muscle myosin.[8] In contrast, delaying ADP release from myosin in smooth muscle[9] and scallop muscle can result in maintained force for an extended period of time without consuming ATP, the so-called latch state and catch state respectively. Yet another specialization of the contractile theme is provided by insect fibrillar flight muscle, which has the ability to contract at very high frequencies (up to 1000 Hz[10,11]). In some cases (e.g., the synchronous muscles), insect flight muscle has evolved an extensive sarcoplasmic reticulum to drive high wing beats. However, the ability of asynchronous (fibrillar) indirect flight muscle (IFM) to sense and transmit force (i.e., its high stiffness) allows high frequency contraction in the absence of extensive calcium cycling machinery.

Insect Flight Systems

To achieve flight many insects must beat their wings at very rapid rates. There are two types of flight muscle that drive insect wing beats: the direct system and the indirect system. Each muscle type can maintain high frequency (e.g., >100 Hz) contractions albeit through very different mechanisms. The two flight systems are described briefly below. For a comprehensive description of flight muscle physiology see the excellent review of comparative insect flight muscle physiology by R.K. Josephson (this volume).

Direct/Synchronous Muscles

In the direct system contraction of the power generating muscles directly move the wings. Contraction of the inner muscles raises the wings while contraction of the outer muscles lowers them thus generating vertical lift. Each contraction of the direct flight muscles is synchronous with nervous stimulation, for this reason they are also referred to as synchronous flight muscles.[1] In addition to direct flight muscles, there are many synchronous muscles that can contract at high frequencies (e.g., the rattle snake shaker muscle[12]). Because contraction and relaxation of synchronous muscles result from rapid calcium release, diffusion and reuptake, these muscles have evolved an extensive sarcoplasmic reticulum (SR) and narrow myofibrils. Narrow myofibrils appear to be the most direct predictor of muscle contraction frequency.[13] These adaptations illustrate the conventional trade off between muscle force and contraction frequency.[14] Despite the limitations on force production and the energetic demands of Ca^{++} cycling synchronous flight systems are used by several insect orders.

Indirect/Asynchronous/Fibrillar Muscles

This muscle type historically has three names based on physiological and ultrastructural observations. They are called indirect muscles because they do not directly move the wings, instead they move the wings indirectly via deformation of the thoracic cuticle. Contraction of the dorsal ventral muscles (DVM) raises the wings while contraction of the dorsal longitudinal muscles (DLM) lowers the wings.

Indirect muscles are also called asynchronous muscles. Unlike synchronous muscles, asynchronous muscles contract without direct (synchronous) nervous stimulation. Stimulation of the muscle at the wing beat frequency results in tetanus,[1] which is readily explained by the limited SR. The myofibril diameter is very large and a much larger proportion of muscle volume is occupied by myofibrils, hence the third name, fibrillar flight muscle. Because such a large proportion of muscle volume is dedicated to myofibrils, it is clear that indirect/asynchronous/fibrillar flight muscles (IFM) are designed for the high power demands of insect flight. The question that remains is how do these muscles achieve such high wing beats without an extensive SR for calcium release and reuptake?

Stretch/Stress Activation

Asynchronous flight muscles contract in an oscillatory manner if they are activated and attached to an under-damped inertial load with a resonant frequency near the wingbeat.[15,16] The oscillatory contractions of the muscle require nervous stimulation; however, the rate of nervous stimulation is much lower than the contraction frequency. It was suggested[17] that these unusual contractions resulted from either a stress sensitive intracellular membrane of the excitation-contraction coupling system or that it might be an intrinsic property of the myofilaments themselves (responding to the length changes imposed by the inertial load). In support of the latter, oscillatory contractions were observed in *Lethocerus* flight muscle where the cell membrane and sarcoplasmic reticulum had been permeabilized with glycerol.[18]

Asynchronous muscle contracts against an inertial load because applying a length change results in a transient tension response as the muscle relaxes toward a new steady state (Fig. 1). Abruptly stretching the muscle results in a rapid increase in force, a decay in force and then a delayed tension rise, generally called stretch activation (Fig. 1). However, it is equally important to note that releasing the muscle results in an opposite and symmetrical response, which results in a delayed decrease in tension. If the timing of the delay matches the characteristic frequency of the resonant system then the muscle will perform oscillatory work. In insects, the resonant system consists of the elasticity of the flight muscles and thorax coupled to the inertia of the wings.[19] Although oscillatory work is often equated with stretch activation it is clear that the phenomenon can be driven by both a delayed tension rise (stretch activation) and delayed tension fall (release deactivation) after a length perturbation.

A transient tension response to a length or, since muscle is viscoelastic, force perturbation is a general property of all muscles.[1,20-22] However, there are important differences that depend on the tissue studied. In skeletal muscle, the delayed tension response is of lower amplitude (Fig. 2)[20,23] than in cardiac muscle, which in turn is of lower amplitude than insect fibrillar muscle.[23,24] The stiffness of relaxed cardiac muscle and IFM are considerably higher than the stiffness of relaxed vertebrate striated muscle (Fig. 2)[25] suggesting that passive stress is linked with the enhanced transient tension response.

The name stretch activation suggests that IFM can be activated by mechanical stretch alone. However, applying stress to relaxed (low calcium) IFM show no signs of activation. Rather, the fiber responds like a passive viscoelastic material. Calcium binding to the thin filament regulatory proteins is required for activation. There is a clear increase in myosin ATPase and force when *Lethocerus* muscle is stretched before[26] or after[27] calcium activation. Phosphate-water oxygen exchange studies also show a direct effect of stress on myosin kinetics.[28] At least two oxygen exchange pathways (two phosphate release pathways) are present in isometrically contracting rabbit psoas muscle. When unstrained, calcium activated IFM

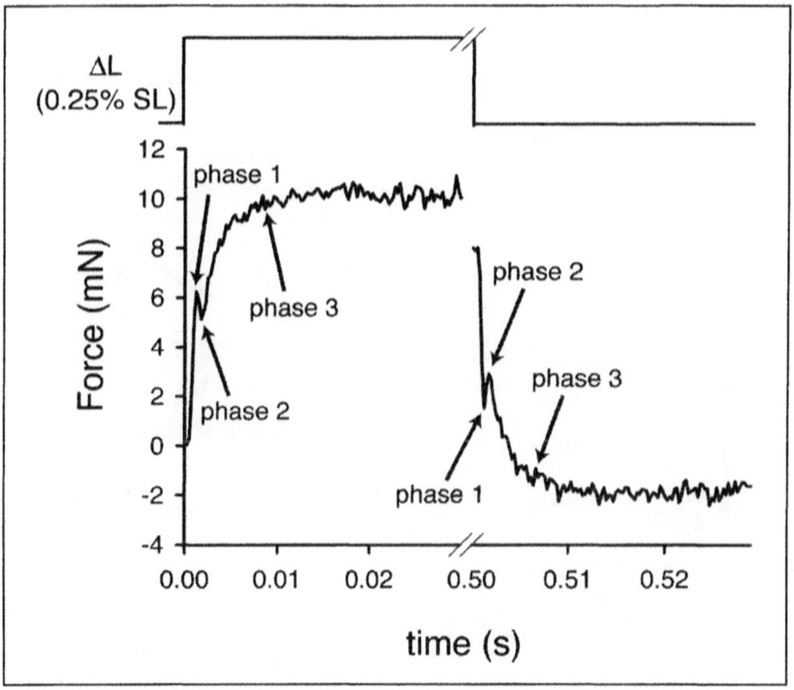

Figure 1. Tension transient of a *Drosophila* IFM fiber in response to a 0.25% step increase followed by a 0.25% step decrease in muscle length at 12 °C. The transient force response after a step stretch exhibits multiple phases. (1) A rapid increase in tension, (2) a decay, and (3) a delayed increase in tension that is generally referred to as stretch activation. The response to a step decrease in length is symmetrical for small length perturbations: phase 1, rapid decay in tension; phase 2, an increase; phase 3, a delayed decrease in tension (J. Moore, unpublished result).

also show two pathways; however, when IFM are stretched, there is a significant increase in ATPase activity,[28] which can be explained by an increase in phosphate release via a single oxygen exchange pathway. In contrast, there is no change in oxygen exchange when skeletal muscle is stretched. Because IFM have a considerable passive stiffness compared to skeletal muscle (Fig. 2), the same amount of stretch would produce a greater stress in IFM than skeletal muscle, again suggesting that changes in muscle stress, not length, influence the number of crossbridges available for cycling in IFM.

Models That Explain the Effect of Stress on Tension

It has been shown that stretch activated tension of IFM results from an increase in the number of crossbridges bound to actin.[29] Several models, which fall into four general categories, have been proposed to explain this crossbridge recruitment: (1) helical matching of thick and thin filaments in the sarcomere (2) connecting filament stress (3) RLC phosphorylation and (4) stress affects the thin filament regulatory system.

Helical Matching

One model of stretch activation is based on the helical structures of flight muscle thick and thin filaments (Fig. 3). Since both thick and thin filament monomers have a long-pitch helical repeat of ~38 nm in fibrillar muscles, stretch induced filament sliding could serve to align actin target sites with myosin heads.[30] During steady muscle lengthening the model predicts that the

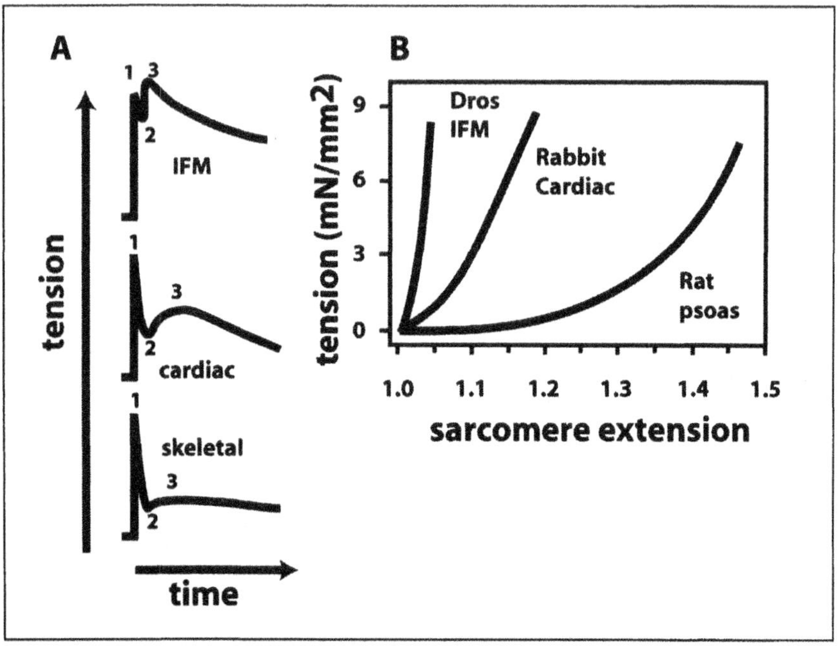

Figure 2. The amplitude of the stretch activation response correlates with passive stiffness. A) Tension transients in response to a step length change are a general property of all striated muscles. However, the amplitude of the delayed increase in tension (phase 3) depends on the source tissue. Conceptual tension transients are shown and phases 1, 2 and 3 are labeled as in Figure 1. B) Passive tension vs. sarcomere length relationship for skeletal muscle, cardiac muscle and IFM myofibrils. Cardiac muscle and IFM have greater passive stiffness (slope) when compared to skeletal muscle. Redrawn from Kulke M, Neagoe C, Kolmerer B et al. Kettin, a major source of myofibrillar stiffness in *Drosophila* indirect flight muscle. J Cell Biol 2001; 154:1045-1057, by copyright permission of The Rockefeller University Press.

probability of acto-myosin interaction changes cyclically as the optimal actin binding sites alternate between being in and out of alignment with myosin crossbridges (Fig. 3). Although this model is attractive and strengthened by experimental evidence in *Lethocerus* muscle,[31] Squire[5] has shown that not all fibrillar muscles possess the helical matching required by the Wray[30] model. Furthermore, cardiac muscle lacks helical matching but displays stretch activation.

Stress via Connecting Filaments

Other models imply a stress-induced change in the rate of crossbridge cycling or recruitment. Thorson and White[32] formulated a mathematical model based on a two-state model like that of Huxley[33] in which stress acts by increasing the number of heads available (recruitment) or by altering either the rate of attachment or detachment (or both). Unfortunately, this model provides little explanation for the molecular basis of changes in crossbridge cycling. It has been proposed that stress in the connecting filaments that connect the Z-band to the thick filament in striated muscle affect myosin activity.[1,34] Indeed, passive stress correlates with the degree of stretch activation (Fig. 2),[25] but what could be the underlying molecular mechanism?

Lattice Spacing Changes

Changes in myofilament lattice spacing have been shown to influence the probability of actomyosin interaction (reviewed in Millman[35]). Garamvolgyi[36] was the first to propose that changes in lattice spacing could be responsible for stretch activation in insect flight muscle;

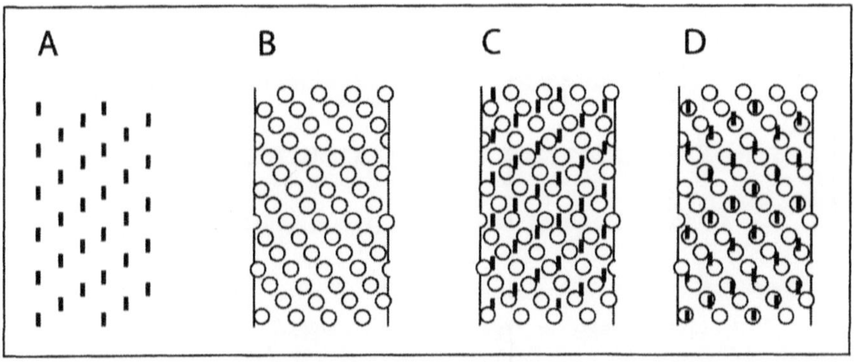

Figure 3. The match mismatch model. A) Actin target sites that would be presented to a myosin thick filament are represented by dashes. The circles in B represent the position of myosin heads on the thick filament. Sliding of the filaments relative to each other places myosin heads in either an inhibited (C) or optimal (D) position for binding to actin target sites. Figure redrawn with permission from Wray J. Filament geometry and the activation of insect flight muscles. Nature 280:325-326, ©1979 Macmillan Publishers Ltd.

however there were no known proteins or structures (besides myosin crossbridges) that could convert a change in length (or force) to a lattice spacing change.

The layout of connecting filaments in the sarcomere suggests a mechanism for how alterations in muscle length could result in lattice spacing changes. In IFM, the connecting filaments are thought to consist primarily of projectin[37] and kettin.[25] Kettin has been proposed to interact with the thin filament near the Z-band on one end and with the thick filament on the other end (Fig. 4A). Therefore, a change in muscle length should result in production of both a longitudinal and a radial force. The radial force would tend to decrease the separation between the thick and thin filaments (Fig. 4A).

In vertebrate cardiac muscle, connecting filaments consist mainly of titin, which, like kettin, binds to actin near the Z-band and associates with the thick filament in the A-band. Titin's layout in the sarcomere was proposed to be responsible for generating passive force when cardiac myocytes are either stretched above or allowed to shorten below their resting sarcomere length.[38] Also, it has been recently proposed that titin-based passive force is responsible for a form of stress-dependent activation in cardiac muscle.[39] Cazorla et al[39] showed that increasing muscle length, and thus passive force, resulted in an increase in calcium sensitivity. As predicted, the increase in calcium sensitivity was correlated with a decrease in the separation between thick and thin filaments.

Lattice spacing changes which were greater than what would be expected from conservation of volume, were also observed in electron micrographs of bee flight muscle fixed at different degrees of extension;[36] however, this may be a fixation artifact.[40] In contrast, Maughan and Irving[41] measured lattice spacing changes in *Drosophila* insect flight muscle during tethered flight in vivo and showed that there was no change in myofilament spacing between wings down (when the DLM were allowed to shorten) and wings up (when the DLM were stretched). Furthermore, effects of passive stress on cardiac muscle Ca^{++} sensitivity have also been observed in the absence of lattice spacing changes.[42] Therefore, although lattice spacing changes can affect the activation state of cardiac muscle, it is not a universal mechanism that can explain stretch activation in IFM.

Thick Filament Stress Affects Myosin Activity

Connecting filaments have been implicated in another mechanism for the effect of stretch (stress) on muscle activity.[1,32,34] The model in Figure 4B shows how passive stress could affect the

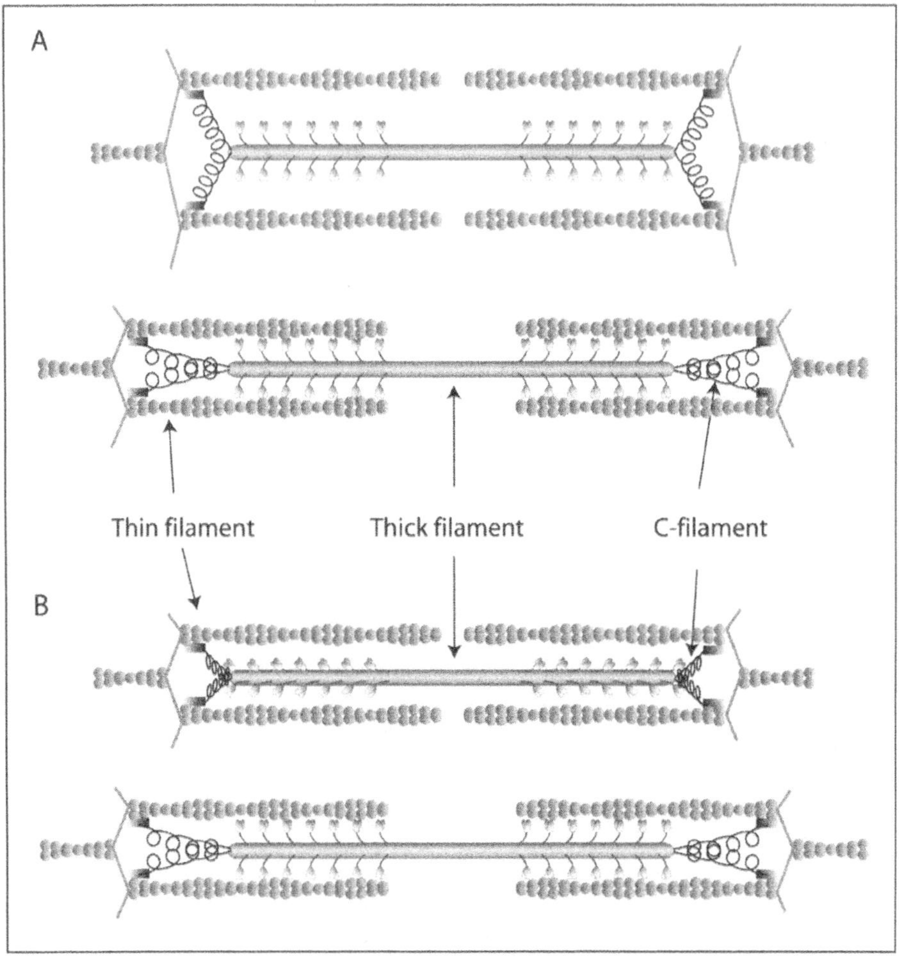

Figure 4. Connecting filament based models for stretch activation. A) Proposed model describing the effects connecting filament based stretch on myofilament spacing.[39] The connecting filament binds to the thin filament near the Z-band and to the ends of the thick filament.[25] In this geometry, stretching the muscle leads to a force vector that has both longitudinal and radial components. The radial forces tend to pull the thick and thin filaments closer thus increasing the probability of acto-myosin interaction. B) Proposed model describing the effects of connecting filament stress on the position of myosin heads. Stress in the thick filament disrupts interactions of the myosin heads with the thick filament backbone allowing them to project toward the thin filament. As in A, the proximity to actin increases the probability of myosin binding to actin. Spacing differences have been exaggerated for illustration purposes.

thick filament in such a way that changes the probability of actomyosin interaction without a lattice spacing change. The effect of passive stress on cardiac muscle myosin activity was recently proposed to result from a specific interaction of the A-band portion of titin with myosin.[43]

The A-band region of cardiac titin contains a super-repeat array of immunoglobulin-like (Ig) and fibronectin-type III (Fn3) domains.[44] Muhle-Goll et al,[43] showed that fibronectin domains representing a particular subset of the A-band super-repeat, bind the S1 proteolytic fragment (globular head) of myosin. Interactions between the myosin head and the thick filament backbone may keep myosin heads in a position that is less favorable for actin binding.

In support of this hypothesis, Muhle-Goll et al[43] showed that incubation of skinned cardiac muscle fibers with an expressed fibronectin domain mimicked the increased calcium sensitivity observed when pre-stressing the fibers.[39] To explain this effect it was proposed that the fibronectin domains in the A-band interact with both the rod and the head domains of the myosin molecules at rest length thus keeping the myosin heads close to the filament backbone. Myosin heads would therefore be in a conformation that has a reduced probability for interacting with actin when there is no stress on the titin molecules. Stress in the titin molecule relieves this inhibition and allows the myosin head to move toward the thin filament.

Although this model was proposed to explain Starling's Law in cardiac muscle, the same phenomenon (with smaller length changes) may give rise to stretch activation and release deactivation. When the muscle is stretched, stress in the connecting filament disrupts the interaction of the connecting filament with the myosin head thus moving the average position of the myosin head toward the actin filament (Fig. 4B). Consequently, the likelihood of actomyosin interaction would be increased resulting in a delayed increase in force following a length perturbation (i.e., stretch activation). In the case of a decrease in connecting filament tension the opposite response would ensue, a delayed fall in tension (release deactivation; Fig. 1). As mentioned above, there are several candidates, acting alone or in combination, for a structure that can transmit length changes to stress on the thick and thin filaments in IFM: projectin, which connects the Z-bands to the thick filaments, and kettin, which connects actin filaments in the Z-band region to the thick filaments.

Kettin contains a repeat pattern of Ig domains with Fn3 domains at the C-terminus[45] and projectin contains a super repeat pattern of Ig and Fn3 domains similar to that found in A-band titin.[46] It has been suggested that projectin function depends on sarcomere location. For example, the *C. elegans* homolog of projectin (twitchin) is located in the A-band and mutant twitchin alleles result in an uncoordinated phenotype. Many of the supressors of twitchin null mutants are mutations in myosin suggesting that twitchin may play a regulatory role in muscle contraction.[47] In tubular muscles of insects, projectin is also located only in the A-band where it is believed to perform a similar function to twitchin. In IFM, there are conflicting reports of the location of projectin. In one, projectin is found in its typical A-band location as well as the I-band.[48] In others it is found only in the I-band,[49,50] where it is thought to fulfill a structural role. If found in the A-band, the fibronectin domains of projectin could interact with the myosin heads in a similar way as has been proposed for titin in cardiac muscle (described above). On the other hand, if projectin is found only in the I-band of IFM, other proteins with myosin binding properties (e.g., kettin,[25] flightin,[51-53] paramyosin[54] and stretchin MLCK[55,56]) or myosin filaments themselves may provide a connection, that can allow the transmission of changes in muscle length from I-band projectin to myosin heads. Recent experiments on the mechanical properties of flight muscle fibers devoid of flightin provide evidence that flightin is necessary for normal passive stiffness. The reduction in passive stiffness was also accompanied by a reduction in oscillatory work (stretch activation) suggesting that flightin is an important component in the pathway that converts muscle stretch to activation.[53]

Flightin could be involved in the interaction of projectin or kettin with the thick filament. Alternatively, flightin has been shown to bind to the LMM portion of the myosin rod[52] and may function to crosslink myosin molecules within the thick filament. Along these lines, Tawada and Kawai[57] have shown that crosslinking of rabbit psoas fibers with the chemical crosslinker, EDC, resulted in an enhanced "insect-like" oscillatory work. Although these results were interpreted as an increase in the crosslinking of actin and myosin, the stability of the filament structures in high salt[57] indicates that the filamentous portion of the thick filament was crosslinked as well. Thus the high stiffness and enhanced stretch activation could be explained, at least partly, by an increase in the thick filament stiffness. Flightin, an IFM specific protein, may function as a natural crosslinker that serves to reinforce the thick filament. Such reinforcements would enhance the ability of the thick filament to respond to length changes and/or transmit myosin generated forces.

The similar responses of IFM and cardiac muscle to stretch suggest that they may act via a similar mechanism. Passive stress could disrupt interactions between myosin and the connecting filament (or other associated proteins). Indeed, stress in the connecting filaments has been shown to elongate *Lethocerus* thick filaments.[58,59] The conformational changes necessary for thick filament lengthening would likely affect the interaction between connecting filament and myosin or between myosin molecules in the thick filament.[59] Although the exact nature of the connecting filaments in flight muscle and an interaction between the connecting filament and myosin needs to be established, flight muscles have several proteins (e.g., kettin, projectin, and flightin) that could perform roles similar to those proposed for titin in vertebrate cardiac muscle.

Regulatory Light Chain (RLC) Phosphorylation

The proposed effect of stress on the position of myosin heads in cardiac muscle and IFM is similar to the observed effects of RLC phosphorylation in striated muscles. Unlike smooth muscle myosin, which is regulated by RLC phosphorylation, striated muscle myosin is enzymatically active unless inhibited by thin filament regulatory proteins. Although myosin is "constitutively on" in striated muscle, RLC phosphorylation has been shown to potentiate muscle activity. RLC phosphorylation increases isometric force production at submaximal calcium concentrations (reviewed in ref. 60), though the effect is minimal. Recently, however, large effects of RLC phosphorylation on myofibrillar ATPase and calcium sensitivity have been observed.[61] A molecular mechanism for force potentiation by RLC phosphorylation was provided by the observation that myosin heads containing dephosphorylated RLC remain close to the thick filament backbone while myosin heads containing phosphorylated RLC swing out away from the filament backbone towards the actin filament.[62] Therefore, the proximity of myosin and actin when the RLC is phosphorylated leads to an increased probability of myosin binding to actin and subsequently, enhanced muscle force.

Skinned flight muscle fibers from transgenic *Drosophila* that lack the myosin light chain kinase substrate serines on the RLC exhibit a marked reduction in oscillatory work.[2,3] It was proposed that the diminished stretch activation resulted from a reduction in the number of myosin heads available to interact with actin without affecting the kinetics of those that can.[3] Consistent with this idea, it has been shown by in vivo X-ray diffraction that there is a reduction in the number of myosin heads bound to the thin filament of transgenic flies expressing the mutant RLC.[41] Thus the reduced stretch activation and oscillatory work may result from a similar mechanism as was proposed for cardiac muscle (see above). That is, an inhibitory interaction between the myosin heads containing dephosphorylated RLC and the thick filament backbone prevents the effect of stress on the recruitment of myosin crossbridges. In the wild-type fly, the interaction between myosin heads containing phosphorylated RLC and the thick filament is weaker and can be overcome by thick filament stress thus leading to stretch activation.

Recent low angle X-ray diffraction patterns of relaxed *Lethocerus* IFM show an apparently inhibitory contact between one head of a myosin dimer and the essential light chain region of an adjacent myosin molecule.[6] Although not the same contact, a similar intramolecular inhibition has been proposed to regulate smooth muscle myosin activity.[6] It is intriguing to propose that this inhibitory contact is influenced by phosphorylation and that mechanical perturbation of the contact could give rise to stretch activation.

Stress Augments Thin Filament Activation

Another model states that IFM at low levels of Ca^{++} concentration can be fully activated by mechanical stretch. Stress was proposed to either (1) change the affinity of muscle regulatory proteins for calcium or (2) affect the ability of the troponin/tropomyosin system to prevent myosin binding to actin.[63]

After an abrupt 2-6% stretch, insect flight muscle[64] and cardiac muscle[64,65] show a rapid increase in tension, a decay, and a delayed increase in tension that is followed by several oscillations in isometric tension. The oscillations, which are thought to be a manifestation of stretch activation, have been proposed to result from an increase in thin filament tension.[66] The molecular mechanism for such an effect has been proposed to involve a flight muscle specific form of tropomyosin, so called troponin heavy (TnH) because it copurifies with the troponin complex.[63] In *Drosophila*, TnH has been suggested to interact with both the thick and thin filaments thus allowing length changes to be directly transmitted to the Ca^{++} regulatory machinery.[63] In *Lethocerus*, homology between TnH and portions of TnI have been proposed to be responsible for the effects of stretch on the troponin complex.[7] However, the normal oscillatory work observed in IFM isolated from *Drosophila*, which are haploid for TnH fail to provide support for its role in stretch activation.[67] Furthermore, TnH is not found in several insect species that have stretch activated IFM thus indicating that TnH is not a universal requirement for stretch activation.[68]

Phosphorylation of thin filament proteins has been shown to modulate thin filament Ca^{++} sensitivity in cardiac muscle.[69] For example, phosphorylation of cardiac TnI decreases the affinity of TnC for $Ca.^{++}$ Projectin from *Locusta migratoria* has been shown to phosphorylate a 30 kD protein believed to be TnI suggesting that projectin may modulate thin filament sensitivity.[70] Interestingly, reduction of projectin kinase in *Drosophila* heterozygous for the projectin mutation, *bt*^{D}, resulted in an altered rate and amplitude of stretch activation.[71] If indeed TnI is the substrate for projectin kinase in *Drosophila*, then reduced TnI phosphorylation could be responsible for the altered IFM mechanical performance. However, phosphorylation of TnI by projectin kinase requires another, yet to be identified, soluble kinase[72] or the distribution of projectin throughout the A-band in IFM, which is a contentious issue (see above). Furthermore, there is no published evidence that *Drosophila* TnI is phosphorylated.

No increase in calcium binding is observed upon stretch[73] thus suggesting that stretch does not directly affect calcium binding by troponin C (TnC). However, an intriguing model involving a flight muscle isoform of TnC has been proposed.[4] In this model, it was hypothesized that IFM are only partially activated by calcium binding to a small number of TnC molecules, which have two calcium binding sites (F2TnC). Full activation was achieved by stress acting on the remaining troponin/tropomyosin complexes that contain a different TnC isoform with only one, nonregulatory, calcium binding site (F1TnC). This model has recently gained experimental support.[7] Agianian et al,[7] show that fibers containing F1TnC exhibit stretch activation with minimal isometric force generation and that stretch activation is greatest at low Ca^{++} levels. On the other hand, fibers containing F2TnC displayed significant Ca^{++} activated isometric tension with minimal stretch activation. Thus, stretch activation and isometric force generation are mutually exclusive and the degree of stretch activation depends on the type of TnC that the IFM contain. *Lethocerus* IFM contain both F1TnC and F2TnC at a ratio of ~5:1[4] and, as expected, skinned fibers with the native complement of TnC behave more like F1TnC fibers than F2TnC fibers.

Whether stretch affects muscle activity via TnH or TnC, it is unclear how stretch applied to the muscle could result in a change in the activation state of the thin filament. It has been suggested that the effect of stretch on tropomyosin position (and ultimately myosin binding) could be transmitted directly through the thin filaments.[1] However, this requires the presence of a mechanical link between adjacent half sarcomeres. There is no evidence for thin filament-thin filament connections crossing the M-line. Another, more reasonable, suggestion is that muscle length changes act indirectly on the thin filament via thick filament-thin filament crosslinks that may include TnH,[63] RLC,[2] slowly cycling crossbridges[1] or kettin[25] (Fig. 5).

Such thick-thin filament crosslinks would be expected to bear a load and may give rise to the IFM's characteristically high relaxed stiffness. Consistent with this proposal, removal of the N-terminus of the *Drosophila* RLC resulted in a decrease in the in-phase stiffness of isolated IFM skinned fibers.[74] On the other hand, treatment of IFM myofibrils with Igase, a protease that preferentially digests one isoform of TnH, suggests that TnH contributes little to passive stiffness.[25] Slowly cycling crossbridges present in relaxed muscle may account for some passive

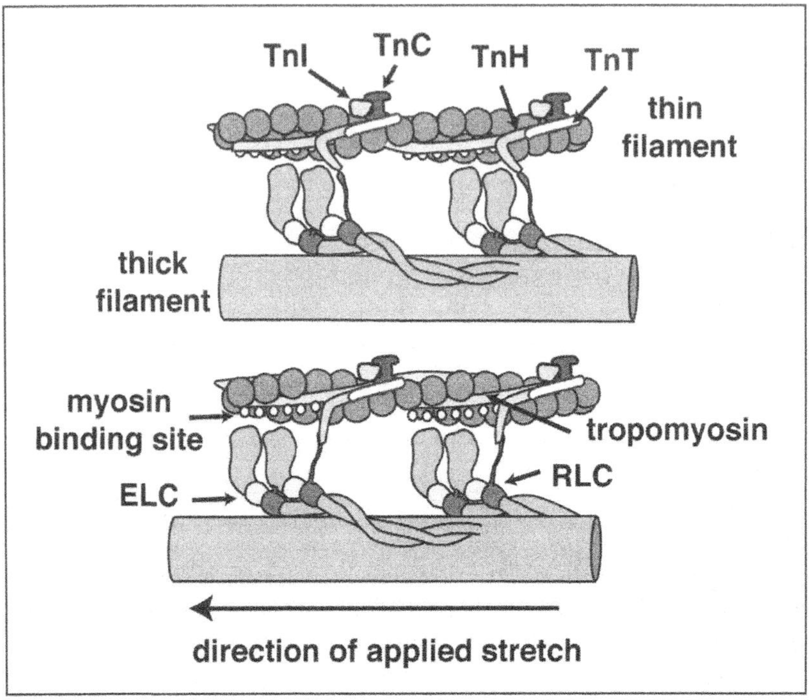

Figure 5. A proposed model to explain stretch effects on thin filament regulatory proteins. Proteins in the insect filament include: actin, the troponin complex (TnC, TnI, TnT, and TnH) and tropomyosin. The myosin heavy chain (MHC) and the myosin light chains (regulatory, RLC, and essential, ELC) are components of the thick filament. In striated muscles regulation of contraction is primarily via the thin filament. Calcium enters the muscle cell and binds TnC. TnI then removes its inhibition of acto-myosin binding. Activation in flight muscle has been proposed to be regulated by another mechanism via a unique form of tropomyosin (TnH), a component of the Tn complex with its hypothesized C-terminal extension interacting with the N-terminal extension of the RLC.[2] When stretch is applied to the thick filament the resulting movement (arrow) pulls the TnH molecule and shifts tropomyosin position thus exposing myosin binding sites on actin (white circles). Binding of myosin crossbridges leads to an increase in tension after the applied stretch (stretch activation). For simplicity other components of the thick and thin filaments (e.g., arthrin, GST-2) are not shown. Based on references 2 and 63.

stiffness; however, their role in stretch activation is questionable since relaxed IFM are not stretch activated (see above).

While there is experimental support for an effect of stretch on thin filament regulation and the unique proteins found in IFM provide possible mechanisms for the effect of stretch on tropomyosin position, evidence against such an effect is provided by electron microscopy and 3-D helical reconstructions of *Drosophila* IFM thin filaments. These studies show normal calcium induced movements of tropomyosin between two distinct positions.[75,76] Therefore, the EM structural studies of IFM thin filaments show the hallmarks of full thin filament activation without the need for stretch. This apparent discrepancy between structural[76] and biochemical/mechanical[7] data can be explained by the relatively high Ca^{++} concentrations used in the EM studies. At the same high Ca,$^{++}$ skinned fibers with the native TnC complement displayed full activation of isometric tension. For the small amount of F2TnC to fully activate the filament the effect of Ca^{++} binding to the single regulatory site of F2TnC must be cooperatively transmitted ~35 actin subunits, which is ~3-fold greater than the cooperative effect observed in skeletal muscle.[77]

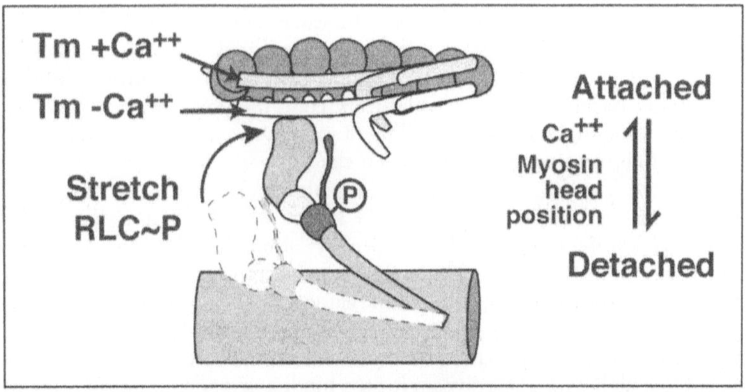

Figure 6. Model to explain the effects of Ca^{++}, stretch and phosphorylation on IFM contractile activity. In this model stretch does not act directly on the thin filament regulatory system as was illustrated in Figure 5. Instead, initial myosin binding to the thin filament is proposed to be regulated by two primary mechanisms: (1) A Ca^{++} induced shift in tropomyosin (Tm) position. (2) The proximity of the myosin heads to the thin filament. High Ca^{++} levels, stretch and phosphorylation all promote acto-myosin formation. The position of tropomyosin is dynamic and the distribution between the two positions depends on Ca^{++} concentration. In the absence of Ca^{++} tropomyosin spends most of its time in a position that blocks myosin binding. At high Ca^{++} levels tropomyosin spends most of its time in a position that allows myosin binding. It is proposed that IFM functions at low Ca^{++} levels where most of the myosin binding sites are blocked. In the absence of stretch and phosphorylation, the tropomyosin position inhibits myosin binding. Activation at this low Ca^{++} is achieved via a stretch and phosphorylation dependent increase in the myosin binding rate via disruption between the myosin head and the thick filament backbone. Tropomyosin position is shown in white in the absence of Ca^{++} and in gray in the presence of saturating Ca^{++}. Most of the troponin complex (see Fig. 5) is not shown. Myosin binding sites are indicated with white circles as in Figure 5.

Bringing It All Together

It appears that components of both thick and thin filaments determine the response to stretch, therefore any model for stretch activation must incorporate components of both (Fig. 6). It is thought that the highly cooperative activation of the thin filament involves both Ca^{++} binding to TnC and myosin binding to actin.

In the absence of Ca^{++}, tropomyosin is in a position that blocks the myosin binding site on actin. However, when TnC binds calcium TnI detaches from actin thus allowing the Tn/Tm complex to slide along the actin filament surface. Tm's new position exposes binding sites necessary for strong myosin binding. The position of the tropomyosin molecule is dynamic. That is, at any given time, a given percentage of the actin binding sites are exposed and a given percentage are blocked. The fraction of time Tm remains in each state (and thus the average positon shown in Figure 6 depends on the Ca^{++} concentration. The higher the Ca^{++} concentration the larger the fraction of time the Tn/Tm regulatory system allows access to the myosin binding sites on actin. This shift increases the probability of myosin binding to actin. Additionally, any other factors, such as stretch and RLC phosphorylation, that modify the ability of the myosin molecule to bind the actin filament will increase the degree of activation for a given Ca^{++} concentration. The initial binding of myosin promotes further myosin binding by (1) reducing tropomyosin mobility and (2) by shifting the Tm molecule further and exposing additional myosin binding sites on actin.

In the model, IFM thin filaments are mostly in the inhibited state, largely as a result of a TnC molecule that lacks the regulatory binding sites for Ca^{++} (F1TnC). When F1TnC is the only isoform of TnC in the fiber isometric contractions are reduced even in the presence of high Ca^{++} concentrations (up to pCa 5.5-4.7).[7] Under normal physiological conditions, where there is a small amount of the calcium binding form of TnC (F2TnC), isometric contractions can be achieved at high Ca^{++} concentrations.[7] This can be consistent with EM structural data,[76] although an exceptionally large cooperative unit is required (see above). Under conditions where the RLC is dephosphorylated or thick filament stress is low the myosin heads are close to the thick filament backbone or in an inhibitory conformation.[6] With myosin in this state, tropomyosin can effectively block the binding of myosin to actin (Fig. 6). When the muscle is stretched, the inhibitory interactions between the thick filament backbone and the myosin heads are disrupted. The degree of disruption (and ultimately stretch activation) would depend on the stiffness of the muscle (i.e., the same % stretch would produce a greater force on the thick filament for a stiffer muscle) and the phosphorylation state of the RLC.[2,3] The movement of heads away from the filament backbone (or out of an inhibited state) would cause the myosin heads to spend more time close to the actin filament. The proximity to actin would result in an increased probability of myosin binding to actin. Relaxed IFM cannot be stretch activated, suggesting that the increased proximity of myosin to actin is not sufficient for myosin to successfully compete with with tropomyosin for the binding site. For activation, the equilibrium between tropomyosin positions must change. The IFM fiber mechanical studies suggest a threshold Ca^{++} concentration is needed before stretch activation can be achieved.[7] Thus, a threshold Ca^{++} concentration causes a shift in the distribution of tropomyosin positions. This shift, coupled with an increased proximity of the myosin head to actin, allows myosin to bind, and cooperatively activate, the thin filament. Teleologically, a threshold calcium requirement for stretch activation seems necessary so that the flight muscle activity can be controlled (albeit asynchronously) by nervous stimulation. It is also possible that mechanical perturbation of the small number of attached heads at threshold calcium influence tropomyosin position and thus cause stretch activation. However, if this were the case, explaning the phenotype of the *Drosophila* phosphorylation site mutants[2] seems more difficult.

Concluding Remarks

Indirect flight muscles are activated by calcium binding to the regulatory proteins of the thin filament and the activation state is augmented by stretch applied by the opposing IFM. While stretch activation is present in all muscles, the extent of augmentation is proportional to the stiffness of the muscle indicating that stress potentiates muscle activity. It follows that stretch activation is enhanced in the stiffer cardiac and indirect flight muscles where it is critical for oscillatory contraction.

While the underlying molecular mechanism for stretch activation is unknown, a large body of information on cardiac muscle and insect flight muscle physiology suggests that a change in sarcomere length transmits stress to the thick filament. Thick filament stress causes a recruitment of force generating myosin crossbridges possibly by decreasing the binding of myosin heads to the thick filament backbone or relieving inhibitory inter-head contacts. Dephosphorylation of RLC may inhibit the effects of stress and phosphorylation may enhance it. In addition to thick filament stress, stretch sensitive calcium regulatory proteins also have been proposed to underly stretch activation. By substituting TnCs containing one or two Ca^{++} binding sites, this model has recently gained significant experimental support.[7]

It appears that components of both thick and thin filaments determine the response to stretch. It is possible that there is no universal mechanism for stretch activation; however, a model is proposed here that incorporates contributions of both thick and thin filament regulation to stretch activation (Fig. 6). It is proposed that flight muscle operates largely with the thin filament in an inhibitory state (at submaximal Ca^{++} activation). Myosin head mobility and proximity to the thin filament (regulated by stretch and phosphorylation) allows myosin to compete for tropomyosin position at low (threshold) calcium levels. This model suggests that IFM operate at submaximal (permissive) Ca^{++} concentrations in vivo. Future measurements of in vivo Ca^{++} levels in contracting IFM may shed light on this hypothesis.

The availability of contractile protein mutants in *Drosophila melanogaster*[78] and cardiac muscle[79-81] provide a unique opportunity to study the myofibrillar interactions necessary for stretch activation. Future studies with both existing and novel contractile protein mutations will illuminate the interactions that transmit the externally applied stress to myosin crossbridges and provide additional details about the role of thin filament activation in stretch activation.

Small angle X-ray diffraction patterns from flight muscles of *Lethocerus* and *Drosophila* seem particularly promising. Extending on the work of Al-Khayat et al,[82] modeling of the myosin head configuration in relaxed-slack and relaxed-pretensioned IFM may confirm the proposed effects of thick filament stress induced myosin head position. In addition, the myosin head configuration in relaxed IFM of the *Drosophila* RLC phosphorylation site mutants should be particularly informative.

Muscle fiber X-ray diffraction data[83,84] and electron micrographic 3-D reconstructions[85,86] have been used to measure shifts in tropomyosin position as a result of myosin or Ca^{++} binding. Changes in tropomyosin position at submaximal Ca^{++} concentrations or in response to stretch using these methods have not been performed but warrant further study. These proposed studies, along with time resolved X-ray diffraction experiments in vitro[29] and in vivo,[41] will likely shed light on the mechanism of stretch activation.

References

1. Pringle JW. The Croonian lecture, 1977. Stretch activation of muscle: Function and mechanism. Proc R Soc Lond B Biol Sci 1978; 201:107-130.
2. Tohtong R, Yamashita H, Graham M et al. Impairment of muscle function caused by mutations of phosphorylation sites in myosin regulatory light chain. Nature 1995; 374:650-653.
3. Dickinson MH, Hyatt CJ, Lehmann FO et al. Phosphorylation-dependent power output of transgenic flies: An integrated study. Biophys J 1997; 73:3122-3134.
4. Qiu F, Lakey A, Agianian B et al. Troponin C in different insect muscle types: Identification of two isoforms in Lethocerus, Drosophila and Anopheles that are specific to asynchronous flight muscle in the adult insect. Biochem J 2003; 371:811-821.
5. Squire JM. Muscle filament lattices and stretch-activation: The match-mismatch model reassessed. J Muscle Res Cell Motil 1992; 13:183-189.
6. Al Khayat HA, Hudson L, Reedy MK et al. Myosin head configuration in relaxed insect flight muscle: X-ray modeled resting cross-bridges in a prepowerstroke state are poised for actin binding. Biophys J 2003; 85:1063-1079.
7. Agianian B, Krzic U, Qiu F et al. A troponin switch that regulates muscle contraction by stretch instead of calcium. EMBO J 2004; 23:772-779.
8. Baker JE, Brosseau C, Joel PB et al. The biochemical kinetics underlying actin movement generated by one and many skeletal muscle myosin molecules. Biophys J 2002; 82:2134-2147.
9. Baker JE, Brosseau C, Fagnant P et al. The unique properties of tonic smooth muscle emerge from intrinsic as well as intermolecular behaviors of Myosin molecules. J Biol Chem 2003; 278:28533-28539.
10. Sotovalta O. The flight tone (wing stroke frequency) of insects. Acta Entomol Fenn 1947; 4:111-117.
11. Sotovalta O. Recordings of high wing-stroke and thoracic vibration frequency in some midges. Biol Bull Woods Hole 1953; 104:439-444.
12. Martin JH, Bagby RM. Properties of rattlesnake shaker muscle. J Exp Zool 1973; 185:293-300.
13. Young D, Josephson RK. 100 Hz is not the upper limit of synchronous muscle contraction. Nature 1984; 309:286-287.
14. Josephson RK, Malamud JG, Stokes DR. Asynchronous muscle: A primer. J Exp Biol 2000; 203(Pt 18):2713-2722.
15. Pringle JW. The excitation and contraction of the flight muscles of insects. J Physiol (Lond) 1949; 108:226-232.
16. Machin KE, Pringle JW. The physiology of insect fibrillar flight muscle. II. Mechanical properties of a beetle flight muscle. Proc R Soc Lond B Biol Sci 1959; 151:204-225.
17. Machin KE, Pringle JW. The physiology of insect fibrillar muscle. III. The effect of sinusoidal changes of length on a beetle flight muscle. Proc R Soc Lond B Biol Sci 1960; 152:311-330.
18. Jewell BR, Ruegg JC. Oscillatory contraction of insect fibrillar flight muscle after glycerol extraction. Proc R Soc Lond B Biol Sci 1964; 164:428-459.
19. Hyatt CJ, Maughan DW. Fourier analysis of wing beat signals: Assessing the effects of genetic alterations of flight muscle structure in Diptera. Biophys J 1994; 67:1149-1154.
20. Huxley AF, Simmons RM. Proposed mechanism of force generation in striated muscle. Nature 1971; 233:533-538.

21. Steiger GJ. Stretch activation and myogenic oscillation of isolated contractile structures of heart muscle. Pflugers Arch 1971; 330:347-361.
22. Warshaw DM, Fay FS. Tension transients in single isolated smooth muscle cells. Science 1983; 219:1438-1441.
23. Abbott RH, Steiger GJ. Temperature and amplitude dependence of tension transients in glycerinated skeletal and insect fibrillar muscle. J Physiol 1977; 266:13-42.
24. Steiger GJ. Tension transients in extracted rabbit heart muscle preparations. J Mol Cell Cardiol 1977; 9:671-685.
25. Kulke M, Neagoe C, Kolmerer B et al. Kettin, a major source of myofibrillar stiffness in Drosophila indirect flight muscle. J Cell Biol 2001; 154:1045-1057.
26. Chaplain RA. Changes in adenosine triphosphatase activity and tension with fiber elongation in glycerinated insect flight muscle. Pflugers Arch 1969; 307:120-126.
27. Guth K, Poole JV, Maughan D et al. The apparent rates of crossbridge attachment and detachment estimated from ATPase activity in insect flight muscle. Biophys J 1987; 52:1039-1045.
28. Lund J, Webb MR, White DC. Changes in the ATPase activity of insect fibrillar flight muscle during calcium and strain activation probed by phosphate-water oxygen exchange. J Biol Chem 1987; 262:8584-8590.
29. Tregear RT, Edwards RJ, Irving TC et al. X-ray diffraction indicates that active cross-bridges bind to actin target zones in insect flight muscle. Biophys J 1998; 74:1439-1451.
30. Wray J. Filament geometry and the activation of insect flight muscles. Nature 1979; 280:325-326.
31. Abbott RH, Cage PE. A possible mechanism of length activation in insect fibrillar flight muscle. J Muscle Res Cell Motil 1984; 5:387-397.
32. Thorson J, White DCS. Distributed representations for actin-myosin interaction in the oscillatory contraction of muscle. Biophys J 1969; 9:360-390.
33. Huxley AF. Muscle structure and theories of contraction. Prog Biophys & Biophys Chem 1957; 7:255-317.
34. Granzier HL, Wang K. Interplay between passive tension and strong and weak binding cross-bridges in insect indirect flight muscle. A functional dissection by gelsolin-mediated thin filament removal. J Gen Physiol 1993; 101:235-270.
35. Millman BM. The filament lattice of striated muscle. Physiol Rev 1998; 78:359-391.
36. Garamvolgyi N. Forces acting between muscle filaments. I. Filament lattice spacing in bee flight muscle. Acta Biochim Biophys Acad Sci Hung 1972; 7:157-164.
37. Saide JD. Identification of a connecting filament protein in insect fibrillar flight muscle. J Mol Biol 1981; 153:661-679.
38. Helmes M, Trombitas K, Granzier H. Titin develops restoring force in rat cardiac myocytes. Circ Res 1996; 79:619-626.
39. Cazorla O, Wu Y, Irving TC et al. Titin-based modulation of calcium sensitivity of active tension in mouse skinned cardiac myocytes. Circ Res 2001; 88:1028-1035.
40. Pringle JW. The mechanical characteristics of insect fibrillar flight muscle. In: Tregear R, ed. Insect Flight Muscle. Elsevier/North Holland Biomedical Press, 1977:177-208.
41. Irving TC, Maughan DW. In vivo X-ray diffraction of indirect flight muscle from Drosophila melanogaster. Biophys J 2000; 78:2511-2515.
42. Konhilas JP, Irving TC, de Tombe PP. Myofilament calcium sensitivity in skinned rat cardiac trabeculae: Role of interfilament spacing. Circ Res 2002; 90:59-65.
43. Muhle-Goll C, Habeck M, Cazorla O et al. Structural and functional studies of titin's fn3 modules reveal conserved surface patterns and binding to myosin S1—a possible role in the Frank-Starling mechanism of the heart. J Mol Biol 2001; 313:431-447.
44. Tskhovrebova L, Trinick J. Role of titin in vertebrate striated muscle. Philos Trans R Soc Lond B Biol Sci 2002; 357:199-206.
45. Bullard B, Linke WA, Leonard K. Varieties of elastic protein in invertebrate muscles. J Muscle Res Cell Motil 2002; 23:435-447.
46. Ayme-Southgate A, Vigoreaux J, Benian G et al. Drosophila has a twitchin/titin-related gene that appears to encode projectin. Proc Natl Acad Sci USA 1991; 88:7973-7977.
47. Benian GM, Kiff JE, Neckelmann N et al. Sequence of an unusually large protein implicated in regulation of myosin activity in C. elegans. Nature 1989; 342:45-50.
48. Royuela M, Fraile B, De Miguel MP et al. Immunohistochemical study and western blotting analysis of titin-like proteins in the striated muscle of Drosophila melanogaster and in the striated and smooth muscle of the oligochaete Eisenia foetida. Microsc Res Tech 1996; 35:349-356.
49. Saide JD, Chin-Bow S, Hogan-Sheldon J et al. Characterization of components of Z-bands in the fibrillar flight muscle of Drosophila melanogaster. J Cell Biol 1989; 109:2157-2167.

50. Vigoreaux JO, Saide JD, Pardue ML. Structurally different Drosophila striated muscles utilize distinct variants of Z-band-associated proteins. J Muscle Res Cell Motil 1991; 12:340-354.
51. Vigoreaux JO, Saide JD, Valgeirsdottir K et al. Flightin, a novel myofibrillar protein of Drosophila stretch-activated muscles. J Cell Biol 1993; 121:587-598.
52. Ayer G, Vigoreaux JO. Flightin is a myosin rod binding protein. Cell Biochem Biophys 2003; 38:41-54.
53. Henkin JA, Maughan DW, Vigoreaux JO. Mutations that affect flightin expression in Drosophila alter the viscoelastic properties of flight muscle fibers. Am J Physiol Cell Physiol 2004; 286:C65-C72.
54. Vinos J, Domingo A, Marco R et al. Identification and characterization of Drosophila melanogaster paramyosin. J Mol Biol 1991; 220:687-700.
55. Champagne MB, Edwards KA, Erickson HP et al. Drosophila stretchin-MLCK is a novel member of the Titin/Myosin light chain kinase family. J Mol Biol 2000; 300:759-777.
56. Patel S, Saide JD. A(225), A novel A-band protein of Drosophila indirect flight muscle. Biophys J 2001; 80:71a.
57. Tawada K, Kawai M. Covalent cross-linking of single fibers from rabbit psoas increases oscillatory power. Biophys J 1990; 57:643-647.
58. Trombitas K, Tigyi-Sebes A. The continuity of thick filaments between sarcomeres in honey bee flight muscle. Nature 1979; 281:319-320.
59. Irving T, Wu Y, Fukuda N et al. Changes in sarcomeric structure as a function of titin-based passive tension in skeletal muscle. Biophys J 2004; 86:188a.
60. Morano I. Tuning the human heart molecular motors by myosin light chains. J Mol Med 1999; 77:544-555.
61. Szczesna D, Zhao J, Jones M et al. Phosphorylation of the regulatory light chains of myosin affects Ca2+ sensitivity of skeletal muscle contraction. J Appl Physiol 2002; 92:1661-1670.
62. Levine RJ, Kensler RW, Yang Z et al. Myosin regulatory light chain phosphorylation and the production of functionally significant changes in myosin head arrangement on striated muscle thick filaments. Biophys J 1995; 68:224S.
63. Reedy MC, Reedy MK, Leonard KR et al. Gold/Fab immuno electron microscopy localization of troponin H and troponin T in Lethocerus flight muscle. J Mol Biol 1994; 239:52-67.
64. Steiger GJ. Stretch activation and tension transients in cardiac, skeletal and insect flight muscle. In: Tregear RT, ed. Insect Flight Muscle. North Holland, 1977:221-268.
65. Schadler M, Steiger GJ, Ruegg JC. Mechanical activation and isometric oscillation in insect fibrillar muscle. Pflugers Arch 1971; 330:217-229.
66. Smith DA. Quantitative model for Schadler's isometric oscillations in insect flight and cardiac muscle. J Muscle Res Cell Motil 1991; 12:455-465.
67. Kreuz AJ, Simcox A, Maughan D. Alterations in flight muscle ultrastructure and function in Drosophila tropomyosin mutants. J Cell Biol 1996; 135:673-687.
68. Peckham M, Molloy JE, Sparrow JC et al. Physiological properties of the dorsal longitudinal flight muscle and the tergal depressor of the trochanter muscle of Drosophila melanogaster. J Muscle Res Cell Motil 1990; 11:203-215.
69. van der Velden, Papp Z, Zaremba R et al. Increased Ca2+-sensitivity of the contractile apparatus in end-stage human heart failure results from altered phosphorylation of contractile proteins. Cardiovasc Res 2003; 57:37-47.
70. Weitkamp B, Jurk K, Beinbrech G. Projectin-thin filament interactions and modulation of the sensitivity of the actomyosin ATPase to calcium by projectin kinase. J Biol Chem 1998; 273:19802-19808.
71. Moore JR, Vigoreaux JO, Maughan DW. The Drosophila projectin mutant, bentD, has reduced stretch activation and altered indirect flight muscle kinetics. J Muscle Res Cell Motil 1999; 20:797-806.
72. Vigoreaux JO, Moore JR, Maughan DW. Role of the elastic protein projectin in stretch activation and work output of Drosophila flight muscles. Adv Exp Med Biol 2000; 481:237-247.
73. Marston S, Tregear RT. Calcium binding and the activation of fibrillar insect flight muscle. Biochim Biophys Acta 1974; 347:311-318.
74. Moore JR, Dickinson MH, Vigoreaux JO et al. The effect of removing the N-terminal extension of the Drosophila myosin regulatory light chain upon flight ability and the contractile dynamics of indirect flight muscle. Biophys J 2000; 78:1431-1440.
75. Ruiz T, Bullard B, Lepault J. Effects of calcium and nucleotides on the structure of insect flight muscle thin filaments. J Muscle Res Cell Motil 1998; 19:353-364.
76. Cammarato A, Hatch V, Saide J et al. Drosophila muscle regulation characterized by electron microscopy and three-dimensional reconstruction of thin filament mutants. Biophys J 2004; 86:1618-1624.

77. Maytum R, Lehrer SS, Geeves MA. Cooperativity and switching within the three-state model of muscle regulation. Biochemistry 1999; 38:1102-1110.
78. Bernstein SI, O'Donnell PT, Cripps RM. Molecular genetic analysis of muscle development, structure, and function in Drosophila. Int Rev Cytol 1993; 143:63-152.
79. Rayment I, Holden HM, Sellers JR et al. Structural interpretation of the mutations in the beta-cardiac myosin that have been implicated in familial hypertrophic cardiomyopathy. Proc Natl Acad Sci USA 1995; 92:3864-3868.
80. Seidman JG, Seidman C. The genetic basis for cardiomyopathy: From mutation identification to mechanistic paradigms. Cell 2001; 104:557-567.
81. Vemuri R, Lankford EB, Poetter K et al. The stretch-activation response may be critical to the proper functioning of the mammalian heart. Proc Natl Acad Sci USA 1999; 96:1048-1053.
82. al-Khayat HA, Yagi N, Squire JM. Structural changes in actin-tropomyosin during muscle regulation: Computer modelling of low-angle X-ray diffraction data. J Mol Biol 1995; 252:611-632.
83. Huxley HE. The structural basis of muscular contraction. Proc R Soc Lond B Biol Sci 1971; 178:131-149.
84. Parry DA, Squire JM. Structural role of tropomyosin in muscle regulation: Analysis of the x- ray diffraction patterns from relaxed and contracting muscles. J Mol Biol 1973; 75:33-55.
85. Lehman W, Craig R, Vibert P. Ca(2+)-induced tropomyosin movement in Limulus thin filaments revealed by three-dimensional reconstruction. Nature 1994; 368:65-67.
86. Xu C, Craig R, Tobacman L et al. Tropomyosin positions in regulated thin filaments revealed by cryoelectron microscopy. Biophys J 1999; 77:985-992.

Section II
Components of the Myofibril

CHAPTER 5

Myosin

Becky M. Miller and Sanford I. Bernstein

Abstract

The molecular motor myosin, composed of two heavy chains and four light chains, is responsible for defining both structural and mechanical properties of insect flight muscle. Myosin polymerizes into thick filaments that are a major component of the sarcomeric units of myofibrils. In the presence of Ca^{2+}, the globular head of myosin interacts with actin-containing thin filaments to generate force and movement in an ATP-dependent fashion. While myosin biochemical properties have been studied in only a few insects to date, the tools of molecular genetics have revealed that multiple isoforms of insect myosin exist in a single species, with specialized isoforms accumulating in flight muscles. In at least some insect species, isoforms of myosin heavy chain and the essential light chain arise from the process of alternative splicing of transcripts from a single gene. Mutations in *Drosophila* myosin, in conjunction with molecular modeling, implicate particular amino acid residues in thick filament assembly, sarcomere stability and ATPase activity. Molecular genetic approaches and transgenic technology in *Drosophila* are proving powerful in demonstrating how structural elements of myosin affect functional properties at the biochemical, fiber and whole organism levels. These integrative studies show that properties of the indirect flight muscle are critically dependent on the specific myosin isoform expressed.

Myosin Structure and Function

The cycle of shortening and lengthening in insect flight muscle is dependent upon transient interactions between myosin-containing thick filaments and actin-containing thin filaments. Myosin II of insect flight muscle, like myosin II's of other invertebrate and vertebrate muscles, is composed of two heavy chains, two essential light chains and two regulatory light chains (Fig. 1). Myosin heavy chain (MHC) (~200 kDa monomer) contains an N-terminal head domain and a C-terminal rod region. Initial dimerization and subsequent multimerization of myosin occurs through the rod region (~1150 amino acid residues). The multimerized rods serve as the main constituent of the thick filament. The head (~850 amino acid residues) contains the nucleotide-binding site, actin-binding site and the lever arm domain that binds the light chains. The head protrudes from the thick filament surface, forming a cross-bridge upon transient strong binding to actin, but spends the remainder of the contraction cycle in a detached or weakly bound state.[1]

Significant conformational changes occur in myosin as it progresses through its biochemical cycle.[2-5] These include reconfiguration of the nucleotide-binding pocket, opening and closing of a long cleft that runs between the actin-binding site and the active site, and movement of the lever arm (Fig. 2). Thermodynamic coupling of the actin and nucleotide-binding domains ensures that the chemical energy released during ATP hydrolysis results in mechanical movement of the lever arm. Power generated from the lever arm stroke promotes muscle

Nature's Versatile Engine: Insect Flight Muscle Inside and Out, edited by Jim Vigoreaux.

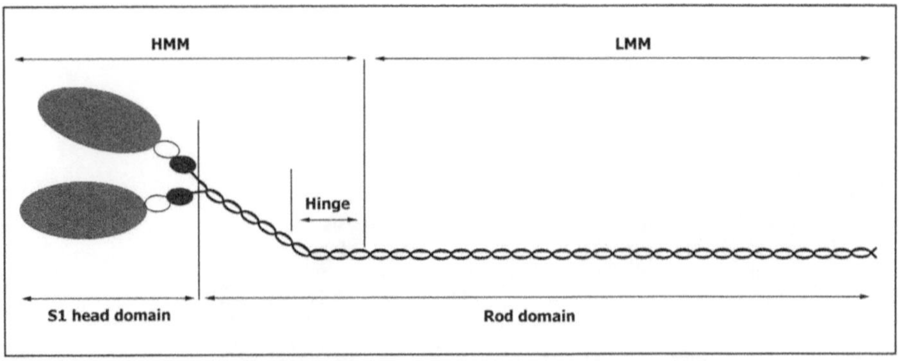

Figure 1. Proteolytic fragments of myosin. Prior to assembling into the thick filament, myosin molecules dimerize via their alpha-helical rod regions to produce the hexameric complex diagrammed here. Partial digestion with α-chymotrypsin or papain cleaves the molecule just C-terminal to the regulatory light chain (black oval) binding domain, producing the S1 head and the rod domain. The RLC can be digested away during α-chymotrypsin incubation, while the essential light chain (white oval) is retained.[14] Biochemical analysis of S1 obviates difficulties associated with full-length myosin where the rod domain causes aggregation and precipitation in low ionic solutions. Treatment of vertebrate myosin dimers with trypsin cleaves the molecule on the C-terminal side of the hinge region resulting in heavy meromyosin (HMM) and light meromyosin (LMM), while prolonged incubation liberates the hinge region. At least some insect myosins lack the rod chymotryptic cleavage sites found in vertebrates.

contraction by thin filament sliding. This force-generating step occurs concurrently with attachment to actin and is generally thought to take place just prior to or simultaneous with phosphate release.[3,6-8] Summation of the forces generated by individual myosin powerstrokes provides the force for sarcomeric shortening. ADP is released after the powerstroke and produces the rigor state; subsequent binding of ATP to the rigor myosin conformation liberates the myosin head from actin and thus governs exit from the attached state.

Myosin ATPase activity has been studied for only a few insect flight muscle isoforms. It is measured in the presence of Mg^{2+} (or Ca^{2+}, K^+, NH_4^+) and can be potentiated by addition of actin. Basal and actin-activated ATPase activities have been reported for *Lethocerus*[9] and *D. melanogaster* myosin isoforms.[10-12] Insect myosin ATPase activity is less stable than that of vertebrate myosins.[11,13] However, production of the S1 head fragment by proteolysis (Fig. 1) results in more stable ATPase activity,[14,15] similar to that seen with vertebrate S1.

Recent evidence in *Drosophila* suggests stage- and tissue-specific myosin isoforms are primary determinants of muscle contractile properties,[10,16] as is the case for vertebrates.[17-19] Isoform-specific kinetic properties drive the differences in muscle mechanical output.[10,20] While extensive homologies among insect and vertebrate muscle myosins imply conservation of structural properties and functional mechanisms, some unique kinetic properties in *Drosophila* are becoming evident.[14]

Structural studies, mostly in noninsect systems, have significantly contributed to visualizing myosin's power stroke. Crystallographic analyses of myosin have identified regions within the head that change conformation depending on the molecule present in the nucleotide-binding pocket.[21-26] These structures, in conjunction with image reconstruction techniques that define the structure of S1 heads bound to actin,[25,27-31] resulted in the development of a detailed mechanism of actomyosin interaction, specifically proposing how conformational changes in the catalytic domain are converted to swinging of the lever arm domain. Further, high resolution reconstructions of cross-bridges in actively contracting[32,33] and relaxed[34] insect flight muscle yield insight into this process in vivo. A discussion of these data and their contributions to our understanding of myosin's function is presented in other chapters.

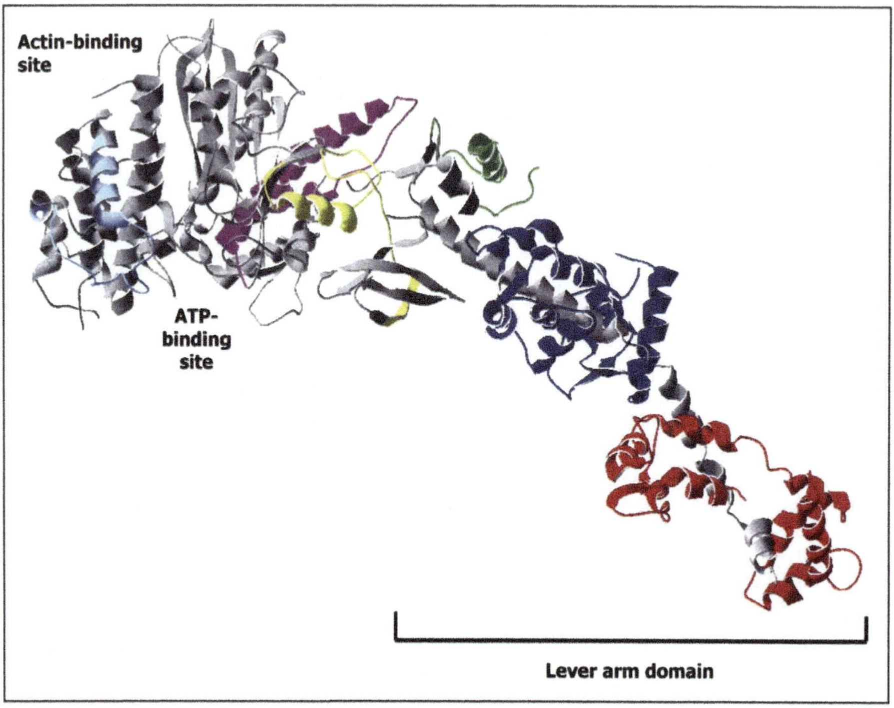

Figure 2. Crystal structure of myosin S1 fragment. Depicted is the 3-dimensional structure of scallop myosin heavy and light chains complexed with MgADP as determined by X-ray crystallography (PDB: 1B7T).[24] The essential light chain (ELC) is blue and the regulatory light chain (RLC) is red. The light chains interact with the long alpha-helix of the heavy chain and this complex is called the lever arm domain. The residues in the myosin heavy chain that are encoded by alternative exons in *Drosophila* are colored to illustrate their position (exon 3 domain is yellow, exon 7 light blue, exon 9 dark pink and exon 11 green). Two other alternative domains (hinge and tail regions) are located in the rod domain of the molecule and are therefore not shown; approximate positions can be seen in Figure 1. Insect flight muscle myosin has not been crystallized, but structural data from other myosins can assist in the interpretation of biochemical and mechanical data obtained from insect studies.

Myosin light chains bind to the alpha-helical lever arm of the heavy chain (Fig. 2). The essential light chain (ELC), also known as MLC1 or the alkali light chain, is a protein of ~20 kDa.[35] The role for the ELC in insect muscle contraction is not clear, and as yet there is no consensus regarding its function in vertebrates. The regulatory light chain (RLC), also designated MLC2, is a protein of ~30-34 kDa[36,37] and is important for regulation of muscle contraction. In *Drosophila,* RLC phosphorylation at Ser 66 and 67[38] is necessary for maximal ATPase activity,[39,40] recruitment of cross-bridges,[41] enhancement of stretch-activation of the IFM[38] and optimal flight function. The RLC has also been implicated in stabilizing the myosin neck[42] and is important for maintaining the appropriate structural separation of the myosin heads of a given dimer.[43] In *Drosophila* there is a unique RLC N-terminal extension of 46 amino acids, similar to the ELC extension in vertebrates that binds actin.[44,45] This extension is thought to be important in maintaining a link between the thick and thin filament parallel to the myosin cross-bridge.[46,47]

The heptad repeat of charged and hydrophobic amino acids within the C-terminal rod domain of MHC plays a role in assembly of monomers into dimers through formation of an α-helical coiled-coil (Fig. 1). Subsequently, these myosin dimers multimerize into hollow thick filaments

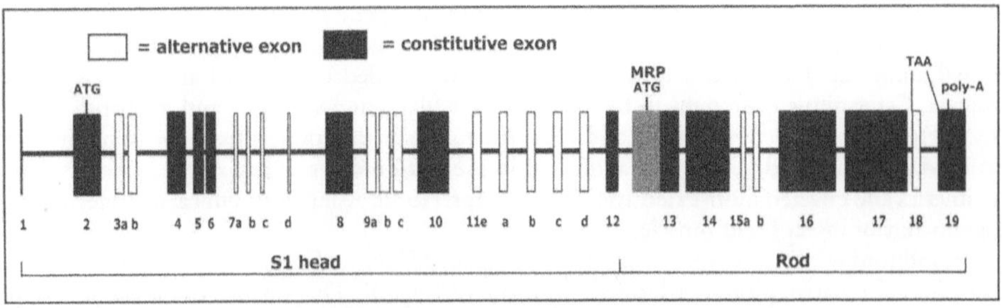

Figure 3. Structure of the *Drosophila* myosin heavy chain gene. The single muscle myosin heavy chain gene in *Drosophila melanogaster* contains 5 alternative exon sets (empty rectangles: 3, 7, 9, 11, 15), 13 constitutive exons (filled rectangles) and a single exon (18) that is either included or excluded.[58] One member of each alternative exon set is included in each *Mhc* transcript. The coding start site (ATG) is located in exon 2 and one of two alternative coding stop sites (TAA) is determined by the presence or absence of exon 18. Exon 19 contains two polyadenylation signals. The first 12 exons code for the S1 head (catalytic domain) and the remaining 7 exons code for the α-helical rod domain. Usage of a promoter upstream of exon 13 (gray rectangle) produces the myosin rod protein (MRP) that contains an identical rod domain to MHC.[70]

through higher order interactions between neighboring rod dimers, α-helical coiled-coil paramyosin core proteins and flightin.[48,49] Studies of the indirect flight muscle (IFM) of *Drosophila* show that the rod domain is critical for assembly[50] and stability[51,52] of the thick filament. Interestingly, the catalytic head domain also serves to fine-tune the assembly of thick filaments and sarcomeres.[50]

Myosin Isoform Expression

Muscle myosins are very well conserved,[53] but few insect myosins or their genes have been fully sequenced. For MHC, *Drosophila melanogaster* and *Drosophila hydei* protein coding potentials are known. Each has a single *Mhc* gene with an identical exon-intron pattern and a high degree of conservation (97%) in their protein coding regions.[54] It appears that a single *Mhc* gene exists in *Drosophila virilis* as well.[55] The *D. melanogaster Mhc* gene gives rise to multiple myosin isoforms through alternative RNA splicing.[56-61] Nineteen exons code for the MHC protein (Fig. 3). Thirteen of these are expressed in all transcripts whereas five exons offer two or more splice variants. The penultimate exon is either included or excluded in the mRNA. Given this gene structure, 480 myosin isoforms are theoretically possible, but only 15 are known to exist.[58,61-65] In situ hybridization analysis of thoracic muscles revealed expression of different myosin isoforms in IFM, direct flight muscles, and jump muscle.[62] The IFM MHC isoform is unique and appears to be critical for proper function of this specialized muscle type.[11,12,16,65]

Localization of the sequence diversity encoded by each of the *Drosophila Mhc* splice variants provides insight into myosin isoform functional diversity. Four of the six variable regions are found in the catalytic domain, one is at the hinge in the rod domain, and one is at the C-terminus of the protein.[52,66] Bernstein and Milligan[66] mapped the four catalytic alternative domains using the crystal structure of chicken skeletal S1. Figure 2 shows the locations of these *Drosophila* alternative domains on the scallop MHC structure. The N-terminal exon 3 splice variants encode residues near the reactive sulfhydryls and lever arm pivot point. The exon 7 series encodes amino acids forming one lip of the nucleotide-binding pocket and a portion of the surface of the myosin head. A kinked α-helical loop, a portion of which is termed the relay loop, is coded for by the exon 9 variants and shows minimal variation. The exon 11 series displays the greatest degree of sequence variation and codes for a portion of the converter domain, a region suggested to be important for amplification of the signals from the nucleotide-binding pocket to the lever arm domain.[67] Residues encoded by alternative exon 15 are found in the central region of the myosin rod hinge (Fig. 1), thought to be subject to melting and proposed to be linked to force production.[68] The presence or absence of the final alternative exon, 18, produces different C-termini. If included, this exon encodes a single terminal amino acid and if excluded, the 27 residues encoded by exon 19 are expressed. The myosin C-terminus is thought to be important for filament assembly[69] and myofibril stability.[52] Experimental determination of the modulatory role that these alternative domains play in IFM myosin function is briefly addressed in the section on transgenic myosin heavy chain chimerics and covered more extensively in the chapters addressing biochemical and mechanical parameters of insect flight muscle.

Standiford et al[70] uncovered a unique isoform of MHC in *Drosophila* that is produced by use of an alternative promoter located in intron 12 (Fig. 3). This myosin rod protein (MRP) has an identical rod domain with MHC but contains a unique N-terminal region of 77 residues that has homology to the *Drosophila* RLC N-terminal extension. MRP is expressed in some somatic, cardiac, and visceral myofilaments. Although it is included in three direct flight muscles of the adult, it is absent from the IFM. The MRP assembles into homodimers and becomes randomly integrated into the thick filament leading to less ordered thick and thin filament packing in muscle myofibrils.[71] It is postulated that lack of the catalytic domain in this isoform would decrease maximal power output in muscles where it is expressed.

Given the exon complexity of the *Drosophila melanogaster Mhc* gene, it is proving useful for analysis of the mechanism of alternative RNA splicing. Thus far, only sequences important for regulating exon 11 and exon 18 have been documented. The inclusion of a particular exon 11 isovariant is governed by local intronic *cis*-acting elements interacting with *trans*-acting factors whose expression varies in a developmental and tissue-specific manner.[72] In addition, nonconsensus 5'-splice sites are essential for the general regulation of exon 11 splicing, but not for specific alternative exon inclusion in a muscle-specific manner.[72] Splicing specificity of IFM-specific exon 11e is controlled by at least three conserved intronic elements.[72,73] Each has a specific function that acts upon the nonconsensus 5' splice site; CIE1 is a splice site repressor, whereas CIE2 and CIE3 behave as splice site enhancers.[73] Thus, inclusion of MHC 11e in IFM is the result of a combination of intronic elements and nonconsensus 5' splice sequences.

Neither exon 18's unusual purine-rich 3' splice site[74] nor extensive noncoding sequences within the exon[55] are critical for regulation of alternative splicing. However, the nonconcensus 5' and 3' splice junctions are necessary for exon 18 exclusion in larval muscle and a distant polypyrimidine tract in intron 17 is an essential positive regulator for exon 18 inclusion in adult muscle.[55] This suggests that inclusion of exon 18 in the IFM is due to the presence of positive *trans*-acting factors.

Myosin light chain expression patterns are far less complex than those for MHC. For *D. melanogaster*, the *Mlc2* (RLC) gene yields two transcripts with the same coding potential.[36,37,75] Phosphorylation of the RLC produces multiple isoforms that are observed on a two-dimensional gel.[76] Transcripts from the *Mlc1* (ELC) gene of *D. melanogaster* are alternatively spliced in their C-terminal coding regions to produce two isoforms, one of which is IFM specific.[77] In a study of related *Drosophila* species, Leicht et al[78] found that this tissue-specific pattern of alternative splicing is conserved in *D. simulans*, *D. pseudoobscura*, and *D. virilis*. Comparison of ELC coding potential among the four species revealed ~94-99% sequence identity, with complete conservation of residue charges.[78] Some noncoding DNA sequences also show significant conservation among these four genes, suggesting these are important *cis*-regulatory elements.[78]

Myosin Mutational Analysis

The *Drosophila melanogaster* system has been exploited to study myosin function through the use of genetic and transgenic approaches. The presence of single genes encoding muscle MHC, ELC and RLC allowed for mutagenesis and gene knockout followed by germline transformation[79] with engineered versions of the gene. The resulting transgenic flies can be tested for their molecular, biochemical, structural, physiological and behavioral defects. As a result of these efforts, a myriad of mutations are yielding insight into the functional contribution of myosin to muscle contraction and specifically to the physiology of insect flight muscle. These include mutations defining particular amino acids critical for myosin multimerization, for myofibril assembly and stability and for myosin's mechanochemical coupling mechanism.

Most myosin mutations have been identified in genetic screens designed to select for or against an IFM phenotype, such as flight ability or hypercontraction.[80-82] Mutants can be broadly distributed into four categories: (1) null/hypomorphic, (2) hypermorphic, (3) missense and (4) suppressor. Descriptions of each type of mutation and the phenotype of representative examples are included below. Detailed descriptions of all myosin mutations are available at the FlyBase website (http://flybase.bio.indiana.edu/), a comprehensive database of *Drosophila* information.

Myosin Heavy Chain Mutants

Null or hypomorphic mutations eliminate or reduce myosin accumulation in all or selective muscle types. In *Mhc1* mutants, the presence of a 101 bp deletion removes exon 5 and most of the intron preceding it, resulting in a stop-codon early in the mRNA.[83] Homozygous embryos fail to accumulate thick filaments, show no muscular contraction and die before reaching the larval stage.[83] Thus, *Mhc1* is a null MHC allele and reveals important insights into the relationship between gene copy number and muscle ultrastructure/function. A single copy of the null allele causes an ~50% reduction in thick filament accumulation in many muscle types (Fig. 4) but only flight and jump ability are severely affected. This is presumably because precise myofilament organization in other muscle types is not as critical for function. While this allele knocks out full-length MHC accumulation, the MRP protein is still produced as a result of its internal transcriptional promoter, yielding accumulation of thick filament-like structures in some embryonic muscles.[71]

Mutations generated by transposable element insertion in an intron of *Mhc* (*Mhc2, Mhc3, Mhc4*) result in premature termination of transcription.[84,85] These recessive lethal mutations appear to be hypomorphic in nature. Some MHC is produced by the mutant alleles when termination of transcription within the transposable elements does not occur and this is followed by splicing of introns containing these transcribed elements.[85]

Mhc10 is a hypomorphic allele causing a null phenotype in specific muscle types. The mutation is at the 3' splice site of exon 15a, producing unstable mRNA transcripts containing this exon.[57] As a result, MHC fails to accumulate in the IFM (Fig. 4) and jump muscles and is reduced in other adult muscle types. However, normal MHC levels are retained in embryonic and some adult muscles.[57,86] Other mutations in this class (e.g., *Mhc7, Mhc9, Mhc11*) affect specific muscle types depending upon the alternative exon that is mutated.

Hypermorphic mutations result in increased protein dosage. Just as optimal functioning of some muscle types is sensitive to reduction in MHC protein dosage (see above), overexpression of MHC also interferes with myosin assembly and function.[87] The IFM is particularly sensitive to changes in gene dosage of several muscle proteins.[80,87-91] Overexpression of MHC in the IFM results in excess misaligned thick filaments without myofibril cracking (Fig. 4).[87] Functional assessment of muscle performance revealed a reduction in jumping ability and a flightless phenotype, presumably due to myosin overexpression in the IFM and other muscles in the thorax important for flight. Sensitivity to MHC overexpression is also found in other muscle types.[87]

Missense mutations cause single amino acid changes and thus can elucidate the role of a particular amino acid in protein function. Determining the biochemical and biophysical properties of these mutant myosins provide insight into myosin's structural properties and its mechanochemical cycle. All of the MHC missense mutations studied to date occur in constitutive exons and therefore affect all muscle types. Generally, the most severe effects are found in the IFM, given its highly organized structure and rigorous mechanical requirements.

Mhc5 is a point mutation that changes residue 200 from glycine to aspartic acid.[92] Myofibril assembly appears to be normal in homozygotes but myofibrils quickly degenerate into a

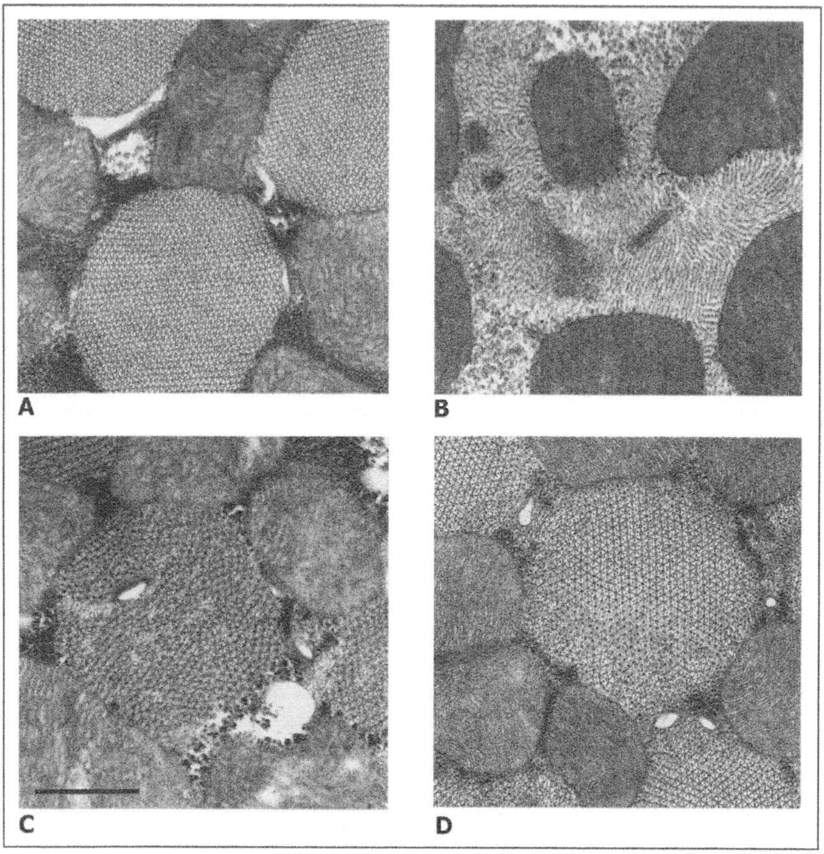

Figure 4. Effect of *Mhc* copy number on thick filament accumulation in the IFM. Expression and accumulation of myosin heavy chain in *Drosophila* is dependent upon the gene copy number. A) cross-section of Canton-S (wild-type) myofibril (two copies of *Mhc*) shows normal hexagonal packing of thick and thin filaments. B) cross-section of *Mhc10* mutant (MHC null mutant in the IFM) shows accumulation of only thin filaments.[86] C) cross-section of *Mhc1/+* mutant (single copy of *Mhc*) reveals ~50% reduction of the thick filaments.[83] D) cross-section of *pWMHC*; + transgenic line (four copies of *Mhc*) shows disruption of thick and thin filament packing at the myofibril peripheries due to excess accumulation of thick filaments.[87] Scale bar is equal to 0.55 μm.

disordered array of myofilaments in the adult (Kronert and Bernstein, unpublished data). The location of the mutated amino acid residue near the ATP entry site suggests it is involved in the ATPase function of MHC.

The *Mhc8* allele is recessive lethal and causes a substitution of histidine for tyrosine at amino acid 823.[92] IFM's of *Mhc8/Mhc10* flies show severely disrupted myofibrils that degenerate with age. However, areas of normal filament packing suggest a problem with stability rather than assembly (Kronert and Bernstein, unpublished data). The mutant residue is located in the RLC binding region of the myosin lever arm domain and may disrupt or destabilize the RLC interaction with myosin, reducing optimal myosin function.

Lastly, *Mhc13* converts a glutamic acid to a lysine at residue 1557 in the myosin rod domain.[51] Conservation of this charged residue in all muscle and nonmuscle myosins and in paramyosins suggests it plays a critical role. IFM myofibrils in homozygous flies appear normal at eclosion but over time the thick and thin filaments become more randomly arrayed and an increase in proteolysis of MHC is observed. The number of shorter and disorganized sarcomeres also increase with age suggesting that the IFM hypercontracts.[51] Recent in vitro evidence shows this mutation abolishes binding of the myosin rod domain to flightin.[93] In addition, in vivo, *Mhc13* flies have reduced flightin accumulation[51] leading to alterations in the passive mechanical properties of the IFM.[94] Taken together, these results suggest this interaction between flightin and the myosin rod domain is important for myosin and myofibril stability in the IFM (for further details, see chapter by Barton and Vigoreaux).

Suppressor mutations are isolated to identify molecular interactions between proteins. In principle, the original mutation and the suppressor mutation reveal direct protein-protein contacts important for myosin assembly or function. The interaction can also be indirect, either mediated by an intermediate protein or through altered function of a general muscle protein component that compensates for the original mutation by partially or completely restoring wild-type function. To isolate suppressor mutants, a mutant exhibiting a specific phenotype is exposed to a mutagen and its progeny are screened for individuals that display a partial or complete wild-type phenotype.

Two large-scale suppressor screens have been performed on the *hdp2* mutant, which has a single amino acid change in the thin filament protein troponin I.[95] *hdp2* flies have severe defects in the IFMs that cause their wings to be held up (*hdp*). Electron micrographic images of their IFMs reveal a hypercontraction phenotype.[92,95] The first screen of *hdp2* suppressors led to the isolation of four mutations (*D1, D41, D45, D62*) that suppressed the wings-up phenotype and mapped to chromosome II.[96] Kronert et al[92] determined that all four suppressor mutations are within a region of the *Mhc* gene coding for the motor domain. A second screen of *hdp2* produced three new *Mhc* alleles (*Su(2)A,B,C, Su(2)D, Su(2)F*) with mutations in the motor domain that fully suppress the hypercontraction phenotype.[82] The location of these mutations, in addition to the observation that flies expressing a headless myosin isoform show a similar phenotype, suggests that hypercontraction is suppressed either by decreasing the force MHC is able to produce or by affecting myosin regulation of muscle contraction.

Transgenic Myosin Heavy Chain Chimerics

Our laboratory has used germline transformation to produce a series of chimeric *D. melanogaster* myosin isoforms, designed to clarify the modulatory role of each alternative domain in MHC. Crossing these transgenes into either the *Mhc1* or *Mhc10* homozygous backgrounds yields *Mhc* gene replacement in either the whole organism or the IFM and jump muscles, respectively. One set of chimeric MHCs has the IFM isoform (IFI) backbone with a single alternative domain normally found in a native embryonic body wall muscle isoform (EMB). A second set of chimeric MHCs has the native EMB backbone with a single alternative domain that is expressed normally in IFI. Several recent studies have begun to define the role(s) for each alternative domain in MHC function with emphasis on insect flight muscle physiology.[10-12,14,16] A description of this class of mutants is included here with further discussion of their contribution to the understanding of insect flight muscle found in the chapter by Maughan.

Studying exon 3, which codes for amino acids near the N-terminus of MHC, Swank et al[12] found that the domain swaps had effects on either ATPase or in vitro actin motility. Thus, this domain independently influences at least two kinetic steps in the actomyosin ATPase cycle. IFM fiber power output was enhanced by the presence of the IFI domain in the EMB backbone, although like all EMB-based transgenics, flies were flightless. IFM fiber power output was reduced when the EMB exon was substituted into the IFI backbone, resulting in decreased flight ability.[97]

Amino acids coded by exon 7 form one outer lip of the ATP-binding pocket and a portion of the adjacent external surface. Actin-activated ATPase activity of both chimerics were similar to that of IFI.[98] These data suggest that inclusion of 7d in the EMB backbone significantly elevates ATPase activity while 7a in the IFI backbone has much less of an effect on ATPase activity. Actin filament velocities generated by the chimeric myosins reveal no differences compared to their backbones.[99] As a result, the exon 7 isovariant is postulated to modulate kinetic transitions of myosin when unbound or weakly bound to actin. Interestingly, in transient kinetic studies of these chimeric isoforms the ADP release rate without actin appears to be modulated by this domain but there is no effect with actin present.[100]

Residues coded for by exon 9 compose the relay helix, a central communication region[3] that responds to conformational changes in the actin and nucleotide-binding sites and propagates the structural changes to the converter domain through a direct interaction.[67] Substitution of EMB exon 9b into IFI reduces IFM power output and flight ability.[101] Studies are in progress to determine the contribution of this domain to myosin mechanochemistry.

Exon 11 residues code for a portion of the converter domain, which is proposed to be responsible for amplifying hydrolysis-induced conformational changes of the catalytic head into the swing of the lever arm domain.[67] Functional studies suggest the converter domain affects MHC kinetics and muscle fiber mechanics by changing the duty ratio, the fraction of time myosin is strongly bound to actin during its ATPase cycle.[10,16] While this domain alters in vitro motility, there was no observed change in step size[10] and very little influence on the rate of ADP release and ATP-induced actomyosin dissociation,[14] suggesting that the converter domain modulates earlier kinetic steps in strong-binding actomyosin interactions. The effects of the exon 11 switches on fiber power output are similar to those seen for the exon 3 swaps. The slowing of muscle kinetics resulting from insertion of the EMB converter into IFI necessitated a reduction in wing beat frequency to sustain flight.[16]

The central 26 amino acid residues of the S2 hinge domain of MHC are coded for by exon 15. Substitution of the EMB exon 15 domain into the IFI backbone yields a dramatic decrease in flight ability, with only modest effects on myofibril structure.[102] The hinge is hypothesized to affect either myofibrillar stiffness or myosin cross-bridge kinetics; its mechanism of action is currently under investigation.

The final alternative domain, exon 18, codes for a single C-terminal amino acid. Transcripts excluding exon 18 result in the addition of 27 C-terminal residues coded by exon 19. Expression of EMB in an IFM null *Mhc* mutant permits normal myofibril assembly, but results in severe degeneration.[65] Myofibrils expressing the embryonic isoform with the IFI C-terminus encoded by exon 18 show enhanced myofibril stability[52] suggesting the C-terminal domain is important for stabilizing the sarcomeric architecture.

Myosin Light Chain Mutants

Mutational analysis of insect myosin light chains has been pursued only in *Drosophila melanogaster* and then, only for the RLC (MLC2). A functionally null allele of the *Mlc2* gene results from a nonsense mutation at the tenth codon position (*Mlc2E38*); homozygotes exhibit a recessive lethal phenotype.[75,89] Heterozygotes, *Mlc2E38/+*, are dominant flightless. They show some ultrastructural differences in the myofibril periphery, have as much as a 50% reduction in the number of thick filaments, and display reduced wing beat frequency. These differences result in decreased contraction kinetics during sinusoidal analysis of skinned IFM fibers from these heterozygotes.[89]

The transgenic approach was used to study the importance of phosphorylation in RLC function.[38] Such phosphorylation enhances ATPase activity of locust, cricket, and *D. melanogaster* myosin.[39,103,104] Mutation of the phosphorylation sites (Ser 66 and 67) of *Drosophila* RLC revealed that light chain phosphorylation markedly enhances the stretch activation response and, therefore, power output, but has little effect on myofibril assembly or maximal isometric force development in the IFM.[38] The enhancement of stretch activation and power output appears to result not from altered kinetics of the cross-bridges themselves, but from phosphorylation-dependent recruitment of cross-bridges into the power-producing pool.[41]

A similar approach was used to study the N-terminal extension of RLC in *Drosophila*. Deletion of the N-terminal 46 amino acids from the RLC affects flight performance of homozygotes but not myofibrillogenesis. A reduction in passive stiffness rather than power-producing properties of the IFM was suggested to explain the observed depression in power-generating ability at the onset of flight.[46] Later studies conducted at more physiological lattice spacing, however, revealed a reduction in power-producing capability of skinned fibers at sub-maximal calcium activation.[105] The light chain extension may help preposition the myosin head for optimum force generation. Removal of this assist element would be especially deleterious when fewer heads are available for power generation, as at submaximal calcium activation. Taken together, these mutant studies suggest the RLC has several important roles in optimal performance of the asynchronous IFM.

Conclusion

Work on insect myosin expression, structure and function has progressed rapidly, particularly at the genetic and whole fiber levels. However, biochemical and biophysical studies have been hampered by the paucity of protein available. Given the fast pace of technological advances to study small quantities of contractile proteins, it is anticipated that future work assessing isolated insect muscle myosin will contribute greatly to our understanding of insect flight muscle function. The field would also be greatly stimulated by the preparation of adequate quantities of protein to determine the three-dimensional structure of insect myosins. Further, studies of insect myosin sequence and structure function relationships will be enhanced by ongoing (*Anopheles*, *Apis*) and future genome sequencing projects. The continued use of the genetically tractable insect, *D. melanogaster*, will provide powerful mutational analyses to further our understanding of the molecular mechanism of myosin mechanochemical coupling and the functions of particular residues in myofibril assembly. Insights from the integrative studies described here and in other chapters will facilitate understanding of myosin's contribution to the unique functional mechanism of asynchronous, oscillatory insect flight muscles.

Acknowledgments

We appreciate comments on the manuscript provided by Drs. David Maughan and Douglas Swank (University of Vermont). In addition, we thank Drs. Michelle Mardahl-Dumesnil and Patrick O'Donnell for providing the micrographs in Figure 4. We acknowledge research support from the National Institutes of Health (GM32443, AR43396) and the Muscular Dystrophy Association. BMM is the recipient of a predoctoral fellowship from the American Heart Association, Western States Affiliate.

References

1. Lymn RW, Taylor EW. Mechanism of adenosine triphosphate hydrolysis by actomyosin. Biochemistry 1971; 10:4617-4624.
2. Bagshaw CR, Trentham DR. The reversibility of adenosine triphosphate cleavage by myosin. Biochem J 1973; 133:323-328.
3. Geeves MA, Holmes KC. Structural mechanism of muscle contraction. Annu Rev Biochem 1999; 68:687-728.
4. Huxley AF. Muscle structure and theories of contraction. Prog Biophys Biophys Chem 1957; 7:255-318.
5. Rayment I, Rypniewski WR, Schmidt-Base K et al. Three-dimensional structure of myosin subfragment-1: A molecular motor. Science 1993; 261:50-58.
6. Fortune NS, Geeves MA, Ranatunga KW. Tension responses to rapid pressure release in glycerinated rabbit muscle fibers. Proc Natl Acad Sci USA 1991; 88:7323-7327.
7. Geeves MA, Goody RS, Gutfreund H. Kinetics of acto-S1 interaction as a guide to a model for the crossbridge cycle. J Muscle Res Cell Motil 1984; 5:351-361.
8. White HD, Taylor EW. Energetics and mechanism of actomyosin adenosine triphosphatase. Biochemistry 1976; 15:5818-5826.
9. White DC, Zimmerman RW, Trentham DR. The ATPase kinetics of insect fibrillar flight muscle myosin subfragment-1. J Muscle Res Cell Motil 1986; 7:179-192.
10. Littlefield KP, Swank DM, Sanchez BM et al. The converter domain modulates kinetic properties of Drosophila myosin. Am J Physiol Cell Physiol 2003; 284:C1031-1038.

11. Swank DM, Bartoo ML, Knowles AF et al. Alternative exon-encoded regions of Drosophila myosin heavy chain modulate ATPase rates and actin sliding velocity. J Biol Chem 2001; 276:15117-15124.
12. Swank DM, Knowles AF, Kronert WA et al. Variable N-terminal regions of muscle myosin heavy chain modulate ATPase rate and actin sliding velocity. J Biol Chem 2003; 278:17475-17482.
13. Tanikawa M, Ueyama A, Maruyama K. Instability of insect myosin ATPase activity and its protection. Comp Biochem Physiol 1987; 86B:63-65.
14. Miller BM, Nyitrai M, Bernstein SI et al. Kinetic analysis of Drosophila muscle myosin isoforms suggests a novel mode of mechanochemical coupling. J Biol Chem 2003; 278:50293-50300.
15. Silva R, Sparrow J, Geeves M. Isolation and kinetic characterisation of myosin and myosin S1 from the Drosophila indirect flight muscles. J Muscle Res Cell Motil 2003; 24:489-498.
16. Swank DM, Knowles AF, Suggs JA et al. The myosin converter domain modulates muscle performance. Nat Cell Biol 2002; 4:312-317.
17. Barany M. ATPase activity of myosin correlated with speed of muscle shortening. J Gen Physiol 1967; 50(Suppl):197-218.
18. Lowey S, Waller GS, Trybus KM. Function of skeletal muscle myosin heavy and light chain isoforms by an in vitro motility assay. J Biol Chem 1993; 268:20414-20418.
19. Marston SB, Taylor EW. Comparison of the myosin and actomyosin ATPase mechanisms of the four types of vertebrate muscles. J Mol Biol 1980; 139:573-600.
20. Tyska MJ, Warshaw DM. The myosin power stroke. Cell Motil Cytoskeleton 2002; 51:1-15.
21. Fisher AJ, Smith CA, Thoden JB et al. X-ray structures of the myosin motor domain of Dictyostelium discoideum complexed with MgADP.BeFx and MgADP.AlF4. Biochemistry 1995; 34:8960-8972.
22. Fisher AJ, Smith CA, Thoden J et al. Structural studies of myosin: Nucleotide complexes: A revised model for the molecular basis of muscle contraction. Biophys J 1995; 68:19S-26S; discussion 27S-28S.
23. Gulick AM, Bauer CB, Thoden JB et al. X-ray structures of the MgADP, MgATPgammaS, and MgAMPPNP complexes of the Dictyostelium discoideum myosin motor domain. Biochemistry 1997; 36:11619-11628.
24. Houdusse A, Kalabokis VN, Himmel D et al. Atomic structure of scallop myosin subfragment S1 complexed with MgADP: A novel conformation of the myosin head. Cell 1999; 97:459-470.
25. Rayment I, Holden HM, Whittaker M et al. Structure of the actin-myosin complex and its implications for muscle contraction. Science 1993; 261:58-65.
26. Smith CA, Rayment I. X-ray structure of the magnesium(II).ADP.vanadate complex of the Dictyostelium discoideum myosin motor domain to 1.9 A resolution. Biochemistry 1996; 35:5404-5417.
27. Flicker PF, Milligan RA, Applegate D. Cryo-electron microscopy of S1-decorated actin filaments. Adv Biophys 1991; 27:185-196.
28. Holmes KC, Angert I, Kull FJ et al. Electron cryo-microscopy shows how strong binding of myosin to actin releases nucleotide. Nature 2003; 425:423-427.
29. Volkmann N, Hanein D, Ouyang G et al. Evidence for cleft closure in actomyosin upon ADP release. Nat Struct Biol 2000; 7:1147-1155.
30. Volkmann N, Ouyang G, Trybus KM et al. Myosin isoforms show unique conformations in the actin-bound state. Proc Natl Acad Sci USA 2003; 100:3227-3232.
31. Whittaker M, Wilson-Kubalek EM, Smith JE et al. A 35-A movement of smooth muscle myosin on ADP release. Nature 1995; 378:748-751.
32. Beinbrech G, Ashton FT, Pepe FA. Orientation of the backbone structure of myosin filaments in relaxed and rigor muscles of the housefly: Evidence for nonequivalent crossbridge positions at the surface of thick filaments. Tissue Cell 1990; 22:803-810.
33. Reedy MC. Visualizing myosin's power stroke in muscle contraction. J Cell Sci 2000; 113:3551-3562.
34. Al-Khayat H, Hudson L, Reedy MK et al. Myosin head configuration in relaxed insect flight muscle: X-ray modeled resting cross-bridges in a prepowerstroke state are poised for actin binding. Biophys J 2003; 85:1063-1079.
35. Falkenthal S, Parker VP, Davidson N. Developmental variations in the splicing pattern of transcripts from the Drosophila gene encoding myosin alkali light chain result in different carboxyl-terminal amino acid sequences. Proc Natl Acad Sci USA 1985; 82:449-453.
36. Toffenetti J, Mischke D, Pardue ML. Isolation and characterization of the gene for myosin light chain two of Drosophila melanogaster. J Cell Biol 1987; 104:19-28.
37. Parker VP, Falkenthal S, Davidson N. Characterization of the myosin light-chain-2 gene of Drosophila melanogaster. Mol Cell Biol 1985; 5:3058-3068.

38. Tohtong R, Yamashita H, Graham M et al. Impairment of muscle function caused by mutations of phosphorylation sites in myosin regulatory light chain. Nature 1995; 374:650-653.
39. Takahashi S, Takano-Ohmuro H, Maruyama K. Regulation of Drosophila myosin ATPase activity by phosphorylation of myosin light chains—I. Wild-type fly. Comp Biochem Physiol B 1990; 95:179-181.
40. Takahashi S, Takano-Ohmuro H, Maruyama K et al. Regulation of Drosophila myosin ATPase activity by phosphorylation of myosin light chains—II. Flightless mfd- fly. Comp Biochem Physiol B 1990; 95:183-185.
41. Dickinson MH, Hyatt CJ, Lehmann FO et al. Phosphorylation-dependent power output of transgenic flies: An integrated study. Biophys J 1997; 73:3122-3134.
42. Vibert P, Craig R. Structural changes that occur in scallop myosin filaments upon activation. J Cell Biol 1985; 101:830-837.
43. Schaub MC, Jauch A, Walzthoeny D et al. Myosin light chain functions. Biomed Biochim Acta 1986; 45:S39-44.
44. Sutoh K. Identification of myosin-binding sites on the actin sequence. Biochemistry 1982; 21:3654-3661.
45. Trayer IP, Trayer HR, Levine BA. Evidence that the N-terminal region of A1-light chain of myosin interacts directly with the C-terminal region of actin. A proton magnetic resonance study. Eur J Biochem 1987; 164:259-266.
46. Moore JR, Dickinson MH, Vigoreaux JO et al. The effect of removing the N-terminal extension of the Drosophila myosin regulatory light chain upon flight ability and the contractile dynamics of indirect flight muscle. Biophys J 2000; 78:1431-1440.
47. Andreev OA, Saraswat LD, Lowey S et al. Interaction of the N-terminus of chicken skeletal essential light chain 1 with F-actin. Biochemistry 1999; 38:2480-2485.
48. Maroto M, Arredondo J, Goulding D et al. Drosophila paramyosin/miniparamyosin gene products show a large diversity in quantity, localization, and isoform pattern: A possible role in muscle maturation and function. J Cell Biol 1996; 134:81-92.
49. Reedy MC, Bullard B, Vigoreaux JO. Flightin is essential for thick filament assembly and sarcomere stability in Drosophila flight muscles. J Cell Biol 2000; 151:1483-1500.
50. Cripps RM, Suggs JA, Bernstein SI. Assembly of thick filaments and myofibrils occurs in the absence of the myosin head. EMBO J 1999; 18:1793-1804.
51. Kronert WA, O'Donnell PT, Fieck A et al. Defects in the Drosophila myosin rod permit sarcomere assembly but cause flight muscle degeneration. J Mol Biol 1995; 249:111-125.
52. Swank DM, Wells L, Kronert WA et al. Determining structure/function relationships for sarcomeric myosin heavy chain by genetic and transgenic manipulation of Drosophila. Microsc Res Tech 2000; 50:430-442.
53. Sellers JR Myosins. 2nd ed. Oxford: Oxford University Press, 1999.
54. Miedema K, Harhangi H, Mentzel S et al. Interspecific sequence comparison of the muscle-myosin heavy-chain genes from Drosophila hydei and Drosophila melanogaster. J Mol Evol 1994; 39:357-368.
55. Hodges D, Cripps RM, O'Connor ME et al. The role of evolutionarily conserved sequences in alternative splicing at the 3' end of Drosophila melanogaster myosin heavy chain RNA. Genetics 1999; 151:263-276.
56. Bernstein SI, Mogami K, Donady JJ et al. Drosophila muscle myosin heavy chain encoded by a single gene in a cluster of muscle mutations. Nature 1983; 302:393-397.
57. Collier VL, Kronert WA, O'Donnell PT et al. Alternative myosin hinge regions are utilized in a tissue-specific fashion that correlates with muscle contraction speed. Genes Dev 1990; 4:885-895.
58. George EL, Ober MB, Emerson Jr CP. Functional domains of the Drosophila melanogaster muscle myosin heavy-chain gene are encoded by alternatively spliced exons. Mol Cell Biol 1989; 9:2957-2974.
59. Rozek CE, Davidson N. Drosophila has one myosin heavy-chain gene with three developmentally regulated transcripts. Cell 1983; 32:23-34.
60. Wassenberg 2nd DR, Kronert WA, O'Donnell PT et al. Analysis of the 5' end of the Drosophila muscle myosin heavy chain gene. Alternatively spliced transcripts initiate at a single site and intron locations are conserved compared to myosin genes of other organisms. J Biol Chem 1987; 262:10741-10747.
61. Zhang S, Bernstein SI. Spatially and temporally regulated expression of myosin heavy chain alternative exons during Drosophila embryogenesis. Mech Dev 2001; 101:35-45.
62. Hastings GA, Emerson Jr CP. Myosin functional domains encoded by alternative exons are expressed in specific thoracic muscles of Drosophila. J Cell Biol 1991; 114:263-276.

63. Kazzaz JA, Rozek CE. Tissue-specific expression of the alternately processed Drosophila myosin heavy-chain messenger RNAs. Dev Bio 1989; 133:550-561.
64. Kronert WA, Edwards KA, Roche ES et al. Muscle-specific accumulation of Drosophila myosin heavy chains: A splicing mutation in an alternative exon results in an isoform substitution. EMBO J 1991; 10:2479-2488.
65. Wells L, Edwards KA, Bernstein SI. Myosin heavy chain isoforms regulate muscle function but not myofibril assembly. EMBO J 1996; 15:4454-4459.
66. Bernstein SI, Milligan RA. Fine tuning a molecular motor: The location of alternative domains in the Drosophila myosin head. J Mol Biol 1997; 271:1-6.
67. Dominguez R, Freyzon Y, Trybus KM et al. Crystal structure of a vertebrate smooth muscle myosin motor domain and its complex with the essential light chain: Visualization of the prepower stroke state. Cell 1998; 94:559-571.
68. Harrington WF, Ueno H, Davis JS. Helix-coil melting in rigor and activated cross-bridges of skeletal muscle. Adv Exp Med Biol 1988; 226:307-318.
69. Kiehart DP, Pollard TD. Inhibition of acanthamoeba actomyosin-II ATPase activity and mechanochemical function by specific monoclonal antibodies. J Cell Biol 1984; 99:1024-1033.
70. Standiford DM, Davis MB, Miedema K et al. Myosin rod protein: A novel thick filament component of Drosophila muscle. J Mol Biol 1997; 265:40-55.
71. Polyak E, Standiford DM, Yakopson V et al. Contribution of myosin rod protein to the structural organization of adult and embryonic muscles in Drosophila. J Mol Biol 2003; 331:1077-1091.
72. Standiford DM, Davis MB, Sun W et al. Splice-junction elements and intronic sequences regulate alternative splicing of the Drosophila myosin heavy chain gene transcript. Genetics 1997; 147:725-741.
73. Standiford DM, Sun WT, Davis MB et al. Positive and negative intronic regulatory elements control muscle-specific alternative exon splicing of Drosophila myosin heavy chain transcripts. Genetics 2001; 157:259-271.
74. Hess NK, Bernstein SI. Developmentally regulated alternative splicing of Drosophila myosin heavy chain transcripts: In vivo analysis of an unusual 3' splice site. Dev Biol 1991; 146:339-344.
75. Warmke JW, Kreuz AJ, Falkenthal S. Colocalization to chromosome bands 99E1-3 of the Drosophila melanogaster myosin light chain-2 gene and a haplo-insufficient locus that affects flight behavior. Genetics 1989; 122:139-151.
76. Takano-Ohmuro H, Takahashi S, Hirose G et al. Phosphorylated and dephosphorylated myosin light chains of Drosophila fly and larva. Comp Biochem Physiol B 1990; 95:171-177.
77. Falkenthal S, Graham M, Wilkinson J. The indirect flight muscle of Drosophila accumulates a unique myosin alkali light chain isoform. Dev Biol 1987; 121:263-272.
78. Leicht BG, Lyckegaard EM, Benedict CM et al. Conservation of alternative splicing and genomic organization of the myosin alkali light-chain (Mlc1) gene among Drosophila species. Mol Biol Evol 1993; 10:769-790.
79. Rubin GM, Spradling AC. Genetic transformation of Drosophila with transposable element vectors. Science 1982; 218:348-353.
80. Mogami K, Hotta Y. Isolation of Drosophila flightless mutants which affect myofibrillar proteins of indirect flight muscle. Mol Gen Genet 1981; 183:409-417.
81. Cripps RM, Ball E, Stark M et al. Recovery of dominant, autosomal flightless mutants of Drosophila melanogaster and identification of a new gene required for normal muscle structure and function. Genetics 1994; 137:151-164.
82. Nongthomba U, Cummins M, Clark S et al. Suppression of muscle hypercontraction by mutations in the myosin heavy chain gene of Drosophila melanogaster. Genetics 2003; 164:209-222.
83. O'Donnell PT, Bernstein SI. Molecular and ultrastructural defects in a Drosophila myosin heavy chain mutant: Differential effects on muscle function produced by similar thick filament abnormalities. J Cell Biol 1988; 107:2601-2612.
84. Davis MB, Dietz J, Standiford DM et al. Transposable element insertions respecify alternative exon splicing in three Drosophila myosin heavy chain mutants. Genetics 1998; 150:1105-1114.
85. Mogami K, O'Donnell PT, Bernstein SI et al. Mutations of the Drosophila myosin heavy-chain gene: Effects on transcription, myosin accumulation, and muscle function. Proc Natl Acad Sci USA 1986; 83:1393-1397.
86. O'Donnell PT, Collier VL, Mogami K et al. Ultrastructural and molecular analyses of homozygous-viable Drosophila melanogaster muscle mutants indicate there is a complex pattern of myosin heavy-chain isoform distribution. Genes Dev 1989; 3:1233-1246.
87. Cripps RM, Becker KD, Mardahl M et al. Transformation of Drosophila melanogaster with the wild-type myosin heavy-chain gene: Rescue of mutant phenotypes and analysis of defects caused by overexpression. J Cell Biol 1994; 126:689-699.

88. Tansey T, Schultz JR, Miller RC et al. Small differences in Drosophila tropomyosin expression have significant effects on muscle function. Mol Cell Biol 1991; 11:6337-6342.
89. Warmke J, Yamakawa M, Molloy J et al. Myosin light chain-2 mutation affects flight, wing beat frequency, and indirect flight muscle contraction kinetics in Drosophila. J Cell Biol 1992; 119:1523-1539.
90. Karlik CC, Fyrberg EA. An insertion within a variably spliced Drosophila tropomyosin gene blocks accumulation of only one encoded isoform. Cell 1985; 41:57-66.
91. Beall CJ, Sepanski MA, Fyrberg EA. Genetic dissection of Drosophila myofibril formation: Effects of actin and myosin heavy chain null alleles. Genes Dev 1989; 3:131-140.
92. Kronert WA, Acebes A, Ferrus A et al. Specific myosin heavy chain mutations suppress troponin I defects in Drosophila muscles. J Cell Biol 1999; 144:989-1000.
93. Ayer G, Vigoreaux JO. Flightin is a myosin rod binding protein. Cell Biochem Biophys 2003; 38:41-54.
94. Henkin JA, Maughan DW, Vigoreaux JO. Mutations that affect flightin expression in Drosophila alter the viscoelastic properties of flight muscle fibers. Am J Physiol Cell Physiol 2003; 286:C65-C72.
95. Beall CJ, Fyrberg E. Muscle abnormalities in Drosophila melanogaster heldup mutants are caused by missing or aberrant troponin-I isoforms. J Cell Biol 1991; 114:941-951.
96. Prado A, Canal I, Barbas JA et al. Functional recovery of troponin I in a Drosophila heldup mutant after a second site mutation. Mol Biol Cell 1995; 6:1433-1441.
97. Swank DM, Kronert WA, Bernstein SI et al. Alternative N-terminal regions of Drosophila myosin heavy chain tune cross-bridge kinetics for optimal muscle power output. Biophys J 2004; 82:191a.
98. Miller BM, Zhang S, Suggs JA et al. Domain near ATP binding pocket of Drosophila muscle myosin influences kinetics. Submitted.
99. Swank DM, Zhang S, Suggs JA et al. Functional analysis of a myosin domain located near the site of ATP entry. Biophys J 1999; 76:A81.
100. Miller BM, Nyitrai M, Swank DM et al. An alternative domain near the nucleotide binding pocket of Drosophila myosin modulates weak-binding states. Biophys J 2004; 86:29a.
101. Swank DM, Kronert WA, Maughan D et al. Alternative versions of the myosin relay loop influence Drosophila muscle kinetics. Biophys J 2004; 86:565a.
102. Suggs JA, Kronert WA, Nikkhoy M et al. Functional importance of the hinge region of the muscle myosin heavy chain rod in Drosophila melanogaster. Biophys J 2001; 80:573a.
103. Winkelman L, Bullard B. Phosphorylation of a light subunit of locust myosin. In: Insect Flight Muscle. Amsterdam: North-Holland Pub. Co., 1977:285-289.
104. Takano-Ohmuro H, Tanikawa M, Maruyama K. Phosphorylation of cricket myosin light chain and Mg^{2+} -activated actomyosin ATPase activity. Zool Sci 1986; 3:715-717.
105. Irving T, Bhattacharya S, Tesic I et al. Changes in myofibrillar structure and function produced by N-terminal deletion of the regulatory light chain in Drosophila. J Muscle Res Cell Motil 2001; 22:675-683.

CHAPTER 6

Paramyosin and Miniparamyosin

Margarita Cervera, Juan Jose Arredondo and Raquel Marco Ferreres

Abstract

In *Drosophila*, paramyosin and miniparamyosin are structural components of thick filaments that have a similar structure to the myosin heavy chain rod tail. Both proteins are rod-like molecules with a high α-helical content in the long central domains, and exist as dimers. While miniparamyosin is mainly located in the M line and at both ends of the thick filaments in *Drosophila* indirect flight muscles (IFM), paramyosin is present all along the thick filaments. The relative amounts of myosin, paramyosin and miniparamyosin vary in the distinct muscles, reflecting the differences in the organization of their thick filaments. Moreover, as for other contractile proteins, the phosphorylation of these two proteins is involved in the acquisition of the capacity to fly. Thus, miniparamyosin has a possible role in the sequential transition of nonfunctional to functional muscle, in general, while the paramyosin transition is more specifically related to the functional onset of IFM.

Paramyosin and miniparamyosin are encoded by the same gene, which contains 10 exons and 9 introns. In *Drosophila*, paramyosin and miniparamyosin share only the last two exons and have a molecular weight of 107 and 60 kDa, respectively. In a similar manner to other *Drosophila* muscle proteins, paramyosin is expressed at two distinct stages of development, while miniparamyosin is present only in the adult musculature. The complex spatio-temporal patterns of paramyosin and miniparamyosin expression depend on two different promoters situated upstream of their transcriptional initiation sites, these two promoters being organized in a modular fashion. In early embryonic development, paramyosin functions as a cytoplasmic protein where it plays an important role in myoblast fusion before its assembly into thick filaments. The properties of thick filaments in invertebrates vary in response to the proportions of myosin, paramyosin and additional proteins such as miniparamyosin, myosin rod protein, and flightin among others.

Introduction: Paramyosin and Miniparamyosin, Components of Invertebrate Thick Filaments

Paramyosin and miniparamyosin are two proteins that are only found in association with thick filaments in invertebrate striated muscles. *Drosophila* paramyosin is a major structural component of thick filaments and it is similar in structure to paramyosin from other invertebrates, having a central α-helical coiled-coil rod flanked by two non α-helical terminal regions.[1-3] The terminal regions of *Drosophila* paramyosin are very short and the regions are much less conserved than the rod portion when the sequences of different species are compared.

Paramyosin has a very similar structure to the myosin heavy chain rod tail. It contains 878 amino acids and is a rod-like molecule in which two 102 kDa monomers interact to form a coiled coil.[2,4] Indeed, it displays the biophysical properties expected of a largely α-helical protein. In *Drosophila melanogaster* three paramyosin isoforms exist, which share a similar molecular mass. In general, the protein accumulates in the adult head, abdomen and thorax

---

Nature's Versatile Engine: Insect Flight Muscle Inside and Out, edited by Jim Vigoreaux.

and in vivo, the appearance of the most acidic phosphorylated isoform of paramyosin has been specifically related to the functioning of the flight-related thoracic musculature.[5] Moreover, as in the case of myosin light chains, and of the flightin protein, it appears that the phosphorylation of the paramyosin isoforms may modulate the assembly and/or function of thick filaments.[6-9]

Paramyosin is produced during two different stages of development. It is detected at the late embryonic stages and remains present through the larval instars. It can then be detected again in mid-pupal stages through adulthood. The protein appears at around ten hours post-fertilization and it accumulates progressively during middle and late embryogenesis, until reaching maximal levels in adults. The protein is present in oocytes and very early embryos (3 hours of development), indicating a possible maternal inheritance. By immunocytochemistry in whole embryos and cryosections of adult flies, paramyosin protein can be seen to accumulate in all muscle groups: pharyngeal, somatic, visceral and specialized thoracic musculature. In electron microscopy images, paramyosin is distributed along the entire A-band of the sarcomere in *Drosophila* IFM and Tergal depressor of the trochanter (TDT) muscles.[5]

The miniparamyosin dimer is a minor component of myofibrillar thick filaments.[1,5] This protein is found mainly in the thoracic tubular jump muscles (TDT) while it is much less predominant in the fibrillar flight muscles. The miniparamyosin monomer has a predicted molecular weight of 54.887 daltons and it shares a common 363 amino acid C-terminal α-helical domain with paramyosin. In addition, miniparamyosin has a unique 114 amino acid N-terminal domain that lacks homology to other known proteins.[1] Although miniparamyosin is present in many invertebrates including arthropods, annelids, mollusks, and echinoderms, it is not found in the nematode *Caeorhabditis elegans*, nor in vertebrates.[10] Six in vivo phosphorylated miniparamyosin isoforms have been identified in the head and thoracic musculature, over a very wide pH range (pH 6 to 8). As with paramyosin, it is possible that the phosphorylation/dephosphorylation of miniparamyosin is involved in the acquisition of flight ability.[5]

In contrast to paramyosin, miniparamyosin is almost exclusively found in the adult musculature. A transient accumulation of miniparamyosin has been detected in 3^d instars larvae but this decreases rapidly during pupation.[5] Furthermore, miniparamyosin is differentially distributed in the sarcomeres of *Drosophila* muscle.[5] In the IFM, the protein is found in the M-line and at the ends of the A-band whereas in the TDT, it is located throughout the A-band.

Paramyosin and Miniparamyosin in Insect Muscles

The paramyosin content of invertebrate muscles varies with the structural organization of the fibers and with the dimensions of the thick filaments.[3,10-12] Thus, a myosin/paramyosin ratio close to or less than 1 has been reported for lamellibranch smooth adductors. In these molluscan "catch" muscles, the thick filaments range from 0.05 to 0.15 μm in diameter and from 10 to 40 μm in length. In *Drosophila*, the myosin/paramyosin ratio of fibrillar muscle is 34:1, while for tubular muscle it is 6:1.[3] Through a biochemical analysis of different muscle-types in distinct species, a wide range of myosin/paramyosin ratios have been established.[3,10-12] Invertebrate muscles were originally classified into 3 groups with respect to the length of the thick filaments and the maximum active tension.[12] Class I muscles, structurally most similar to vertebrate striated muscles, have short thick filament lengths (1.9 μm) and the lowest paramyosin/myosin ratios (less than 0.2). Class II muscles, intermediate in structural type between Class I and smooth catch muscles (Class III), have longer thick filament lengths (3 μm) and ratios (0.3-0.6), and Class III muscles have the highest ratios (greater than 2.0). When filament length and the myosin/paramyosin ratios are compared in muscles of different species, it appears that paramyosin is involved in determining the length and the active tension generated.[12] This was later confirmed in a study of *C. elegans* mutants.[13] When correlated to the classification set out previously[12], *Drosophila* fibrillar and tubular muscles can be included in class I, fibrillar muscles having one of the highest myosin/paramyosin ratios and tubular muscles one of the lowest in the group. However, in agreement with the differences in paramyosin content, the thick filaments are shorter in fibrillar than in tubular muscles.

Drosophila muscles can be grouped into four categories based on the relative amount of myosin, paramyosin and miniparamyosin: (i) miniparamyosin and paramyosin are abundant protein components, as seen in adult tubular muscle (TDT), with a paramyosin/miniparamyosin ratio near to 1; (ii) paramyosin is expressed at higher levels than miniparamyosin and there is a relatively high ratio of paramyosin to myosin as found in leg and abdomen muscles; (iii) there is a low ratio of paramyosin and miniparamyosin to myosin, as in asynchronous indirect flight muscles; (iv) there is no miniparamyosin and relatively large amounts of paramyosin as in embryonic supercontractile muscles.[5] Furthermore, the relatively large number of paramyosin and miniparamyosin isoforms adds to the diversity of thick filaments in adult invertebrate muscles.

Paramyosin, Myosin and Thick Filament Structure

Paramyosin and myosin are the most abundant proteins in invertebrate thick filaments, and it has been proposed that the alpha-helical coiled-coil paramyosin dimer interacts with its homologous counterpart, the myosin rod dimer.[14-18] Analysis of the paramyosin and myosin heavy chain rod sequences has revealed a remarkable pattern of alternating groups of charged residues associated with a 28 residue repeat.[19] Interactions between these oppositely charged segments are thought to play a significant role in the assembly of these two proteins into thick filaments.[20,21] The C-terminal domains of both molecules are critical for both solubility and assembly. These domains appear to function as modulators of assembly in both proteins.[19] Although previous studies demonstrated that myosin and paramyosin possessed the ability to self-assemble,[3,22] the filaments formed in vitro lacked important features of thick filaments in vivo. Nevertheless, the organization and exact location of paramyosin in the structure of the insect thick filaments remains unclear. It is proposed that paramyosin, together with additional proteins, is assembled in the core of the thick filament. This disposition will facilitate the attachment of myosin, the functional motor protein of the thick filament, at the periphery of the filament.

Accumulated evidence suggests that in invertebrates, the assembly of thick filaments of distinct length, diameter, electron density and rigidity requires the presence of myosin, paramyosin and several additional proteins, such as miniparamyosin, myosin rod protein and flightin, among others.[1,5,23-26] These proteins are present in distinct amounts according to the length, diameter and electron density of the fibres.[3,5] Thus, in various invertebrates a correlation has been established between the properties of the muscles and the amount of paramyosin.[3,5,27,28] In *C. elegans* paramyosin mutants, the length and diameter of thick filaments are also affected by the paramyosin content.[13] A model of thick filament structure has been proposed wherein the tubular thick filament core in *C. elegans* is formed by seven paramyosin subfilaments supported by an internal sleeve of filagenins, each paramyosin subfilament containing four strands of paramyosin.[29,30] Since filagenins have not been identified in *Drosophila* and many more distinct types of fibres are present in insects, different mechanisms of molecular assembly may exist in different organisms.

Paramyosin and Miniparamyosin Function in Myofibril Formation

The differences in the functional role of paramyosin and miniparamyosin in each type of insect muscle probably reflect the characteristics and specialized requirements of each muscle.

Paramyosin functions as a cytoplasmic protein in early embryonic development and is important for myoblast fusion and myofibril formation.[31] Homozygous paramyosin mutants, *prm1*, obtained by mobilizing a P element located in the paramyosin promoter, are strongly hypomorphic alleles of the paramyosin gene (paramyosin is reduced to 1% of wild type level). They die at the late embryo stage and display defects in both myoblast fusion and myofibril assembly in muscles of the embryonic body wall (Fig. 1). Sarcomeres do not assemble properly and muscle contractility is impaired. Paramyosin was shown to be important for the production of an adequate number of morphologically normal thick filaments. Although thin filaments are correctly assembled in these mutants, thick filament containing paramyosin are important for the organization of the regular sarcomeric patterns (Fig. 1). An abnormal interaction between thin and thick filaments might be the cause of this phenotype. As in these paramyosin

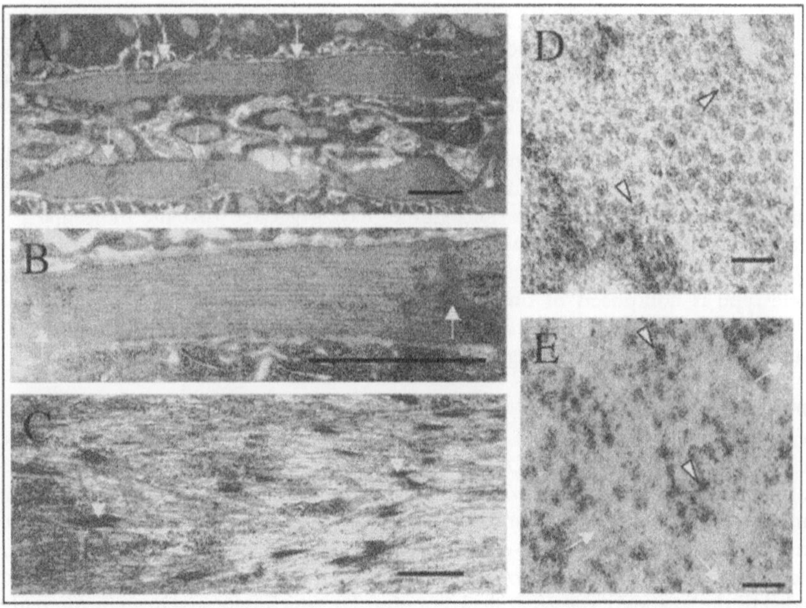

Figure 1. Electron microscopy of wild-type and homozygous *prm1* embryonic body wall muscles. Thin and thick filaments in embryonic body wall muscles are well organized. In longitudinal sections (A,B), thin and thick filaments are parallel and arrange into smooth filament bundles. Z-bodies align to form Z-band (A,B, arrows), which mark the myofibrils into sarcomeres. In a cross section (D), each thick filament is composed of several dense particles and they are arranged into a circular entity (arrowheads). Thin and thick filaments are organized into regular arrays. Mutation of paramyosin in *prm1* causes a reduction in thick filament number and disrupts myofibril organization (C,E). *prm1* myofibrils are shorter and wavy (C). Z-band material is poorly organized (C, arrows). In areas where thick filaments are formed, they are structurally abnormal. Dense particles in a cross–section of thick filaments are no longer organized into circular, hollow structures (E, arrowheads). Bars: (A-C) 1 μm; (D,E) 0.1 μm. Reproduced from The Journal of Cell Biology, 2003, 160, 904, by copyright permission of The Rockefeller University Press.

mutants, most of the *null* mutations affecting *Drosophila* contractile proteins disrupt sarcomere organization, highlighting the influence that the interaction between thin and thick filaments has on the sarcomere.[32-37]

The role of miniparamyosin in muscle assembly, development, and function has been assessed by analyzing the phenotypic perturbations produced by protein overexpression in *Drosophila*.[38] Such analyses were focused on the IFM, since IFM myofibril assembly and flight ability are quite sensitive to the stoichiometry of muscle proteins.[39] In contrast to expectations, overexpression of miniparamyosin has little impact on the assembly of IFM myofibrils, neither thick filament electron density nor sarcomere length is affected. Nevertheless, overexpression of miniparamyosin does cause IFM dysfunction and age-dependent myofibril degeneration. Transgenic flies undergo progressive deterioration of the myofibrils, producing a gradual loss of flight muscle functionality. These observations indicate that the correct stoichiometry of miniparamyosin is important to maintain integrity of myofibrils, and for the proper function of the flight musculature.

Regulation and Control

The differentiation of muscle types is a complex, multistep developmental process involving multiple gene regulatory mechanisms.[40-42] This process is largely controlled by the transcriptional regulation of a large battery of genes encoding muscle-specific proteins.[43]

The differences in the contractile and biochemical properties of muscle types originate in the selective expression of genes encoding specific contractile protein isoforms. The mechanisms controlling the expression of each of these genes are highly tissue specific, and they are rapidly and strongly activated only in muscle lineages. However, protein accumulation varies in each type of muscle according to their specific properties and functions.[39]

The protein stoichiometry during myofibril assembly must be maintained and as such, the activation or down-regulation of a muscle protein-encoding gene must be precisely offset by the regulation of other genes. Previous studies have suggested that the molecular pathways controlling muscle formation are ancient and evolutionary conserved in flies and vertebrates.[44] In *Drosophila*, in contrast to mammals, few specialized muscle types are generated and each muscle-type is composed of only one fiber type.[45,46] In this sense, the paramyosin/miniparamyosin gene in *Drosophila melanogaster* represents a good model system to study the regulatory mechanisms controlling expression of muscle specific genes.

General Organization of the Gene

The structure of the *Drosophila melanogaster* paramyosin/miniparamyosin gene has been determined by DNA sequencing of cDNA and genomic clones.[1,2] The paramyosin/miniparamyosin gene is located in region 66D14 on the left arm of the third chromosome, and it spans 12. 8 kb, organized into 10 exons and 9 introns (Fig. 2). Through the use of different promoters and alternative exon splicing, this gene encodes both paramyosin and miniparamyosin, which share the last two exons of the gene (exons 8 and 9). Paramyosin uses an upstream promoter and is encoded by the 9 exons of the gene, all except exon 1B, which is located in an intron that follows exon 7 and encodes the 5′ end of miniparamyosin. Indeed, miniparamyosin is produced by the use of an alternative promoter, and it includes exon 1B joined to the last two exons of the paramyosin transcript by RNA splicing. The two overlapping transcriptional units of the paramyosin/miniparamyosin gene act independently, thus, the two promoters of the gene separately regulate the expression of the transcripts. Indeed, during pupal myogenesis, both transcripts are expressed in the same fibers. This type of genomic organization has also been described for the *Drosophila* tropomyosin and myosin heavy genes. Internal promoters in these genes produce transcripts encoding cytoplasmic tropomyosin and the myosin rod protein, respectively.[26,47-50]

Temporal and Spatial Control of the Gene

The paramyosin/miniparamyosin genes from *D. melanogaster, D. virilis* and *D. pseudoobscura* share a high degree of similarity in their open reading frames. Despite the fact that *D. pseudoobscura* and *D. virilis* diverged from *D. melanogaster* more than 30 and 50 millions years ago, the two genes display identical patterns of expression in each of the three species.The complex spatio-temporal regulation of paramyosin and miniparamyosin expression depends on two regulatory regions situated upstream of the transcriptional initiation sites of the paramyosin and miniparamyosin mRNAs, each with distinct properties. Through a transgenic approach complemented with sequence comparison, we have established that these two promoters present a modular organization, supporting and providing a theoretical foundation for our experimental findings in vivo.[51]

Both promoters contain regions extending 90-100 nucleotides upstream of the paramyosin and miniparamyosin transcriptional initiation sites. These regions are over 90% conserved in the three *Drosophilidae*, indicating that they may correspond to RNA polymerase complex binding domains. Moreover, apart from paramyosin expression in IFM, the spatial and temporal patterns of transgene expression driven by both promoters depend on discrete regions located between -0.9 and -2 kb of the paramyosin and miniparamyosin initiation sites. Thus, besides the basal promoter, two discrete regions in the paramyosin promoter have been identified.[51] A MEF2-E region, [52]containing a group of conserved E boxes and a MEF2 site, is located -1400bp upstream of the start site and seems to act as a distal muscle activator that differently regulates paramyosin expression in embryonic/larval and adult muscles (Figs. 2, 3).

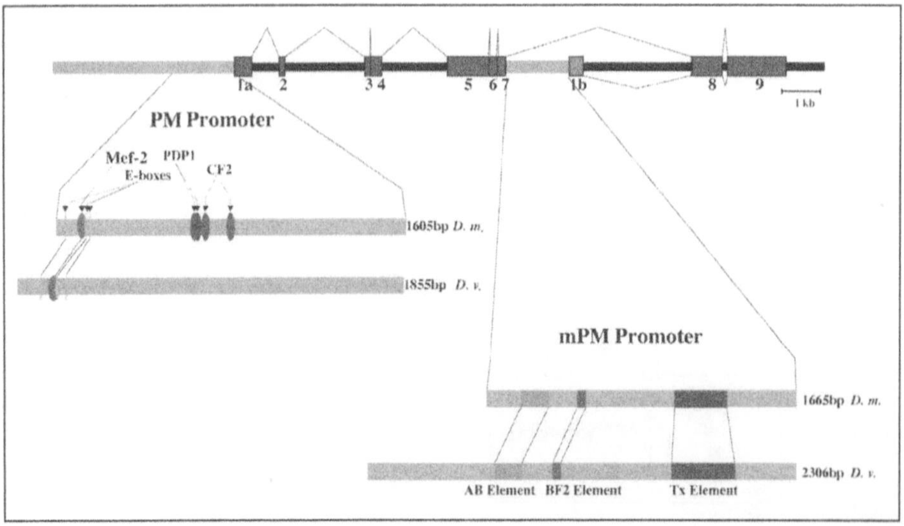

Figure 2. Conserved regions in the sequences upstream of paramyosin/ miniparamyosin gene in *D. melanogaster* and *D. virilis*. A schematic representation of the *D. melanogaster* paramyosin/miniparamyosin gene. The exons are shown as blue boxes and are specified by number. The lower part shows the alignment of the sequences upstream of the start sites of the paramyosin and miniparamyosin transcription units in *D. melanogaster* and *D. virilis* genes. Proximal regions of both transcriptional units are conserved.[10] The site for MEF2 at -1488 and three E boxes at -1587, -1461 and -1436 in the paramyosin promoter are conserved in *D. virilis*. In the same region, two CF2 sites and two PDP1 sites are present in *D. melanogaster* but not in *D. virilis*. In the upstream sequences of the miniparamyosin transcription unit, three regions are conserved in sequence and position. These are located at -477 to -740 (TX element), -1173to -1207 (BF2 element) and -1342 to -1469 (AB element).

This region is essential for the high levels of expression in larval muscles. A second region, containing several PDP1 and CF2 sites,[24,53] seems to be responsible for the low basal levels, and the temporal- and muscle-specific expression. Indeed, the expression of the transgene is not completely abolished until the region containing the PDP1 sites is eliminated. These MEF-E and PDP1/CF2 regions act coordinately to reproduce the complete spatio-temporal pattern of paramyosin expression.

Miniparamyosin expression is regulated by the basal promoter and three highly conserved elements, which have been named AB (127bp), BF2 (34bp), and TX (263bp). These are located at -477 to -740 (TX element), -1173 to -1207 (BF2 element) and -1342 to -1469 (AB element). The AB element specifically drives high levels of expression in IFM, whereas the TX element drives high levels of expression in TDT and low expression in larval body wall muscles. The presence of the two elements, AB and TX, is needed to recapitulate transgene expression in adult hypodermic muscles. The BF2 element is unable to direct detectable reporter expression by itself. However, when transgene transcription is directed by a combination of BF2 with either, TX or AB, trangene expression decreases in IFM and TDT. BF2 might acts as an expression modulator in adult muscles.

An interesting aspect of paramyosin/miniparamyosin expression is how the overlapping transcriptional units regulate transcription of the two mRNAs during adult myogenesis. Despite the fact that the two proteins are expressed at the same stage of adult development, the transcriptional units act independently.[51] One possible explanation for how these two promoters can function without influencing one another may be that the promoters don't act in the same nucleus. Muscle fibers are a syncytium, and in the mouse it has been demonstrated that

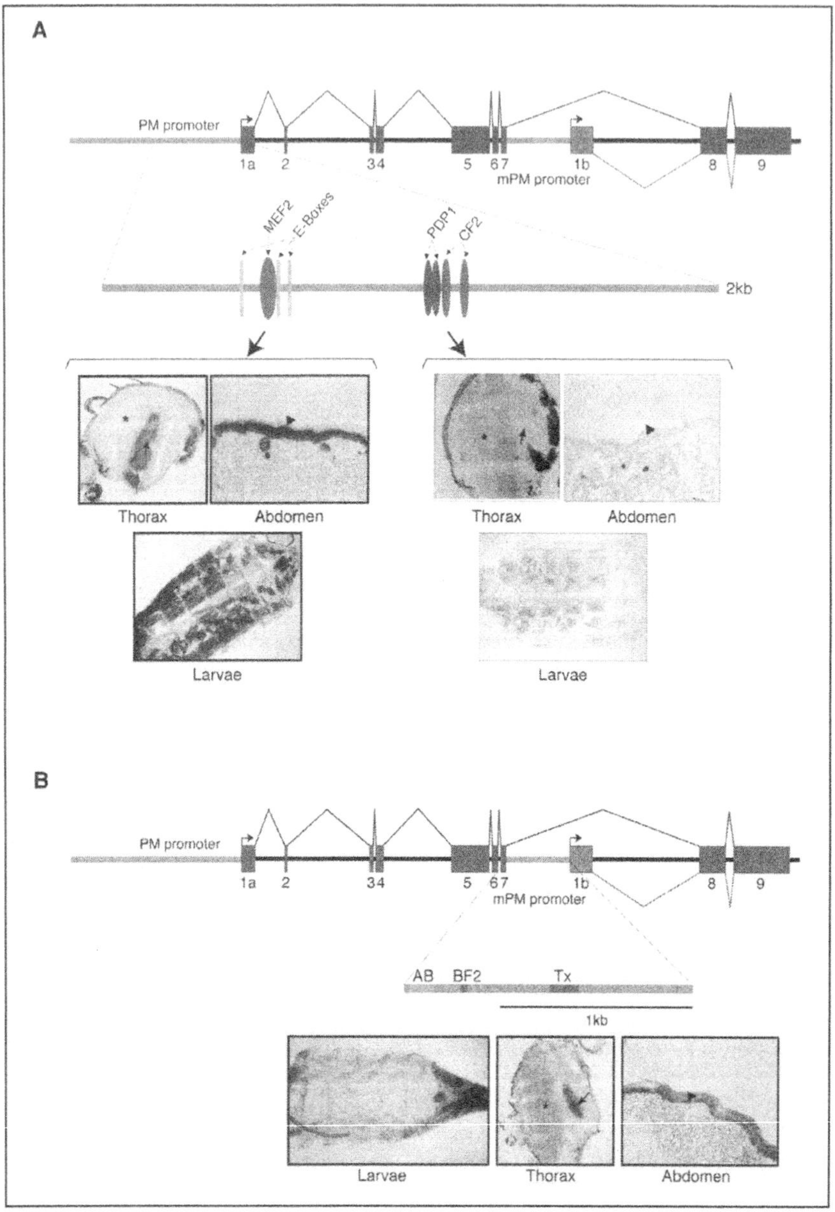

Figure 3. Comparison of β-galactosidase gene expression driven by selected sequences upstream from the paramyosin and miniparamyosin transcription start sites. A) Transgene expression in abdomen and thorax thin sections, and dissected 3rd instar larvae transformed with constructs containing the 5′ sequences 2.2 Kb upstream of (in the lower part of the MEF2-E region) and 1.4Kb upstream of (in the lower part of the PDP1-CF2 region) the paramyosin transcription start site, inserted in the pCasper β-gal plasmid. B) Transgene expression in abdomen and thorax thin sections and dissected 3rd instar larvae transformed with a construct containing the 5′ sequences 2.7 Kb upstream of the miniparamyosin transcription start site inserted in the pCasper β-gal plasmid. Hypodermic muscles, black arrowhead; IFM, black asterisk; and TDT, black arrow.

individual nuclei present in the same muscle cell do not necessarily transcribe the same genes at the same time.[54] Indeed, genes may or may not be transcribed in a particular nucleus, but the number of nuclei that transcribe a specific gene is constant. This might occur in the case of the paramyosin/miniparamyosin gene. A further possible explanation invokes the presence of an insulator (a region that defines independent domains of gene function throughout the genome) separating the two promoters.

Evolutionary Aspects

The evolutionary diversification of paramyosin in invertebrates, including the presence of miniparamyosin, has been investigated by using antibodies specific to the two proteins.[10] Both types of proteins have been found in all the invertebrate species studied except *C. elegans*. Paramyosin shows slight variations in its molecular mass but these are minor compared to the wide variation in molecular mass of miniparamyosin (from 50 to 80 kDa). A specific antiserum against *Drosophila* miniparamyosin recognizes a single protein with a similar mobility to miniparamyosin only in *Diptera*. In contrast, the antiserum recognizes two proteins, paramyosin and the putative miniparamyosin, in all the other species analyzed, including *Sphaerechinus, Mytilus, Helix, Scalopendra, Locusta, Formica* and *Coccinella*. This suggests that the protein domain specific to *D. melanogaster* miniparamyosin (exon 1B) may also be expressed in paramyosin in other species. In the *Annelida* and *Mollusca*, this cross-reaction with paramyosin was stronger with the anti-miniparamyosin than with a paramyosin specific antisera (exon 5), suggesting that the function fulfilled by the protein domain encoded by exon 1B is common to both proteins in some species. This is in agreement with the much greater variability in exon organization of the paramyosin/miniparamyosin genes in different species.[5] In light of this, it is interesting that the anti-exon 5 antiserum specific to paramyosin, not only recognizes paramyosin, but also myosin in several invertebrate species. *Coccinella* provides an extreme example because this serum only cross-reacts with myosin, that a similar exchange of functions may occur, in part, between myosin and paramyosin in these species. Furthermore, in the case of the deuterostome invertebrates, such as *Echinodermus*, the data suggest that a miniparamyosin isoform may be present in addition to the paramyosin described. In conclusion, we suggest that the patterns of antibody reactivity reflect the distinct specialization of the paramyosin gene complex such that, in *Diptera* (possibly through the evolution of a separate regulatory control) miniparamyosin has acquired a structure and function distinct to that of paramyosin. In any case, the widespread occurrence of miniparamyosin in invertebrate muscles and the complexity of the transcriptional regulation of the two distinct proteins encoded by a single gene in *Drosophila* species point to the importance of the functional role of paramyosin and miniparamyosin in producing the structural and functional diversity of invertebrate muscles.

Concluding Remarks

Elucidating the detailed organization and the function of the proteins found in distinct invertebrate thick filaments has proved to be a complex task, even for the same functional types of muscle. For example, multiple levels of complexity have been described in the flight muscle thick filaments of different organisms. Indeed, filament length and diameter show wide variations when compared with the relative uniformity found in vertebrate thick filaments. Despite the fact that several models have been proposed, they fail to fully explain the complex situation in invertebrate muscles. Nevertheless, genetic and biochemical studies have begun to reveal a clearer picture of how thick filaments are organized in distinct muscle types. One fact that emerges is the importance of the accessory proteins in these filaments and the influence that the relative amount of such proteins has on establishing the normal length, diameter and density of the filament. Furthermore, the relative large number of isoforms and the phosphorylation of some of these isoforms can play a vital role in the development and function of muscle. We expect that even greater insight into muscle function will come from studies of the different components of thick filaments.

References

1. Becker KD, O'Donnell PT, Heitz JM et al. Analysis of Drosophila paramyosin: Identification of a novel isoform which is restricted to a subset of adult muscles. J Cell Biol 1992; 116:669-681.
2. Vinos J, Maroto M, Garesse R et al. Drosophila melanogaster paramyosin: Developmental pattern, mapping and properties deduced from its complete coding sequence. Mol Gen Genet 1992; 231:385-394.
3. Vinos J, Domingo A, Marco R et al. Identification and characterization of Drosophila melanogaster paramyosin. J Mol Biol 1991; 220:687-700.
4. Cohen C, Lanar DE, Parry DA. Amino acid sequence and structural repeats in Schistosome paramyosin match those of myosin. Biosci Rep 1987; 7:11-16.
5. Maroto M, Arredondo J, Goulding D et al. Drosophila paramyosin/miniparamyosin gene products show a large diversity in quantity, localization, and isoform pattern: A possible role in muscle maturation and function. J Cell Biol 1996; 134:81-92.
6. Vigoreaux JO. Alterations in flightin phosphorylation in Drosophila flight muscles are associated with myofibrillar defects engendered by actin and myosin heavy-chain mutant alleles. Biochem Genet 1994; 32:301-314.
7. Sparrow J. Flight and phosphorylation. Nature 1995; 373:592-593.
8. Tohtong R, Yamashita H, Graham M et al. Impairment of muscle function caused by mutations of phosphorylation sites in myosin regulatory light chain. Nature 1995; 374:650-653.
9. Bárány M, Bárány K. Phosphorylation of the myofibrillar proteins. Ann Rev Physiol 1980; 42:275-292.
10. Maroto M, Arredondo JJ, San Roman M et al. Analysis of the paramyosin/miniparamyosin gene. Miniparamyosin is an independently transcribed, distinct paramyosin isoform, widely distributed in invertebrates. J Biol Chem 1995;270:4375-4382.
11. Beinbrech G, Meller U, Sasse W. Paramyosin content and thick filament structure in insect muscle. Cell Tissue Res 1985; 241:607-614.
12. Levine RJ, Elfvin M, Dewey MM et al. Paramyosin in invertebrate muscles. II. Content in relation to structure and function. J Cell Biol 1976; 71:273-279.
13. Mackenzie Jr JM, Epstein HF. Paramyosin is necessary for determination of nematode thick filament length in vivo. Cell 1980; 22:747-755.
14. Bennett PM, Elliott A. "Splicing" of paramyosin filaments. J Mol Biol 1984; 175:103-109.
15. Epstein HF, Aronow BJ, Harris HE. Myosin-paramyosin cofilaments: Enzymatic interactions with F-actin. Proc Natl Acad Sci USA 1976; 73:3015-3019.
16. Harris HE, Epstein HF. Myosin and paramyosin of Caenorhabditis elegans: Biochemical and structural properties of wild-type and mutant proteins. Cell 1977; 10:709-719.
17. Hoppe PE, Waterston RH. Hydrophobicity variations along the surface of the coiled-coil rod may mediate striated muscle myosin assembly in Caenorhabditis elegans. J Cell Biol 1996; 135:371-382.
18. Szent-Gyorgyi AG, Cohen C, Kendrick-Jones J. Paramyosin and the filaments of molluscan "catch" muscles. II. Native filaments: Isolation and characterization. J Mol Biol 1971; 56:239-258.
19. Cohen C, Parry DA. A conserved C-terminal assembly region in paramyosin and myosin rods. J Struct Biol 1998; 122:180-187.
20. McLachlan AD, Karn J. Periodic charge distributions in the myosin rod amino acid sequence match cross-bridge spacings in muscle. Nature 1982; 299:226-231.
21. Kagawa H, Gengyo K, McLachlan AD et al. Paramyosin gene (unc-15) of Caenorhabditis elegans. Molecular cloning, nucleotide sequence and models for thick filament structure. J Mol Biol 1989; 207:311-333.
22. Huxley HT. Electron microscope studies on the structure of natural and synthetic protein filaments from striated muscle. J Mol Biol 1963; 7:281-308.
23. Barral JM, Epstein HF. Protein machines and self assembly in muscle organization. Bioessays 1999; 21:813-823.
24. Reddy KL, Wohlwill A, Dzitoeva S et al. The Drosophila PAR domain protein 1 (Pdp1) gene encodes multiple differentially expressed mRNAs and proteins through the use of multiple enhancers and promoters. Dev Biol 2000; 224:401-414.
25. Vigoreaux JO, Saide JD, Valgeirsdottir K et al. Flightin, a novel myofibrillar protein of Drosophila stretch-activated muscles. J Cell Biol 1993; 121:587-598.
26. Standiford DM, Davis MB, Miedema K et al. Myosin rod protein: A novel thick filament component of Drosophila muscle. J Mol Biol 1997; 265:40-55.
27. Bullard B, Luke B, Winkelman L. The paramyosin of insect flight muscle. J Mol Biol 1973; 75:359-367.
28. Bullard B, Dabrowska R, Winkelman L. The contractile and regulatory proteins of insect flight muscle. Biochem J 1973; 135:277-286.

29. Epstein HF, Lu GY, Deitiker PR et al. Preliminary three-dimensional model for nematode thick filament core. J Struct Biol 1995; 115:163-174.
30. Muller SA, Haner M, Ortiz I et al. STEM Analysis of Caenorhabditis elegans muscle thick filaments: Evidence for microdifferentiated substructures. J Mol Biol 2001; 305:1035-1044.
31. Liu H, Mardahl-Dumesnil M, Sweeney ST et al. Drosophila paramyosin is important for myoblast fusion and essential for myofibril formation. J Cell Biol 2003; 160:899-908.
32. Roulier EM, Fyrberg C, Fyrberg E. Perturbations of Drosophila alpha-actinin cause muscle paralysis, weakness, and atrophy but do not confer obvious nonmuscle phenotypes. J Cell Biol 1992; 116:911-922.
33. Sparrow J, Drummond D, Peckham M et al. Protein engineering and the study of muscle contraction in Drosophila flight muscles. J Cell Sci 1991; 14(Suppl):73-78.
34. Beall CJ, Fyrberg E. Muscle abnormalities in Drosophila melanogaster heldup mutants are caused by missing or aberrant troponin-I isoforms. J Cell Biol 1991; 114:941-951.
35. Beall CJ, Sepanski MA, Fyrberg EA. Genetic dissection of Drosophila myofibril formation: Effects of actin and myosin heavy chain null alleles. Genes Dev 1989; 3:131-140.
36. Fyrberg E, Fyrberg CC, Beall C et al. Drosophila melanogaster troponin-T mutations engender three distinct syndromes of myofibrillar abnormalities. J Mol Biol 1990; 216:657-675.
37. O'Donnell PT, Bernstein SI. Molecular and ultrastructural defects in a Drosophila myosin heavy chain mutant: Differential effects on muscle function produced by similar thick filament abnormalities. J Cell Biol 1988; 107(6 Pt 2):2601-2612.
38. Arredondo JJ, Mardahl-Dumesnil M, Cripps RM et al. Overexpression of miniparamyosin causes muscle dysfunction and age-dependant myofibril degeneration in the indirect flight muscles of Drosophila melanogaster. J Muscle Res Cell Motil 2001; 22:287-299.
39. Bernstein SI, O'Donnell PT, Cripps RM. Molecular genetic analysis of muscle development, structure, and function in Drosophila. Int Rev Cytol 1993; 143:63-152.
40. Hughes SM, Salinas PC. Control of muscle fibre and motoneuron diversification. Curr Opin Neurobiol 1999; 9:54-64.
41. McKinsey TA, Zhang CL, Olson EN. Signaling chromatin to make muscle. Curr Opin Cell Biol 2002; 14:763-772.
42. Stockdale FE. Mechanisms of formation of muscle fiber types. Cell Struct Funct 1997; 22:37-43.
43. Buckingham M, Houzelstein D, Lyons G et al. Expression of muscle genes in the mouse embryo. Symp Soc Exp Biol 1992; 46:203-217.
44. Scott MP. Intimations of a creature. Cell 1994; 79:1121-1124.
45. Bate M. The embryonic development of larval muscles in Drosophila. Development 1990; 110:791-804.
46. Baylies MK, Bate M, Ruiz Gomez M. Myogenesis: A view from Drosophila. Cell 1998; 93:921-927.
47. Karlik CC, Fyrberg EA. Two Drosophila melanogaster tropomyosin genes: Structural and functional aspects. Mol Cell Biol 1986; 6:1965-1973.
48. Hanke PD, Storti RV. The Drosophila melanogaster tropomyosin II gene produces multiple proteins by use of alternative tissue-specific promoters and alternative splicing. Mol Cell Biol 1988; 8:3591-3602.
49. Gremke L, Lord PC, Sabacan L et al. Coordinate regulation of Drosophila tropomyosin gene expression is controlled by multiple muscle-type-specific positive and negative enhancer elements. Dev Biol 1993; 159:513-527.
50. Polyak E, Standiford DM, Yakopson V et al. Contribution of Myosin rod protein to the structural organization of adult and embryonic muscles in Drosophila. J Mol Biol 2003; 331:1077-1091.
51. Arredondo JJ, Ferreres RM, Maroto M et al. Control of Drosophila paramyosin/miniparamyosin gene expression. Differential regulatory mechanisms for muscle-specific transcription. J Biol Chem 2001; 276:8278-8287.
52. Black BL, Olson EN. Transcriptional control of muscle development by myocyte enhancer factor-2 (MEF2) proteins. Annu Rev Cell Dev Biol 1998; 14:167-196.
53. Gogos JA, Hsu T, Bolton J et al. Sequence discrimination by alternatively spliced isoforms of a DNA binding zinc finger domain. Science 1992; 257:1951-1955.
54. Newlands S, Levitt LK, Robinson CS et al. Transcription occurs in pulses in muscle fibers. Genes Dev 1998; 12:2748-2758.

Chapter 7

Novel Myosin Associated Proteins

Byron Barton and Jim O. Vigoreaux

Abstract

Asynchronous insect flight muscle (IFM) relies on high frequency operation to achieve higher power output than a comparable synchronous muscle. The biochemical, ultrastructural, and mechanical adaptations that define the performance of this muscle type are not completely understood. IFM is characterized by its high stiffness, a property that influences the magnitude of stretch activation and ability to deliver high power to the wings. IFM is also characterized by the presence of unique (novel) myofibrillar proteins, and atypical modifications of conventional proteins, that likely affect the functional properties of this muscle. Here we review the properties of three thick filament associated proteins that may be important in stabilizing the myofilament lattice and impart rigidity to the myofibril. *Drosophila* flightin is an ~20 kDa myosin rod binding protein found exclusively in IFM that exists as two unphosphorylated and nine phosphorylated isovariants in adult muscle. Genetic analyses have shown that flightin is essential for IFM development, structural integrity and function. Zeelins are found in *Lethocerus* leg (zeelin 1) and flight muscle (zeelin 1 and zeelin 2). Although the exact role of zeelins is not known, it has been speculated that they play a role in maintaining the structural organization of the thick filament and the regularity of the myofilament lattice. Stretchin-MLCK is a conceptual *Drosophila* protein that has been studied through the virtual translation of its coding region. The transcription unit is hypothesized to express seven different transcripts. None of the conceptual protein products have been shown to be expressed in IFM, although preliminary studies have identified a protein that may correspond to one of the small kinase isoforms and a second, kettin-like isoform that does not correspond to any of the predicted products.

Introduction

Flight muscle sarcomeres are comprised of at least three filament types, the most predominant of which are the thick filaments. Myosin II, the highly conserved molecular motor, is the main component of the thick filament and one of the most abundant proteins in insect flight muscle. Aside from myosin, only a handful of other proteins have been identified in the thick filaments of insect flight muscle by traditional biochemical and/or genetic approaches. These include the ubiquitous invertebrate muscle proteins paramyosin and mini-paramyosin (see chapter by Cervera et al in this volume), and the novel proteins flightin (*Drosophila*), and zeelins (*Lethocerus*). This chapter will review the characteristics and possible functions of these novel proteins, focusing on their potential role in defining the structural and functional properties of insect indirect flight muscles (IFM). In addition, we will discuss the hypothetical *Drosophila* protein stretchin-MLCK, a member of the titin/myosin light chain kinase family, and one of its newly identified isoforms, myostrandin.

Nature's Versatile Engine: Insect Flight Muscle Inside and Out, edited by Jim Vigoreaux.

Flightin

Gene Structure and Expression

Drosophila flightin is an ~20 kDa protein that is expressed exclusively in IFM. Northern blot and microarray analyses showed that transcripts are present from late pupal stages through adulthood, a pattern consistent with flightin being specific to adult muscle.[1,2] In situ hybridization and southern blot analyses indicated that flightin is encoded by a single gene located in polytene region 76 D/E,[1] results further corroborated by whole genome sequence. Like many other genes encoding myofibrillar proteins, the first intron in the flightin gene separates a small noncoding exon from the open reading frame. Alternative start sites give rise to two transcripts that differ in their 5'noncoding region but share a single open reading frame. The functional significance of the two 5'noncoding regions is not known. There is no evidence of alternatively spliced exons or differential exon exclusion/inclusion in flightin. Instead, multiple isoelectric variants are generated by post-translational modifications.[3] Two small introns of 66 base pairs and 62 base pairs interrupt the coding region. The flightin gene in *D. virilis* shares a similar structure to the *D. melanogaster* gene, including location, but not the size, of introns (our unpublished results).

A recent microarray study investigating the expression of *Drosophila* genes during aging found that flightin was down-regulated with increasing age and after feeding with paraquat, a free radical generator. More significantly, flightin was one of only 42 genes, of more than 4500 tested, whose expression was regulated with both age and oxidative stress.[4] The significance of this regulation is not known. As discussed below, the expression of flightin (as determined by western blot analysis) also is affected by mutations in several contractile protein genes.

Association with the Myofibril and Myosin Binding

Immunolocalization studies and biochemical fractionation of IFM fibers have provided evidence that flightin is a myofibrillar protein.[1] In the sarcomere, flightin is distributed rather homogeneously throughout the A band except at the M line and the edge of the A/I junction.[5] This distribution suggests that flightin is associated with the thick filament, a conclusion that is further supported by genetic and biochemical studies. Flightin is present in the myosin-enriched fraction obtained by high ionic strength extraction of skinned IFM fibers[5] and in the cytomatrix fraction obtained from the IFM of *act88F*^{KM88}, an actin null mutant that lacks thin filaments.[6] A null mutation in the myosin heavy chain (MHC) gene, *Mhc*^{7}, which does not permit myosin expression prevents assembly of thick filaments and accumulation of flightin. In contrast, flightin is present in *act88F*^{KM88}IFM.[6]

Studies of the MHC mutants *Mhc*^{13} and *Mhc*^{6} have permitted more detailed mapping of the flightin binding site in the thick filament. Both of these alleles are single nucleotide changes in exon 16 that alter one amino acid in zone 27 of the light meromyosin (LMM) region of the MHC rod.[7] *Mhc*^{13} changes glutamic acid 1554 to lysine and *Mhc*^{6} changes arginine 1559 to histidine. Although these mutations are in a constitutive exon (i.e., expressed in all muscle myosins), they appear to seriously affect only the IFM. Two-dimensional gel electrophoresis (2DE) analysis indicates that both mutations affect flightin accumulation in the IFM. In *Mhc*^{6} IFM, low levels of flightin (primarily the nonphosphorylated variant, N1) are initially present in the pupa, but in the adult the full complement of phosphovariants is absent. *Mhc*^{13} prevents the accumulation of all but a minute amount of nonphosphorylated flightin.[7]

In contrast to these MHC rod mutations, the accumulation of flightin is not affected by mutations in the MHC motor domain. In particular, transgenic flies that express a recombinant myosin lacking the motor domain ('headless' myosin[8]) accumulate the full complement of flightin isovariants (our unpublished result). One interpretation of these results is that flightin interacts with the myosin rod, at or near the region defined by the *Mhc*^{13} and *Mhc*^{6} mutations.

To test the hypothesis that flightin binds the myosin rod, Ayer and Vigoreaux performed in vitro binding studies using full length wild-type and *Mhc*^{13} myosin, as well as several recombinant constructs representing sequentially shorter regions of the myosin rod.[9] They demonstrated that flightin binds full length myosin, headless myosin, and a recombinant fragment that includes the C terminal 600 amino acids of the rod. Flightin failed to bind *Mhc*[13] myosin and recombinant fragments from the C-terminal 400 amino acids and shorter. The results suggest that MHC aspartic acid 1554 is necessary but not sufficient for flightin binding to the myosin rod.

Sequence Features and Phosphorylation

Flightin bears no sequence homology to any known protein and it also lacks motifs or modular domains that may allow its assignation to a protein family. There are several interesting features in the sequence of flightin that may be relevant to its functional properties. The N-terminal one third of the protein is highly acidic (pI ~3.8) while the C-terminal two thirds are basic (pI ~10.6). Separating these two regions is a stretch of 5 prolines. The tertiary structure of flightin is not known, but secondary structure prediction programs suggest the protein has two α-helical regions (residues 88-114 and 149-180), with the rest of the protein having an extended or random conformation. The highly charged nature of flightin (pI ~5.2) combined with its low predicted hydrophobicity (Fig. 1) are features that are characteristic of unstructured or natively unfolded proteins.[10]

Preliminary DNA sequence analysis of flightin from several *Drosophila* species revealed that the central part of the protein is highly conserved but the N-terminal one third of the protein shows an unusually high degree of evolutionary divergence.[11] These studies suggest that flightin is a hybrid protein, raising the possibility that it performs a dual function, one conserved across species and one species-specific. The availability of sequenced genomes for *Anopheles*, *Apis*, and other insects with IFM will allow further testing of this hypothesis.

Eleven isoelectric variants of flightin have been resolved by 2DE of adult *D. melanogaster* IFM. Nine of the isoelectric variants are phosphorylated in vivo, as demonstrated by ^{32}P incorporation and treatment of skinned fibers with alkaline phosphatase.[3] Initial mass spectrometry analysis by MALDI-TOF identified residues Ser 139, Ser 141, Ser 145, Thr 158, and Ser 162 as likely phosphorylation sites.[12] Interestingly, two of these sites (Ser 141 and Ser 145) are not conserved in *D. virilis*, consistent with the observation that only nine isoelectric variants are detected by 2DE for this species.[11] At least three conditions have been shown to affect the extent of flightin phosphorylation: (i) stage of development (ii) mutations in contractile protein genes, and (iii) physiological state.

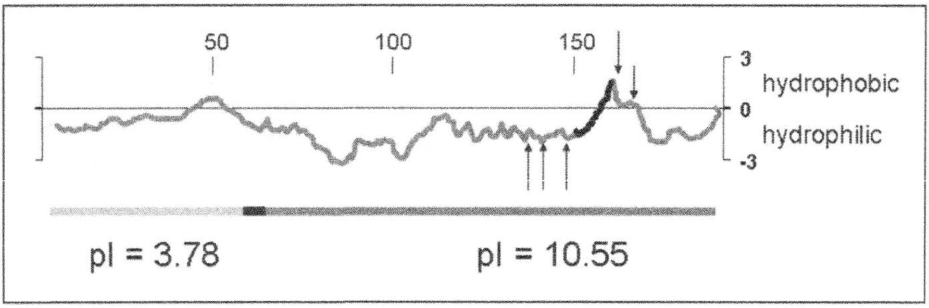

Figure 1. Hydrophilicity plot of flightin. The N-terminal one third of the protein is highly acidic (pI ~ 3.8), while the C-terminal two thirds are basic (pI ~ 10.6). These two regions are separated by 5 proline residues (black rectangle). The arrows point to five phosphorylation sites. The darkened section between the third and fourth arrows corresponds to amino acids 150-156 that are similar to part of the F-actin binding site of villin (see Fig. 4). Note that this sequence extends into the most hydrophobic region of flightin. Phosphorylation of the surrounding sites may be necessary to expose the motif and make it available for F-actin binding.

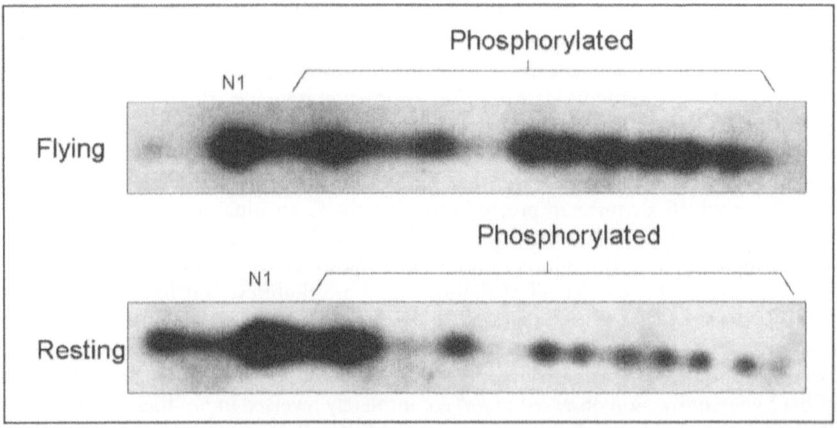

Figure 2. Phosphorylation profile of flightin during flight and rest. Note the shift in the ratio of phosphorylated to nonphosphorylated (N1) flightin between flying and resting flies. Changes in the proportion of phosphovariants may influence the dynamic stiffness of the IFM.

Phosphorylation of flightin commences during the final stages of pupal (IFM) development and continues throughout eclosion and the initial hours of adult life. While mid-stage pupa contain predominantly nonphosphorylated flightin, the number of phosphovariants progressively increases from zero to nine by the time the fly develops flight competency (~2-3 hours post-eclosion).[3] Mature adults have nearly a 1:1 ratio of nonphosphorylated to phosphorylated flightin.

Unlinked mutations also interfere with the pattern of flightin phosphorylation, as described above for the MHC rod mutants. Of particular interest is *Mhc*^{6}, which appears to affect accumulation of phosphorylated flightin, but not the unphosphorylated form.[7] An opposite effect is seen in the actin null mutant *act88F*^{KM88} which causes premature phosphorylation of flightin (as early as pupal stage P9) and preferential accumulation of the phosphorylated forms.[6] The reason for these changes are not known but it is evident from these results that phosphorylation of flightin is sensitive to changes in the thick filament as well as the thin filament.

Little is known about how the levels of flightin phosphorylation are controlled, the identity of the protein kinases involved, or of the kinases' regulation. Recent studies, however, have shown that the levels of flightin phosphorylation in the adult are not static. In particular, we note a shift in the ratio of nonphosphorylated to phosphorylated flightin between nonflying and flying flies (Fig. 2). In the latter, there tends to be a greater abundance of the more acidic variants, suggesting an increased in phosphorylation with flight. The functional significance of this difference is discussed below.

Genetic Analysis of Flightin Function

Genetic studies have shown that flightin is essential for proper IFM development and function. *Drosophila* that are heterozygous for a genetic deficiency spanning the flightin gene, *Df(3L)fln*^{1}, demonstrate a 20% reduction in flightin abundance. This reduction impairs flight ability and results in slight ultrastructural defects in the myofibrils.[13] Myofibrils from *Df(3L)fln*^{1} have an intact central core that resembles wild type, but the myofilaments along the myofibril periphery are loosely organized. When treated with nonionic detergent, the core remains intact but the peripheral myofilaments are washed away, indicating that they were not structurally connected to the lattice. This observation lends insight into the function of flightin, as it suggests that flightin is essential for maintaining the structural integrity of the myofibril by linking or stabilizing the thick filaments in the lattice.

Mechanical studies of skinned IFM fibers showed that the reduction in flightin expression has little effect on net power output but increases f_{max}, the frequency required to achieve maximum power.[13] This increase reflects a change in crossbridge kinetics, in particular the transition from a nonforce producing state to a force producing state (i.e., the apparent rate constant $2\pi b$). It is unlikely that flightin has a direct effect on crossbridge kinetics; instead the increased frequency may compensate for the reduced number of myosin motors that form an integral part of the myofibril. Nevertheless, the observation that heterozygotes of the projectin mutant *bent*^{dominant}, in which a truncated projectin missing the C-terminal protein kinase domain is expressed, show a similar increase in $2\pi b$, raise the possibility that projectin kinase may modulate crossbridge function through flightin phosphorylation.[14]

Flies that are altogether devoid of flightin (*fln*^{0}) are flightless and have severe fiber and myofibrillar defects.[5] IFM fibers progressively shorten during the initial hours of adult life as a result of contractile activity. This 'hypercontraction' causes the fibers to pull away from one end of the cuticle and compress themselves against the other end, or to rip apart along their length. The fiber hypercontraction observed in *fln*^{0} is completely reversed in *fln*^{0} flies that also carry an MHC transgene encoding a headless myosin or MHC alleles with mutations in the motor domain that were isolated as suppressors of the wings up phenotype of the TnI mutant *heldup*^{2}.[15] These results demonstrate that fiber hypercontraction in *fln*^{0} results from actomyosin generated contractile force and suggest that flightin is an important structural protein in maintaining sarcomere integrity in active muscle.

At the myofibrillar level, sarcomeres in late stage pupa are longer in *fln*^{0} than in wild type (~3.8 µm versus 3.1), but otherwise normal.[5] However, sarcomere structure is not preserved in adults as massive degeneration ensues several hours after eclosion. Fraying of myofilaments is evident as well as disintegration of Z bands and M lines to the point where individual sarcomeres are no longer discernable. These results suggest that flightin is not only informative in thick filament length determination during development, but it is also essential for maintaining the structural integrity of the myofibril in the working muscle. The lack of myofibrillar integrity in *fln*^{0} adults is partly explained by the instability of thick filaments. In the absence of flightin, myosin is more susceptible to proteolysis at or near the S2 hinge, a feature that is also seen in the myosin rod mutant *Mhc*^{13}. Solubility studies also showed that myosin, but not actin, is more easily dissociated from the fiber when flightin is not present.[5]

Sinusoidal analysis of skinned IFM fibers from *fln*^{0} and myosin rod mutants (*Mhc*^{6} and *Mhc*^{13}) that affect flightin expression revealed that flightin contributes to the viscoelastic properties of the flight muscle fiber.[16] Passive stiffness was significantly decreased in *fln*^{0} and *Mhc*^{13} but not in *Mhc*^{6}; this latter mutation has little effect on the accumulation of nonphosphorylated (N1) flightin. In contrast, maximally Ca^{2+} activated fibers from all three mutants exhibited deficits in viscous moduli, the amplitude of which is proportional to the work produced by the fiber. Because the absence of flightin has no effect on isometric tension, the decrease in work and power output evident in the mutant fibers likely arises from defects in force transmission. In the next section, we will discuss a model for how flightin may fulfill its role as a structural protein that stiffens the thick filament.

Current studies in our lab aimed at elucidating the function of flightin phosphorylation have focused on the characterization of transgenic lines expressing mutant flightin genes in which two (Thr 158 and Ser 162), three (Ser139, Ser141, and Ser145) or all five phosphorylated residues were changed to alanine. All three transgenes result in structural and functional IFM defects when expressed in *fln*^{0} background.[17] Remarkably, transgenes with two (fln^{3TSA}) or five (fln^{5STA}) mutated residues exhibit a dominant negative effect. When expressed in a wild-type background, these transgenes result in flightlessness and ultrastructural defects (Table 1). These preliminary studies indicate that the phosphorylation of flightin is important for IFM structure and function. Further studies, however, are needed to define the functional role of phosphorylation.

Table 1.* Drosophila *IFM mutations that affect flightin expression

Name	Mutation	Effect on Flightin Expression*	Phenotype*
Df(3L) fln^{1}	Deficiency deletes entire flightin gene	~20% reduction	-Modest flight impairment -Peripheral disruption of myofibril
fln^{0}	Null (codon 8 to stop)	No flightin expressed	-Flightless -Severe fiber hypercontraction -Myofibrillar degeneration
fln^{5STA}	Transgene with five phosphorylation sites changed to alanine (Ser139, Ser141, Ser145, Thr 158 and Ser 162)	Reduced accumulation and fewer phosphovariants	-Flightless, even in presence of wild type flightin -Severe fiber hypercontraction -Myofibrillar degeneration
fln^{5SA}	Transgene with three phosphorylation sites changed to alanine (Ser139, Ser141 and Ser145	Slight reduction in adult accumulation	-Flightless, but can fly in presence of wild type flightin -Myofibrillar degeneration
fln^{3TSA}	Transgene with two phosphorylation sites changed to alanine (Thr 158 and Ser 162)	Reduced accumulation and fewer phosphovariants	-Flightless, even in the presence of wild type flightin -Myofibrillar degeneration
Mhc^{7}	MHC null	No flightin expressed	-Flightless
Mhc^{6}	R to H in LMM zone 27 of MHC rod	Phosphovariants not present	-Flightless -Fiber hypercontraction -Myofibrillar degeneration
Mhc^{13}	E to K in LMM zone 27 of MHC rod	Greatly reduced expression of all isovariants	-Flightless -Severe fiber hypercontraction -Myofibrillar degeneration
Act88F^{KM88}	Actin null	Premature phosphorylation	-Flightless

*All results are for homozygotes except for *Df(3L)fln*^{1}.

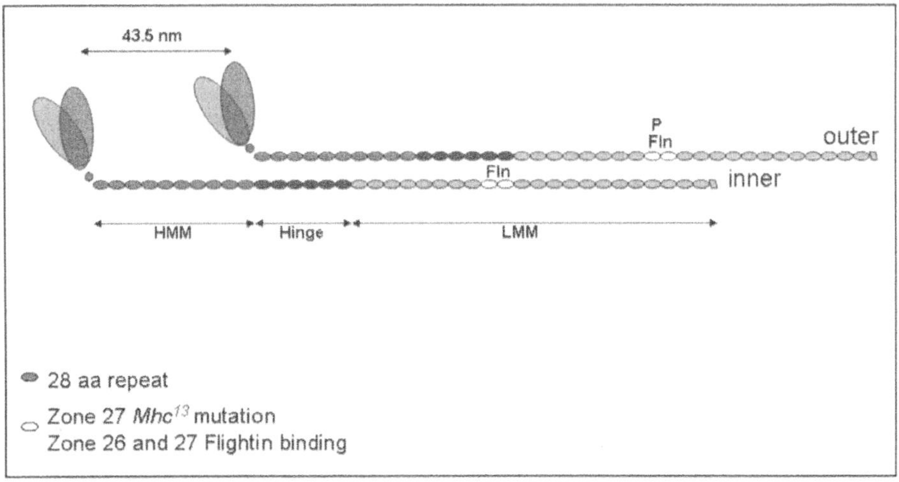

Figure 3. Model of flightin-myosin interaction. The two myosin molecules shown represent one subfilament.[20] Note that flightin (Fln) bound to zone 26-27 of the inner myosin is juxtaposed to the S2 hinge of the outer myosin. An interaction between flightin and S2 hinge may explain why this myosin region is susceptible to proteolysis in the absence of flightin and why IFM fibers that lack flightin show reduced stiffness.[5,7,16] In adult IFM, the ratio of phosphoflightin to dephosphoflightin is ~1:1.[3] Lattice constraints may allow flightin on the outer myosin to be more readily phosphorylated than flightin on the inner myosin. Phosphorylation (P) would expose the villin-like F-actin binding site and allow interaction of flightin with the thin filament.

Possible Functions for Flightin

The studies summarized above provide evidence that flightin is essential for proper thick filament and myofibril assembly, sarcomere integrity in active muscle, and normal contractile function. The manner in which flightin fulfills these various functional roles is not known, but our studies indicate that phosphorylation may be a key mechanism for modulating flightin's dual function in muscle development and contractile activity. Previously, Reedy et al had proposed that hyperphosphorylation of flightin during the pupal-to-adult transition may be part of a mechanism for arresting thick filament elongation and sarcomere growth.[5] According to this 'charge repulsion' model, phosphorylation of flightin would overcome the electrostatic attractions of the 28 residue charge repeat along the LMM which are largely responsible for myosin self-assembly properties.[18] Phosphorylation may add enough negative charges to the surface of the filament so that incoming myosin molecules are repelled and filament growth stops (see Fig. 12 in Reedy et al).[5] A second possibility is that phosphorylation of flightin on the surface of the thick filament may serve to align thick and thin filaments as part of the mechanism for establishing final sarcomere length.[5] Yet, a third possibility is that flightin binding to LMM may interfere with myosin sequences that confer assembly properties to myosin.[19]

In adult muscle, flightin is an important contributor to fiber stiffness. One manner in which flightin may fulfill this role is by cross-linking myosin dimers in a subfilament, as the model in Figure 3 shows. IFM thick filaments consists of 12 subfilaments, and each subfilament is formed by one pair of myosin molecules staggered axially by ~43.5 nm (3x14.5 nm)[20] (see also chapter by Beinbrech and Ader). The model accounts for the observation that myosin is susceptible to proteolysis in the S2 hinge in the absence of flightin. Flightin bound on the LMM heptad repeat 27 is ideally placed to interact with the hinge region of its neighboring myosin molecule. The model also accounts for the observed ratio of myosin to flightin of ~2:1. If flightin interacts with the myosin hinge, the most flexible part of the coiled-coil, it may additionally stiffen the thick filament by stabilizing the helical conformation.

cv	**KE** -x(25)- **K**	QQNL	**KKEK**	G	**LF**
hv	KE -x(25)- K	-x(4)-	KKEK	G	LF
hs	LE -x(25)- K	-x(4)-	KKAK	G	LF
hd	LE -x(25)- K	-x(4)-	KKKA	S	LF
dv	RE -x(25)- K	-x(4)-	KKQF	K	LF
fn	RE -x(14)- R	-x(13)-	KRDK	Q	LI

Figure 4. Alignment of partial headpiece domain sequences and flightin sequence. Amino acids implicated in F-actin binding are shown in bold in the chicken villin (cv) sequence. Hv, human villin; hs, human supervillin; hd, human dematin; dv, Drosophila villin; fn, flightin. The number of residues separating the amino acids shown are indicated by –x(). Adapted from Vardar et al.[22]

A second, more speculative, possibility for how flightin may influence myofibril stiffness is that it forms, or is part of, an interfilament link between the thick and thin filaments, a role analogous to that proposed for vertebrate myosin binding protein C. While binding of flightin to actin has not been tested, there is evidence of genetic interaction between mutants in flightin and mutants in actin and troponin I (our unpublished results). Interestingly, flightin has sequence similarity ($_{150}$KRDKQLI$_{156}$) with part of the F-actin binding motif found in villin[21] (KKEKGLF) and other F-actin bundling proteins[22] (Fig. 4). The five identified phosphorylation sites in flightin flank this potential actin binding site, suggesting that the phosphorylation could somehow be involved in the interaction of actin with flightin. Likewise, the vicinity of the phosphorylation sites might indicate that actin modulates phosphorylation. Consistent with this idea, phosphorylation occurs prematurely in actin null flies.[6] Flightin will need to be largely unfolded to cover the ~20 nm distance between myofilaments. This model is attractive because the extent of phosphorylation may dictate the strength of the flightin-thin filament interaction (by dictating the extensibility of the spring), which in turn may influence muscle stiffness. Changes in muscle stiffness may be important for modulating wing beat frequency during flight.

Zeelins

Zeelins are proteins found in *Lethocerus* leg and flight muscle. Zeelin 1 (35 kDa) is found in both leg and flight muscle, while zeelin 2 (23 kDa) is found only in flight muscle. Immunofluorescence microscopy of IFM myofibrils show that zeelin 1 is found predominantly at the ends of the A band (towards the H zone and Z disc) while zeelin 2 has a more broad distribution throughout the A band, on either side of the H zone.[23] Immunoelectron microscopy of oblique sections showed differences in the labeling by anti-zeelin 1 and anti-zeelin 2 antibodies. Zeelin 1 appears confined to the thick filament interior while zeelin 2 is found on the outside of the thick filament. Removal of myosin left an intact skeleton of zeelin 1. The association of zeelins with thick filaments was further demonstrated by EM of muscle fibers that were stretched so that thick and thin filaments no longer overlapped. Antibodies to zeelin 1 and zeelin 2 labeled the thick filaments but not the thin filaments. The zeelins also were found to cofractionate with thick filaments from which thin filament proteins had been removed by treatment with gelsolin. However, zeelin 2 does not bind myosin in vitro so its association with the thick filament is likely mediated by another protein.[23]

Two dimensional gel electrophoresis revealed that zeelin 2 separates into two major and one minor isoform while the more basic zeelin 1 separates into two isoforms. Lower molecular mass forms of zeelin 1 were also found. The proportion of zeelin isoforms differ in leg muscle and IFM. The isoforms of zeelin 2 were phosphorylated in vitro with the catalytic subunit of cAMP dependent protein kinase, resulting in an uniform shift of the three isoforms. This suggests that the differences among the in vivo isoforms, which are of the same molecular weight, are not due to phosphorylation.[23]

In a low ionic strength solution zeelin 2 formed short filaments 10 nm wide while zeelin 1 formed irregular micelles about 20 nm in diameter, suggesting that zeelin 1 is amphiphilic in nature. The presence of myosin had no effect on zeelin 2 filament formation, but when zeelin 1 and zeelin 2 were combined, clumped, elongated micelles were formed, but no filaments. One interpretation of these results is that zeelin 2 is prevented from forming filaments via an interaction with zeelin 1.[23]

The role of zeelins is not known. Ferguson et al proposed that zeelins may be involved in maintaining the ordered structure of the thick filament and the regularity of the myofibrillar lattice.[23] Given the similarities between zeelin 2 and flightin, these proteins are likely to play analogous roles (see below and Table 2).

Stretchin-MLCK

Stretchin-MLCK is reported to be a novel member of the Titin/Myosin light chain kinase family that has been studied conceptually through virtual translation of its coding region and by RT-PCR.[24] The transcription unit contains thirty-three exons and is hypothesized to express seven different transcripts, the largest of which is stretchin-MLCK at 926 kDa (8,295 amino acids). This protein features the four hallmark motifs associated with the Titin/MLCK family: immunoglobulin domains (32), fibronectin type-III domains (2), PEVK (proline, glutamic acid, valine, lysine) repeats (2), and an MLCK-like kinase domain.[24] The first PEVK-rich domain is similar to the one in titin which has been shown to impart the properties responsible for passive tension and elasticity during a physiological stretch to the sarcomere.[25] The second PEVK repeat contains the amino acid sequence serine, alanine, isoleucine, aspartic acid, glutamic acid, and is therefore referred to as SAIDE repeat. Champagne et al[24] suggest that together PEVK and SAIDE should provide stretchin-MLCK with titin's spring-like properties.

Of the remaining six polypeptides predicted to be encoded by *stretchin-MLCK*, four include the kinase domain and two do not. The four putative kinases range in size from 86 kDa to 497 kDa. Stretchin, the second largest putative protein encoded by the gene, has the PEVK and SAIDE repeats as well as the 24 Ig domains and the unique 1706 residue segment, but does not contain the kinase domain. A kettin-like protein of 269 kDa consists of 16 Ig domains and unique sequences only.

None of the conceptual protein products have been shown to be expressed in the IFM. Preliminary studies in our lab have identified a 165 kD protein as stretchin-MLCK, (Henkin, Maughan and Vigoreaux, in preparation). The molecular weight of this protein closely matches that of one of the predicted kinases.[24] Other recent studies have identified myostrandin (A225), a novel splicing variant that consists of 16 Ig domains preceded by a unique 449 residue region.[26] Myostrandin is expressed in the IFM as two molecular weight variants of ~225 and 231 kD. The heavier one appears to be IFM-specific and is absent in the IFM of the myosin null *Mhc*^{7}. Immuno EM revealed that myostrandin is distributed throughout the A band, except at the H zone and towards the end of the thick filament facing the Z band, a distribution very similar to that described for flightin (see above). Other similarities between flightin and myostrandin is that both proteins are first detected at pupal stage P8 and their RNAs are down-regulated during aging and after paraquat treatment.[4] These results raise the possibility that flightin and myostrandin may be functionally linked in the IFM.

Table 2. Comparison of* Drosophila *flightin and* Lethocerus *zeelin 2

	Flightin	Zeelin 2
Tissue expression	IFM	IFM
Sarcomere distribution	A band	A band
Molecular weight	20 kDa	23 kDa
Isoelectric point*	5.2	5.7
Isoforms (phosphoisoforms)	11 (9)	3 (0)
Hydrophobicity	Low	High
% Pro-Ala	20.5	19.7
Mole per mole myosin	1 : 2	1 : 2.5

* for flightin, calculated pI for the unmodified translation product; for zeelin 2, estimated from 2DE.

Concluding Remarks

Table 2 summarizes some of the properties of zeelin 2 and flightin. The most notable similarity is that both proteins are found exclusively in the A band of the IFM. Both proteins are homogeneously distributed throughout the A band except for the very tip of the A band facing the Z band and the H zone. This distribution suggests that flightin and zeelin 2 associate with the myosin rod, flightin through direct binding and zeelin indirectly. Flightin remains associated with thick filaments in the absence of thin filaments and zeelin 2 remains associated with thick filaments in fibers stretched beyond the overlap (A band) region. Both proteins cofractionate with thick filaments after fiber extractions. Despite these similarities, antibodies to the two proteins do not cross react and comparisons of short peptides indicate that the proteins do not appear to share sequence homology.[23] It is possible that flightin and zeelin 2 fulfill the same functional need, that of stiffening the thick filament, via different mechanisms. It is also possible that flightin, via its multiple phosphorylated and nonphosphorylated isovariants, can perform the role of zeelin 1 and zeelin 2 combined.

The advent of genomes from more insect species with asynchronous IFM, including mosquito and honeybee, may help establish the diversity in myosin associated proteins. Determination of the atomic structure would reveal if flightin and/or zeelin 2 share structural homology with other known myosin binding proteins such as MyBP-C. This information will help address the question of whether these proteins, while different in sequence, share functional attributes.

Note Added in Proof

A new study by Qiu et al[27] showed that zeelin 2 is similar to flightin.

Acknowledgements

This publication was made possible by support from NSF MCB-0090768 and MCB-0315865, and by the Vermont Genetics Network through the NIH Grant Number 1 P20 RR16462 from the BRIN program of the NCRR.

References

1. Vigoreaux JO, Saide JD, Valgeirsdottir K et al. Flightin, a novel myofibrillar protein of Drosophila stretch-activated muscles. J Cell Biol 1993; 121(3):587-598.
2. Arbeitman MN, Furlong EE, Imam F et al. Gene expression during the life cycle of Drosophila melanogaster. Science 2002; 297(5590):2270-2275.
3. Vigoreaux JO, Perry LM. Multiple isoelectric variants of flightin in Drosophila stretch-activated muscles are generated by temporally regulated phosphorylations. J Muscle Res Cell Motil 1994; 15:607-616.

4. Zou S, Meadows S, Sharp L et al. Genome-wide study of aging and oxidative stress response in Drosophila melanogaster. Proc Natl Acad Sci USA 2000; 97(25):13726-13731.
5. Reedy MC, Bullard B, Vigoreaux JO. Flightin is essential for thick filament assembly and sarcomere stability in Drosophila flight muscles. J Cell Biol 2000; 151:1483-1499.
6. Vigoreaux JO. Alterations in flightin phosphorylation in Drosophila flight muscles are associated with myofibrillar defects engendered by actin and myosin heavy chain mutant alleles. Biochem Genet 1994; 32:301-314.
7. Kronert WA, O'Donnell PT, Fieck A et al. Defects in the Drosophila myosin rod permit sarcomere assembly but cause flight muscle degeneration. J Mol Biol 1995; 249:111-125.
8. Cripps RM, Suggs JA, Bernstein SI. Assembly of thick filaments and myofibrils occurs in the absence of the myosin head. EMBO J 1999; 18(7):1793-1804.
9. Ayer G, Vigoreaux JO. Flightin is a myosin rod binding protein. Cell Biochem Biophys 2003; 38:41-54.
10. Uversky VN, Gillespie JR, Fink AL. Why are "natively unfolded" proteins unstructured under physiologic conditions? Proteins 2000; 41(3):415-427.
11. Cajigas IJ, Valsky E, Gorrochategui M et al. Phylogenetic analysis of Drosophila flightin reveals a hybrid protein with conserved and rapidly evolving sequences. Mol Biol Cell 2002; 13S:40a.
12. Barton BE, Ayer G, Cajigas IJ et al. Defects in flight muscle ultrastructure and function in transgenic Drosophila with mutations of phosphorylation sites in flightin. Mol Biol Cell 2002; 13S:319a.
13. Vigoreaux JO, Hernandez C, Moore J et al. A genetic deficiency that spans the flightin gene of Drosophila melanogaster affects the ultrastructure and function of the flight muscles. J Exp Biol 1998; 201:2033-2044.
14. Vigoreaux JO, Moore JR, Maughan DW. Role of the elastic protein projectin in stretch activation and work output of Drosophila muscles. In: Granzier H, Pollack G, eds. Elastic Filaments of the Cell. New York: Kluwer Academics/Plenum Publishers, 2000:237-250.
15. Nongthomba U, Cummins M, Clark S et al. Suppression of muscle hypercontraction by mutations in the myosin heavy chain gene of Drosophila melanogaster. Genetics 2003; 164(1):209-222.
16. Henkin JA, Maughan DW, Vigoreaux JO. Mutations that affect flightin expression in Drosophila alter the viscoelastic properties of flight muscle fibers. Am J Physiol Cell Physiol 2004; 286:C65-C72.
17. Barton BE, Ayer G, Maughan D et al. Mutations of phosphorylation sites in flightin are responsible for loss of function and ultrastructural defects in Drosophila indirect flight muscle. Mol Biol Cell 2003; 14S:309a.
18. McLachlan AD, Karn J. Periodic charge distributions in the myosin rod amino acid sequence match cross-bridge spacings in muscle. Nature 1982; 299:226-231.
19. Sohn RL, Vikstrom KL, Strauss M et al. A 29 residue region of the sarcomeric myosin rod is necessary for filament formation. J Mol Biol 1997; 266(2):317-330.
20. Beinbrech G, Ashton FT, Pepe FA. Orientation of the backbone structure of myosin filaments in relaxed and rigor muscles of the housefly: Evidence for nonequivalent crossbridge positions at the surface of thick filaments. Tissue & Cell 1990; 22(6):803-810.
21. Doering D, Matsudaira P. Cysteine scanning mutagenesis at 40 of 76 positions in villin headpiece maps the F-actin binding site and structural features of the domain. Biochemistry 1996; 35:12677-12685.
22. Vardar D, Buckley DA, Frank BS et al. NMR structure of an F-actin-binding "headpiece" motif from villin. J Mol Biol 1999; 294(5):1299-1310.
23. Ferguson C, Lakey A, Hutchings A et al. Cytoskeletal proteins of insect muscle: Location of zeelins in Lethocerus flight and leg muscle. J Cell Science 1994; 107:1115-1129.
24. Champagne MB, Edwards KA, Erickson HP et al. Drosophila stretchin-MLCK is a novel member of the Titin/Myosin light chain kinase family. J Mol Biol 2000; 300(4):759-777.
25. Linke WA, Granzier H. A spring tale: New facts on titin elasticity. Biophys J 1998; 75:2613-2614.
26. Patel SR, Saide JD. A(225), an A-band associated protein of Drosophila IFM, has a myosin dependent and a myosin independent isoform. Biophys J 2003; 84:564a.
27. Qiu F, Brendel S, Cunha PM et al. Myofillin, a protein in the thick filaments of insect muscle. J Cell Sci 2005; 118:1527-1536.

Chapter 8

Structure of the Insect Thick Filaments

Gernot Beinbrech and Gereon Ader

Abstract

Myosin filaments of insect indirect flight muscles (IFM) are 17 to 19 nm thick and 1.9 to 3.6 μm long structures with probably 4 cross-bridges per level (= crown). These crowns repeat in periods of 14.5 nm along the longitudinal axis of the filament. The cross-bridges are located at 4 helical tracks with axial spacings of 38.7 nm and a true axial repeat of 116 nm on the surface of the filaments. Twelve myosin subfilaments, arranged in pairs, run parallel to the longitudinal filament axis and form a wall around a myosin-free filament core. The core may be filled by additional elements, the number of which is related to the paramyosin content of the filaments.

Aggregates of peptide fragments representing the C-terminal two-thirds of *Drosophila* light meromyosin with the exon-19 encoded C-terminus display 116 nm repeating units with substructured elements of 43.5-29-14.5-29 nm in width. Optical diffraction patterns of these aggregates show layer line characteristics that resemble those of negatively stained, isolated thick filaments. These results suggest that the IFM specific cross-bridge pattern results from the aggregation properties of insect myosin. Filament models, based on these properties are consistent with electron microscopy and X-ray diffraction data.

Introduction

Myosin filaments of insect indirect flight muscles (IFM) have diameters of 17 nm to 19 nm (Table 1). Their lengths are less uniform but are close to the sarcomere length, making I-bands barely visible (Fig. 1). The filament lengths vary from 1.9 μm to 2.7 μm in the flight muscles of honeybees, to 2.8 μm to 3.6 μm in the flight muscles of flesh flies (Table1).[1] In some insect orders that have been studied these values are associated to the capability of the animals to move the wings at frequencies of up to 1000 Hz by small amplitude oscillations of the flight muscles.[2] This is possible because the up and down strokes of the wings can be performed by one synchronous stroke of the cross-bridges of the dorso-ventral and the dorso-longitudinal flight muscles, respectively (see chapter by Moore in this volume). The percent fiber shortening resulting from the powerstroke is proportionally greater the shorter the sarcomere. Therefore, the length of the thick filament must be fine tuned to the properties of the different components of the flight machinery: wings, wing hinges, structure and elasticity of the thorax and of the contractile apparatus.

Another structural preposition for oscillating cross-bridges is a well organized arrangement of the contractile proteins of IFM within their sarcomeres and their myofilaments. It is unique that in asynchronous IFM (as seen in the rigor state), actin and myosin molecules are arranged within the actin and the myosin filaments, respectively, along helices spaced about 38.7 nm (Fig. 1a).[3,4] The matching of the repeat lengths of both helical tracks has the functional advantage that the activated cross-bridges and their target areas along the actin

Nature's Versatile Engine: Insect Flight Muscle Inside and Out, edited by Jim Vigoreaux.

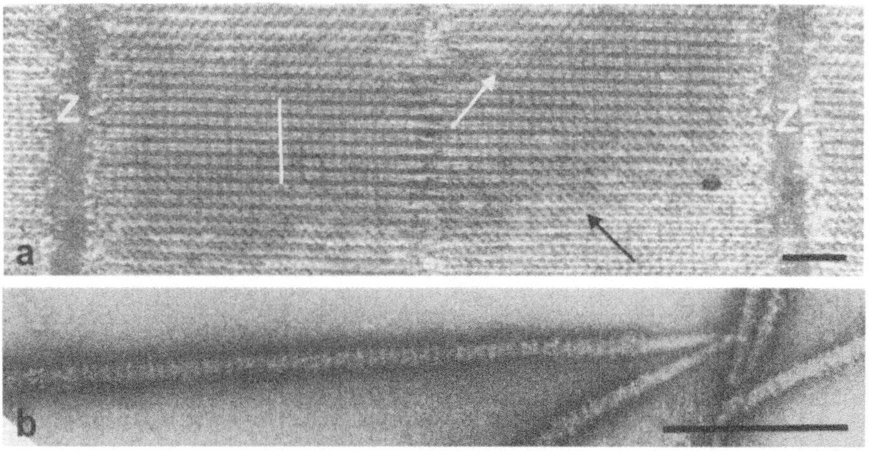

Figure 1. Electron micrographs showing (a) a longitudinal section of a sarcomere of an asynchronous flight muscle fiber of a waterbug, (*Lethocerus spec*) which had been fixed in rigor solution: (b) a negatively stained myosin filament (prepared in an ATP containing buffer) from an asynchronous flight muscle of a waterbug. Note the 14.5 nm periodicities corresponding to cross-bridges. The arrows in (a) point to lines formed by the origins of the cross-bridges at the surface of the filaments in 38.7 nm distances. The lines are better seen if the micrograph is viewed in an acute angle along the shaft of the arrows. The white line indicates the lines formed by the attachment of the cross-bridges in distances of 38.7 nm along the actin filaments. Scale bars 250 nm. b) From Reedy MK, Leonard KR, Freeman R, Arad T. Thick myofilament mass determination by electron scattering measurements with the scanning transmission electron microscope. J Muscle Res Cell Motil 1981; 2: 45-64. Copyright ©1981 Kluwer Academic Publishers. Reprinted with kind permission of Kluwer Academic Publishers.

filaments[3,5] may be directed periodically toward each other at the same levels. This should allow the cross-bridges to attach to the actin filaments and to perform their synchronous, oscillatory power strokes in a very efficient way. This particular arrangement of the cross-bridges at the surface of the myosin filaments, however, makes the understanding of the backbone structure of the thick filaments, i.e., the stagger of the myosin molecules, more difficult. The packing of the myosin molecules in the filament backbone must be different than that of myosin in vertebrate thick filaments. In fact, as in vertebrate skeletal muscles, the myosin heads of insect muscles are arranged in "crowns"[6] that repeat regularly along the filament axis at distances of 14.5 nm at either side of the M-line (Fig. 1b). There are three myosin molecules per crown in vertebrate muscles[7] but four to six double heads per crown have been calculated for insect muscles.[1,6,7] In IFM the number seems to be closer to four, in insect leg muscles the number is closer to six (Table 1). Models for the backbone structure of IFM myosin filaments, which try to relate the arrangement of cross-bridges at the surface of the filaments[8,9] to that of the rods in the filament backbone, therefore, have to take into account the following:

1. the number of cross-bridges per crown
2. the 14.5 nm periodicity of the crowns along the filament length
3. the arrangement of the cross-bridges on the filament surface along helices with 38.7 nm axial spacing
4. the specific self assembly properties of insect myosin molecules
5. the specific distribution and possible role of myosin associated proteins in the assembly process.

Table 1. Comparison of structural, biochemical and functional parameters of different muscles

Muscle	Filament Type	Wing Beat Frequency (s^{-1})	Diameter of Myosin Filaments (nm)	Length of Myosin Filaments (mm)	Myosin Molecules per Crown	Paramyosin Percentage of Filament Mass
rabbit (psoas)	●	–	12-13	1.6	3.0	–
Melolontha (IFM)	●	50	17-19	2.8-3.3	4.9	9.5
Apis mellifera (IFM)*	●	200	17-19	1.9-2.7	3.8	24
Lethocerus (IFM)	●	20-38**	17-19		4.2***	11***
Locusta (IFM)	●	20	17-19	2.4-3.3	4.5	8.6
Locusta (leg muscle)	●	–	17-19	7.7	6.2	6.4
Phormia (IFM)	° /●	180	17-19	2.8-3.6	4.5	3.8
Musca (IFM)	°	165	17-19	2.4-3.4	4.6	2.7
Drosophila (IFM)	°	200	17-19	2.8-3.3		2.6

*From Hinkel-Aust S, Hinkel P, Beinbrech G. Four cross-bridge strands and high paramyosin content in the myosin filaments of honey bee flight muscles. Experientia 1990; 46:872-874. **From Barber SB, Pringle JWS. Functional aspects of flight in belostomid bugs (Hemiptera). Proc Royal Soc B 1966; 164:21-39. ***From Reedy MK, Leonard KR, Freeman R et al. Thick myofilaments mass determination by electron scattering measurements with the scanning transmission electron microscope. J Muscle Res Cell Motil 1981; 2:45-64. Other data from Beinbrech G, Meller U, Sasse W. Paramyosin content and thick filament structure in insect muscles. Cell Tissue Res 1985; 241:607-614. ● = solid and ° = tubular myosin filaments.

Substructures in Transverse Sections of Myosin Filaments

IFM thick filaments consist mainly of three proteins: myosin, paramyosin and projectin (mini-titin). The bulk protein is myosin. Projectin is a minor component with 1 molecule per 35 myosin molecules as estimated for locust flight muscle.[10] The paramyosin content is either about 3% (tubular filaments of the flesh-fly *Phormia*) or close to multiples of 3%, i.e., about 9% (locust), 11%, waterbug or 18 to 24% (honey-bee).[11,12] The paramyosin content of the flight muscles of *Drosophila* and of *Musca* (house-fly) is lower than 3 % (Table 1). In these cases cross sections of filaments show a well defined solid wall and a hollow core and it is reasonable to assume that the substructures of the walls are myosin aggregates. The results of averaging and filtering procedures (Fig. 2) suggest that the filament walls consist of 12 subfilaments, as proposed by Wray.[13] The subfilaments, however, seem to be arranged in pairs rather than in equivalent positions. The center to center distance between the two subfilaments in a pair is about 3 nm, while the center to center distance between two subfilaments of neighboring pairs, i.e., across the gap, is 4 nm (Fig. 2h). These values are similar to those of subfilaments of vertebrate skeletal muscles.[14]

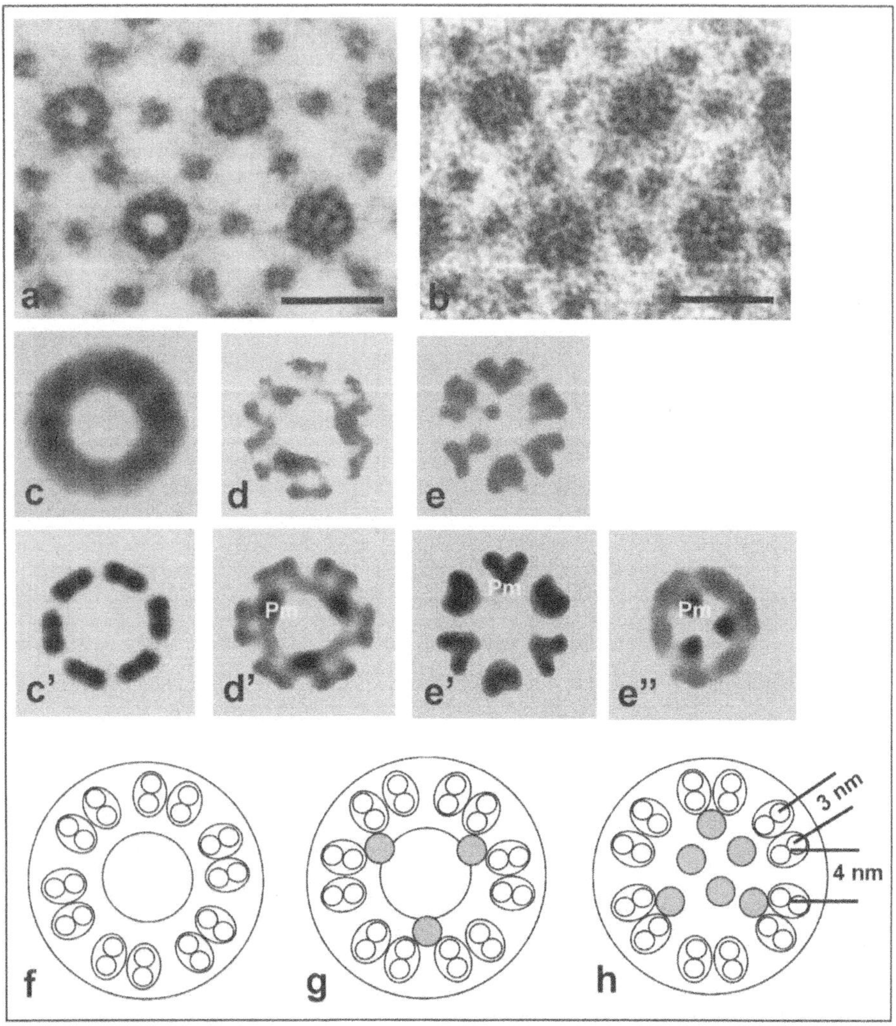

Figure 2. Cross-section electron micrographs of: a) the IFM of *Phormia terrae-novae* (flesh-fly) showing tubular and solid myosin filaments, and b) *Apis mellifera* (honey-bee) IFM showing exclusively solid myosin filaments. c-e) Computer processed images obtained by superimposing and averaging 50 to 60 digitized individual filaments. The 6-fold rotational centers of the filament walls have been used for the superimposition. c'-e") The corresponding images obtained by filtering the averaged images: c') is the result of filtering the averaged image (c) of the tubular flesh-fly filaments for 6-fold rotational symmetry. d') Obtained from the averaged image (d) by filtering for 3-fold symmetry also using the center for 6-fold symmetry of the filament wall. The same procedure has been applied to (e), the average image of solid filaments of bee muscles, to get (e'). e") A filtered image obtained by filtering the core region of the average image of the same filaments as in (e), but superimposed by using the center for 3-fold symmetry of the filament cores. f-h) Schematic drawings of the arrangement of subunits within cross-sectioned myosin filaments. They are deduced from the filtered images (c'), (d') and the combined (e') and (e"), respectively. The shaded circles indicate subunits presumably consisting of paramyosin (Pm). The distances between subfilaments in (h) are also valid for (f) and (g). Scale bars for (a) and (b) 30 nm.

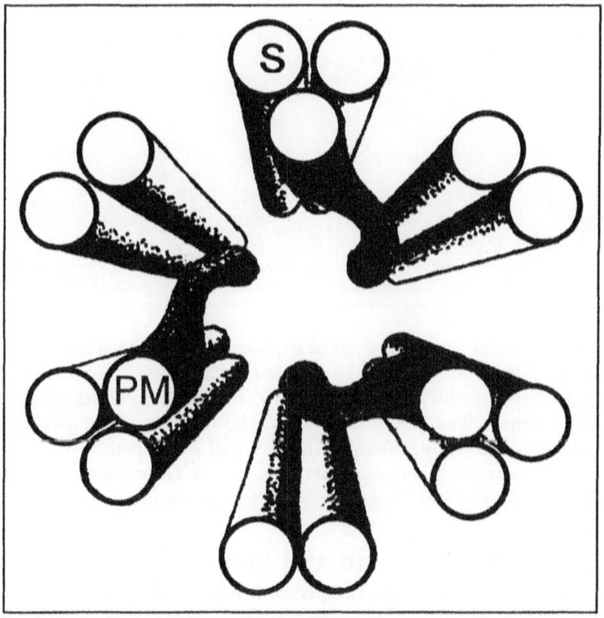

Figure 3. Model of solid thick filaments of the flesh-fly, *Phormia terrae-novae*, representing the location of 6 pairs of myosin subfilaments (S) and 3 substructures consisting of paramyosin (PM). The paramyosin structures are located at the inner edge of the filament wall and wobble between two pairs of myosin subfilaments. Reprinted from: Schmitz H, Ashton FT, Pepe FA, Beinbrech G. Substructures in the core of thick filaments: arrangement and number in relation to the paramyosin content of insect flight muscles. Tissue Cell 1994; 26: 83-100. ©1994 Elsevier, with permission.

If the paramyosin content is higher than 3 %, e.g., about 9 % or 11 % as in the thick filaments of locusts, cock-chafers or waterbugs, we may assume that additional substructures seen in cross-sections of filaments consist of paramyosin. The sarcomeres of flesh-fly flight muscles seem to be an exception. There are two different types of filaments (Fig. 2a). About one third are solid, with electron dense cores, whereas the other two thirds are tubular with solid walls in their periphery and hollow cores.[1] The average paramyosin content in these filaments is about 3.8 % (Table 1). These observations suggest that tubular filaments have ~3% paramyosin (e.g., *Drosophila* and *Musca* flight muscles) and solid filaments have ~9 % paramyosin (e.g., flight muscles of locusts and cock-chafers). In these solid flesh-fly filaments 3 paramyosin substructures seem to be located at the inner edge of the wall[15] (Fig. 2d,d',g). The higher paramyosin content in the bee muscles is correlated with extra elements in the core (Fig. 2h) in addition to those of the inner edge of the wall (Fig. 2 e",g). Figure 2g and h only show one possible location (see below).[15-17]

We may conclude more from electron micrographs of transverse sections. The results of tilt series of 140 nm thick sections[16] and reconstructions of serial sections of the walls[17,18] suggest that the 6 pairs of myosin subfilaments in the wall are arranged parallel to the longitudinal axis of the filaments. The paramyosin subunits in the solid flesh-fly filaments (with 3 paramyosin subunits), on the other hand, seem to wobble from one pair of myosin subfilaments to the next, thereby, crossing the gap (Fig. 3), whereas the 5 paramyosin subunits of the waterbug filaments (11 % paramyosin[12]) seem to keep their position along or between the myosin structures.[18]

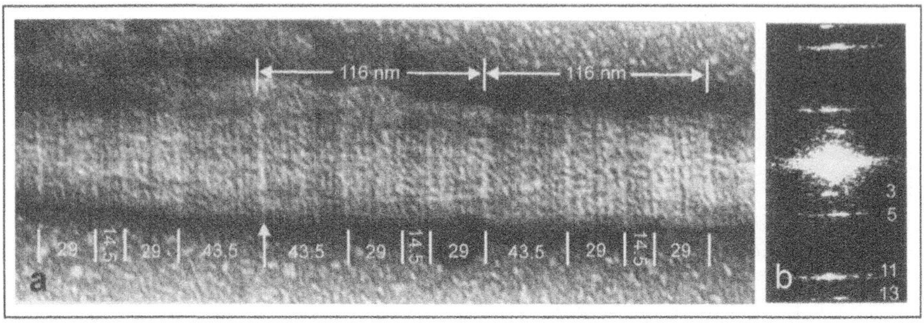

Figure 4. a) Paracrystal formed by slow dialysis of fragments of *Drosophila* light meromyosin consisting of the C-terminal two-thirds with the exon 19 encoded C-terminus.[19] The vertical arrow points to the mirror plane of the paracrystals with a characteristic 116 nm periodicity at either side. The periods consist of 43.5-29-14.5-29 nm units. The fine lines of the striation pattern of the paracrystal are about 4 nm apart. They are more clearly seen by looking at the wide side of the paracrystal in an acute angle. b) A typical optical diffraction pattern of these aggregate preparations. The numbers indicate the 3rd (38.7nm), 5th (23.2 nm), 11th (10.5 nm), and 13th order (8.9 nm) of 116 nm.

Assembly Properties of Myosin and Light Meromyosin Fragments

Myosin is able to self assemble into filaments. It has also been demonstrated that the myosin associated proteins paramyosin, mini-paramyosin and the proteins of the Titin/myosin light chain kinase family, are important for formation of functional filaments during muscle development (see chapters by Cervera et al and Ayme-Southgate and Southgate in this volume). To determine if the IFM specific arrangement of cross-bridges at the filament surface is defined by the properties of the whole myosin molecule, the light meromyosin (LMM) part of myosin, or by the presence of myosin-associated proteins, we studied the aggregation properties of myosin and LMM fragments in vitro, separately and in combination with myosin-associated proteins.[10,19,20,21]

The most interesting results have been obtained with the C-terminal two-thirds of LMM (residues 1541 to 1935 + 1 amino acid encoded by exon 18 or 27 amino acids encoded by exon 19, respectively[22]). This tail piece forms paracrystals with remarkable properties. The paracrystal of Figure 4a shows distinct dark and bright cross striations with periods of about 4 nm that may result from the distribution of positive and negative charges along the LMM coiled-coil, as originally described by McLachlan and Karn.[23] The stripes form a pattern of 116 nm repeats, each one consisting of 43.5-29-14.5-29 segments (from left to right in Fig. 4a). This nonuniformity suggest they are not caused by a positive staining effect of this particular thin paracrystal. Optical diffraction diagrams of these paracrystals (Fig. 4b) show layer line orders of 116 nm: off-meridional intensity on $l = 3, 5, 11$ and 13. The 8th order (14.5 nm) of 116 nm is barely visible, the 16th order, however, is strong but outside the figure. These diagrams resemble optical diffraction patterns from isolated myosin filaments with respect to the layer line characteristics.[8] We conclude, therefore, that the aggregation pattern of these LMM fragments in the paracrystal is strongly related to the pattern of their corresponding regions in the myosin molecules that form the backbone of the thick filaments.

The roles of myosin-associated proteins in filament assembly have been studied by codialysis experiments. Myosin and paramyosin dissolved in high ionic strength buffers were dialysed against low ionic strength solutions. This kind of experiment yielded thick filaments with defined lengths and in vivo filament-like diameters if projectin was present in sufficient amounts.[10] At low projectin concentrations, occasional thin aggregates with diameters of 6- 7 nm were observed (Fig. 5). These diameters correspond to those of the subfilament pairs seen in cross-sectioned and averaged myosin filaments (Fig. 2). As seen in Figure 5, pairs of thin

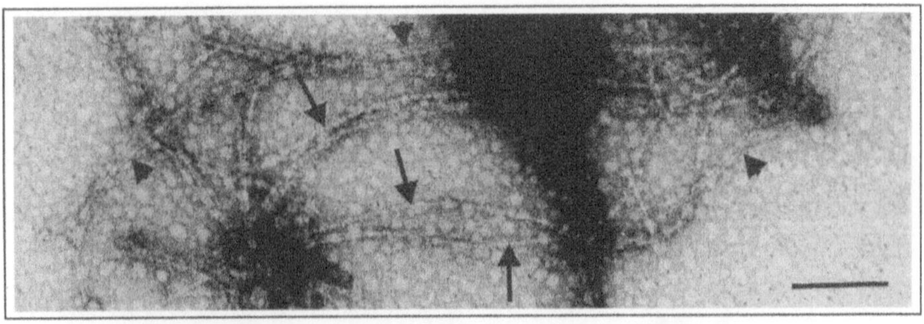

Figure 5. Negatively stained aggregates obtained by codialysis of locust flight muscle myosin purified by DEAE Sepharose chromatography (with traces of contaminating projectin) and paramyosin extracted from ether dried powder of *Mytilus* byssus retractor muscle.[30] The molar ratio of the proteins was 1 mole myosin to 0.5 mole paramyosin. The arrows point to 6 to 7 nm wide filaments, the arrowheads to pairs of such filaments. Cross-bridge-like protrusions may also be recognized along the filaments. Scale bar 200 nm.

aggregates of 14 nm to 17 nm width sometimes run side by side. These values are close to the widths of two subfilament pairs, including the gap between the pairs, within the cylinder of the filament wall (Fig. 2). Single subfilaments could not be resolved. This observation suggests that the packing of the subfilaments of a pair is very close.

Filament Models

Mike Reedy[3,24] was the first to propose that the cross-bridge origins of IFM thick filaments are located along helical strands with intervals of 38.7 nm. A basic helical track with 8 elements in 3 turns is demonstrated in Figure 6A. According to Reedy's model with two helical strands, a basic helix has a pitch of 77.3 nm (p = nc, where p is the pitch, the number of helical tracks and c the axial repeat of the basic helices which is 38.7 nm like in Figure 6A. Two cross-bridges repeat every 14.5 nm along the filament. This cross-bridge arrangement also causes a helix with an opposite screw sense and a period of 23.2 nm. Regardless of the number of cross-bridge strands, models of the filament backbone have to meet, at least, the following demands: diffraction diagrams must show layer lines based on 116 nm repeats like the 3rd order (38.7 nm), the 5th order (23.2 nm), the 8th order (14.5 nm), the 11th order (10.5 nm) and the 13th order (8.9 nm) of 116 nm. The 8th order line is supposed to be on the meridian, the other orders are off-meridionals.

The first approach that related the cross-bridge lattice at the filament surface to substructures of the backbone of insect muscles was performed by Squire.[25] Squire arranged the myosin molecules of a thick filament in equivalent positions in the wall of the filaments with 6 molecules per crown and within 6 helical tracks. Later, Wray[13] assumed that the myosin molecules (4 molecules per crown) were arranged within 12 subfilaments. In Wray's model, the subfilaments were localized in equivalent positions around the core of the filaments. Wray then, in an elegant way, tilted the subfilaments in the filament wall by a few degrees (Fig. 6B). The result was an axial separation of the helical tracks by 38.7 nm with supposedly little change in the size of the axial repeats of the cross-bridge crowns. Wray carefully demonstrated that this model is in good agreement with Reedy's suggestion[3] and the relevant X-ray data. However, evidence for tilted subfilaments have not yet been found in IFM. Instead, computer processing of images of cross-sectioned filaments indicate the existence of subfilaments in nonequivalent positions and running parallel to the filament axis.[15-18] It is not likely, therefore, that the model proposed for the structure of crustacean muscles also applies to IFM.

Another approach to elucidate the backbone structure of IFM thick filaments was to check whether the IFM myosin molecules aggregate in the same pattern as those of skeletal myosin.

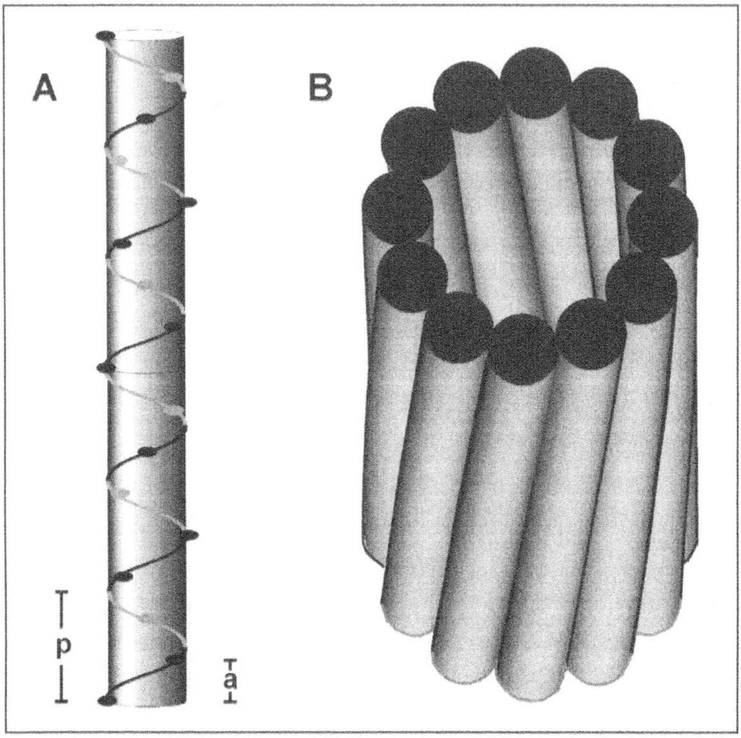

Figure 6. A) Cylinder representing a segment (2 x 116 nm in length) of a putative single stranded thick filament. The cross-bridge origins on the filament surface repeat every 14.5 nm (a) along the axis of the filament. They are rotated azimuthally by 135° causing the helical track to twist through 38.7 nm[24] (p = 38.7 nm) instead of 43 nm (3 x 14.3 nm) as in skeletal muscles. The black ovals are at the near side of the cylinder, the gray ovals are at the far side. B) Wray's[13] filament model consisting of 12 subfilaments in equivalent positions and tilted by a few degrees to explain the azimuthal rotation of the cross-bridge origins in (A).

It has been demonstrated (Fig. 4) that paracrystals of insect LMM fragments show the IFM specific 116 nm periodicity,[3] here apparently emphasizing -43.5-29-14.5-29- nm units. As mentioned above, the paracrystals also showed the features of the diffraction diagrams of isolated myosin filaments of IFM.[8] An attempt to propose a filament model based on these properties has to make additional assumptions. With respect to the axial repeat period of 116 nm within the paracrystal, we favoured an end to end overlap of the fragments within the same fragment strand. Since the fragments are about 60 nm long, a short end-to-end overlap results in a 58 nm periodicity of the individual fragment strands (Fig. 7, left). Any larger overlap of the fragments would cause too many cross-bridges per crown in the model of Figure 8. The second assumption was a 14.5 nm stagger of the two fragment strands within one subfilament (a in Fig. 7, left) and a 29 nm stagger of the second subfilament (b in Fig. 7, left). A vice versa stagger would lead to the same result. Eventually, we assumed a coiling of the two fragment strands of a subfilament because it would enable the cross-bridges of either strand to be located on the filament surface. A more speculative assumption is that the 2 subfilaments (a and b in Fig. 7). of a subfilament pair form super coiled coils (ab in Fig 7, center). This assumption is supported by the presence of aggregates (Fig. 5) with diameters corresponding to those of subfilament pairs in cross sections.

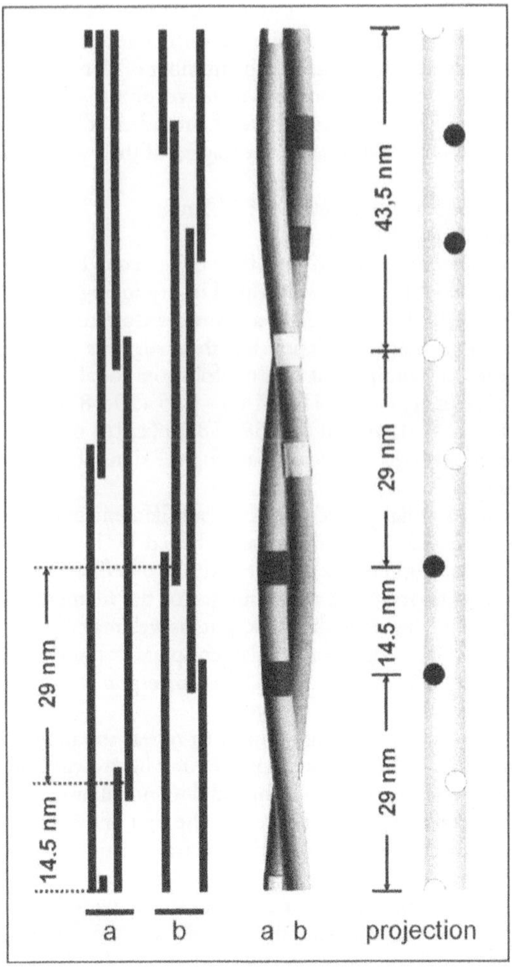

Figure 7. Molecular modelling of a 116 nm period of the paracrystal shown in Figure 4a. One 'subfilament', a or b, respectively, consists of two LMM fragment strands with an end-to- end packing of the 60 nm long molecules. The second strand of 'subfilament' a is shifted relative to the first strand by 14.5 nm. 'Subfilament' b is shifted relative to 'subfilament' a by 29 nm. The center of the figure shows a 3D diagram of a hypothetical coiled coil structure of the two 'subfilaments' with a pitch of 116 nm. The ends of the molecules of 'subfilament' a are indicated by white squares, the ends of 'subfilament' b by black squares. At the right side of the figure is a projection of the 3D structure of the center. The white and black circles again indicate the ends of the molecules of 'subfilament' a and b, respectively. This projection of the subfilament pair has been used as the basic unit for Figure 8A.

An electron micrograph corresponds to a projection of the particle density of the specimen by the electron beam of the microscope. If we project, likewise, the 3-dimensional structure of the double helix shown in the center of (Fig. 7ab) to a 2-dimensional image ("projection" in Fig. 7, right), we can see that the overlaps of the fragments at either side of the projection now show 116 nm repeats with the units of 43.5-29-14.5-29 nm as observed in the paracrystal of (Fig. 4a). A 38.7 nm repeat is not yet visible at this level of organization. This will become evident after the next steps of model building.

Any combination of these subfilaments to form a filament wall necessarily has to take into account the layer lines of the helical diffraction pattern of an IFM and fulfill the selection rule $l = 8m \pm 3$ where l is the layer line number, 8 the number of elements (in this case cross-bridge origins), m is an integer which can be positive, negative, or zero, and 3 is the result of t times n (t is the number of turns of a basic helix, $n \cong N$, the number of helical tracks in the simplified case of less ordered objects).[26] Additionally, the stagger of the 6 subfilament pairs has to reflect the following:

- a step width of 14.5 nm or a multiple of 14.5 nm
- six steps that add up to 0 or 116 nm.

Among the numerous possibilities to combine coiled coils Longley's[27] "knobs-into-hole" model of closely packed double helices was helpful in producing a cylinder of 6 closely packed subfilament coils. This cylinder would have a diameter close to 20 nm as in IFM thick filaments. The underlying geometrical relationship then suggests the stagger pattern "aabaab". This reflects the minimum requirement of a twofold rotational symmetry of IFM filaments. Four combinations of steps are possible: (a) –14.5 –14.5 +29 (-87) nm; (b) –29 –29 +58 (-58) nm; (c) –43.5 –43.5 +87 (-29) nm and (d) –58 –58 ±0 (±116) nm. The values in parentheses are equivalent possibilities for the 3rd step. Changing the signs of the individual steps alters the handedness of the helices.

Figure 8A represents a radial projection of six subfilament pairs (Fig. 7) modelled into a filament. The ends of the LMM fragments are labelled by black spots. If we consider the fragments as parts of myosin molecules in the backbone of a filament, the double heads of the myosin molecules would be staggered at the surface of the filaments in the same way as the fragments in the backbone. Therefore, the black spots as well may represent areas of cross-bridge origins. The projections (Fig. 7, right) of subfilament pairs 1 through 4 are staggered in Figure 8A by –14.5, –14.5 and +29 nm, respectively. This corresponds to possibility (a) from above. The stagger pattern repeats itself from pair 4 to 6.

The model provides four cross-bridge origins to repeat in crowns separated by 14.5 nm along the longitudinal axis of the filament (Fig. 8A). The crowns consist of 6 cross-bridges which are located along 4 helical tracks. The model does not allow conclusions on the handedness of the helices: the screw sense is influenced by the sign of the stagger values as mentioned above. It is possible, however, to analyse other characteristics of the surface lattice of the cross-bridge origins of the filament model by optical diffraction (Fig. 8). The inset of Figure 8B shows a diffraction pattern with layer lines numbered as beat periods of 116 nm. We would like to point particularly to the lines corresponding to 38.7 nm, 23.2 nm, 14.5 nm, 10.5 nm, 8.9 nm and 7.25 nm (3rd, 5th, 8th, 11th, 13th, and 16th order of 116 nm, respectively) which are also present in X-ray and optical diffraction diagrams of sarcomeres[24] and isolated thick filaments.[8] The off-meridional positions and intensities on the 3rd (38.7 nm), 5th (23.3 nm), 11th (10.5 nm) and 13th (8.9 nm) order line reflections suggest the existence of two coexisting sets of helices within the same filament. One set being 4-stranded and the other set 2-stranded and running in the opposite direction. Both of them fulfill the IFM-specific helical selection rule. The 8th (14.5 nm) and 16th (7.25 nm) order lines are meridional reflections that are the result of the repeat of the cross-bridge crowns along the longitudinal axis of the model.

The proposed model has been deduced from the aggregation properties of *Drosophila* LMM fragments. It is consistent with X-ray and electron microscopy data in regards to strandedness and the layer lines of the optical diffraction diagrams. It is interesting that this consistency is achieved by a model with 6 myosin double heads per crown. However, the positions on the strands are taken by either one or two myosin molecules.

At present, most people working on IFM favor the idea of 12 subfilaments and 4 helical cross-bridge strands with 4 cross-bridges per crown instead of 6. If two cross-bridge origins per level are systematically eliminated in Figure 8B, e.g., one from each of the two double taken lattice points, one obtains 4 cross-bridges per crown and the pattern shown in Figure 8C. This pattern shows a more uniform distribution of the cross-bridge origins because all lattice points

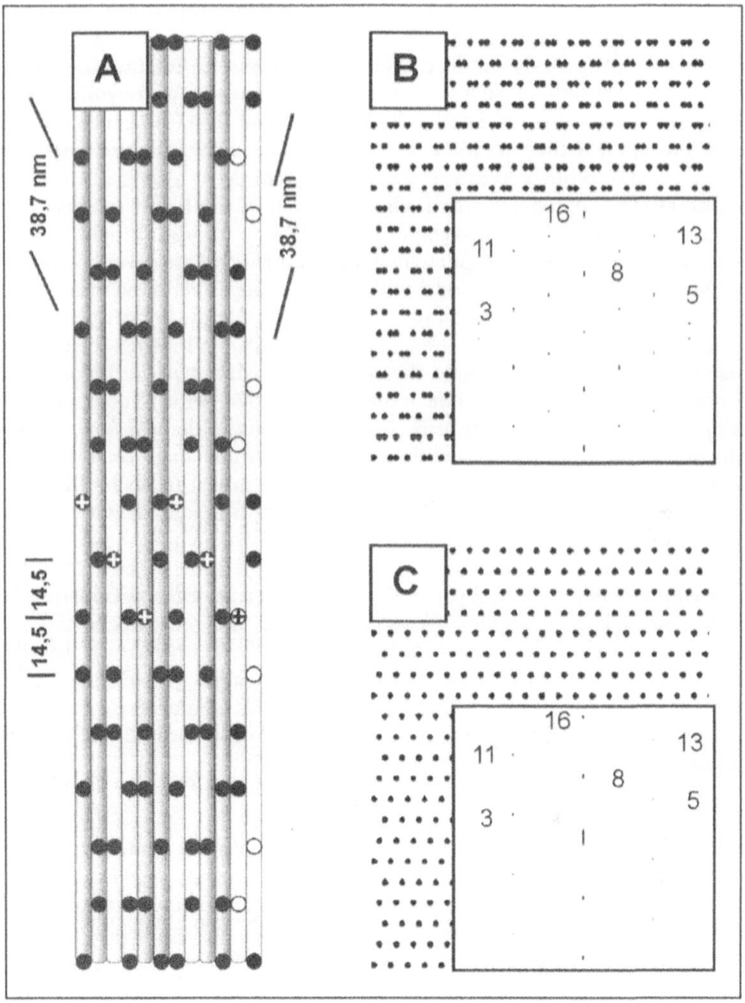

Figure 8. Unrolled cylindrical wall of a filament model with a length of 2 x 116 nm (A). The model is composed of 6 projections (gray or bright, respectively, consisting of two cylinders) of coiled coil subfilament pairs. They correspond to the "projection" of the coiled coil subfilament pair in Figure 7; as an example, the projected subfilament pair at the right side of Figure 8A is labelled with black and white spots like the projection of Figure 7. The projected subfilament pairs are shifted relative to each other by -14.5 nm; -14.5 nm; +29.0 nm. The shifts are indicated by plus signs in the spots at the "equator" of the model. The spots represent the positions of cross-bridges on the surface of the model. Horizontal lines with 6 cross-bridge origins and a periodicity of 14.5 nm may be observed along the longitudinal axis of the model. Each periodicity contains 2 single black spots and 2 pairs of black spots representing the 6 cross-bridge origins per crown. B,C) Diagrams of lattices with 6 cross-bridges (B) and 4 cross-bridges (C) per crown, respectively, represented as black dots. The insets in (B) and (C) are the corresponding optical diffraction patterns. The numbers within the diffraction patterns indicate order lines of 116 nm.

are occupied by single cross-bridges. This pattern reflects the distribution pattern of cross-bridge positions proposed by Wray[13] but slightly distorted. If there are 12 subfilaments forming the filament wall and 4 cross-bridges per crown, this would result in a nonintegral average number of 2.6 myosin molecules per 116 nm period of every subfilament. In other words, gaps within the formerly continuous myosin strands would be caused by missing LMMs. We get nonequivalent 116 nm periods within the subfilaments and an increase in disorder of the backbone structure. It would be highly speculative to assume that paramyosin (which is only about 3 % of the filament mass in tubular thick filaments of IFM, though considerably more in solid ones[1]) would take these positions to stabilize the filament structure. Other candidates to fill that role could be proteins like the myosin rod protein (MRP), a 155 kD headless myosin which is coexpressed and copolymerized with myosin in direct flight, somatic, cardiac and visceral muscles of *Drosophila* but it is absent in the IFM.[28,29] In direct flight muscle, the 1:3 ratio of MRP to myosin results in regions along the filament that are devoid of crossbridges.[28]

The diffraction diagram of the model with 4 cross-bridges per crown (inset in Fig. 8C) shows the layer lines already discussed for the model with 6 cross-bridges per crown. Again, strong 8th and 16th order lines of 116 nm are on the meridian. However, the off-meridionals show characteristic changes. For example, the innermost maximum of the 3rd order line of 116 nm (38.7 nm) is only seen on the left side of the meridian, that of the 5th order line (23.2 nm) only at its right side. This indicates the helix characteristics at the filament surface have changed by the reduction in the number of cross-bridges per crown from 6 to 4 (i.e., the set of 2 helices disappeared). This results in a single set of 4 parallel helices spaced axially by 38.7 nm, thereby matching one set of the actin target area helices.[5] Thus, a 6-headed crown with two sets of cross-bridge helices would have the functional advantage of matching the target areas of the two start helix set on the actin filaments.[5]

A model with 6 myosin molecules per crown with two sets of helices of opposite screw sense might structurally facilitate the matching of cross-bridges and their attachment sites at the actin filaments. A model with the favoured 4 myosin molecules per crown would have the disadvantage of disordered myosin strands. Our data do not allow us to distinguish between 4 or 6 myosin molecules per crown. Our models, however, demonstrate that four helical cross-bridge strands at the filament surface do not exclude the possibility of 6 myosin molecules per crown.

Concluding Remarks

The filament model presented here is based on the aggregation properties of a *Drosophila* light meromyosin fragment and several assumptions: (a) the end by end aggregation of the fragments, (b) the coiling of 2 fragment strands to form 'subfilaments', and (c) the coiling of coiled subfilaments to form subfilament pairs. It is encouraging that the diffraction pattern of Figure 4. matches those of isolated filaments and further studies on the aggregation properties of these fragments should be pursued.

The fragments are efficiently expressed in E. coli (up to ~80 % of total bacterial protein) and their ability to self assemble into large crystalline structures within the bacterial cell should permit their analysis in situ, i.e., by electon microscopy and computer tomography, or 'in vivo' by X-ray diffraction of bacterial suspensions or even of single bacteria. Another promising method would be the production of 2-dimensional crystals of the fragments on lipid layers and their analysis by electron microscopy, image reconstruction and molecular modelling.

References

1. Beinbrech G, Meller U, Sasse W. Paramyosin content and thick filament structure in insect muscles. Cell Tissue Res 1985; 241:607-614.
2. Sotavalta O. Recordings of the high wing-stroke and thoracic vibration frequency in some midges. Biol Bull 1953; 104:439-444.
3. Reedy MK. Ultrastructure of insect flight muscle. I. Screw sense and structural grouping in the rigor cross-bridge lattice. J Mol Biol 1968; 31:155-176.
4. Trombitas K, Tigyi-Sebes A. Structure of thick filaments from insect flight muscle. Acta Biochim Biophys Hungaricae 1986; 21:115-128.

5. Wray JS. Filament geometry and the activation of insect flight muscles. Nature 1979; 280:325-326.
6. Reedy MK, Bahr GF, Fishman DA. How many myosins per cross-bridge? Flight muscle myofibrils from the blowfly, Sarcophaga bullata. Cold Spring Harbor Symp Quant Biol 1973; 37:397-422.
7. Tregear RT, Squire JM. Myosin content and filament structure in smooth and striated muscle. J Mol Biol 1973; 77:279-290.
8. Morris EP, Squire JM, Fuller GW. The 4-stranded helical arrangement of myosin heads on insect (Lethocerus) flight muscle thick filaments. J Struct Biol 1991; 107:237-249.
9. Schmitz H, Lucaveche C, Reedy MK et al. Oblique section 3-D reconstruction of relaxed insect flight muscle reveals the cross-bridge lattice in helical registration. Biophys J 1994; 67:1620-1633.
10. Kölsch B, Ziegler C, Beinbrech G. Length determination of synthetic thick filaments by cooperation of two myosin-associated proteins, paramyosin and projectin. Naturwiss 1995; 82:239-241.
11. Bullard B. Contractile proteins of insect flight muscle. TIBS 1983; 8:68-70.
12. Reedy MK, Leonard KR, Freeman R et al. Thick myofilaments mass determination by electron scattering measurements with the scanning transmission electron microscope. J Muscle Res Cell Motil 1981; 2:45-64.
13. Wray JS. Structure of the backbone in myosin filaments of muscle. Nature 1979; 277:37-40.
14. Pepe FA, Ashton FT, Street C et al. The myosin filament X. Observation of nine subfilaments in transverse sections. Tissue Cell 1986; 18:499-508.
15. Beinbrech G, Ashton FT, Pepe FA. Invertebrate myosin filament: Subfilament arrangement in the wall of the solid filaments of insect flight muscles. Biophys J 1992; 61:1495-1512.
16. Beinbrech G, Ashton FT, Pepe FA. Invertebrate myosin filament: Subfilament arrangement in the wall of tubular filaments of insect flight muscles. J Mol Biol 1988; 201:557-565.
17. Schmitz H, Ashton FT, Pepe FA et al. Invertebrate myosin filament: Parallel subfilament arrangement in the wall of solid filaments from the honeybee, Apis mellifica. Tissue Cell 1993; 25:111-119.
18. Schmitz H, Ashton FT, Pepe FA et al. Substructures in the core of thick filaments: Arrangement and number in relation to the paramyosin content of insect flight muscles. Tissue Cell 1994; 26:83-100.
19. Ader G, Pepe FA, Beinbrech G. Functional domains of Drosophila LMM affecting self assembly and paramyosin binding. J Muscle Res Cell Motility 1997; 18:246-247.
20. Ziegler C, Jurk K, Weitkamp B et al. In vitro interactions of proteins from insect myosin filaments. Biophysics 1996; 41:79-87.
21. Fährmann M, Fonk I, Beinbrech G. The kinase activity of the giant protein kinase projectin of the flight muscle of Locusta migratoria. Insect Biochem Mol Biol 2002; 32:1401-1407.
22. George EL, Ober MB, Emerson CPJ. Functional domains of the Drosophila melanogaster muscle myosin heavy-chain gene are encoded by alternative spliced exons. Mol Cell Biol 1989; 9:2957-2974, (Published erratum Mol Cell Biol 1989; 9:4118).
23. McLachlan AD, Karn J. Periodic charge distributions in the myosin rod amino acid sequence match cross-bridge spacings in muscle. Nature 1982; 299:226-231.
24. Reedy MK. Cross-bridges and periods in insect flight muscle. Am. Zoologist 1967; 7:465-481.
25. Squire JM. General model for the structure of all myosin-containing filaments. Nature 1971; 233:457-462.
26. Stewart M. Computer image processing of electron micrographs of biological structures with helical symmetry. J Electron Microsc Techn 1988; 9:325-358.
27. Longley W. The packing of double helices. J Mol Biol 1975; 93:111-115.
28. Standiford DM, Davis MB, Miedema K et al. Myosin rod protein: A novel thick filament component of Drosophila muscle. J Mol Biol 1997; 265:40-55.
29. Polyak E, Standiford DM, Yakopson V et al. Contribution of myosin rod protein to the structural organization of adult and embryonic muscles in Drosophila. J Mol Biol 2003; 331:1077-1091.
30. Dufhues G, Philipp L, Ziegler C et al. The ATPase activity of actomyosin in the presence of C-protein and low paramyosin concentrations. Comp Biochem Physiol 1991; 99B:871-877.

Chapter 9

Actin and Arthrin

John C. Sparrow

Dedication

This chapter is dedicated to the memory of the late Dr. Eric Fyrberg, collaborator and colleague of many authors of chapters in this volume, who pioneered the study of actin molecular genetics with his research on the flight muscle-specific actin gene, *Act88F*, of *Drosophila melanogaster*.

Abstract

Filamentous actin forms the core of all muscle thin filaments and is an integral part of the acto-myosin motor system that powers muscle contraction. Muscle actin isoforms show considerable sequence conservation compared to all actins, but insect actins form a distinct group. Within insect actins the flight muscle actin sequences do not form a statistically distinct group. The flight muscle actins have biochemical properties and post-translational modifications almost indistinguishable from those of vertebrate muscle actins. The major exception is the specific mono-ubiquitination of some flight muscle-specific actins through a single isopeptide bond at a specific actin lysine residue to form arthrin. Though the ability to carry out this modification seems to have arisen de novo at least twice in insect evolution it is not required for flight and a specific function remains elusive. The close conservation of insect actins with all other actins, the presence of a single indirect flight muscle-specific isoform encoded by the *Act88F* gene in *Drosophila*, that is not required for any other vital functions (and thus does not affect viability), has allowed this particular insect flight muscle actin to play a major role in the molecular genetic study of actin function.

Introduction

Actin is a major constituent of all muscles, including insect flight muscles. Stoichiometrically it is the most prevalent myofibrillar protein. As filamentous F-actin it is the structural core of the sarcomeric thin filaments. Within the thin filament actin serves both as part of the actomyosin motor complex to produce, with myosin, the crossbridge activity that generates the force and movement that drives muscle contraction, and as a structural component that transmits the forces produced along the thin filament to the Z-disc, in which its ends are embedded. In all insect flight muscles F-actin is associated with other thin filament proteins, including tropomyosin and the three components of the troponin complex (See Ferrus chapter) that have roles in the regulation of muscle contraction.

Evolutionarily actin is a highly conserved protein found in all eukaryotic cells where it exists in two forms, the soluble monomeric G -actin and the filamentous polymeric F-actin. It contributes as a major cytoskeletal component to the structure of cells, and as actomyosin to generate force with members of the different families of the so-called nonconventional myosins. In all these activities the roles of F-actin are determined by its interactions with a vast array of cellular

Nature's Versatile Engine: Insect Flight Muscle Inside and Out, edited by Jim Vigoreaux.

proteins known as actin-binding proteins (ABP). In the mature myofibril the structure and function of the thin filament and the Z-disc are largely determined by the interaction of F-actin with a set of muscle-specific proteins or protein isoforms. In addition to the thin filament proteins these also include specific Z-disc proteins such as the actin cross-linking protein α-actinin.

During myofibrillogenesis sarcomeric G-actin polymerises to F-actin to form the core of the thin filament and the Z-disc. In vertebrate muscles there is a continual remodelling of myofibrils through sarcomeric protein turnover, including thin filaments, post muscle differentiation. In *Drosophila* indirect flight muscles (IFM) messenger RNA levels decline rapidly after eclosion and the high levels of IFM protein synthesis do not continue much beyond the first day or so after eclosion (Sparrow, unpublished); so any remodelling of the muscle, if it occurs, cannot involve de novo synthesis. This might not be universal for all insect flight muscles. A number of insects, including the giant waterbug, *Lethocerus spp.*, are known to undergo seasonal changes in their flight muscles suggesting that imaginal flight muscles can show post-maturational remodelling.

G-Actin Properties

Monomeric G-actin consists of a single polypeptide chain of 375 residues with a mass of 42,000 that binds a single adenosine nucleotide (ATP or ADP) and a single divalent cation, usually magnesium in vivo. Actin has an ATPase activity. G-actin.ATP monomers are added to the 'barbed' end (see below) of the F-actin during polymerisation and, shortly afterwards, the actin.ATP is converted to actin.ADP.Pi with the subsequent release of Pi. Actin shows structural homology to some other ATPases such as cell cycle proteins, hexokinase and hsp70.[1]

The atomic structure of G-actin was initially determined from cocrystals of rabbit skeletal muscle actin with the actin monomer ABP DNaseI[2] and subsequently from cocrystals with profilin,[3] gelsolin subfragment-1[4] and vitamin D binding protein.[5] Most recently a structure has been determined from crystals of chemically modified actin.[6] Despite small, but significant, differences between these structures they all reveal that actin consists of four subdomains in which pairs of subdomains (1 and 2) and (3 and 4) form larger domains of equivalent sizes separated by a cleft and connected by two passes of the polypeptide chain known as the hinge (see Fig. 1A). At the base of the cleft, and involving residues on either side of the cleft, are the nucleotide and divalent cation binding sites. With respect to the F-actin filament helix (see below) subdomains 1 and 2 form the outer domain and subdomains 3 and 4 the inner domain of the molecule (see Fig. 1B).

F-Actin Properties

Polymerised, filamentous actin, F-actin, can be visualised in the electron microscope.[7] It is helical with a diameter of about 10nm. Decoration of F-actin with myosin subfragment 1 (S1) reveals a 'herringbone' pattern, which indicates that F-actin filaments are polar from one end to the other. All the actin monomers must lie in similar orientations (see Fig. 1B) within a filament. Polymerisation (monomer addition) is much more rapid at one end (defined as the + end) than the other (-). The 'herringbone' appearance of S1-decorated F-actin has led to the terms 'barbed' for the + end and 'pointed' for the – end. In all striated muscles, including IFM, it is the barbed ends which are inserted into the sarcomeric Z-discs, so that during contraction thick and thin filament sliding is consistent with the universal active movement of members of the myosin II family (includes all sarcomeric myosins) towards the F-actin barbed end.

The atomic structure of F-actin has not been determined by direct experimental methods. Rather, data from EM reconstructions into which the G-actin atomic structure was modelled led to low resolution F-actin structures;[8] such models were further refined by altering the orientation of the monomers within the structure until a best fit was achieved between the low angle X-ray diffraction pattern data from oriented F-actin gels and that predicted from the models.[9,10]

F-actin is stabilised by interactions between neighbouring actin monomers and the amino acids involved have been deduced from the F-actin models.[2] A protein loop with a hydrophobic

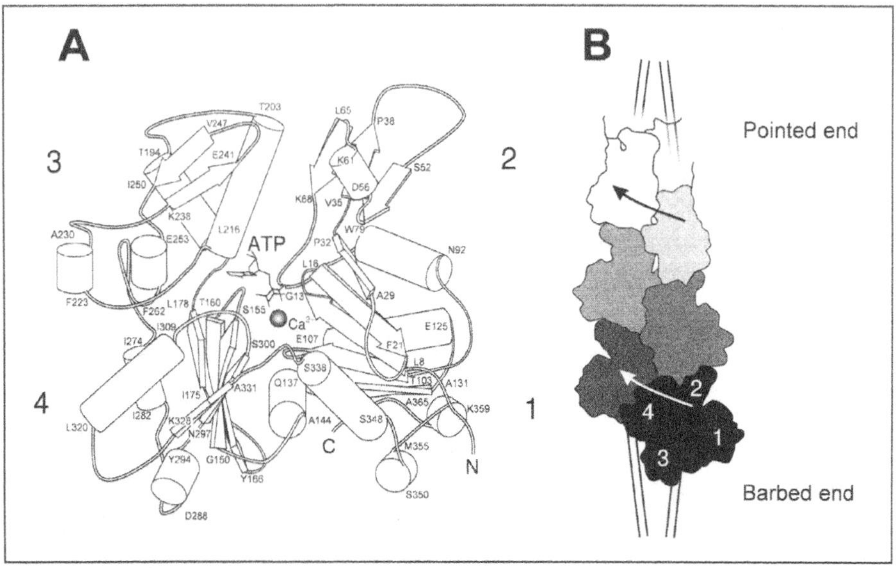

Figure 1. Molecular structure of actin. A) The monomeric structure of actin (G-actin) derived from actin images generated using the 1ATN.pdb file coordinates of DNase 1/actin cocrystals.[2] Cylinders represent α-helices; small numbers represent actin residues, numbered from N- to C-termini, at which the α-helices begin and stop. Note at the bottom of the 'cleft' between the two domains (consisting of subdomains 1+2 and 3+4) a single nucleotide (ATP in G-actin, and ADP in F-actin) and a divalent cation (usually Mg^{++} in vivo, but Ca^{++} can be bound in vitro) indicated by the shaded sphere. The polypeptide chain passes twice between the two major domains, forming the hinge about which actin is flexible, allowing nucleotide exchange during polymerization/depolymerisation of actin. B) Diagram of F-actin to show the overall structure of the F-actin helix. This can be considered as either a one- or two-start helix. The curved arrows show the path of the one-start helix. Note that subdomains 1 and 2 are on the outer edge of the helix; subdomains 3 and 4 are closer to the axis of the helix. (Diagrams redrawn from Sheterline et al[24]).

apex, seen in the G-actin structure, has been proposed[9] to form a 'plug', with the potential to increase cross helix stability by inserting into a hydrophobic pocket formed between two actin monomers in the opposite strand. Supporting evidence for the proposed plug function in F-actin comes from yeast[11,12] where mutations reducing plug hydrophobicity led to cold-sensitivity of in vivo growth and in vitro to effects on actin polymerisation kinetics. Intriguingly, single amino acid differences between homologous isoforms in *D. melanogaster* and *D. virilis* cluster at the base of the 'plug'.[13]

Thin Filament Helical Repeat

The F-actin helix can be considered as either a single start left-handed helix with a rise of 2.75 nm/monomer which gives a pitch of 36.5 nm for 13 monomers and six turns or as a two-start right-handed double helix with a half-pitch of 36.5 nm.

The insect flight muscle thin filament actin helix has an axial repeat length of 38.5 nm,[14-17] compared to the 36.5 nm of vertebrate thin filaments and of F-actin polymerised in vitro. The larger axial repeat in insect thin filaments seems to be a property of the thin filament, rather than insect F-actin itself as this repeat is not found in isolated insect flight muscle F-actin preparations.[18-22] Undisrupted bundles of isolated insect thin filaments have a 38.5 nm axial repeat. Individual thin filaments isolated in the absence of calcium (+EGTA) have an axial repeat of 36.5 nm, whereas more than half of the single filaments have an axial repeat of 38.5 nm following isolation at pCa equivalent to those present during muscle contraction.[23] These data strongly support the notion that the thin filament axial repeat is due to an F-actin helix distortion associated with its interactions with other thin filament proteins. Whether the interactions are with components of the tropomyosin-troponin (Tm-Tn) complex of insects is not clear. The individual components, although showing significant homology to their vertebrate thin filament homologues, contain extra novel protein domains (see chapters by Ferrus, Bullard et al). A structural consequence of the 38.5 nm repeat is that the Tm-Tn complexes on either side of the filament follow a straight, parallel path along the thin filament rather than a helical path, as seen in vertebrate thin filaments.

Actin Isoforms

Most multicellular organisms contain multiple actins[24] and in animals these can usually be distinguished as muscle and cytoplasmic isoforms. In insects (as in all arthropods) actins can usually be classified by their patterns of expression into adult muscle actins, larval-adult actins and cytoskeletal actins. In the context of this chapter the focus is on the actin isoforms expressed in the indirect flight muscles. The best-studied insect actin genes are the six contained within the *Drosophila melanogaster* genome. Four of these genes (known as *Act57B*, *Act79B*, *Act87E* and *Act88F* from their positions on the polytene chromosome map) encode sarcomeric actins and one, *Act88F*, encodes all of the IFM sarcomeric actin,[25,26] though more recently[27] it has been found that its expression is not limited to these muscles.

An increasing number of actin sequences are now known from other insects, especially with the recent completion of the genome sequence of the mosquito *Anopheles gambiae*.[28] Various actin gene nomenclatures have been used and these (including their usage within this chapter) are shown in Table 1 where an attempt has been made to classify the insect actins into indirect flight muscle, nonIFM, and cytoplasmic actins, at least by their major sites of expression. The muscle/cytoplasmic isoform separation is usually very clear, but given the presence of two such distinct muscle types in adult insects the larval-adult classification seems inappropriate in this context. However, the known sites of expression are often defined by Northern blots on RNA isolated from thoraces or whole pupae, so it is not possible to unambiguously assign all sequences to IFM/nonIFM classes on the information available. In the case of *D. virilis* the E2 isoform shows only two amino acid differences from the *D. melanogaster* ACT88F isoform (Table 2) and is expressed solely in the IFM.[13] The amino acid sequences of a *D. simulans* IFM actin were identical to *Act88F*[29] as is a muscle actin sequence (ACT2_BACDO) from the oriental fruitfly *Batrocera dorsalis*[30] and a recent sequence from the *D. pseudoobscura* (within contig8255_contig717 of the *D. pseudoobscura* genome at Human Genome Sequence Center, Baylor College of Medicine). It is therefore assumed (Table 1) that these, as apparent close homologues of *Act88F*, are all IFM-specific isoforms. All of these data suggest a very high degree of sequence conservation of the IFM-specific actin in the Drosophilidae, a conservation maintained outside this group to the Tephritidae with the observation of the identical sequence of the IFM-specific actin of *B. dorsalis*,[30] but still confined to the Muscomorpha, within the Diptera.

In addition to the *AeAct-1* muscle actin gene from the yellow-fever mosquito *Aedes aegypti*[31] three more muscle actin genes have been found recently and studied. The *AeAct-4* gene appears to be a muscle isoform, that is expressed specifically in the flight muscles of the female mosquitoes, a strange and so far unique observation for any species (Dr. A. James, personal communication).[32] Why the *AeAct-4* isoform expression occurs only in the females remains unanswered, but could conceivably reflect the greater muscle power output required for the larger female that must take on a blood meal before laying many eggs. Two other muscle actin isoforms have been detected in *A. aegypti*, *AeAct-2* and *AeAct-3*.[33] Expression of *AeAct-2* begins in the last (fourth) larval instar, is found in the pupae and at higher levels in the adult where it is detected in the body wall, gut and head tissues; in Table 1 it has been classified as a nonIFM muscle isoform. The *AeAct-3* gene is expressed in the pupae and at higher levels only in young adult male mosquitoes. This is a pattern consistent with expression in the male IFM, especially as the *AeAct-4* gene is female IFM-specific but it has not yet been determined directly if *AeAct-3*

Table 1. Nomenclature of insect actin genes

Usage	Swissprot/ Genbank	Gene Name	Genome	Species
IFM-specific				
Act6_DROME#	P2575	Act88F	CG5178	*Drosophila melanogaster*
Act6_DROPSE	##			*D. pseudoobscura*
Act1_BACDO	P45893			*Bactrocera dorsalis*
Act6_DROSIM	P2575			*D. simulans*
ActE2_DROVIR	AF358263			*D. virilis*
AeAct-4	AY531222	AeAct-4		*Aedes aegypti*
AeAct-3	AY289765			*A. aegypti*
Non-flight muscle				
Act3_DROME	P53501	Act57B	CG10067	*D. melanogaster*
Act4_DROME	P02572	Act79B	CG7478	*D. melanogaster*
Act5_DROME	P10987	Act87E	CG18290	*D. melanogaster*
ActC2_DROVIR	AF358264			*D. virilis*
ActD1_DROVIR	AF358266			*D. virilis*
ActE1_DROVIR	AF358265			*D. virilis*
Act2_BACDO	P45885			*B. dorsalis*
Act3_BACDO	P45886			*B. dorsalis*
Act5_BACDO	P45887			*B. dorsalis*
Act1_AEDAE	P49128			*A. aegypti*
AeAct-1	U20287			*A. aegypti*
AeAct_2	AY289764			*A. aegypti*
Act_MANSE	P49871			*Manduca sexta*
Act1_BOMMO	P07836			*Bombyx mori*
Act2_BOMMO	P07837			*Bo. mori*
Cytoplasmic				
Act1_DROME	P10987	Act42A	CG12051	*D. melanogaster*
Act2_DROME	P02572	Act5C	CG4027	*D. melanogaster*
Act3_BOMMO	P04829			*Bo. mori*
Act4_BOMMO*	Q27250			*Bo. mori*
Act3A_HELAM	Q25010			*Helicoverpa armigera*
Act_SPOLI (partial)	Q11212			*Sodoptera littoralis*

#identical with *D. simulans* (P45893), *Ba. dorsalis* (P45885) and *D. pseudoobscura*. ## *D. pseudoobscura* sequences from http:/www.hgsc.bcm.tmc.edu/drosophila/. *identical with Act1D from *Anopheles gambiae*.

gene expression is IFM-specific (Dr Q Lan, personal communication). On balance for the purposes of this chapter (Table 1) it is assumed that this will be the case. Whether the sexual dimorphism in the expression of these two actin genes is a unique feature of this mosquito species alone or whether functional homologues will be found in the *Anopheles gambiae* genome (in which 8 actin sequences are predicted), and others, remains to be determined. Whether the *Anopheles gambiae* genes also encode flight muscle-specific actins will require direct molecular determination.

All the insect species for which we have the IFM actin sequences (Table 2) have asynchronous indirect flight muscles (and are all Diptera). A partial sequence of actin from the flight muscle of the giant waterbug, *Lethocerus griseus* (Hemiptera), also with asynchronous flight

Table 2. Statistical table from the alignment of insect actins

	%Identity to Act6_DROME	Amino Acid Differences*	Conserved Residues+
Flight muscle			
ACT6_DROME	100	0	376
ACT6_DROPSE	100	0	376
ACTE2_DROVIR	99	2	374
AeAct-3	93	25	365
AeAct-4	95	17	365
Non-flight muscle			
ACT3_DROME	97	10	373
ACT4_DROME	97	11	372
ACT5_DROME	97	9	373
ACTC2_DROVIR	97	10	373
ACTD1_DROVIR	97	11	372
ACTE1_DROVIR	97	9	373
ACT2_BACDO	97	10	372
ACT3_BACDO	97	10	372
ACT5_BACDO	97	8	373
AeAct-1	93	24	357
AeAct-2	98	4	375
ACT_MANSE	98	7	374
ACT1_BOMMO	97	8	374
ACT2_BOMMO	98	6	374
Cytoplasmic			
ACT1_DROME	96	5	372
ACT2_DROME	95	7	372
ACT3_BOMMO	95	7	370
ACT4_BOMMO	96	15	372
ACT3_HELAM	95	16	371
ACT_SPOLI	-	-	-

*number of non-identical residues compared to the *Act6_DROME* (ACT88F) sequence. +total residues that are identical or at which conserved substitutions have occurred.

muscle has been determined by mass spectrometry.[34] The 40 strongest de-isotoped peaks gave a best match against the flight muscle actin sequence from *B. dorsalis, Act3_BACDO* (P45886) and covered about 54% of the sequence. Where necessary single, specific amino acid changes gave complete agreement between the *Lethocerus* peptide fragments and the *in silico* predicted masses of tryptic fragments from *Act3_BACDO.* A small number (6) of amino acid substitutions were detected between this partial *Lethocerus* sequence and that of *Drosophila* ACT88F.

An alignment from a compilation of all known insect actin sequences (Table 3) shows the extreme conservation of actin sequences, a feature typical of actin sequences from most groups.[24] Within the sequences used (Table 1) the greatest divergence is between *AeAct-1* and *AeAct-3* (88% identity). Even in comparison with the *Act6_DROME* sequence (Table 2) the *Ae. aegypti* sequences show some of the largest number of amino acid differences (25; *AeAct-3*) within the IFM-specific actins and the smallest (4; *AeAct-2*) within the much more diverse nonIFM muscle actins. This suggests that the *Ae. aegypti* actin sequences are especially diverse. Due to the very small number of amino acid differences between the insect actin sequences used, none of the phylogenetic tree branchpoints are statistically significant and the trees are therefore not presented.

Table 3. Alignment of insect actin sequences to the sequence of the *Drosophila ACT88F* IFM-specific actin. Residue numbering given is that produced by the alignment software (ClustalX). For comparison to the complete actin protein family the first residue of the mature protein is Asp2. (Met, M, is numbered –1)

```
                      *        20         *        40         *        60         *
ACT6_DROME : MCDDDAGALVIDNGSGMCKAGFAGDDAPRAVFPSIVGRPRHQGVMVGMGQKDSYVGDEAQSKRGILTLKY
ACT6_DROPS : ......................................................................
ACTE2_DROV : ......................................................................
AeAct-3    : .....Q..............................................A.................
AeAct-4    : ....................................................A.................
ACT3_DROME : ....EVA...............................................................
ACT4_DROME : ...EE.S.............................................C.................
ACT5_DROME : ....EVA...............................................................
ACTC2_DROV : ....EVA...............................................................
ACTD1_DROV : ...EE.S.............................................C.................
ACTE1_DROV : ....EVA...............................................................
ACT2_BACDO : ....EVA...............................................................
ACT3_BACDO : ....EVA...............................................................
ACT5_BACDO : ....E.S...............................................................
ACT1_BOMMO : .....VR...............................................................
ACT2_BOMMO : ......................................................................
AeAct-1    : .....VRRS.................V......................N........V...........
AeAct-2    : ......................................................................
ACT_MANSE  : .....VA...............................................................
ACT1_DROME : ...EEVA...............................................................
ACT2_DROME : ...EEVA...............................................................
ACT3_BOMMO : ...EEVA...............................................................
ACT4_BOMMO : ...EEVA...............................................................
ACT3_HELAM : ...EEVA...............................................................
ACT_SPOLI  : ----------............................................................

                     80         *       100         *       120         *       140
ACT6_DROME : PIEHGIITNWDDMEKIWHHTFYNELRVAPEEHPVLLTEAPLNPKANREKMTQIMFETFNSPAMYVAIQAV
ACT6_DROPS : ......................................................................
ACTE2_DROV : ......................................................................
AeAct-3    : ............................................S.............AA..A.......
AeAct-4    : ............................................S.............A...........
ACT3_DROME : ......................................................................
ACT4_DROME : ......................................................................
ACT5_DROME : ...........................................................A..........
ACTC2_DROV : ......................................................................
ACTD1_DROV : ......................................................................
ACTE1_DROV : ...........................................................A..........
ACT2_BACDO : ......................................................................
ACT3_BACDO : ......................................................................
ACT5_BACDO : ......................................................................
ACT1_BOMMO : ......................................................................
ACT2_BOMMO : ...........................................................C..........
AeAct-1    : .N....................................................................
AeAct-2    : ......................................................................
ACT_MANSE  : ......................................................................
ACT1_DROME : ......................................................................
ACT2_DROME : ......................................................................
ACT3_BOMMO : ......................................................................
ACT4_BOMMO : ......................................................................
ACT3_HELAM : ......................................................................
ACT_SPOLI  : ......................................................................
```

continued on next page

Table 3. Continued

```
                   *       160         *       180         *       200         *
ACT6_DROME : LSLYASGRTTGIVLDSGDGVSHTVPIYEGFALPHAILRLDLAGRDLTDYLMKILTERGYSFTTTAEREIV
ACT6_DROPS : ......................................................................
ACTE2_DROV : .......................................X..............................
AeAct-3    : ......................................................................
AeAct-4    : ......................................................................
ACT3_DROME : ......................................................................
ACT4_DROME : ......................................................................
ACT5_DROME : ......................................................................
ACTC2_DROV : ......................................................................
ACTD1_DROV : ......................................................................
ACTE1_DROV : ......................................................................
ACT2_BACDO : ......................................................................
ACT3_BACDO : ......................................................................
ACT5_BACDO : ......................................................................
ACT1_BOMMO : ......................................................................
ACT2_BOMMO : ......................................................................
AeAct-1    : ......................................................................
AeAct-2    : ......................................................................
ACT_MANSE  : ......................................................................
ACT1_DROME : ......................................................................
ACT2_DROME : ......................................................................
ACT3_BOMMO : ......................................................................
ACT4_BOMMO : ......................................................................
ACT3_HELAM : ......................................................................
ACT_SPOLI  : ..................................------------------------------------

                    220         *       240         *       260         *       280
ACT6_DROME : RDIKEKLCYVALDFEQEMATAAASTSLEKSYELPDGQVITIGNERFRCPEALFQPSFLGMESCGIHETVY
ACT6_DROPS : ......................................................................
ACTE2_DROV : ..............................................................S.......
AeAct-3    : ..................QA......S....................A.............AT.......
AeAct-4    : ..................QA......S....................A..............T.......
ACT3_DROME : ..................................................S...................
ACT4_DROME : ...............................................T......................
ACT5_DROME : ..................................................S...................
ACTC2_DROV : ..................................................S...................
ACTD1_DROV : ...............................................T......................
ACTE1_DROV : ..................................................S...................
ACT2_BACDO : ..................................................S...........S.......
ACT3_BACDO : ..................................................S...........S.......
ACT5_BACDO : ...............................................T......................
ACT1_BOMMO : ......................................................................
ACT2_BOMMO : ..................Q...................................................
AeAct-1    : ...............................................AQKLSSSHPS.............
AeAct-2    : ....................................L.................................
ACT_MANSE  : ......................................................................
ACT1_DROME : ......................S......................................A......T.
ACT2_DROME : ......................S...........................S..........A......T.
ACT3_BOMMO : ......................S......................................AN.....T.
ACT4_BOMMO : ......................S......................................A......T.
ACT3_HELAM : ......................S......................................AN.....T.
ACT_SPOLI  : ----------------------------------------------------------------------
```

continued on next page

Table 3. Continued

		* 300 * 320 * 340 *
ACT6_DROME	:	NSIMKCDVDIRKDLYANSVLSGGTTMYPGIADRMQKEITALAPSTIKIKIIAPPERKYSVWIGGSILASL
ACT6_DROPS	:	....
ACTE2_DROV	:	....
AeAct-3	:	......................K................S..............................
AeAct-4	:	....S..............................
ACT3_DROME	:	.................I.....................S..............................
ACT4_DROME	:	Q................N..
ACT5_DROME	:	.................I..
ACTC2_DROV	:	.................I.....................S..............................
ACTD1_DROV	:	Q................N..
ACTE1_DROV	:	.................I..
ACT2_BACDO	:	.................I..
ACT3_BACDO	:	.................I..
ACT5_BACDO	:	Q................N..
ACT1_BOMMO	:	....
ACT2_BOMMO	:	....
AeAct-1	:	....
AeAct-2	:	....
ACT_MANSE	:	....
ACT1_DROME	:	....
ACT2_DROME	:	....
ACT3_BOMMO	:	....R..............................
ACT4_BOMMO	:	....
ACT3_HELAM	:	....
ACT_SPOLI	:	--

		360 *	
ACT6_DROME	:	STFQQMWISKQEYDESGPGIVHRKCF	:
ACT6_DROPS	:	....	:
ACTE2_DROV	:	....	:
AeAct-3	:	....T.....H....G..........	:
AeAct-4	:	....A..........G..........	:
ACT3_DROME	:	....	:
ACT4_DROME	:	....	:
ACT5_DROME	:	....	:
ACTC2_DROV	:	....	:
ACTD1_DROV	:	....	:
ACTE1_DROV	:	....	:
ACT2_BACDO	:	....	:
ACT3_BACDO	:	....	:
ACT5_BACDO	:	....	:
ACT1_BOMMO	:	....	:
ACT2_BOMMO	:	....	:
AeAct-1	:	....A.............A.......	:
AeAct-2	:	....	:
ACT_MANSE	:	....	:
ACT1_DROME	:	..................S.......	:
ACT2_DROME	:	..................S.......	:
ACT3_BOMMO	:	..................S.......	:
ACT4_BOMMO	:	..................S.......	:
ACT3_HELAM	:	..................S.......	:
ACT_SPOLI	:	--------------------------	:

In the comparisons of actins from the alignment which follow please note the following convention. As the first actins were sequenced as mature protein from vertebrate cells and muscles[35] the numbering of the actin protein family begins with Aspartate as #1; post-translational processing (see below) removes the first two N-terminal amino acids in all higher eukaryotic actins. In addition all cytoskeletal and a number of nonvertebrate muscle actins, including all the insect actins, are shorter by one residue than the vertebrate muscle actin. Thus the first residue (invariably aspartate - D) in mature insect actins corresponds to residue #2 of the rabbit striated muscle actin from which the first sequence and actin atomic structure[2] were derived. All specific residues are referred to in this chapter using the actin family numbering system.

Alignment of all known actin sequences produces a phylogenetic tree which shows that insect actins form a statistically significant separate group.[24] In the alignment of all the insect actin sequences compared to the *Act6_DROME* (ACT88F) sequence (Table 3) there appears to be a cluster of amino acid differences at the N-terminus of actin. Amino acid variation at the N-terminus is a general finding within the actin sequence family.[24] With the exception of a glutamine at residue 5 the IFM-specific actins show no amino acid differences (Table 3) at the N-terminus. All the other actins (muscle and cytoplasmic), except for *AeAct-2*, show a number of shared amino acid substitutions in residues 3-6 with overall a much larger fraction of glutamate (E) residues than aspartate (D). Other than these the rest of the alignment does not reveal any consistent substitutions between the flight muscle and other muscle actin isoforms, with the possible exception of a fairly common substitution in the nonflight muscle actins of a serine at residue 260 and substitutions at residue 297 by isoleucine (I) or asparagine (N). In the rather small sample of insect cytoplasmic actins currently available there seem to be diagnostic substitutions compared to muscle actins of EEVA at residues 3-4, a substitution of Ala231 with serine (S), at Ser271 with alanine (A), Val278 with threonine (T) and at Gly368 with serine (S). These are at variance with the previously identified insect muscle/nonmuscle actin diagnostic substitutions.[36,37]

IFM-Specific Actin

Overall very little in the primary sequences distinguishes these insect actin isoforms from the nonIFM isoforms. The largest difference concerns the N-terminus. These differences may affect myosin binding as this region of the molecule has been modelled as a major, probably early contact, in the binding of myosin to F-actin,[38,39] but direct comparison of specific actin residues predicted in the myosin binding site in what is believed to be the stereospecific-binding show few differences (see later). Whether the amino acid differences between insect actins effect specific functional significance in the peculiar physiology of asynchronous insect flight muscles remains to be determined. However, studies with transgenic *Drosophila* in which in vitro genetic changes were made in the *Act88F* gene followed by its genomic insertion into flies lacking a functional copy of this gene (*Act88F*^{KM88}) showed that there is not a simple answer.[40] Single amino acid substitutions at 10 of the 27 residues at which the six *Drosophila* actins differ were made. Only one of these disrupted myofibrillar structure or function from which it was argued that most isoform residue replacements are of minor functional importance. Substitution of portions of ACT88F with those from the other actin genes (involving more than one residue change) produced functional muscle with only one construct (from 5) and the severity of the effects correlated with the number of amino acid substitutions. Complete conversion of ACT88F into a specific nonmuscle isoform, involving 18 amino acid substitutions, severely disrupted flight muscle structure and function. The overall picture seems to be that many isoform specific substitutions acquired during insect actin evolution are relatively neutral but divergent. As more substitutions are introduced this leads to isoform-specific function.

Earlier claims that the *Drosophila Act88F* gene expression is restricted to the IFM[25,26,41,42] had to be modified following detection of *Act88F* gene expression in other adult muscles in transgenic flies containing the *Act88F* gene promoter linked to GFP (green fluorescent protein) or β-galactosidase.[27] In this study *Act88F* gene expression was detected in the cells of the developing wing blade, in uterine muscles, one of the femoral muscles and a number of minor muscles, including those responsible for raising the head. Flies homozygous or heterozygous for a 'null' mutation of this gene are flightless as expected,[26] but the homozygotes also showed reduced oviposition (by uterine retention of fertilised eggs), though the flies are otherwise normal. This suggests that alternate actin isoforms are expressed in the developing wing blade and these other muscles that are able to produce normal functions in the absence of ACT88F actin. It may be that no actin isoforms can truly be described as indirect flight muscle actin isoforms, but little detail is yet known of actin expression in insects other than the Drosophilidae. Until the specificity of the actins expressed in insect IFM are better known, the term 'IFM-specific actin' remains a useful collective term for them and has been used throughout this chapter with this implicit caution.

Post-Translational Modifications of Insect Flight Muscle Specific Actins

N-Terminal Processing

In most eukaryotes the N-terminus of actin is post-translationally processed, in common with many cytosolic proteins. The N-terminal methionine is proteolytically removed, followed by acetylation of the N-terminal amino group.[43,44] For many actins, including all muscle-specific actins, the second amino acid is cysteine. In these actins the acetyl-cysteine is proteolytically removed by a specific actin N-acetylaminopeptidase[45] followed by acetylation of the new N-terminal amino group,[46] invariably the first amino acid is acidic, either aspartate or glutamate. The functions of this processing are not clear.

In vitro transcription/translation using rabbit reticuloyte lysate has shown the sequential removal of the first two amino acids of the *Drosophila* ACT88F isoform.[47-49] Inhibition of processing affected actin polymerization, probably at the filament nucleation phase.[49] Investigation of this in vivo processing of the ACT88F actin in *Drosophila* flight muscles showed that while removal of the N-terminal methionine and cysteine occurs, acetylation of the terminal amino group does not.[50] This muscle actin, so far uniquely, has a free N-terminal amino group. The *mod*⁻ mutation, in an unidentified gene on the *Drosophila* third chromosome, affects N-terminal processing, leaving the ACT88F actin with an acetyl-cysteine at its N-terminus. This has small, but significant, effects on the flight ability of the flies, confirming that the nature of the actin N-terminus is important for actomyosin interactions and raising further questions about the functional importance of N-terminal processing.

Methylation

Complete methylation of His73 occurs in all actins, except for a few fungal actins. Mass spectrometry of *Drosophila* ACT88F flight muscle actin showed the presence of the methyl group on the appropriate tryptic fragment.[34]

Arthrin

A novel 'heavy' actin, given the name arthrin, was first described in actin preparations from the IFM of the waterbug *Lethocerus*.[18] This protein has very similar properties to actin: it polymerises to form F-arthrin, filaments which can be decorated with myosin S1 subfragments. F-arthrin will activate the myosin Mg.ATPase activity and this activity is regulatable by tropomyosin-troponin complexes from insect or mammalian muscle sources in the same way as F-actin. Arthrin has now been detected in a large number of insects[34,51] including all the higher Diptera, except the primitive *Tipula spp.*, all Hymenoptera and a number of the Hemiptera. In all cases examined it has been found to be restricted to the IFM.

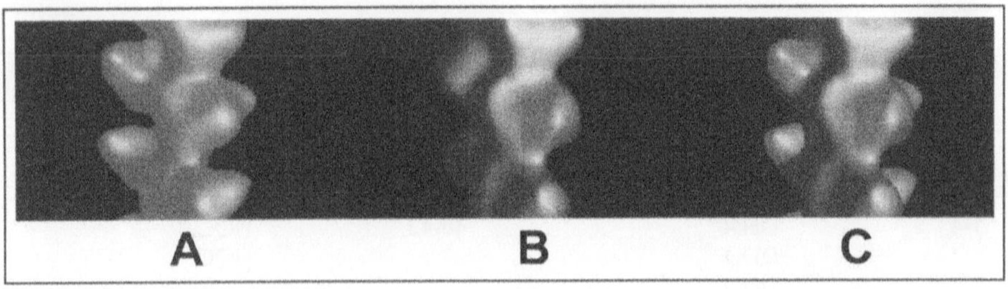

Figure 2. Reconstructions of F-actin and F-arthrin from electron micrographs of purified frozen-hydrated protein. A) F-arthrin; B) F-actin; C) F-arthrin image rendering changes to show the location of the major part of the ubiquitin mass on the F-actin. The pictures were taken of material preserved in vitreous ice and processed using a single particle analysis.[21] Images kindly provided by Prof. John Trinick and Dr. Stan Burgess, University of Leeds, U.K.

Studies of arthrin in *Drosophila* showed that it is not encoded by a novel gene but is a post-translationally modified form of the IFM-specific ACT88F actin isoform.[51] Western blots using anti-ubiquitin antibodies showed that the modification in mass (from 42000 to 55000) and charge was due to addition of a single molecule of ubiquitin. The difference in apparent mass on denaturing gels (55000) and the combined mass of actin and ubiquitin (approx. 51000) could be explained if the conjugation of ubiquitin occurred through an isopeptide bond. Recent mass spectrometry studies have confirmed that arthrin is formed through conjugation of the carboxyl group of the C-terminal glycine of ubiquitin to one of the ε-amino groups of lysine118 of actin.

Comparisons of electron micrograph reconstructions of F-actin and F-arthrin (Fig. 2) show that the extra mass contributed by the ubiquitin lies 'behind' subdomain 1, on the opposite side to the major myosin binding site;[19-22] this is consistent with the covalent bond formed between the ε-amino group of actin lysine118 and the terminal carboxyl group of ubiquitin.

Electron micrographs of *Drosophila* IFM thin filaments labelled with anti-ubiquitin antibodies shows substantial added mass (not seen in the absence of antibodies) at about 40nm intervals. This demonstrates that every seventh actin monomer along the thin filament long-pitch helix is ubiquitinated[21] in line with earlier stoichiometric measurements[51] which showed a ratio of 1 arthrin to 6 actins in flight muscle myofibrils i.e., one per each thin filament repeat. Specification of which actin monomers are ubiquitinated is probably related to the position of the troponin complex spacing (one per 7 actins) as, although actin accumulates, no arthrin is formed in the IFM of the *hdp³* troponin I null mutant.[52]

Modelling the F-arthrin with tropomyosin present leads to differences of opinion between the Burgess and Egelman groups as to whether the ubiquitin will interfere with tropomyosin movement in vivo, a motion that has a major role in activating muscle contraction. On balance the evidence (see below) makes an effect of arthrin on this unlikely.

Transgenic *Drosophila* containing the K118R mutation in their only functional copy of the IFM-specific *Act88F* gene fly almost as well as wild type though they show some reluctance to do so (Schankin, Schmitz and Sparrow, unpublished). The obvious conclusion is that arthrin is not required for asynchronous flight muscle function. This is consistent with the absence of arthrin in the asynchronous IFM of many insects but begs the question as to why the IFM of so many insects contain it and why this specific actin modification appears to have evolved twice.[34] There are no correlations between the presence of arthrin and insect size, wing beat frequencies, life style etc.,[34] though recent results (Nongthomba and Sparrow unpublished) indicate that *Act88F* ^{K118R} flies can fly well at eclosion, but cease flying at younger ages than wild-type; of course there is no such thing as a neutral amino acid substitution and these effects could arguably be due to the lysine to arginine substitution rather than the absence of ubiquitination.

Table 4. Comparison of amino acids in the myosin binding sites on actin in rabbit skeletal actin and insect muscle actins

AA No.	Primary	AA No.	Monomer Below
1-4	DEDE	**40-42**	HQG
	DDD		HQG
	DDQ		
24,25	DD	**332-334**	PPE
	DD		PPE
(99,100)	EE		**Secondary**
	EE		
144-148	ASGRTT	**(79-92)**	WDDMEKIWHHTFYN
	ASGRTT		WDDMEKIWHHTFYN
341-354	IGGSILASLSTFQQ	**91-100**	YNELRVAPEE
	IGGSILASLSTFQQ		YNELRVAPEE
	IGGSILASLSTFQT		
	IGGSILASLSTFQA		

The upper sequence row in each 'cell' of the table is the rabbit skeletal actin sequence; the next one down is the major insect sequence and the lower rows any other variants. Actin residues in the myosin binding site are from reconstructions with chicken S1[38] or, in parentheses, with *Dictyostelium* myosin.[39]

Actomyosin Interactions

Actin and arthrin have been purified and separated using standard muscle actin protocols from the flight muscles of *Lethocerus*[18] and *Drosophila*.[50,53,54] These actins have been used in biochemical kinetics experiments and behave almost exactly as their vertebrate striated muscle homologues. The only major difference is the observation that the IFM myosins have lower actin affinities than vertebrate myosins irrespective of the source of the actin (see Sparrow and Geeves Chapter).

The myosin-binding site on actin appears highly conserved, in line with insect actomyosin kinetics experiments[55,56] that show that differences in IFM acto-myosin kinetics reside in the source of IFM myosin not the actin (see Sparrow and Geeves chapter). The actin residues involved in actomyosin binding have not been directly observed but deduced from docking the atomic structure of rabbit skeletal actin with structures obtained from chicken skeletal muscle subfragment 1 (S1)[38] or a myosin II from the slime mould, *Dictyostelium discoideum*[39] guided by 3-D reconstructions from electron micrographs of S1-decorated F-actin. These showed that the myosin head binds two neighbouring actin monomers in the F-actin, but in two discrete areas—a primary site largely involving residues of the first monomer but also some others in the second, and a secondary site involving residues only of the second monomer. The two models give slightly different assignments of residues to a large extent because of small differences in the S1 structures. Comparison of residues in the binding site between rabbit skeletal actin (Table 4) and the insect flight muscle actins (from Table 1) show almost no differences. The assignment of amino acids in the two myosin binding sites on actin should be treated with caution since the exact binding conformation of these two proteins is not resolved sufficiently well nor is it known if these change as the actomyosin complex undergoes conformational change during the power part of the crossbridge cycle.

Actin Molecular Genetics

The *D. melanogaster Act88F* gene has been used for the study of actin molecular genetics, because, unlike the vast majority of known actin genes, its IFM-specific expression is dispensable, except for flight. *Act88F* mutations are readily recoverable by selecting for a dominant flightless phenotypes.[57-59] A null allele,[26,60] $Act88F^{KM88}$, also allows the study, following transgenesis, of actin genes mutated in vitro.[26,40,61] Changes in muscle function[62] (flight ability and in vitro skinned fibre mechanics), muscle structure,[40,60-64] protein accumulation[41] and heat shock protein synthesis[26,65] have all then been used to assess the effects of the mutations. More recently the development of protocols for the isolation and purification of the IFM-specific isoform and/or arthrin[50,54] in sufficient quantities for biochemical and biophysical studies have led to insights into actomyosin interactions,[53,54,66] N-terminal processing of actin,[50] its methylation[50] and investigations of the structures of arthrin and F-arthrin.[19-21,34] *Act88F* remains the only muscle actin gene that is amenable to genetic manipulation, whose mutant phenotypes can be characterised in muscle and whose product can be purified for molecular characterisation. The recent discovery of human cardiac[67-69] and skeletal actin[70] myopathies, together with the recent demonstration that *Act88F* mutants homologous to those in the human *ACTA1* gene, produce a nemaline myopathy phenotype similar that found in human skeletal muscle (Sparrow unpublished) mean that this flight muscle-specific actin will continue to be the focus of experimentation.

Acknowledgements

I would like to thank specifically Drs. Tony James and Que Lan for providing prepublication manuscripts on their recent work on *Aedes aegypti* actin genes. Also to Drs. Stan Burgess and John Trinick for providing Figure 2.

References

1. Bork P, Sander C, Valencia A. An ATPase domain common to prokaryotic cell cycle proteins, sugar kinases, actin and hsp70 heat shock proteins. Proc Natl Acad Sci (USA) 1992; 89:7290-7294.
2. Kabsch W, Mannherz HG, Suck D et al. Atomic structure of the actin: DNase I complex. Nature 1990; 347:37-44.
3. Schutt CE, Myslik JC, Rozycki MD et al. The structure of crystalline profilin-β-actin. Nature 1993; 365:810-816.
4. McLaughlin PJ, Gooch JT, Mannherz HG et al. Structure of gelsolin segment 1-actin complex and the mechanism of filament severing. Nature 1993; 364:685-692.
5. Otterbein LR, Cosio C, Graceffa P et al. Crystal structures of the vitamin D-binding protein and its complex with actin: Structural basis of the actin-scavenger system. Proc Natl Acad Sci (USA) 2002; 99:8003-8008.
6. Otterbein LR, Graceffa P, Dominguez R. The crystal structure of uncomplexed actin in the ADP state. Science 2001; 293:708-711.
7. Egelman EH. The structure of F-actin. J Muscle Res Cell Motil 1985; 6:129-151.
8. Milligan RA, Whittaker M, Safer D. Molecular structure of F-actin and location of surface binding sites. Nature 1990: 348:217-221.
9. Holmes KC, Popp D, Gebhard W et al. Atomic model of the actin filament. Nature 1990; 288:44-49.
10. Lorenz M, Popp D, Holmes KC. Refinement of the F-actin model against X-ray fiber diffraction data by the use of a directed mutation algorithm. J Mol Biol 1993; 234:826-836.
11. Chen X, Cook RK, Rubenstein PA. Yeast actin with a mutation in the hydrophobic plug between subdomain-3 and subdomain-4 (L(266)D) displays a cold-sensitive polymerization defect. J Cell Biol 1993; 123:1185-1195.
12. Kuang B, Rubenstein PA. Beryllium fluoride and phalloidin restore polymerizability of a mutant yeast actin (V266G,L267G) with severely decreased hydrophobicity in a subdomain 3/4 loop. J Biol Chem 1997; 272:1237-1247.
13. Lovato TL, Meadows SM, Baker PW et al. Characterization of muscle actin genes in Drosophila virilis reveals significant functional complexity in skeletal muscle types. Insect Mol Biol 2001; 10:333-340.

14. Reedy M. Ultrastructure of insect flight muscle. I. Screw sense and structural grouping in the rigor cross-bridge lattice. J Mol Biol 1968; 31:155-176.
15. Miller A, Tregear RT. Structure of insect fibrillar flight muscle in the presence and absence of ATP. J Mol Biol 1972; 70:85-104.
16. Reedy MK, Reedy MC. Rigor crossbridge structure in tilted single filament layers and flared-X formation from insect flight muscle. J Mol Biol 1985; 185:145-176.
17. Schmitz H, Lucaveche C, Reedy M et al. Oblique section 3-D reconstruction of relaxed insect flight muscle reveals cross-bridge lattice in helical registration. Biophys J 1994; 67:1620-1633.
18. Bullard B, Bell J, Craig R et al. Arthrin: A new actin-like protein in insect flight muscle. J Mol Biol 1985; 182:443-454.
19. Burgess SA, Knight PJ, Walker M et al. Real-space 3-D reconstruction of frozen-hydrated arthrin and actin filaments at 2 nm resolution. Biophys J 2000; 78:A47.
20. Burgess SA, Walker M, Knight PJ et al. Structural studies of arthrin: Mono-ubiquitinated actin. Biophys J 2003: 84:480A.
21. Burgess S, Walker M, Knight PJ et al. Structural studies of arthrin: Monoubiquitinated actin. J Mol Biol 2004; 341:1161-1173.
22. Galkin VE, Orlova A, Lukoyanova N. The location of ubiquitin in Lethocerus actin. J Mol Biol 2003; 325:623-628.
23. Ruiz T, Bullard B, Lepault J. Effects of calcium and nucleotides on the structure of insect flight muscle thin filaments. J Muscle Res Cell Motil 1998; 19:353-364.
24. Sheterline P, Clayton JD, Sparrow JC. Actin. In: Sheterline P, ed. Protein Profile. 4th ed. Oxford: Oxford University Press, 1998.
25. Fyrberg EA, Mahaffey JW, Bond BJ et al. Transcripts of the six Drosophila actin genes accumulate in a stage- and tissue-specific manner. Cell 1983; 33:115-123.
26. Hiromi Y, Hotta Y. Actin gene mutations in Drosophila: Heat shock activation in the indirect flight muscles. EMBO J 1985; 4:1681-1687.
27. Nongthomba U, Pasalodos-Sanchez S, Clark S et al. Expression and function of the Drosophila ACT88F actin isoform is not restricted to the indirect flight muscles. J Muscle Res Cell Motil 2001; 22:111-119.
28. Holt A, Subramian GM, Halpern A et al. The genome sequence of the malaria mosquito Anopheles gambiae. Science 2002; 298:129-149.
29. Beifuss MJ, Durica DS. Sequence analysis of the indirect flight muscle actin encoding gene of Drosophila simulans. Gene 1992; 118:163-170.
30. He M, Haymer DS. The actin gene family in the oriental fruit fly Bactrocera dorsalis. Insect Biochem Mol Biol 1994; 24:891-906.
31. Ibrahim MS, Eisinger SW, Scott AL. Muscle actin gene from Aedes aegypti (Diptera:Culicidae). J Med Entomol 1996; 33:955-962.
32. Munoz-Labiano D, Jimenez A, James AA. The Aeact-2 gene is expressed specifically in the developing dorsal longitudinal (indirect) flight muscles of female Aedes aegypti. Insect Mol Biol 2004; in press.
33. Vyazunova I, Lan Q. Stage-specific expression of two actin genes in the yellow fever mosquito, Aedes aegypti. Insect Mol Biol 2004; 13:241-249.
34. Schmitz S, Schankin S, Prinz H et al. Molecular evolutionary convergence of the flight muscle protein arthrin in Diptera and Hemiptera. Mol Biol Evol 2003; 20:2019-2033.
35. Vanderkerckhove J, Weber K. Actin amino-acid sequences. Comparison of actins from calf thymus, bovine brain and SV40-transformed mouse 3T3 cells with rabbit skeletal muscle actin. Eur J Biochem 1987; 90:451-462.
36. Mounier N, Gaillard J, Prudhomme J-C. Nucleotide sequence of the coding region of two actin genes in Bombyx mori. Nucl Acids Res 1987; 15:2781-2781.
37. Mounier N, Sparrow JC. Structural comparisons of muscle and nonmuscle actins give insights into the evolution of their functional differences. J Mol Evol 1997; 44:89-97.
38. Rayment I, Holden HM, Whittaker CB et al. Structure of the actin-myosin complex and its implications for muscle contraction. Science 1993; 262:58-65.
39. Schroder ER, Manstein DJ, Jahn W et al. Three-dimensional atomic model of F-actin decorated with Dictyostelium myosin. Nature 1993; 364:171-174.
40. Fyrberg EA, Fyrberg CC, Biggs JR. Functional nonequivalence of Drosophila actin isoforms. Bioch Genet 1998; 36:271-287.
41. Mahaffey JW, Coutu MD, Fyrberg EA et al. The flightless Drosophila mutant raised has two distinct genetic lesions affecting accumulation of myofibrillar proteins in flight muscles. Cell 1985; 40:101-110.

42. Hiromi Y, Okamoto H, Gehring WJ et al. Germline transformation with Drosophila mutant actin genes induces constitutive expression of the heat-shock genes. Cell 1986; 44:293-301.
43. Redman K, Rubenstein PA. NH_2-terminal processing of Dictyostelium discoideum actin in vitro. J Biol Chem 1981; 256:13226-13229.
44. Rubenstein PA, Martin DJ. NH_2-terminal processing of actin in mouse L-cells in vivo. J Biol Chem 1983; 258:3961-3966.
45. Sheff DR, Rubenstein PA. Isolation and characterization of the rat-liver actin N-acetylaminopeptidase. J Biol Chem 1992; 267:20217-20224.
46. Sheff DR, Rubenstein PA. Amino-terminal processing of actins mutagenized at the Cys-1 residue. J Biol Chem 1992; 267:2671-2678.
47. Rubenstein PA, Martin DJ. NH_2-terminal processing of Drosophila melanogaster actin. Sequential removal of two amino acids. J Biol Chem 1983; 258:11354-11360.
48. Solomon TL, Solomon LR, Gay LS et al. Study on the role of actin's aspartic acid 3 and aspartic acid 11 using oligonucleotide-directed site specific mutagenesis. J Biol Chem 1988; 263:19662-19669.
49. Hennessey ES, Drummond DR, Sparrow JC. Post-translational processing of the amino terminus affects actin function. Eur J Biochem 1991; 197:345-352.
50. Schmitz S, Clayton J, Nongthomba U et al. Drosophila ACT88F indirect flight muscle-specific actin is not N-terminally acetylated: A mutation in N-terminal processing affects actin function. J Mol Biol 2000; 295:201-210.
51. Ball E, Karlik CC, Beall CJ et al. Arthrin, a myofibrillar protein of insect flight muscle, is an actin-ubiquitin conjugate. Cell 1987; 51:221-228.
52. Nongthomba U, Clark S, Cummins M et al. Troponin I is required for myofibrillogenesis and sarcomere formation in Drosophila flight muscle. J Cell Sci 2004; 117:1795-1805.
53. Anson M, Drummond DR, Geeves MA et al. Actomyosin kinetics and in vitro motility of wild type Drosophila actin and the effects of 2 mutations in the Act88F gene. Biophysical J 1995; 68:1991-2003.
54. Razzaq A, Schmitz S, Veigel C et al. Actin residue E93 is identified as an amino acid affecting myosin binding. J Biol Chem 1999; 274:28321-28328.
55. Swank DM, Bartoo ML, Knowles AF et al. Alternative-exon-encoded regions of Drosophila myosin heavy chain modulate ATPase rates and actin sliding velocity. J Biol Chem 2001; 276:15117-15124.
56. Iliffe C. The kinetics and mechanics of insect flight muscle. PhD thesis. University of York, 2002.
57. Mogami K, Hotta Y. Isolation of Drosophila flightless mutants which affect myofibrillar proteins of indirect flight muscle. Mol Gen Genet 1981; 183:409-417.
58. An HS, Mogami K. Isolation of 88F actin mutants of Drosophila melanogaster and possible alterations in the mutant actin structures. J Mol Biol 1996; 260:492-505.
59. Cripps RM, Ball E, Stark M et al. Dominant flightless mutants of Drosophila melanogaster and identification of a new gene required for normal muscle structure and function. Genetics 1994; 137:151-164.
60. Karlik CC, Coutu MD, Fyrberg EA. A nonsense mutation within the Act88F actin gene disrupts myofibril formation in Drosophila indirect flight muscles. Cell 1984; 38:711-719.
61. Drummond DR, Hennessey ES, Sparrow JC. Characterisation of missense mutations in the Act88F gene of Drosophila melanogaster. Mol Gen Genet 1991; 226:70-80.
62. Drummond DR, Peckham M, Sparrow JC et al. Actin mutants causing changed muscle kinetics. Nature 1990; 348:440-442.
63. Reedy MC, Beall C, Fyrberg E. Formation of reverse rigor chevrons by myosin heads. Nature 1989; 339:481-483.
64. Sparrow J, Reedy M, Ball E et al. Functional and ultrastructural effects of a missense mutation in the indirect flight muscle-specific actin gene of Drosophila melanogaster. J Mol Biol 1991; 222:963-982.
65. Karlik CC, Saville DL, Fyrberg EA. 2 Missense alleles of the Drosophila melanogaster Act88F gene are strongly antimorphic but only weakly induce synthesis of heat-shock proteins. Mol Cell Biol 1987; 7:3084-3091.
66. Bing W, Razzaq R, Sparrow JC et al. Tropomyosin and troponin regulation of wild-type and E93K mutant actin filaments from Drosophila flight muscle: Charge reversal on actin changes actin-tropomyosin from ON to OFF state. J Biol Chem 1998; 273:15016-15021.
67. Mogensen J, Klausen IC, Pedersen AK et al. Alpha-cardiac actin is a novel disease gene in familial hypertrophic cardiomyopathy. J Clin Invest 1999; 103:R39-43.
68. Olson TM, Michels VV, Thibodeau SN et al. Actin mutations in dilated cardiomyopathy, a heritable form of heart failure. Science 1998; 280:750-752.
69. Olson TM, Doan TP, Kishimoto NY et al. Inherited and de novo mutations in the cardiac actin gene cause hypertrophic cardiomyopathy. J Mol Cell Cardiol 2000; 32:1687-1694.
70. Sparrow JC, Nowak K, Durling HJ et al. Muscle disease caused by mutations in the skeletal muscle alpha-actin gene, ACTA1. Neuromuscular Disord 2003; 13:519-531.

Chapter 10

Troponin, Tropomyosin and GST-2

Alberto Ferrús

Abstract

Troponin, Tropomyosin and GST are generic names of protein families that play a variety of cellular roles in the biology of uni- and multicellular organisms. In muscles, specific family members are associated to the actin based thin filament where they contribute to the sarcomere contraction/relaxation cycle. Different muscles have different regulatory mechanisms; in most muscles, including the vertebrate cardiac and insect flight muscles, regulation is mediated by the binding of Ca^{2+} to the troponin C component of the troponin-tropomyosin complex of the thin filament. This switch triggers a fast cascade of allosteric events involving all the thin-filament proteins whose modulatory properties change as a function of Ca^{2+} concentration and phosphorylation status. The corresponding genes in *Drosophila* and other species of insects and vertebrates show relatively well conserved regulatory mechanisms that are likely to sustain the coordinated physiology of the encoded proteins in each cellular context. In general, the diversity of protein isoforms is attained by differential exon splicing (insects) or independent gene expression (vertebrates). Flight is a well studied manifestation of the insect muscle system. To sustain flight in small insects, wings must beat at frequencies well above the firing capability of the corresponding motorneurons. This is achieved by a mechanism known as stretch activation in which muscles contract when subject to a small deformation and Ca^{2+} concentrations below the activation level. The phenomenon is also present in vertebrate heart. The underlying molecular mechanisms of stretch activation are not yet known, although models based on troponin/tropomyosin protein isoforms specific to these muscles have been proposed.

Troponin Complex Components

The regulation of the sliding of thin over thick filaments during muscle contraction depends to a great extent on the proper activity of the troponin complex.[1-4] This is a tetrameric protein ensemble composed of three troponins: I, C and T, and tropomyosin. The troponin acronyms stand for their respective roles within the complex. In thin filament regulated muscle contraction, troponin I (TnI) inhibits the interaction between actin and the myosin head at rest. Troponin C (TnC) is the calcium binding component when this ion is released from the storage compartments at the muscle activation point. Troponin T (TnT) anchors the complex to tropomyosin (Tm) providing a regular spacing pattern along the actin based thin filament. In addition, tropomyosin (along with tropomodulin) is known to play a role in thin filament capping by preventing actin monomer depolymerization at the pointed ends.[5] The interactions between these proteins have been elucidated at the level of peptide fragments and are summarized in Figure 1. Genes encoding troponin complex components appear similarly organized in insects and humans suggesting evolution under similar constraints. Thus, TnI and TnT genes appear as single units yielding multiple isoforms (insects) or grouped in at least three clusters yielding one or very

Nature's Versatile Engine: Insect Flight Muscle Inside and Out, edited by Jim Vigoreaux.

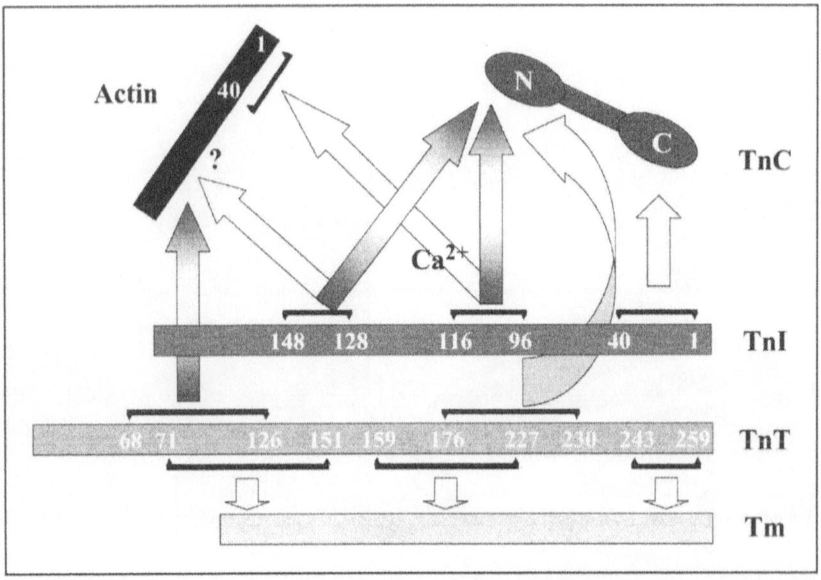

Figure 1. Summary of interactions between vertebrate thin filament components as detected by physicochemical methods using peptide fragments. The five main components of thin filaments, actin, tropomyosin and troponins I, C and T are shown schematically. Numbers indicate amino acids that define each fragment tested, indicated by thin lines. Arrows between fragments and protein domains indicate interactions. Empty arrows show interactions at the relaxed state and shaded arrows indicate interactions following muscle Ca^{2+} activation. Data summarized from various authors.[3,16,45,92-94]

few isoforms each (human 1q32, 11p15.5 and 19q13.4),[6] while TnC genes, which are derived from a calmodulin-like ancestor, are found in multiple copies in both insects and humans.[7]

Troponin I

TnI is a globular protein about 269 amino acids long depending on isoforms or species. In *Drosophila*, it is encoded by a single gene, *wings up A (wupA)*, located in chromosomal band 16F7, that yields up to ten protein isoforms by differential splicing.[8] In the mouse, however, there are at least three independent genes, fast, slow, and cardiac. The cardiac isoform has a characteristic Pro-Ala rich extension towards its N-terminus that is present also in exon 3 containing insect isoforms. All of them evolved from a single ancestor, close to that found in today's ascidians, by means of alternative and mutually exclusive exon splicing from a single gene (insects) or by duplication and specialization of up to three genes (vertebrates).[9] TnI isoforms are muscle type specific. A given muscle, however, can express several TnI isoforms whose specific roles are still unknown. In vertebrates, isoform replacement during embryo development has been reported as part of muscle maturation and fiber diversification.[10] In the adult heart, the TnI isoform repertoire seems constant throughout aging,[11] and only under severe cardiac damage TnI C-terminus hydrolysis is detected.[12] In *Drosophila*, a similar phenomenon seems to occur, since some TnI isoforms are mainly embryonic while others are mostly adult.[8] Also, muscles show differential sensitivity to the same mutation or protein isoform. The structure of the indirect flight muscles (IFM) and, consequently flight ability, are the most sensitive features in the biology of these muscles. The *wupA*^{hdp2} mutation A116V, which affects all TnI isoforms because it is located in a constitutive exon, yields viable adults with normal motor activity except flight[13] and, to a minor extent, walking.[14] All muscles develop normally but only the IFM collapse upon use, showing a sarcomere structure that can be interpreted as

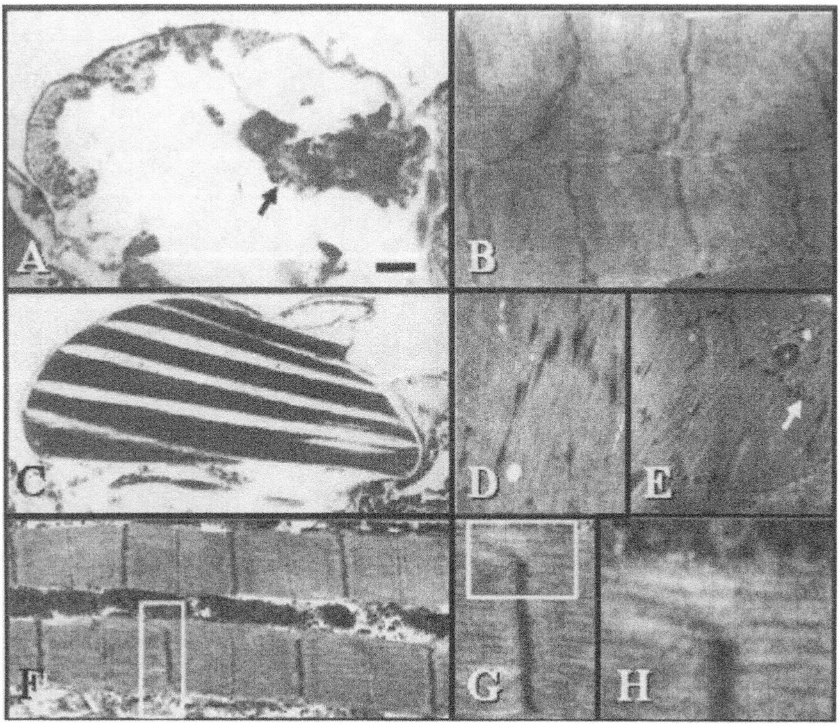

Figure 2. TnI mutant effects in *Drosophila* IFM. A) Lateral view of an adult thorax showing the IFM remnants (arrow) due to the *hdp*^{2} (A116V) mutation (Bar = 200 μm). B) In the IFM remnants, sarcomeres appear hypercontracted. C) The suppressed phenotype due to a second site (L118F) mutation in TnI allows a normal IFM structure and, in some cases, flight. D) A deficiency that eliminates the TnI regulatory region URE, reduces the level of TnI to such an extent that thin filaments of somatic muscles in the embryo are scant, poorly oriented and accumulate disorganized Z-like deposits. E) A similar mutation that separates IRE from URE regulatory regions (see Fig 3) produces a less severe reduction of TnI, still incompatible with viability, but enough as to organize Z protodisks (arrow). Sarcomeres, however, are never observed. F) Thin filaments of IFM in near normal sarcomeres from *hdp*^{2}, *Su(hdp*^{2}*)* double mutants do not necessarily anchor to Z disks. G) Detail of inset in F. H) Further detail of inset in G.

"clamped in the contracted mode"[15] (Fig. 2). It appears that this mutated form of TnI prevents normal relaxation. The mutated residue (A116V) is equivalent to rabbit's TnI position 25, near the TnI/TnC interaction site[16] which is now known also to include the TnT binding site (see below). Similarly, the depletion of a whole set of isoforms due to the *wupA*^{hdp3} mutation at the splicing acceptor site for exon 6d also yields almost normal motile adults, albeit with severe IFM abnormalities preventing flight.[17] The repertoire of TnI isoforms in *Drosophila* and other insects is provided by a set of 4 mutually exclusive exons known as the exon 6 family and 1-3 alternative exons located at either end of the gene sequence. It is worth pointing out that this exon structure is very similar to that of TnT encoding genes in several insect species (Herranz et al unpublished result) (see below). The conserved gene structure and mechanism for generating isoform diversity between TnI and TnT encoding genes is further underlined by a conserved array of regulatory sequences (see below).

Beyond the unknown role of various TnI isoforms within the same muscle cell, it is becoming evident that TnI isoform stoichiometry may play a crucial role in muscle development and function. The insertional mutations *wupA*^{PL87} and *wupA*^{PG31} disrupt the regulatory regions of *wupA* (see below) and are embryonic lethals. The mutations result in reduced levels of all isoforms below a critical point that is insufficient for muscle development, although thin filaments do assemble to a large extent, including the formation of proto-Z discs (Fig 2). Sarcomeres, however, are not observed at these low TnI levels.[18] Based on quantitative RT-PCR data, it has been shown that absolute levels of TnI transcripts do not correlate with phenotype severity. Instead, it is more plausible that the phenotype severity results from inbalance among the levels of expression of certain TnI isoforms. This is also the case in vertebrates.[19]

The most characteristic role of TnI is its inhibitory effect of muscle contraction by preventing the actin-myosin head interaction, that is, the acto-myosin MgATPase enzymatic activity.[20] However, the TnT/TnC interaction plus the Tm binding to actin are also contributing factors to this enzymatic inhibition. Recent data using peptidic fragments of these proteins and NMR methods indicate that the binding of TnI to actin causes a conformational change in the latter that result in allosteric effects at additional sites of the actin monomer.[21] A further modulatory process on TnI, at least the vertebrate cardiac isoform, is phosphorylation first at Ser24 and later at Ser 23 by cAMP-dependent protein kinase.[22] This modulation affects the Ca^{2+}-binding activity by TnC and, perhaps, relates to the phenomenon of stretch activation (see below). In *Drosophila*, the putative phosphorylation of TnI has not been investigated. The IFM of this insect, however, express high levels of a TnI isoform containing exons 3 and 9 (Herranz unpublished result).[8,17] This isoform exhibits a long Pro-Ala rich extension very similar to that of two heavy tropomyosin 1 isoforms, TnH33 and TnH34, which are also IFM specific. Although interactions between all these isoforms have been proposed as mechanisms sustaining stretch activation (see below), their specific role and structural relationships remain to be elucidated.

It is likely that TnI will have additional functions outside the sarcomere as already indicated by the role of the vertebrate homologue as a potent angiogenesis inhibitor through its interaction with polycystin-2.[23] In that context, the *Drosophila* TnI gene harbors an unusual class of mutations, the dominant lethal (DL) alleles, all of which are rearrangements affecting the exon 7-8 interval.[24] These mutant alleles are lethal, even in heterozygotes, and show severe neural phenotypes whose molecular foundations are not yet known.

Troponin C

TnC is a Ca^{2+}-binding protein about 155 amino acids long that represents a specialized form of calmodulin, to which it is evolutionary related.[25] The crystal structure of vertebrate fast skeletal version shows two globular domains bound by an α-helix. Two EF-hand motifs for Ca^{2+} binding are present in each globular domain. Depending on muscle type and species, however, these four sites may be active, inactive, or mutated to a Mg^{2+}-binding activity.[26-28] Also, their affinity constant for these cations differ according to isoforms[29] (see Table 1). Contrary to vertebrates, in most insects and molluscs the two N-terminal EF-hand sites are either inactive, due to changes in cation coordinating key amino acids (site 1), or show low affinity for Ca^{2+} (site 2).[30,31] The two C-terminal sites bind Mg^{2+} (site 3) and exhibit low affinity for Ca^{2+} (site 4), in contrast to vertebrates where both sites have high affinity for Ca^{2+}. It is through the C-terminus that the interaction between TnC and TnI takes place at rest, when the free Ca^{2+} concentration is very low. Thus, the two EF-hand sites at the C-terminus play a structural role in vertebrates (because of their high cation affinity) while they are regulatory in invertebrates (because of their low affinity). Also at rest, the N-terminus of vertebrate TnC interacts with TnT. When mollusc muscle is activated, by transient elevation of free Ca^{2+} concentration, these interactions change drastically due, in part, to the binding of Ca^{2+} to the low affinity EF-hand sites at the N terminus of TnC (see Fig. 1).[32,33] At this stage, the interaction between TnC and TnI shifts towards the N-terminus of TnC and the C-terminus of TnI liberating the inhibition on the actin-myosin head interaction.[34]

In addition to the structural and functional diversity of the four EF-hand motifs, insect TnC isoforms differ in the α-helix separating the two globular domains. Single amino acid

Table 1. Types and properties of* Drosophila *Troponin C

Crystal Structure	EF Hand Sites				α Helix	Expression			Gene	Class
	1	2	3	4		IFM	Tub	E		
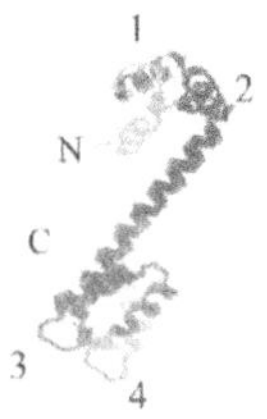	o	o	m	●		+	+	+	*DmTnC25D*	II
	o	●	m	●		+	+	–	DmTnC41C (*DmTnC1*)	IIIa
	o	o	m	●	+E	+	+	–	DmTnC41F (*DmTnC4*)	IIIb
	o	●	o	●		–	+	+	DmTnC47D (*DmTnC2*)	Ib
	o	●	o	●		+	+	+	DmTnC73F (*DmTnC3*)	Ia

Diagram shows the crystal structure of rabbit TnC. Note the two globular domains at the amino (N) and carboxy (C) termini. Numbers 1-4 indicate the EF hand motifs. The active (●), inactive (o) or Mg^{2+} (m) binding condition of each EF hand site is indicated. The affinity constant of site 4 is always low. Insertions (+) of glutamic acid in the α-helix domain is indicated. Expression based on RT-PCR or in situ data;[35] thus, the information shown is qualitative. Expression data on the bottom row is from cDNAs and immunoblots.[37] Gene names are according to Herranz et al.[35] with corresponding names in parentheses from Qiu et al.[37] The class of gene refers to intron/exon position.[35,37,91] Isoform 41F (a.k.a. 4) has only one Ca^{2+} binding site active (site 4) as demonstrated by direct assay on the expressed protein and site directed mutagenesis of the *Lethocerus* homolog.[37] Isoforms 47D (a.k.a. 2) and 25D are not yet identified in *Lethocerus*. Other insect TnCs differ in the α-helix sequence or the activity of the EF-hands, mainly site 2.

insertions in this helix are frequent in certain isoforms of *Drosophila, Anopheles, Apis* and *Lethocerus* (see Table 1). Considering their restricted muscle type expression, these structural peculiarities are likely to sustain specialized functional properties, such as stretch activation (see chapter by Moore for a description of possible role of TnC in stretch activation).

In *Drosophila*, recent data[35] indicate that TnC isoforms are encoded by five independent genes: *DmTnC25D, DmTnC41C, DmTnC41F, DmTnC47D* and *DmTnC73F*, the same number of genes as in mosquito (*Anopheles*) or bee (*Apis*), but three more than in the nematode (*C. elegans*). Considering their developmental profile of expression, activity of the four EF-hand motifs and protein sequence, the putative evolutionary history of TnC genes can be traced back to a common ancestor predating the radiaton of metazoans.[25,35] In *Drosophila*, each gene has a characteristic developmental profile of expression. Thus, TnC25D is the main isoform throughout development and it becomes restricted to the head in the adult. TnC47D and TnC73F are also expressed during early embryonic development but their expression in the adult is restricted almost exclusively to the abdomen. TnC41C and TnC41F are specific to the adult thorax.[35-37] In *C. elegans*, mutational analysis of the TnC gene *pat-10* has revealed two important features: the D64N mutation abolishes Ca^{2+} binding, and a truncated form at position W153 is unable to bind TnI.[38,39]

Troponin T

TnT is a long molecule (18 nm) that interacts with Tm for most of its length and, as a result, serves to space the Tn/Tm complex at 38.5 nm intervals along the thin filament. Many different isoforms are expressed throughout development either from different genes (vertebrate skeletal) or by differential splicing (insects and vertebrate cardiac). The main functional difference among TnT isoforms is their effect on the actomyosin MgATPase sensitivity to Ca^{2+} concentrations.[40] TnT is a 396 amino acid long protein that, in *Drosophila*, is encoded by the

upheld (up) gene at chromosome band position 12A7. TnT mutants yield phenotypes that are informative on IFM myofibrillar morphogenesis.[41] Thus, up^2 and up^3 alter RNA splicing and virtually eliminate TnT which result in no thin filament assembly except near the proto-Z discs, a very similar effect to TnI deprivation (see above). Mutant alleles up^{101} and up^{int3}, however, produce single amino acid substitutions that yield apparently normal thin filaments that collapse upon use, a phenotype very similar to that of the missense TnI alleles described earlier. Finally the missense mutation up^{whu} yield myofibrils of half the normal diameter.[41] There is now evidence, based on comparative genome analysis of several *Drosophila* species, for a new exon, 10A, that is mutually exclusive with the previously known exon 10, now renamed exon 10B (Herranz et al, unpublished result). The new exon is adult specific, expressed in the IFM and TDT muscles. Exon 10B, by contrast, is not detected in these muscles. The predicted amino acid sequence of the new exon 10A is, as with exon 10B, 26 amino acids long.

In contrast to vertebrates, *Drosophila* TnT has a negatively charged, Glu-rich, C-terminal extension that has been proposed to bind Ca^{2+} and influence the Ca^{2+} activation of the muscle.[42,43] This will suggest that the mechanism by which insect TnT influences Ca^{2+} activation differs from that of vertebrate TnT where the Glu-rich extension is very short.[44] Recent crystallographic and residue substitution data of vertebrate cardiac troponin indicate that the C-terminus of TnT interacts with the N-terminus of TnI / C-terminus of TnC interface resulting in a tripartite structure ($TnT^{Ct}/TnI^{Nt}/TnC^{Ct}$) that is critical for muscle Ca^{2+} activation.[45,46] How insect TnT, with its extended C-terminus, would fit into the troponin complex remains to be established. At least in *Drosophila* and bees, the role of TnT seems to be modulated by phosphorylation at an unknown residue.[42]

Tropomyosin

Tropomyosins exhibit several (6-7 in vertebrates) actin-binding domains, each composed of about 40 amino acids.[47] Two Tm monomers bind in opposite orientation to form the functional unit. Its two α helices arrange into a coiled-coil structure that is able to bind along the two grooves of the actin filament through clusters of Ala residues that produce local destabilization of the coiled-coil rigid structure, allowing the required flexibility to adapt to the filament grooves.[48] Vertebrate α-tropomyosin cooperates with the actin-binding protein tropomodulin 1 in the process of thin filament capping and stabilization.[49,50] The crystal structure of the chicken capping Z protein has been solved recently.[51] At the other end of the thin filament, where it binds to the Z disk lattice, tropomyosins are competed out by kettin[52] (see chapter by Bullard, Leake and Leonard).

While vertebrate tropomyosins are encoded by four independent genes (α, β, γ and δ) yielding more than 20 protein isoforms,[47] the *Drosophila* counterparts are encoded by two genes: *Tm1* and *Tm2*, located at chromosomal band positions 88E12 and 88E13 respectively with shared regulatory elements.[53-55] Tropomyosin 1 is a 518 amino acids long protein that has been referred to as Tropomyosin II in previous literature.[55,56] In addition to a regular muscle (mTm1) and a cytoskeletal (cTm1) protein isoforms, the gene *Tm1* encodes two IFM specific isoforms, TnH33 and TnH34, characterized by a Pro-Ala rich extension of about 200 amino acids at the C-terminus.[53] These extensions are produced by alternatively spliced exons 15-16 and include a repeated motif, APPAEGA, which has been proposed to form a coiled-coil structure with a similarly repeated motif, PAANGKA, found in exon 3 containing, IFM specific, TnI isoforms. It has been proposed also, that these Pro-Ala extensions, following phosphorylation, could form connections between thin and thick filaments and be involved in the phenomenon of stretch activation (Mateos et al, unpublished result) (see below). Tm1 is expressed also in the oocytes (cTm1), where it helps to localize the *oskar* mRNA, and in the nervous system.

Tropomyosin 2 is a 284 amino acids long protein that has been referred to as Tropomyosin I and Ifm(3)3 in former publications.[55] Genetic interactions have been reported between Tm2 and Mhc and TnI (see suppressor approach, below). Based also on genetic interaction studies, mutants in *Drosophila* tropomyosin 2 enhance phenotypes of *broad* mutant alleles, suggesting that Tm2 plays a role in cell shape changes during embryogenesis and metamorphosis through an ecdysone and rho-1GTPase mediated signaling cascades.[57] The null mutants are embryonic lethals. The human homologue, β-tropomyosin, maps to 9p11.3 and is associated with some forms of the distal arthrogryposis syndrome.[58]

Tm1 and Tm2 play distinct roles in muscle biology. While a single copy of the *Tm2* gene yields flightless adults with some defects in myofibrillogenesis, a single copy of *Tm1* does not.[55] In the first case, muscle power output is 32% of wild type and in the second is 73%. Also, flies nearly deprived of Tm2 yield 1% of power output only, though the core myofibrillar structure is rather normal. These data do not seem to support the key role suggested for Tm1 (TnH isoforms) unless some compensatory effects between both genes takes place, perhaps through the common regulatory elements. This speculation, however, remains to be tested.

Protein isoforms of unusually high mass (about 80 kD) (hence the H acronym) were first isolated from *Lethocerus* IFM[59] and later from *Drosophila*.[53] Their initial characterization was based on immunological criteria that only years later could be contrasted with sequencing data.[60] As a consequence, the different names have become somewhat misleading. Since *Lethocerus* has no regular 20 kD TnI and the 80 kD protein had an inhibitory role on the acto-myosin interaction and coimmunoprecipitated with the troponin complex, it was named heavy troponin (TnH). Subsequent sequencing of *Lethocerus* TnH revealed that it is indeed a TnI isoform with a long Pro-Ala rich extension (Bullard pers. comm.). The 80 kD protein from *Drosophila* IFM is recognized by the monoclonal antibody raised against the 80 kD *Lethocerus* protein. Its sequence, however, revealed that the fly heavy proteins (TnH33 and TnH34) are tropomyosin isoforms with long Pro-Ala rich extensions.[53] For some time, it was thought that *Drosophila* IFM had no TnI and that its role would have been played by this large Tm1. However, the discovery of a bona fide TnI gene in *Drosophila*, one of whose isoforms has a Pro-Ala rich extension, changed the scenario.[8] *Drosophila* TnH isoforms are phosphorylated at the Pro-Ala extension as part of the muscle maturation process that takes place posteclosion prior to flight (Mateos et al, unpublished result).

Regulation of Thin Filament Encoding Genes

Studies on the regulatory elements of genes encoding TnI, TnT and Tm2 in *Drosophila*, followed by *in silico* comparisons with the corresponding orthologs and homologs derived from Genome Sequence Programs, reveal a rather common trend. Two regulatory regions, URE and IRE, located up and down-stream of the promoter respectively, have been identified in these genes.[18,61] They include a particular array of transcription factor binding sites for MEF2, TINMAN, CF2 and BINIOU (Fig. 3). These regions are synergistic, rather than redundant, in producing the proper levels and ratios of protein isoforms required for normal muscle morphogenesis.[18,54,56,61] This conserved regulatory scenario[62] most likely arose as a consequence of the functional constraints imposed by thin filament mechanics. That is, the requirement of specific stoichiometry for myofibril assembly and efficient muscle contraction as shown for TnI in *Drosophila*[18] and vertebrates,[19] as well as for TnT.[63] Also, it is likely to be the mechanism responsible for the down-regulation of several thin-filament gene expression observed in flies mutant for TnI or TnT[64] as well as in zebra fish TnT mutants.[63] It is worth pointing out, however, that other genes that do not encode thin filament proteins, such as β-3-tubulin, but are expressed in somatic muscles and the dorsal vessel, also contain a similar regulatory array.[65,66] These studies are beginning to unravel a number of common trends in gene regulatory networks that, ultimately, could determine the exquisite coordination in the physiology of the corresponding proteins.[67-70]

GST-2

Glutathione S-transferase is a family of enzymes that play a general role in oxidative and detoxification processes.[71] In insects, GSTs can be classified in three types (I-III). While types

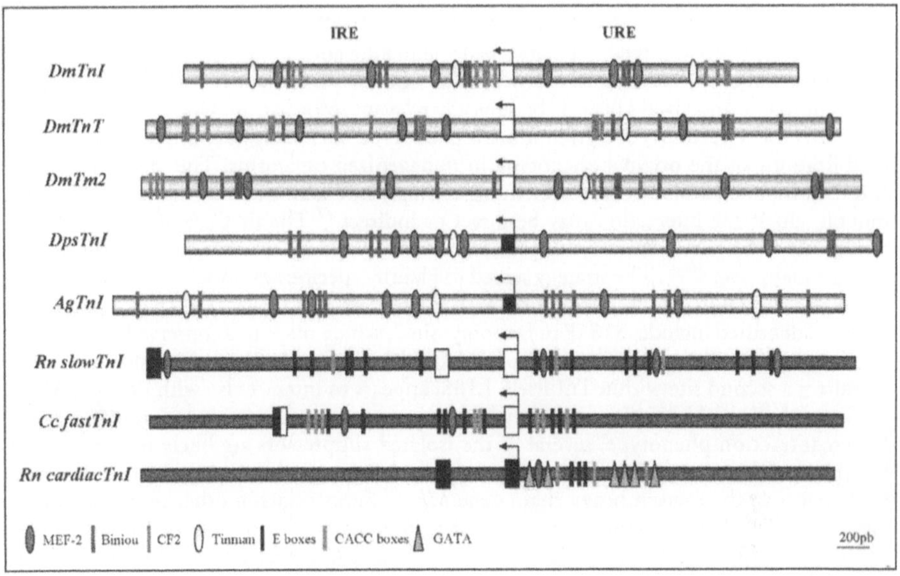

Figure 3. Comparison of regulatory regions in TnI, TnT and Tm2 encoding genes across species. Two regulatory regions URE and IRE are identified up- and down stream, respectively, of the transcription initiation site (arrow) in insects *D. melanogaster* (*Dm*), *D. pseudoobscura* (*Dps*), *Anopheles gambiae* (*Ag*) and vertebrates *Ratus norvegicus* (*Rn*) and *Coturnix coturnix* (*Cc*). Note the possible lost of IRE in the cardiac case. Known transcription binding sites are color coded.[18] A color version of this figure is available online at http://www.Eurekah.com.

I and III (a.k.a. δ and ε classes respectively) are involved in insecticide resistance, type II (a.k.a. σ class) remain poorly studied. Currently known as DmGstS1-1 (E.C.:2.5.1.18), *Drosophila* GST-2 is a type II GST enzyme that is abundant in the indirect flight muscles.[72]

Since GST-2 has been shown to conjugate 4-hydroxynonenal, an aldehyde product from lipid peroxidation,[73] it is thought to protect from oxidative stress during flight. Its crystal structure at 1.75Å resolution is known.[74] It forms dimers showing the canonical GST fold with glutathione bound to only one of the two binding sites. In addition, the alpha6 helix domain provides a novel electrophilic substrate-binding site (H-site) which is similar to that of other GSTs and may explain the selectivity for peroxidation products thanks to the directionality of its polar and hydrophobic residues.[74] The gene is also expressed in the central nervous system and wing disks, two structures without a significant role in oxidative detoxification, which suggests additional functional roles for GST-2. The recently isolated mutants are embryonic lethal.[75]

Proteome Data for *Drosophila* Thin Filament Proteins

Based on a systematic yeast-two-hybrid study, many *Drosophila* gene products have been found to interact in this semi-in vivo assay.[76] The full set of results can be found at the web address: http://portal.curagen.com/cgi-bin/interaction/flyGene.pl. Among the thin filament proteins described here, the interactions between the following pairs were reported to occur with high confidence: TnI with TnC47D, Androcan, and Gbeta76C; TnC41C with CG11456; TnC47D with TnI; Tm1 with CG5376, CG5338, CG15468 and CG4213. All others failed to show significant interactions in this assay. Given that most of these putative interactions correspond to still non described gene products (CG numbers), a substantial number of in vivo and in vitro experiments are still required to certify the biological relevance of these yeast two-hybrid interactions.

Searching for Protein-to-Protein Interactions in Vivo

To identify protein interacting partners, let alone the corresponding regions of interface, is a very long and costly task that rely mostly on in vitro assays. Genetic methods, by contrast, are more affordable and always provide biologically relevant information. One of these methods is the search for suppressors of a given mutant phenotype. Generally, it consists of the search for modifications of the original phenotype in mutagenized organisms. The identified proteins and their mutated amino acids represent functionally interacting sites with that of the original mutant, albeit the interaction may be direct or indirect.[77] The first use of this approach in *Drosophila* IFM was to identify suppressors of the "wings up" phenotype of a TnI missense mutant allele, $wupA^{hdp2}$. The strategy served to identify specific amino acids in other sarcomere proteins as functionally interacting sites with the mutated A116V of TnI (Fig. 4). Some suppressors identified include, S185F in tropomyosin 2, which maps to a conserved TnT interacting domain,[14] and sites near the entry of the ATP binding pocket in the myosin head.[78] Most revealing, a second site within TnI itself, L188, appears to interact also with position A116.[15]

Since the initial mutation against which suppressors were screened, $wupA^{hdp2}$, causes a hypercontraction phenotype, several of the isolated suppressors are likely to correspond to modifiers of the acto-myosin force generating mechanism. This has proven to be the case for suppressors in the myosin heavy chain gene *Mhc*.[79] Suppressors in other proteins, however, are more difficult to explain on the basis of acto-myosin force modifiers, in particular the second site suppressors within TnI itself. In any case, it is evident that the suppressor approach has revealed functional interactions that were unsuspected hereto. Future crystallographic studies will benefit from these data and eventually will provide a full view of the interaction mechanisms.

Stretch Activation

To sustain flight, small insects must beat their wings at high frequencies (about 200 Hz in *Drosophila*). Being small animals, however, implies also small neurons whose thin axons can not conduct action potentials at that frequency. As a result, flight in small animals would be impossible unless a neuron independent mechanism for muscle contraction would have appeared earlier in evolution. This is the case for stretch activation, a process whereby muscles enter in contraction/relaxation cycles when a small distortion in cell shape (>1% muscle length) occur. In vivo, this phenomenon is accompanied also by a small increase in free Ca^{2+} concentration, perhaps an indirect result from membrane depolarization due to the activity of stretch-activated K^+ channels that are insufficient to fully activate the muscle. Since contractions of the IFM are not coincident with motorneuron firing, these muscles are also referred to as asynchronous (see chapters by Josephson and Moore for a full description of stretch activation). It is still a matter of further debate whether stretch activation is a result of convergent evolution or if it originated from a single mechanism.[80-82]

The molecular bases of stretch activation are still poorly known. However, since IFM exhibit thin filament regulation, thin filament proteins with IFM specific isoforms are good candidates to sustain the molecular mechanism of this phenomenon. In this context, *Drosophila* and *Lethocerus* are reported to exhibit a one Ca^{2+}-binding site (site 4) TnC isoform predominantly expressed in IFM, isoform 41F (Table 1) (a.k.a. TnC4).[37] In addition to this major isoform, these muscles express also a minor TnC isoform with two Ca^{2+}-binding sites (sites 2 and 4). It has been suggested that initial muscle activation is mediated by the minor isoform while full activation depends on stretch and the major TnC isoform.[37] Indeed, based on TnC isoform substitution experiments in *Lethocerus* IFM, the TnC isoform with only one, high affinity, Ca^{2+}-binding site has been shown to elicit stretch activation, while the isoform with two Ca^{2+}-binding sites generates isometric tension mainly.[83] It should be noticed, however, that other proteins or protein isoforms are also specific to asynchronous muscles. This is the case of TnI and Tm1 isoforms with the Pro-Ala rich extensions at their N- and C-termini respectively. It has been speculated that these extensions will form a coiled-coil structure that

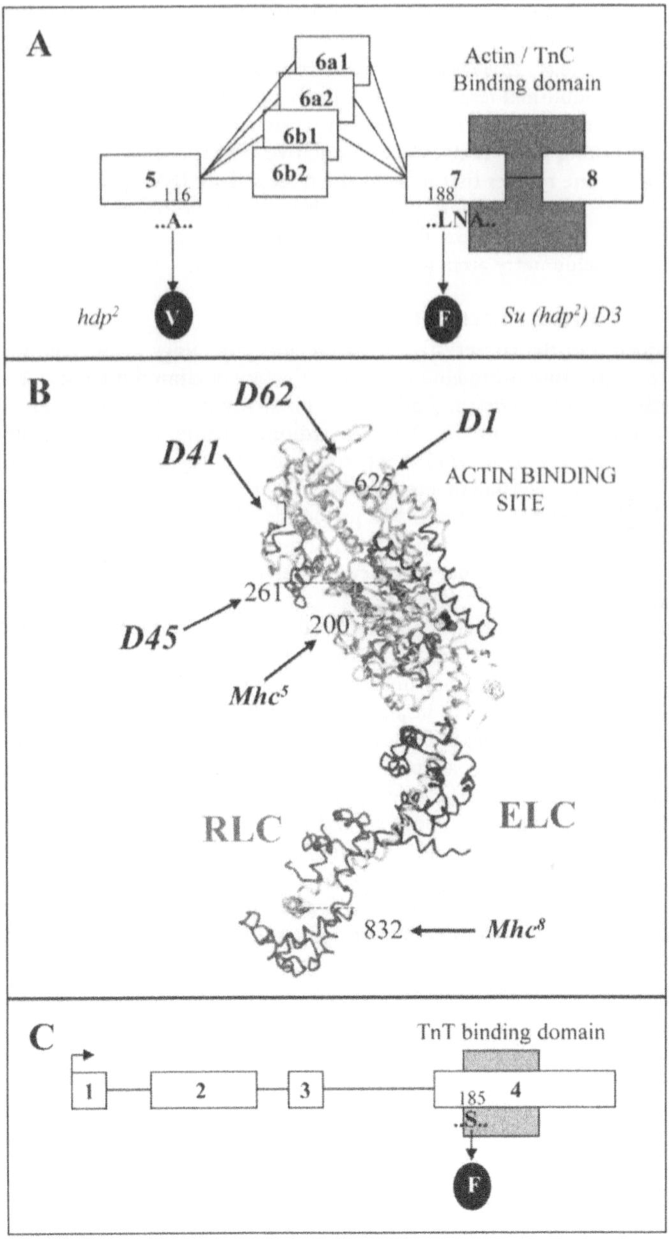

Figure 4. Functional interactions of *Drosophila* TnI detected as genetic suppressors of the *hdp*^{2} (A116V) mutant allele. A) Second site suppressor (L188F) located adjacent to the actin and TnC binding domain encoded in exons 7 and 8.[15] B) Suppressor mutations (D#) in the myosin heavy chain (MHC) protein placed on the crystal structure of chicken MHC.[78] Note that all D mutations are located at or near the actin binding site, or the ATP entering mouth, while mutations MhC^5 and MhC,^{8} located slightly further away, do not suppress. ELC = essential light chain, RLC = Regulatory light chain. C) Suppressor (S185F) in tropomyosin 2 located within the TnT binding domain. This mutation also suppresses[14] the TnT allele *up*^{101}.

would link with the thick filament (Mateos et al, unpublished result). How such a rigid structure will contribute to stretch activation is not clear. In addition, other flying insects with asynchronous IFM (e.g., *Anopheles*) do not encode a TnI isoform with a Pro-Ala extension (Herranz et al, unpublished result). It seems likely that the molecular bases for stretch activation will result from a peculiar multiprotein structure rather than from a single protein species. Moreover, contrasting the available data on putative IFM specific expression of thin filament protein isoforms, one realizes that none of them are strictly IFM specific (TDT expression is frequently detected as well). More properly, they should be referred as "very abundant" or "major isoforms". In this context, the phenomenon of stretch activation is likely to emerge from a critical stoichiometry among several muscle proteins.

Features of Clinical Interest

A consequence of the conservative structure and physiology across animal systems is the possibility to study protein-protein interactions that are of clinical interest in humans. In that context, findings in *Drosophila* and other experimentally tractable organisms can provide relevant information at very low cost. Various cardiomyopathies result from mutations in contractile proteins. One of them, familial hypertrophic cardiomyopathy (FHC) can be elicited by mutations in any of seven sarcomere proteins.[84-86] Those in α-tropomyosin tend to cluster around the TnT interacting domain.[87] As previously demonstrated in *Drosophila* through the suppressor approach (see above), the fly's counterpart, tropomyosin 2, may exhibit mutations in the TnT interacting region that cause severe muscle phenotypes but can be suppressed by a specific mutation inTnI.

Congenital abnormal limb twitching in distal arthrogryposis (DA) may result from mutations in the C terminus of fast TnI. As in the case of FHC, DA is genetically heterogeneous meaning that a mutant subject may not show the symptoms. Once more, the suppressor approach in *Drosophila* has served to inspire the search for an explanation of this genetic heterogeneity and an additional gene, β-Tropomyosin, was identified as alternative origin for DA.[58]

Under myocardial stunning, a type of reversible ischemia, a C-terminal fragment of TnI is characteristically cleaved serving as index of heart failure severity.[12,88] The remaining product, human cardiac TnI 1-193, still appears to perform the regular TnI functions, albeit with increased Ca^{2+} sensitivity and diminished sliding velocity during sarcomere contraction, as deduced from experiments using the mouse and rat homologues of the human fragment.[89] It is worth noting that changes in Ca^{2+} sensitivity is also a result of mutations in the α Tm interacting domain of TnC that often cause hypertrophic cardiomyopathies.[90] It is evident that finding the mechanisms that control Ca^{2+} sensitivity in *Drosophila* will provide clues and tools to direct research in vertebrates.

Open Questions and Future Prospects

Muscles are among the best known multicellular structures in biology. Yet, we are still short from being able to offer a mechanistic explanation of their activity that could have predictive value. More precise measurements of in situ protein concentrations, binding activity and force generation are needed. Technical limitations will end up determining the order in which the remaining questions are finally solved, some of which are listed below:

Crystal Structure for All Thin Filament Components

This is likely to be achieved in the immediate future. Even though TnI appears to be a globular protein, the described varieties among species, including the Pro-Ala rich extensions, indicate structural peculiarites that, in all likelihood, will be functionally relevant. For TnC, the available crystal structure from vertebrate fast skeletal is being used as standard reference. Nevertheless, the existence of mutated versions eliminating Ca^{2+}-binding activity or affinity implies a diversity of functional roles that must be sustained by specific structures.

High Resolution Structural Details of the Thin Filament

This is, perhaps, the most urgent type of information that is needed. In particular, the suspected structural heterogeneities along the thin filament will be most interesting to know. It is not clear to what extent their structure may be modified by current methods of EM sample preparation and, in any event, how faithful it is with respect to in vivo. Also, the thin filament structure at their anchorage to the Z discs remains obscure for the most part. These features seem critical to correctly interpret future data on force generation measurements.

Functional Data in Vivo

The physicochemical methods used so far to elucidate interactions between fragments of thin filament proteins have served to frame future crystallographic experiments. The final test, however, is always the physiology of the animal. In this context, molecular engineering of IFM in *Drosophila* is, by far, the most promising avenue of experimental inquiry. Muscles are, perhaps, the most suitable system where to assay the upcoming methods of proteomic studies (see chapter by Henkin and Vigoreaux). We are close to knowing the full repertoire of protein components of muscle and to having a yeast-two-hybrid based menu on the interactions that these components might exhibit. In addition to descriptive procedures, the study of muscle biology requires a substantial experimental effort on its normal and pathological physiology. Independently of whether these novel techniques are successful or not, it is clear that the classical rationale of linear interactions among proteins must yield to network methods if a comprehensive understanding of muscle function is ever to be attained.

References

1. Farah CS, Reinach FC. The troponin complex and regulation of muscle contraction. FASEB J 1995; 9:755-767.
2. Geeves MA, Lehrer S. The muscle thin filament as a classical cooperative/allosteric regulatory system. J Mol Biol 1998; 277:1081-1089.
3. Filatov VL, Katrukha AG, Bulargina TV et al. Troponin: Structure, properties, and mechanism of functioning. Biochemistry (Mosc.) 1999; 64:969-985.
4. Gordon AM, Homsher E, Regnier M. Regulation of contraction in striated muscle. Physiol Rev 2000; 80:853-924.
5. Kostyukova AS, Hitchcock-DeGregori SE. Effect of the structure of the N-terminus of tropomyosin on tropomodulin function. J Biol Chem 2004; 279:5066-5071.
6. Barton PJ, Cullen ME, Townsend PJ et al. Close physical linkage of human troponin genes: Organization, sequence, and expression of the locus encoding cardiac troponin I and slow skeletal troponin T. Genomics 1999; 57:102-109.
7. Kawasaki H, Nakayama S, Kretsinger RH. Classification and evolution of EF-hand proteins. Biometals 1998; 11:277-295.
8. Barbas JA, Galceran J, Krah-Jentgens I et al. Troponin I is encoded in the haplolethal region of the Shaker gene complex of Drosophila. Genes Dev 1991; 5:132-140.
9. Hastings KE. Molecular evolution of the vertebrate troponin I gene family. Cell Struct Funct 1997; 22:205-211.
10. Schiaffino S, Reggiani C. Molecular diversity of myofibrillar proteins: Gene regulation and functional significance. Physiol Rev 1996; 76:371-423.
11. Siedner S, Kruger M, Schroeter M et al. Developmental changes in contractility and sarcomeric proteins from the early embryonic to the adult stage in the mouse heart. J Physiol 2003; 548:493-505.
12. Marston SB, Redwood CS. Modulation of thin filament activation by breakdown or isoform switching of thin filament proteins: Physiological and pathological implications. Circ Res 2003; 93:1170-1178.
13. Beall CJ, Fyrberg E. Muscle abnormalities in Drosophila melanogaster heldup mutants are caused by missing or aberrant troponin-I isoforms. J Cell Biol 1991; 114:941-951.
14. Naimi B, Harrison A, Cummins M et al. A tropomyosin-2 mutation suppresses a troponin I myopathy in Drosophila. Mol Biol Cell 2001; 12:1529-1539.
15. Prado A, Canal I, Barbas JA et al. Functional recovery of troponin I in a Drosophila heldup mutant after a second site mutation. Mol Biol Cell 1995; 6:1433-1441.

16. Vassylyev DG, Takeda S, Wakatsuki S et al. Crystal structure of troponin C in complex with troponin I fragment at 2.3-Å resolution. Proc Natl Acad Sci USA 1998; 95:4847-4852.
17. Barbas JA, Galceran J, Torroja L et al. Abnormal muscle development in the heldup[3] mutant of Drosophila melanogaster is caused by a splicing defect affecting selected troponin I isoforms. Mol Cell Biol 1993; 13:1433-1439.
18. Marín MC, Rodríguez JR, Ferrús A. Transcription of Drosophila Troponin I gene is regulated by two conserved, functionally identical, synergistic elements. Mol Biol Cell 2004; 15:1185-1196.
19. Metzger JM, Michele DE, Rus EM et al. Sarcomere thin filament regulatory isoforms: Evidence of a dominant effect of slow skeletal troponin I on cardiac contraction. J Biol Chem 2003; 278:13118-13123.
20. Perry SV. Troponin I: Inhibitor or facilitator. Mol Cell Biochem 1999; 190:9-32.
21. Patchell VB, Gallon CE, Hodgkin MA et al. The inhibitory region of troponin-I alters the ability of F-actin to interact with different segments of myosin. Eur J Biochem 2002; 269:5088-5100.
22. Keane NE, Quirk PG, Gao Y et al. The ordered phosphorylation of cardiac troponin I by the cAMP-dependent protein kinase: Structural consequences and functional implications. Eur J Biochem 1997; 248:329-337.
23. Li Q, Shen PY, Wu G et al. Polycystin-2 interacts with troponin I, an angiogenesis inhibitor. Biochemistry 2003; 42:450-457.
24. Prado A, Canal I, Ferrús A. The haplolethal region at the 16F gene cluster of Drosophila melanogaster: Structure and function. Genetics 1999; 151:163-175.
25. Yuasa HJ, Takagi T. The genomic structure of the scallop, Patinopecten yessoensis, troponin C gene: A hypothesis for the evolution of troponin C. Gene 2000; 245:275-281.
26. Garone L, Theibert JL, Miegel A et al. Lobster troponin C: Amino acid sequences of three isoforms. Arch Biochem Biophys 1991; 291:89-91.
27. Allhouse LD, Guzman G, Miller T et al. Characterisation of a mutant of barnacle troponin C lacking Ca^{2+}-binding sites at positions II and IV. Pflugers Arch 1999; 438:30-39.
28. Allhouse LD, Li Q, Guzman G et al. Investigating the role of Ca^{2+}-binding site IV in barnacle troponin C. Pflugers Arch 2000; 439:600-609.
29. Collins JH, Theibert JL, Francois JM et al. Amino acid sequences and Ca^{2+}-binding properties of two isoforms of barnacle troponin C. Biochemistry 1991; 30:702-707.
30. Kobayashi T, Takagi T, Konishi K et al. Amino acid sequences of the two major isoforms of troponin C from crayfish. J Biol Chem 1989; 264:18247-18259.
31. Kobayashi T, Kagami O, Takagi T et al. Amino acid sequence of horseshoe crab, Tachypleus tridentatus, striated muscle troponin C. J Biochem (Tokyo) 1989; 105:823-828.
32. Ojima T, Koizumi N, Ueyama K et al. Functional role of Ca^{2+}-binding site IV of scallop troponin C. J Biochem (Tokyo) 2000; 128:803-809.
33. Ojima T, Ohta T, Nishita K. Amino acid sequence of squid troponin C. Comp Biochem Physiol B Biochem Mol Biol 2001; 129:787-796.
34. Grabarek Z, Tao T, Gergely J. Molecular mechanism of troponin-C function. J Muscle Res Cell Motil 1992; 13:383-393.
35. Herranz R, Díaz-Castillo C, Nguyen T et al. Expression patterns of the whole Troponin C gene repertoire during Drosophila development. Mechanisms of Dev 2004; 168:183-190.
36. Fyrberg C, Parker H, Hutchison B et al. Drosophila melanogaster genes encoding three troponin-C isoforms and a calmodulin-related protein. Biochem Genet 1994; 32:119-135.
37. Qiu F, Lakey A, Agianian B et al. Troponin C in different insect muscle types: Identification of an isoform in Lethocerus, Drosophila and Anopheles that is specific to asynchronous flight muscle in the adult insect. Biochem J 2003; 371:811-821.
38. Terami H, Williams BD, Kitamura S et al. Genomic organization, expression, and analysis of the troponin C gene pat-10 of Caenorhabditis elegans. J Cell Biol 1999; 146:193-202.
39. Ueda T, Katsuzaki H, Terami H et al. Calcium-bindings of wild type and mutant troponin Cs of Caenorhabditis elegans. Biochim Biophys Acta 2001; 1548:220-228.
40. Perry SV. Troponin T. Genetics, properties and function. J Muscle Res Cell Motil 1998; 19:575-602.
41. Fyrberg E, Fyrberg CC, Beall C et al. Drosophila melanogaster troponin-T mutations engender three distinct syndromes of myofibrillar abnormalities. J Mol Biol 1990; 216:657-675.
42. Domingo A, Gonzalez-Jurado J, Maroto M et al. Troponin-T is a calcium-binding protein in insect muscle: In vivo phosphorylation, muscle-specific isoforms and developmental profile in Drosophila melanogaster. J Muscle Res Cell Motil 1998; 19:393-403.
43. Benoist P, Mas JA, Marco R et al. Differential muscle-type expression of the Drosophila troponin T gene. A 3-base pair microexon is involved in visceral and adult hypodermic muscle specification. J Biol Chem 1998; 273:7538-7546.

44. Schaertl S, Lehrer SS, Geeves MA. Separation and characterization of the two functional regions of troponin involved in muscle thin filament regulation. Biochemistry 1995; 34:15890-15894.
45. Takeda S, Yamashita A, Maeda K et al. Structure of the core domain of human cardiac troponin in the Ca(2+)-saturated form. Nature 2003; 424:35-41.
46. Burkart EM, Arteaga GM, Sumandea MP et al. Altered signaling surrounding the C-lobe of cardiac troponin C in myofilaments containing an alpha-tropomyosin mutation linked to familial hypertrophic cardiomyopathy. J Mol Cell Cardiol 2003; 35:1285-1293.
47. Perry SV. Vertebrate tropomyosin: Distribution, properties and function. J Muscle Res Cell Motil 2001; 22:5-49.
48. Singh A, Hitchcock-DeGregori SE. Local destabilization of the tropomyosin coiled coil gives the molecular flexibility required for actin binding. Biochemistry 2003; 42:14114-14121.
49. Mudry RE, Perry CN, Richards M et al. The interaction of tropomodulin with tropomyosin stabilizes thin filaments in cardiac myocytes. J Cell Biol 2003; 162:1057-1068.
50 Mardahl-Dumesnil M, Fowler VM. Thin filaments elongate from their pointed ends during myofibril assembly in Drosophila indirect flight muscle. J Cell Biol 2001; 155:1043-1053.
51. Yamashita A, Maeda K, Maeda Y. Crystal structure of CapZ: Structural basis for actin filament barbed end capping. EMBO J 2003; 22:1529-1538.
52. van Straaten M, Goulding D, Kolmerer B et al. Association of kettin with actin in the Z-disc of insect flight muscle. J Mol Biol 1999; 285:1549-1562.
53. Karlik CC, Fyrberg EA. Two Drosophila melanogaster tropomyosin genes: Structural and functional aspects. Mol Cell Biol 1986; 6:1965-1973.
54. Gremke L, Lord PC, Sabacan L et al. Coordinate regulation of Drosophila tropomyosin gene expression is controlled by multiple muscle-type-specific positive and negative enhancer elements. Dev Biol 1993; 159:513-527.
55. Kreuz AJ, Simcox A, Maughan D. Alterations in flight muscle ultrastructure and function in Drosophila tropomyosin mutants. J Cell Biol 1996; 135:673-687.
56. Tansey T, Schultz JR, Miller RC et al. Small differences in Drosophila tropomyosin expression have significant effects on muscle function. Mol Cell Biol 1991; 11:6337-6342.
57. Ward RE, Evans J, Thummel CS. Genetic modifier screens in Drosophila demonstrate a role for rho1 signaling in ecdysone-triggered imaginal disc morphogenesis. Genetics 2003; 165:1397-1415.
58. Sung SS, Brassington AM, Grannatt K et al. Mutations in genes encoding fast-twitch contractile proteins cause distal arthrogryposis syndromes. Am J Hum Genet 2003; 72:681-690.
59. Bullard B, Leonard K, Larkins A et al. Troponin of asynchronous flight muscle. J Mol Biol 1988; 204:621-637.
60. Lakey A, Ferguson C, Labeit S et al. Identification and localization of high molecular weight proteins in insect flight and leg muscle. EMBO J 1990; 9:3459-3467.
61. Mas JA, Garcia-Zaragoza E, Cervera M. Two functionally identical molecular enhancers in Drosophila Troponin T gene establish the correct protein levels in different muscle types. Mol Biol Cell 2004; 15:1931-1945.
62. Banerjee-Basu S, Buonanno A. Cis-acting sequences of the rat troponin I slow gene confer tissue- and development-specific transcription in cultured muscle cells as well as fiber type specificity in transgenic mice. Mol Cell Biol 1993; 13:7019-7028.
63. Sehnert AJ, Huq A, Weinstein BM et al. Cardiac troponin T is essential in sarcomere assembly and cardiac contractility. Nat Genet 2002; 31:106-110.
64. Nongthomba U, Clark S, Cummins M et al. Troponin I is required for myofibrillogenesis and sarcomere formation in Drosophila flight muscle. J Cell Sci 2004; 117:1795-1805.
65. Damm C, Wolk A, Buttgereit D et al. Independent regulatory elements in the upstream region of the Drosophila beta 3 tubulin gene (beta Tub60D) guide expression in the dorsal vessel and the somatic muscles. Dev Biol 1998; 199:138-149.
66. Kremser T, Hasenpusch-Theil K, Wagner E et al. Expression of the beta3 tubulin gene (beta Tub60D) in the visceral mesoderm of Drosophila is dependent on a complex enhancer that binds Tinman and UBX. Mol Gen Genet 1999; 262:643-658.
67. Baylies MK, Bate M, Ruiz-Gomez M. Myogenesis: A view from Drosophila. Cell 1998; 93:921-927.
68. Maytum R, Lehrer SS, Geeves MA. Cooperativity and switching within the three-state model of muscle regulation. Biochemistry 1999; 38:1102-1110.
69. Cripps RM, Olson EN. Control of cardiac development by an evolutionarily conserved transcriptional network. Dev Biol 2002; 246:14-28.
70. Davidson EH, McClay DR, Hood L. Regulatory gene networks and the properties of the developmental process. Proc Natl Acad Sci USA 2003; 100:1475-1480.

71. Sawicki R, Singh SP, Mondal AK et al. Cloning, expression and biochemical characterization of one Epsilon-class (GST-3) and ten Delta-class (GST-1) glutathione S-transferases from Drosophila melanogaster, and identification of additional nine members of the Epsilon class. Biochem J 2003; 370:661-669.
72. Clayton JD, Cripps RM, Sparrow JC et al. Interaction of troponin-H and glutathione S-transferase-2 in the indirect flight muscles of Drosophila melanogaster. J Muscle Res Cell Motil 1998; 19:117-27.
73. Singh SP, Coronella JA, Benes H et al. Catalytic function of Drosophila melanogaster glutathione S-transferase DmGSTS1-1 (GST-2) in conjugation of lipid peroxidation end products. Eur J Biochem 2001; 268:2912-2923.
74. Agianian B, Tucker PA, Schouten A et al. Structure of a Drosophila sigma class glutathione S-transferase reveals a novel active site topography suited for lipid peroxidation products. J Mol Biol 2003; 326:151-165.
75. Feng YP, Mondal A, Robinson M et al. Isolation and characterization of mutants in the Drosophila Glutathione S-Transferase S1 (GST S1) gene. A Dros Res Conf 2003; 44:500B.
76. Giot L, Bader JS, Brouwer C et al. A protein interaction map of Drosophila melanogaster. Science 2003; 302:1727-1736.
77. Ferrús A, Acebes A, Marín MC et al. A genetic approach to detect muscle protein interactions in vivo. Trends in Cardiovasc Med 2001; 10:293-298.
78. Kronert WA, Acebes A, Ferrús A et al. Specific myosin heavy chain mutations suppress troponin I defects in Drosophila muscles. J Cell Biol 1999; 144:989-1000.
79. Nongthomba U, Cummins M, Clark S et al. Suppression of muscle hypercontraction by mutations in the myosin heavy chain gene of Drosophila melanogaster. Genetics 2003; 164:209-222.
80. Pringle JW. The evolution of fibrillar muscle in insects. J Exp Biol 1981; 94:1-14.
81. Dudley R. The biomechanics of insect flight: Form, Function, Evolution. Princeton University Press, 2000.
82. Josephson RK, Malamud JG, Stokes DR. Asynchronous muscle: A primer. J Exp Biol 2000; 203:2713-2722.
83. Agianian B, Krzic U, Feng Q et al. A troponin switch that regulates muscle contraction by stretch instead of calcium. EMBO J 2004; 23:772-779.
84. Carrier L, Hengstenberg C, Beckman JS et al. Mapping of a novel gene for a familial hypertrophic cardiomyopathy to chromosome 11. Nature Genet 1993; 4:311-313.
85. Coonar AS, McKenna WJ. Molecular genetics of familial cardiomyopathies. Adv Genet 1997; 35:285-324.
86. Erdmann J, Daehmlow S, Wischke S et al. Mutation spectrum in a large cohort of unrelated consecutive patients with hypertrophic cardiomyopathy. Clin Genet 2003; 64:339-349.
87. Prabhakar R, Petrashevskaya N, Schwartz A et al. A mouse model of familial hypertrophic cardiomyopathy caused by a alpha-tropomyosin mutation. Mol Cell Biochem 2003; 251:33-42.
88. Lewinter MM, Vanburen P. Myofilament remodeling during the progression of heart failure. J Card Fail 2002; 8:271-275.
89. Foster DB, Noguchi T, VanBuren P et al. C-terminal truncation of cardiac troponin I causes divergent effects on ATPase and force: Implications for the pathophysiology of myocardial stunning. Circ Res 2003; 93:917-924.
90. Lu QW, Morimoto S, Harada K et al. Cardiac troponin T mutation R141W found in dilated cardiomyopathy stabilizes the troponin T-tropomyosin interaction and causes a Ca(2+) desensitization. J Mol Cell Cardiol 2003; 35:1421-1427.
91. Nakayama S, Kretsinger RH. Evolution of EF-hand calcium-modulated proteins. III. Exon sequences confirm most dendrograms based on protein sequences: Calmodulin dendrograms show significant lack of parallelism. J Mol Evol 1993; 36:458-476.
92. Tobacman LS, Nihli M, Butters C et al. The troponin tail domain promotes a conformational state of the thin filament that suppresses myosin activity. J Biol Chem 2002; 277:27636-27642.
93. Reinach FC, Farah CS, Monteiro PB et al. Structural interactions responsible for the assembly of the troponin complex on the muscle thin filament. Cell Struct Funct 1997; 22:219-223.
94. Wendt T, Leonard K. Structure of the insect troponin complex. J Mol Biol 1999; 285:1845-1856.

Chapter 11

The Thin Filament in Insect Flight Muscle

Kevin R. Leonard and Belinda Bullard

Abstract

In this chapter we describe the special properties of insect muscle thin filament proteins and the way in which they differ from those in vertebrates. As in the vertebrate, the repeating unit of the muscle fibre (sarcomere) contains interdigitated thick (myosin containing) and thin (actin containing) filaments which generate the contractile force. The backbone of the insect muscle thin filament is provided by the helical F-actin polymer. Other proteins along the thin filament are modified versions of proteins present in the vertebrate thin filament. These include arthrin (ubiquitinated actin) and a heavy troponin subunit (TnH). The latter differs in the two insects studied, *Drosophila* and *Lethocerus*, and is absent in the vertebrate. The main functional difference in the insect thin insect filaments is between those in indirect flight muscles (IFM) and all other muscles. The IFM can be regulated at much higher frequencies by the process known as "stretch activation". The mechanism of stretch activation is still not completely understood but it now appears that troponin-C, a normal regulatory component of the thin filaments, is involved. A small amount of high resolution information is now available for the vertebrate troponin complex and we have tried to incorporate this into what other structural data is available for the insect thin filament. The partial X-ray structure for the vertebrate troponin complex would fit within the density envelope found for insect troponin but, as expected, does not account for all the density.

Introduction

Insect muscle thin filaments are similar to those found in vertebrate muscle, but there are some important differences. Perhaps the greatest functional difference between the two types, which has been studied most thoroughly, is found in the indirect flight muscles (IFM) of some insects. A special mechanism for control of muscle contraction has evolved to enable these insects to contract their muscles at the high frequency (up to 1000 Hz in small insects) required for flight. In vertebrate striated muscle and many insect flight muscles, contraction occurs as a result of Ca^{2+} binding to the troponin/tropomyosin complex and relaxation follows dissociation of Ca^{2+} from the complex. At low Ca^{2+} concentration, tropomyosin (Tm) is in a position that blocks the binding of myosin crossbridges to actin; at higher Ca^{2+} concentration, the tropomyosin moves, allowing the crossbridges to attach. In these muscles, contractions are synchronous with nerve impulses. In the special case of IFM, contraction is produced by rapid stretch at constant Ca^{2+} concentrations and is coupled to the natural resonance of the thorax, so that once the contraction cycle begins, opposing groups of muscles are reciprocally activated independently of nerve impulses (asynchronous contraction).[1] For a more detailed description see the chapters by Josephson and by Moore in this volume. We now know that a major factor leading to this phenomenon lies in a modification of a component (troponin-C, TnC) of the regulatory complex in the thin filament.[2]

Nature's Versatile Engine: Insect Flight Muscle Inside and Out, edited by Jim Vigoreaux.

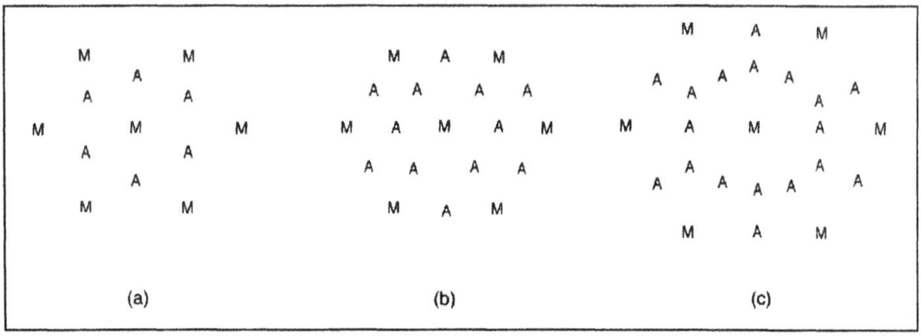

Figure 1. Schematic cross sections showing distribution of thick (M) and thin (A) filaments in (a) vertebrate, (b) insect flight and (c) insect leg muscle. Adapted from Squire J. The Structural Basis of Muscle Contraction. New York: Plenum Press; 1981.

The work on the structure and function of the IFM thin filament, which is reviewed here, began using muscles from the giant water bug, *Lethocerus*. More recently the fruit fly, *Drosophila* has increasingly been used. The availability of the genome sequence and extensive mutant studies now makes *Drosophila* the organism of choice; although *Lethocerus*, because of its greater size is still useful for physiological, biochemical and structural studies of insect muscle.

Most biochemical and structural studies have been carried out on thin filament preparations isolated from fresh or glycerinated muscles. We have found that there is an advantage in using *Drosophila* IFM myosin null mutants, for example Mhc^7, which have no thick filaments in the flight muscle, to avoid the problem of thin filament contamination by thick filaments.[3]

Muscle Lattice Parameters

The arrangement of thick and thin filaments in the lattices of vertebrate skeletal muscle IFM is different (Fig. 1); in vertebrate muscle there are on average 2 thin filaments per thick filament, whereas in IFM, there is a ratio of 3 thin filaments to each thick filament. In insect non-IFM thoracic and leg muscle, the ratio of thin to thick filaments can be even higher (Fig. 1c). The length of thin filaments also varies in different insect muscle types (Fig. 2). In flight muscle, the sarcomere length is about 2.5 µm in *Lethocerus* and 3.2 µm in *Drosophila*. The thin filaments are about 1.1 µm and 1.4 µm long, respectively. In some insect leg muscles, the sarcomeres can be as long as 8 µm and the thin filaments more than 3 µm long.

Thin Filament Proteins

A brief description of the major thin filament associated proteins is given here. See also other chapters in this volume for additional details.

Actin

The principal protein of the thin filament is actin with a monomer molecular weight (MW) of 42 kDa, which polymerises as helical F-actin. F-actin is a flexible polymer and there is always some local variation in the helical twist.[4] The average helical parameters however, can be measured by X-ray fibre diffraction or electron microscopy (EM). For vertebrate thin filaments, the axial helical repeat is close to 13 actin monomers in 6 turns of the basic helix giving a repeat distance of 36 nm. The basic or "genetic" helix is the shallow one-start actin helix. For insect thin filaments in the intact *Lethocerus* IFM lattice, the average axial repeat is 28 monomers in 13 turns, corresponding to an approximate repeat of 38.7 nm.[5] Isolated *Drosophila* thin filaments also show this repeat.[3] The difference in repeat is small but it has the consequence that in the insect thin filament, the axial repeat of 38.7 nm is the same as the repeat of the troponin/tropomyosin complex (see below) which is arranged periodically along

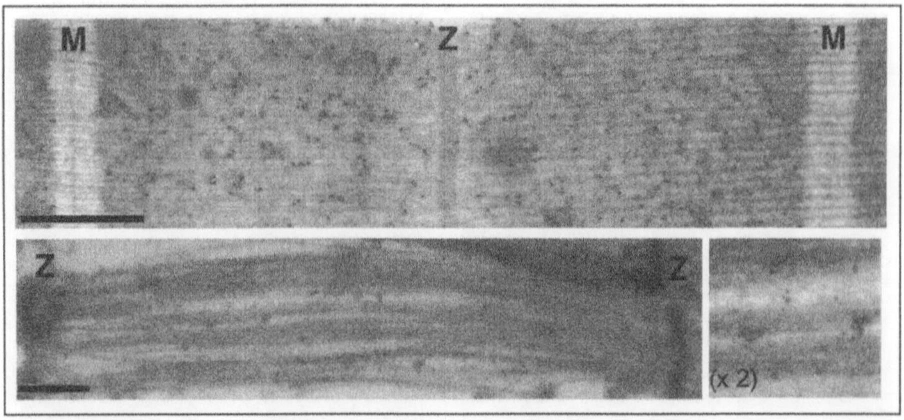

Figure 2. Frozen sections of *Drosophila* muscle sarcomeres. The flight muscle (upper panel) is highly regular and has clear Z and M lines. The section has been labelled with troponin-H antibody and 10 nm protein-A gold. The leg muscle (lower left panel) is much less regular and the sarcomere is almost twice as long as that in the flight muscle. In this case it has been labelled with troponin-C antibody and 10 nm protein-A gold. The lower right panel is part of this image 2x magnified to show the gold particles more clearly. The scale bars are 0.5 μm.

the filament. This matching of the troponin repeat to the thin filament helical repeat results in all troponin complexes along a thin filament having the same azimuthal orientation, which may be important for stretch activation.

Arthrin

In some insect flight muscles, in addition to actin there is a post-translationally modified form of actin, known as arthrin, which has one ubiquitin adduct per molecule, giving a molecular weight of approximately 50 kDa.[6,7] The position of ubiquitin on actin has been determined by mass spectrometric analysis of *Drosophila* and *Lethocerus* arthrin peptides. It is found to be covalently linked to lysine 118.[8,9] This is in agreement with 3D reconstructions of F-arthrin filaments from *Lethocerus* [10] and *Drosophila*.[8] F-arthrin filaments can be decorated with the myosin head subfragment (S1),[6] indicating that the ubiquitin adduct does not block the myosin binding site, which is also consistent with its position on the filament. Electron microscope images of *Drosophila* thin filaments treated with anti-ubiquitin antibodies (Fig. 3e) show cross-links arranged regularly at about 38nm intervals. The ratio of actin to arthrin in the thin filament is about 6:1, which corresponds to one arthrin molecule per troponin/tropomyosin repeat.[6] These data might that there may be an interaction between troponin/tropomyosin and arthrin. Synthetic filaments composed of F-arthrin alone bind tropomyosin/troponin and the interaction with myosin is regulated by Ca^{2+} in the normal way (see below).

Tropomyosin

As in the vertebrate thin filament, IFM tropomyosin dimers (MW 70 kDa) are linked end-to-end to form a continuous α-helical coiled-coil that follows the steeper (long-pitch) helix in F-actin.[3] There are two tropomyosin strands, one on each side of the actin filament, which control muscle contraction by moving between two positions. In the "on" position the myosin heads can bind to actin in the thin filament, in the "off" position, the myosin heads are sterically blocked and unable to bind to actin. It is possible that arthrin (see above) could influence the binding of myosin heads in the "on position" of the thin filament. Although the position of the ubiquitin found in the 3D reconstructions is away from tropomyosin, modelling studies [8] indicate that there is enough flexibility in the ubiquitin linkage for it to make

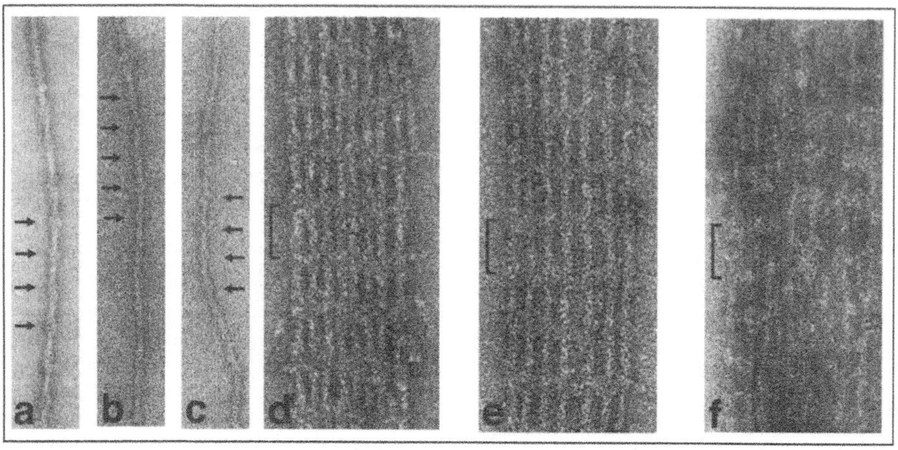

Figure 3. Electron micrographs of negatively stained thin filament preparations. a) *Lethocerus* IFM thin filament. b,c) *Drosophila* IFM thin filaments. The arrows point to troponin protruberances which are separated by approx. 38nm. d) Raft of *Drosophila* IFM thin filaments. The filaments contact through the troponin protruberances. The filaments are separated laterally by about 8 nm. e) Raft of *Drosophila* filaments crosslinked by anti-ubiquitin Igs. The appearance is similar to (d) but the inter-filament spacing increases to about 15 nm. Again the cross-links are at regular intervals of 38 nm. f) *Drosophila* thin filaments crosslinked by antibody (IgG) to the PA extension of TnH (see below). Again there is a 38 nm periodicity along the filaments but the lateral separation between filaments is more variable. This may be due to flexibility of the PA extension which is exposed on the surface of the thin filament.[11] The vertical scale bar in (d), (e), and (f) is 38 nm.

contact with tropomyosin running along the other long pitch actin helix when Tm is in the "on" position.

Troponin

Each tropomyosin dimer has a regulatory troponin complex bound to it, and in insect flight muscle, the complex is attached at intervals of 38.7 nm along the thin filament. The vertebrate skeletal troponin complex is made up of 3 components

1. Troponin-C (TnC) binds Ca^{2+}, normally the first step in Ca^{2+} activation of the muscle.
2. Troponin-I (TnI) inhibits the interaction of myosin and actin by holding Tm in the blocking position.
3. Troponin-T (TnT) attaches the complex to tropomyosin.

In insect flight muscle there are significant differences in these proteins compared with those found vertebrate skeletal muscle. The sizes of the insect troponin components and the corresponding vertebrate subunits are listed in Table 1.

In *Lethocerus*, there is a heavy TnI (referred to as TnH) which is TnI fused to a long PA-rich (proline, alanine) sequence.[11] In *Drosophila*, there is a heavy component (TnH) which is a fusion of tropomyosin and a PA-rich sequence; approximately equal amounts of two closely related isoforms (TnH-33 and TnH-34) are present.[12] A survey of flight muscles from a number of different insect species[13] showed that all contained proteins with a MW of about 50-70 kDa that reacted with an antibody to the TnH PA sequence; therefore all the flight muscles may have proteins with similar PA extensions.

An additional component of the troponin complex in *Drosophila* is a glutathione-S transferase type 2 (GST-2).[14] This binds to the PA-rich sequence of TnH and can be removed by treating the thin filaments with the protease Igase, which selectively cleaves the PA extension. The GST-2 (dimer MW 54 kDa) is thought to play an anti-oxidant role in the flight muscle.[15]

Table 1. Molecular weights of troponin subunits

	TnC	TnI	TnT	TnH
Drosophila	18kDa	30kDa (flight) 24kDa (leg)	47kDa	55kDa[1]
Lethocerus	18kDa	24kDa (leg)	45kDa	52kDa[2]
Vertebrate	18kDa	21kDa	33kDa	

1—*Drosophila* TnH is a fusion of tropomyosin and a PA rich sequence. 2—*Lethocerus* TnH is a fusion of TnI and a similar PA sequence. The sequence of *Lethocerus* is deposited in the EMBL DNA database under accession no, AJ621044 and that of Lethocerus TnI2 under no. AJ621045. Molecular weights of insect troponins are for IFM except where indicated for leg muscle TnI.

It is clear that the insect muscle troponins can be considerably larger than the vertebrate equivalents (Table 1). It is possible to visualise the troponin complex by EM as a "bump" on the outside of the filament every 38 nm (Fig. 3a-c). Insect thin filaments also tend spontaneously to form "rafts" where they are aligned in register and contact through the troponin sites (Fig. 3d). The rafts are flat sheets and should not be confused with the three-dimensional paracrystalline bundles formed by actin at high Mg^{2+} concentrations. Thin filaments can also be cross-linked by antibodies to arthrin and troponin (Fig. 3e-f).

In the case of rabbit troponin, it has been shown by binding a fragment of TnT to tropomyosin crystals, that the most likely position of the troponin complex is about 16nm from the C-terminal end of the tropomyosin rod (Fig. 4).[16] For *Lethocerus*, it has been shown, by electron microscopy of isolated tropomyosin/troponin complexes and Ca^{2+}-induced tropomyosin/troponin paracrystals,[17] that the globular troponin appears to be at one end of the 38 nm

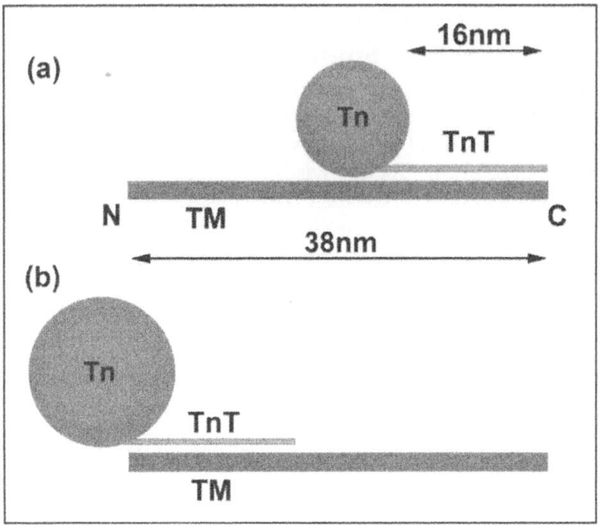

Figure 4. Schematic cartoons of troponin (Tn) binding to tropomyosin (TM) in (a) vertebrate and (b) insect flight muscle thin filaments. The size ratio of the green troponin "spheres" is proportional to the ratio of the molecular weights of the complexes. In (b) the orientation N-C of the tropomyosin is not known. A color version of this figure is available online at http://www.Eurekah.com.

tropomyosin rod. This positional difference may not affect the function of troponin, since the tropomyosin associates end-to-end to form an essentially continuous coiled-coil along the actin filament. However, the position of troponin in *Lethocerus* IFM would mean there could be an interaction between the regulatory region of troponin and the overlap of tropomyosin molecules. The C-terminal end of vertebrate skeletal tropomyosin is composed of parallel rather than coiled-coil α-helices.[18] It is possible that the action of stretch on troponin could be transmitted to tropomyosin via this region if it is also present in the insect.

Troponin-C Isoforms

It has recently been shown[19] that two different isoforms of TnC are present in IFM thin filaments. The major isoform has only one high affinity Ca^{2+} binding site and the bound Ca^{2+} would not be exchangeable during oscillatory contraction. This isoform appears to account for the ability of these muscles to undergo stretch activation. The other minor isoform has two Ca^{2+} binding sites, one of which is exchangeable under physiological conditions and can act as a Ca^{2+} switch to activate non-oscillatory contraction at higher Ca^{2+} concentrations. The different functions of the two isoforms have been demonstrated by measuring the mechanics of *Lethocerus* muscle fibres after replacing the endogenous TnC by the individual recombinant TnC isoforms.[2]

Modular Proteins

Titin (MW 3 MDa), which stretches across the whole length of the half-sarcomere in vertebrate muscle and which interacts with thin filaments in the Z-disk and in a short region of the I-band close to the Z-disk, does not appear to have a homologue in insect muscle. In IFM there are two modular proteins, projectin and kettin with some similarities to titin (for more information on projectin and kettin, see the section on Connecting Filaments in this volume). Kettin (a 540 kDa isoform from the *Drosophila sls* gene) binds to thin filaments in the Z-disk and I-band regions[20] and also links them to the thick filaments.[21] Nebulin, an 800 kDa filamentous protein that binds along the whole length of vertebrate skeletal muscle thin filaments,[22] is not found in insect muscle.

High Resolution Structural Studies

Complete atomic resolution structures of thin filament proteins are known only for vertebrate actin[23] and TnC.[24] An atomic resolution structure has been determined for the *Drosophila* GST-2 dimer which is found associated with TnH in IFM.[15]

Although no high resolution structure has been solved for the actin helix in F-actin, the actin monomer has been modelled into the F-actin helix using X-ray fiber diffraction data.[25] Some conformational modification of the G-actin structure was necessary in order to model the X-ray fiber patterns. This structure is based on the 13:6 geometry of vertebrate F-actin. However, it probably represents equally well the insect F-actin structure.

Data for the structure of vertebrate tropomyosin is limited to X-ray and EM studies, which confirm that the structure is a coiled–coil dimer. Early work showed that the two helices are parallel and in register.[26,27] More recently, a 7Å resolution structure of tropomyosin has been obtained by X-ray diffraction of spermine-induced crystals.[28] Tropomyosin can also be seen in EM 3D helical reconstructions of actin-tropomyosin as a continuous thread of density running along the outside of the filament, following the long-pitch actin helices.[29]

In the case of TnC, the original X-ray crystal structure showed the vertebrate protein to be "dumb-bell" shaped with globular N and C terminal Ca^{2+} binding domains separated by a straight α-helical rod.[24] However, subsequent structural determinations, particularly those carried out by NMR suggest that other conformations are possible, with some flexibility of the central helix.[30]

Recently, a partial X-ray structure for the vertebrate cardiac troponin complex has been described.[31] This contains all of TnC, most of TnI and about half of TnT. Figure 5a,b shows a

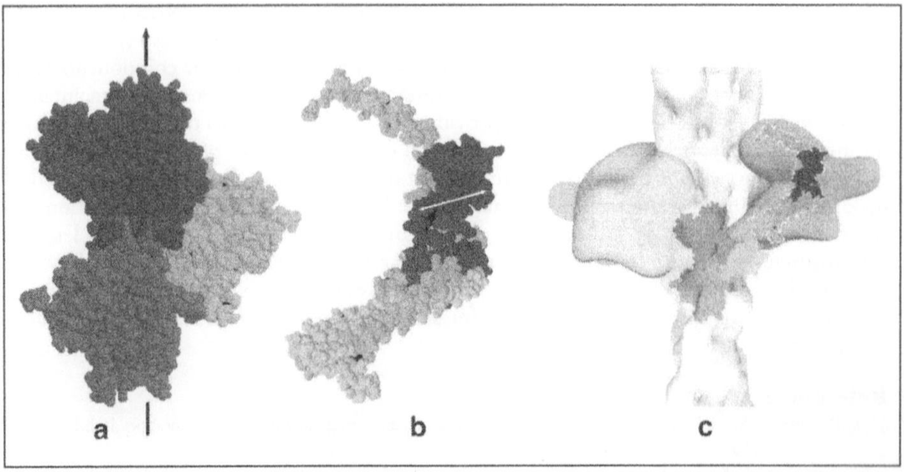

Figure 5. a) 3 monomers of F-actin—the arrow indicates the axis of the helical filament. b) Partial structure for vertebrate troponin taken from the pdb entry.[31] TnC is dark grey, TnI and TnT are light grey. This is at the same scale as (a) and it can be seen that the core of troponin is large enough to span 3 actin monomers. In (b) the TnC D-helix (white arrow) has been used to orient the troponin structure relative to actin using the result of Ferguson et al[32] described above. Because the polarity of the actin filament is not known, there is an ambiguity (±) in the angle of the D-helix. We have chosen the angle which makes the two ends of TnI come closer to actin. c) EM reconstruction of 2 *Lethocerus* troponin complexes docked on each side of actin-tropomyosin.[33] For comparison, the actin trimer and partial troponin X-ray structures have been scaled and superimposed in the same relative orientation on the low resolution troponin 3D reconstruction. A color version of this figure is available online at http://www.Eurekah.com.

comparison of the size of this structure with the size of three actin subunits from the F-actin helix. Another recent development has been the use of fluorescent probes to determine the in vivo orientation of TnC in the vertebrate thin filament.[32] This suggests that the D α-helix in TnC is oriented approximately at right angles to the axis of the thin filament in relaxed muscle and changes in orientation by about 32 degrees on activation. We have used this orientation of the TnC D-helix to align the partial troponin X-ray structure to the F-actin helix in Figure 5b.

A lower resolution 3D reconstruction of the *Lethocerus* troponin complex has been obtained by electron microscopy of negatively stained paracrystals of isolated *Lethocerus* troponin-tropomyosin.[33] This reconstruction is shown in Figure 5c combined with the F-actin-tropomyosin model of Lehman et al.[29] It is interesting to see that the partial X-ray structure for vertebrate troponin, which has a "two-pronged" appearance, fits comfortably within the two main lobes of the EM troponin envelope, although there is clearly a large amount of density unaccounted for. The mass of the partial X-ray model (PDB entry is 375 residues, ~40 kDa) is about 1/3 of that expected for *Lethocerus* troponin complex (~115kDa). There is a projecting lobe on the outside of the EM model which could perhaps be the additional 26kDa PA sequence in *Lethocerus* TnH. This sequence is susceptible to proteolysis suggesting that it is exposed on the thin filament.[11] In addition, immunolabelling has shown that the PA sequence extends towards the rear crossbridge in *Lethocerus* IFM fibres in rigor.[34]

Concluding Remarks

As we have seen, there are many differences between the thin filament proteins in insects and vertebrates. Although the mechanism of regulation of contraction in IFM still occurs via the troponin complex on the thin filament, it has been modified to allow insect muscles to operate at higher frequencies. Little structural information is available yet to determine the exact mechanism of stretch activation. However, we now know that it involves a special isoform of TnC which is controlled by stretch. The role of other thin filament components is still unclear. The PA rich extensions on the heavy troponin subunits may take part in maintaining the geometry of the lattice, or they may provide positional feedback to the troponin complex during stretch. The presence of arthrin in the thin filament does not prevent myosin binding but it could have a subtle effect on the movement of tropomyosin during activation, perhaps making the filament more easily switchable to the "off" state during stretch activation. The absence of nebulin, which is thought to act as a "protein ruler" in determining the length of thin filaments in vertebrate muscle,[22] leaves unclear how the very exact length of IFM thin filaments is regulated. These questions will only be answered when we have more information, including high resolution structural data, for all the components in the insect thin filament assemblage.

References

1. Pringle J. Stretch activation of muscle: Function and mechanism. Proc Roy Soc Lond B 1978; 201:107-130.
2. Agianian B, Krzic U, Qiu F et al. A troponin switch that regulates muscle contraction by stretch instead of calcium. EMBO J 2004; 23(4):772-779.
3. Ruiz T, Bullard B, Lepault J. Effects of calcium and nucleotides on the structure of insect flight muscle thin filaments. J Muscle Res Cell Motil 1998; 19(4):353-364.
4. Egelman EH, Francis N, DeRosier DJ. F-actin is a helix with a random variable twist. Nature 1982; 298(5870):131-135.
5. Reedy MK. Ultrastructure of insect flight muscle. I. Screw sense and structural grouping in the rigor cross-bridge lattice. J Mol Biol 1968; 31(2):155-176.
6. Bullard B, Bell J, Craig R et al. Arthrin: A new actin-like protein in insect flight muscle. J Mol Biol 1985; 182(3):443-454.
7. Ball E, Karlik CC, Beall CJ et al. Arthrin, a myofibrillar protein of insect flight muscle, is an actin-ubiquitin conjugate. Cell 1987; 51(2):221-228.
8. Burgess S, Walker M, Knight PJ et al. Structural studies of arthrin: Monoubiquitinated actin. J Mol Biol 2004; In Press.
9. Schmitz S, Schankin CJ, Prinz H et al. Molecular evolutionary convergence of the flight muscle protein arthrin in diptera and hemiptera. Mol Biol Evol 2003; 20:2019-2033.
10. Galkin VE, Orlova A, Lukoyanova N et al. The location of ubiquitin in Lethocerus arthrin. J Mol Biol 2003; 325(4):623-628.
11. Bullard B, Leonard K, Larkins A et al. Troponin of asynchronous flight muscle. J Mol Biol 1988; 204(3):621-637.
12. Karlik CC, Fyrberg EA. Two Drosophila melanogaster tropomyosin genes: Structural and functional aspects. Mol Cell Biol 1986; 6(6):1965-1973.
13. Peckham M, Cripps RM, White DCS et al. Mechanics and protein content of insect flight muscles. J Exp Biol 1992; 168:57-76.
14. Clayton JD, Cripps RM, Sparrow JC et al. Interaction of troponin-H and glutathione S-transferase-2 in the indirect flight muscles of Drosophila melanogaster. J Muscle Res Cell Motil 1998; 19(2):117-127.
15. Agianian B, Tucker PA, Schouten A et al. Structure of a Drosophila sigma class glutathione S-transferase reveals a novel active site topography suited for lipid peroxidation products. J Mol Biol 2003; 326(1):151-165.
16. White SP, Cohen C, Phillips Jr GN. Structure of co-crystals of tropomyosin and troponin. Nature 1987; 325(6107):826-828.
17. Wendt T, Guenebaut V, Leonard KR. Structure of the Lethocerus troponin-tropomyosin complex as determined by electron microscopy. J Struct Biol 1997; 118(1):1-8.
18. Greenfield NJ, Swapna GV, Huang Y et al. The structure of the carboxyl terminus of striated alpha-tropomyosin in solution reveals an unusual parallel arrangement of interacting alpha-helices. Biochemistry 2003; 42(3):614-619.
19. Qiu F, Lakey A, Agianian B et al. Troponin C in different insect muscle types: Identification of two isoforms in Lethocerus, Drosophila and Anopheles that are specific to asynchronous flight muscle in the adult insect. Biochem J 2003; 371(Pt 3):811-821.

20. van Straaten M, Goulding D, Kolmerer B et al. Association of kettin with actin in the Z-disc of insect flight muscle. J Mol Biol 1999; 285(4):1549-1562.
21. Kulke M, Neagoe C, Kolmerer B et al. Kettin, a major source of myofibrillar stiffness in Drosophila indirect flight muscle. J Cell Biol 2001; 154(5):1045-1057.
22. Labeit S, Gibson T, Lakey A et al. Evidence that nebulin is a protein-ruler in muscle thin filaments. FEBS Lett 1991; 282(2):313-316.
23. Kabsch W, Mannherz HG, Suck D et al. Atomic structure of the actin: DNase I complex. Nature 1990; 347(6288):37-44.
24. Herzberg O, James MN. Common structural framework of the two Ca2+/Mg2+ binding loops of troponin C and other Ca2+ binding proteins. Biochemistry 1985; 24(20):5298-5302.
25. Lorenz M, Poole KJ, Popp D et al. An atomic model of the unregulated thin filament obtained by X-ray fiber diffraction on oriented actin-tropomyosin gels. J Mol Biol 1995; 246(1):108-119.
26. Stewart M. Tropomyosin: Evidence for no stagger between chains. FEBS Lett 1975; 53:1-5.
27. Bullard B, Mercola DA, Mommaerts WF. The origin of the tyrosyl circular dichroism of tropomyosin. Biochim Biophys Acta 1976; 434(1):90-99.
28. Whitby FG, Phillips Jr GN. Crystal structure of tropomyosin at 7 Angstroms resolution. Proteins 2000; 38(1):49-59.
29. Lehman W, Craig R, Vibert P. Ca(2+)-induced tropomyosin movement in Limulus thin filaments revealed by three-dimensional reconstruction. Nature 1994; 368(6466):65-67.
30. Slupsky CM, Sykes BD. NMR solution structure of calcium-saturated skeletal muscle troponin C. Biochemistry 1995; 34(49):15953-15964.
31. Takeda S, Yamashita A, Maeda K et al. Structure of the core domain of human cardiac troponin in the Ca(2+)-saturated form. Nature 2003; 424(6944):35-41.
32. Ferguson RE, Sun YB, Mercier P et al. In situ orientations of protein domains: Troponin C in skeletal muscle fibers. Mol Cell 2003; 11(4):865-874.
33. Wendt T, Leonard K. Structure of the insect troponin complex. J Mol Biol 1999; 285(4):1845-1856.
34. Reedy MC, Reedy MK, Leonard KR et al. Gold/Fab immuno electron microscopy localization of troponin H and troponin T in Lethocerus flight muscle. J Mol Biol 1994; 239(1):52-67.

Chapter 12

The Insect Z-Band

Judith D. Saide

Abstract

The Z-band is an electron dense structure that borders sarcomeres in striated muscle. It is a complex assembly of proteins that organizes and stabilizes both thick and thin filament arrays in the contractile apparatus. By anchoring actin filaments and protein extensions of myosin filaments, the Z-band ensures that active as well as passive tensions are transmitted from one sarcomere to the next along the length of muscle.

Although Z-bands serve these common functions in all muscles, in insects Z-band ultrastructure is strikingly varied among different types of fibers. It appears fragmented and amorphous in some fibers, solid and geometrically ordered in others.[1,2] This architectural diversity suggests that the Z-band has adapted to specific physiological requirements of different muscle fibers.

Although it is not yet possible to describe how proteins are arranged to construct an insect Z-band, there have been significant advances in the field, aided in large part by use of three-dimensional reconstruction techniques, the sequencing of the *Drosophila* genome, and genetic manipulation of proteins. This chapter will review research that has expanded our understanding of the structure of the insect Z-band and the nature of proteins that are assembled to form it. While the focus will be on the Z-band of indirect flight muscle (IFM), an overview of Z-bands in other muscle types will be presented for comparison.

Z-Band Anatomy

Unstructured Z-Bands

Perhaps the least well organized Z-bands in insects are the larval body wall muscles[3,4] and larval and adult visceral muscles.[5-7] These fibers require extensive shortening, and the Z-band is designed to allow the penetration of thin and thick filaments of adjacent sarcomeres during contraction. The fibers are described as "supercontracting" [8,9] because they can shorten to less than 40% of rest length, like vertebrate smooth muscle. In fact, normalized length-tension curves of blowfly larval body wall muscle and vertebrate teania coli smooth muscle are nearly identical.[10] In longitudinal sections the Z-bands in these insect fibers appear as discontinuous, poorly aligned, spindle shaped densities that collect and anchor groups of thin filaments. Although in transverse views, cross-sections of thin filaments can sometimes be seen within the Z-band density, they do not have noticeable order.[5] The dense Z material itself is discontinuous and irregularly shaped in cross-section.[3] In some muscles the densities appear continuous and form a plate with numerous, large perforations.[4-6] Hardie and Hawes,[11] who examined very thick cross sections through Z-bands with high voltage electron microscopy, implied that the perforated Z-structure is more likely to be the rule than the exception in supercontracting insect muscles. Since Z-band material is usually not perfectly planar in these fibers, thin sections of perforated Z-discs would be expected to appear discontinuous. Sections thick enough to include the entire width of the Z-band allow the interconnectedness of the Z material to be seen (Fig. 1).

Nature's Versatile Engine: Insect Flight Muscle Inside and Out, edited by Jim Vigoreaux.

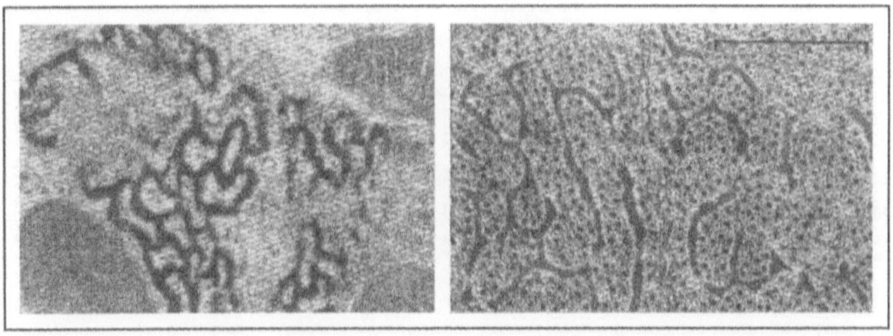

Figure 1. Cross-sections through the perforated Z-band of blowfly larval body wall muscle, stretched (left) and supercontracted (right). Note the expansion of the perforations and their penetration by thin and thick filaments in the supercontracted fiber. Scale for both images = 0.5 μ. Modified from: Osborne MP. J Insect Physiol 1967; 13:1471-1482, ©1967 Pergamon Press Ltd, with permission.

Several investigators have observed that in supercontracted fibers the Z-band becomes noticeably thinner in longitudinal section and that perforations enlarge significantly to accommodate the penetrating thick and thin filaments.[4,5,12,13] This suggests that Z-band components and their interactions must have significant flexibility to support these dynamic changes.

In adult somatic tissues, such as leg, intersegmental, and flight muscles,[1,2] as well as certain visceral muscles,[6] Z-bands are solid and restrict the extent of sarcomere shortening. They are also unstructured. They appear dense and amorphous, binding thin filaments that are distributed without apparent regularity.[14] In all muscles with such Z-bands, the ratio of thin and thick filaments exceeds 3:1 and may be as great as 6:1. It is interesting that in flight muscles, filament ratios have an inverse relationship to wing beat frequencies; the lower the thin/thick filament ratio the higher the frequency.[14,15] In rapidly contracting synchronous fibers and the remarkable, oscillatory asynchronous muscles of the IFM, thin/thick filament ratios drop to 3:1, and filament lattices and Z-bands become precisely ordered.

Structured Z-Bands

In the high performance IFM thin and thick filaments are organized into regular arrays. Thick filaments form a hexagonal lattice in which thin filaments are positioned midway between every two thick filaments.[15] The dense Z-band is modified into a highly ordered structure to stabilize the filament lattices from adjacent sarcomeres. The IFM Z-band has been the subject of several electron microscopy studies, the earliest of which by Auber and Couteax[16] described the IFM Z-band of *Calliphora* in cross section as a plate perforated by numerous equidistant holes surrounded by dense rims that embedded up to six filaments. The authors suggested that thin filaments within the Z-band split in two and were recombined with bifurcated thin filaments from the adjoining sarcomere. Whether the lattices of adjacent sarcomeres were aligned or displaced across the Z-line was unresolved. Later, Ashhurst[17-19] and Saide and Ullrick[20] examined the Z-bands of *Lethocerus* and *Apis*, respectively. They inferred from micrographs of oblique sections that lattices of adjacent sarcomeres were displaced with respect to each other such that a thick filament of one sarcomere projected to the trigonal position of three thick filaments on the opposite side of the Z-band (cf., ref. 21). Thin filaments in the shifted lattices of adjacent sarcomeres passed straight into the Z-band where they interdigitated without any noted change of lattice spacing (Fig. 2). They formed circular arrays of six evenly spaced filaments alternating in polarity, and the arrays themselves were arranged in a hexagonal pattern. In the honeybee Z-band the rims embedding the six filaments in each grouping had a triangular profile (Fig. 3). When viewed in a central transverse plane through the Z-band, these triangles were oriented such that the apex of one pointed to the middle of the base of an adjacent triangle. The orientation of the triangles appeared to shift by several degrees in opposite directions on opposite sides of the central plane (Fig. 3).[20]

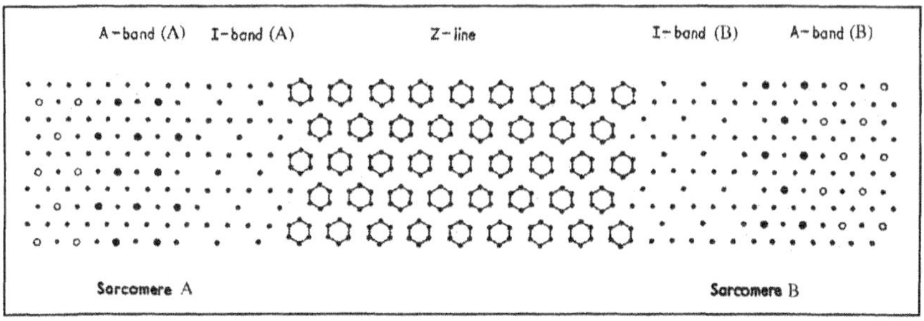

Figure 2. Diagram that illustrates the interdigitation of thin filaments from adjacent sarcomeres to produce the hexagonal lattice of the flight muscle Z-band. Note that lattices on opposite sides of the Z-line are displaced from one another. Connecting filaments that link thick filaments to the Z-band are not indicated. Modified from Ashhurst DE. J Mol Biol 1967; 27(2):385-389, ©1967 Academic Press Inc, with permission.

In 1989 Cheng and Deatherage published a three dimensional reconstruction of the honeybee Z-band from electron micrographs of tilted, thin transverse sections and correlated the findings with locally averaged oblique sections to develop a detailed model of this structure.[22,23] Their technique provided significantly better resolution than that of previous work, and the studies explained and extended earlier observations.

In the central, transverse plane of the Z-band each thin filament in the six filament circular array is associated with two types of connecting densities, C1 and C2 that alternate around the array (Fig. 4). C1 joins a given thin filament to its anti-parallel neighbor on one side; C2 joins it to an anti-parallel neighbor on the other side. The highest density of C1 is equidistant between two anti-parallel filaments and shifted radially slightly outside the circle of filaments away from the center of the array. C2 connection densities are weaker than those of C1 and shifted somewhat in the opposite direction. The result is that the connecting material creates a triangular profile with the greatest density of C1 positioned at the apices and that of C2 along the bases. C1 and C2 form cross connections between thin filaments, as described, and extend axially along them, as well.[22]

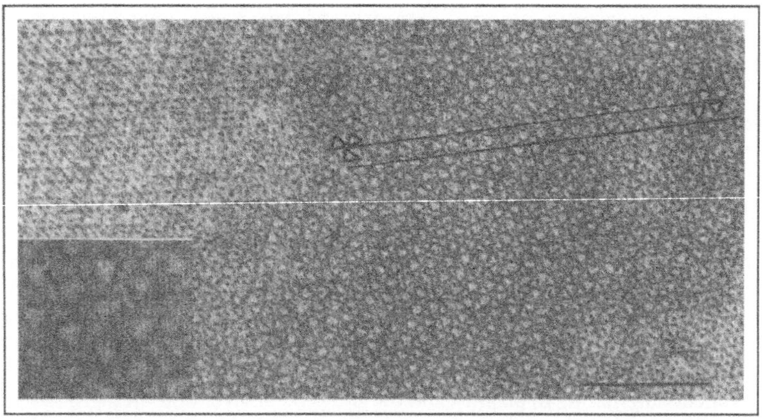

Figure 3. Oblique section through the honeybee IFM Z-band with superimposed outlines to show the orientation of the triangular Z-tubes at opposite sides of the Z-band and to emphasize their rotation. Inset is an enlarged view of the Z-tubes at the center of the Z-band. Scale bar = 0.5 μ. Modified from Saide JD, Ullrick WC. J Mol Biol 1973; 79(2):329-337, ©1973 Academic Press Inc, with permission.

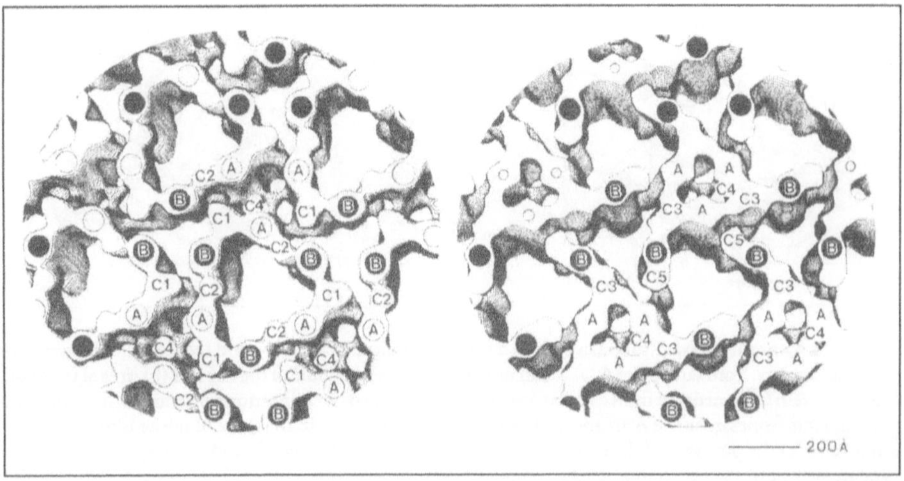

Figure 4. Contour surfaces of the three-dimensional map of the Z-band. Open circles represent thin filaments from sarcomere A, filled circles, those from adjacent sarcomere B. Both the central plane (left) and a plane closer to the I-band of sarcomere B (right) are viewed from sarcomere B. At a level in the Z-band near sarcomere B (right), the position of thin filaments from sarcomere A that do not penetrate the plane of view are marked by small circles. Connections between filaments are labeled C. Modified from Cheng NQ, Deatherage JF. J Cell Biol 1989; 108:1761-1774, ©1989 Rockefeller University Press, with permission.

Thin filaments from adjacent sarcomeres overlap by 80 nm and terminate within the Z-band, which is 120 nm thick in the honeybee. If a transverse view of the Z-band is shifted axially, away from the central plane toward one of the sarcomeres, the ends of thin filaments from the far sarcomere come into view. Here, there is an additional pair of cross-connections, C3 and C5, which appear to have continuity with C1 and C2. C5 is axially displaced from C2 and joins the same two filaments in the array as C2. Since it extends from the thin filament of the near sarcomere to the terminal region of the adjacent filament from the far sarcomere, it may serve as a capping protein.[22]

C3, which is axially continuous with C1 and forms the apices of the triangular profiles at this level, also connects a thin filament from the near sarcomere with the end of a thin filament from the opposite sarcomere. That connection, however, extends most C3 density not to a neighboring filament in the same array, but to the end of a nearby filament of opposite polarity in an adjacent array (Fig. 4). This serves to shift the position of the apical densities of the triangles and makes them appear to have rotated (Fig. 5). Beyond the central plane, on the other side of the Z-band, C3' and C5' , like C3 and C5, appear to have axial continuity with C1 and C2 and make thin filament connections that are related by symmetry to those of C3 and C5. All such connections are between anti-parallel filaments.[22]

The Z-band lattice integrates not only thin filaments from adjacent sarcomeres but also extensions of thick filaments, called connecting filaments (for review see ref. 24). A connecting filament, if extended axially, straight into the Z-band, would meet the hub of a fifth structure, C4, which is positioned at the center of three thin filaments that pass into the Z-band from the opposite sarcomere and, consequently, are of the same polarity. C4 densities are linked to the three thin filaments surrounding them by spoke-like densities that radiate from the hub (Fig. 4). Of interest is that the three filaments in each bundle are drawn closer to the hub as they approach their termination within the Z-band, so that lattice spacing between filaments is, in fact, altered. C4 extends axially toward, but does not cross the central plane of the Z-line. Among the five identified Z-band connecting densities, C4 is unique in that it joins together

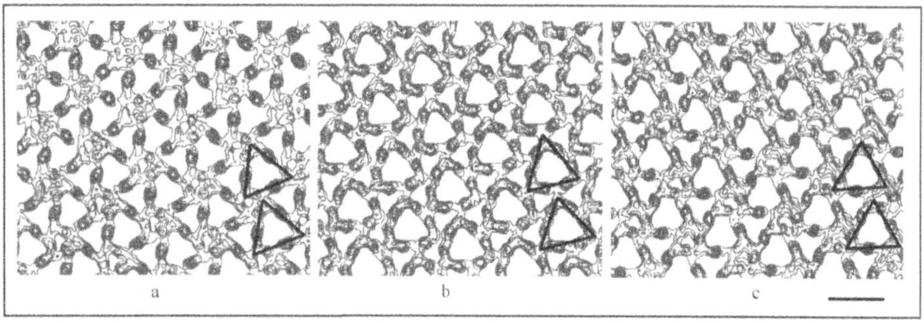

Figure 5. Contour maps at successive levels across the honeybee IFM Z-band. Maps were derived from oblique sections from sarcomere A (a) to sarcomere B (c). At the middle of the Z-band (b) rims of triangular tubes contain 6 filaments (3 from each sarcomere) that are joined by connections C1 and C2. In (a) densities of filaments from sarcomere A are more prominent than those from B. In (c) the opposite is true. Note that in (a) and (c) the positions of the connecting densities, C3' and C5', and C3 and C5, respectively, cause an apparent shift in the orientation of the triangles. See superimposed outlines. Bar scale = 500Å. Modified from Deatherage JF, Cheng NQ, Bullard B. J Cell Biol 1989; 108:1775-1782, with permission. ©1989 Rockefeller University Press.

thin filaments from the same sarcomere and links them to thick, rather than thin, filaments of the opposite sarcomere.[22] To date no proteins have been assigned with certainty to the specific densities (C1 to C5) in the Z-band.

Inter-Z Bridges

Filamentous connections between Z-bands of adjacent myofibrils were first described by Garamvölgyi who observed them in electron micrographs of honeybee fibers[25] and demonstrated by micro-dissection that they were elastic.[26] The structures were resistant to glycerol, and they survived in solutions of high ionic strength which dissolved most of the contractile proteins and left insoluble networks of interconnected isolated Z-disks.[27] More recently, Trombitas and Pollack[28] showed by scanning electron microscopy that inter-Z cables not only link neighboring Z-bands to one another, but that every Z-band in peripheral myofibrils is anchored to the sarcolemma, as well. The composition of inter-Z bridges is not known, but they are electron dense and are associated with granules, vesicles and mitochondria. They may serve to position these organelles along the fibril, but by aligning and interconnecting myofibrils to one another and to the membrane, inter-Z bridges offer a mechanism by which Z-bands can propagate forces transversely across the fiber as well as along its length.[28]

Isolated Z-Discs

A unique property of structured Z-bands from insect IFM is their ease of isolation. Treatment of IFM myofibrils with dilute acids or concentrated salt solutions dissolves thick and thin filaments and liberates Z-disc backbones.[27,29,30] These resistant, isolated Z-discs, prepared from honeybee IFM, have the hexagonal symmetry of the intact Z-band and retain remnants of connecting filaments that project from the Z-disc surfaces (Fig. 6).[31] The fine structural order of the parent structure is disturbed, however, and it is possible that some Z-band components are lost during extraction procedures, or that extraneous proteins may bind to the Z-disc scaffold (e.g., zeelins,[32] flightin[33]). Isolated Z-disc preparations have been exploited to identify Z-band structural proteins. Preparations have been purified from *Drosophila*,[34] *Apis*,[35] and *Lethocerus*[36] and used as immunogens to generate monoclonal antibodies against constituent proteins. These antibodies have been extremely useful in identifying insect Z-band components.

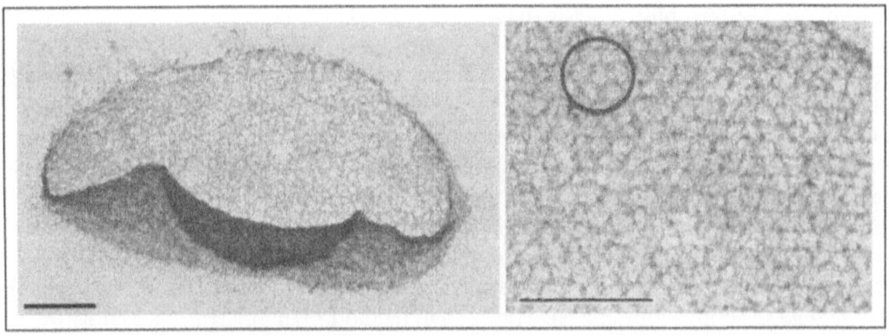

Figure 6. Electron micrographs of isolated Z-discs expanded in distilled water and stained with 1% phosphotungstic acid. Note honeycomb appearance in the enlarged Z-disc surface (right) and the connecting filaments extending from the edge of the fold in the Z-disc on the left. Scale bars = 1 μ. Modified from Saide JD, Ullrick WC. J Mol Biol 1974; 87:671-683, ©1974 Academic Press Inc, with permission.

Z-Band Proteins

α-Actinin

Alpha actinin was discovered in 1964 by Ebashi[37] and localized in vertebrate Z-bands by Masaki et al in 1967.[38] It has been found in Z-bands or related structures that anchor thin filaments in every muscle studied,[39,40] and it is retained in the isolated insect Z-disc backbone.[34-36] Though vertebrates and insects diverged early in evolution, *Drosophila* and vertebrate muscle α-actinin sequences have an average 68% identity.[41] The α-actinin monomer has two N-terminal calponin homology domains that bind actin, a central rod segment with four spectrin-like repeats and a C-terminal calmodulin-like domain with two EF-hand motifs.[39] Since a recombinant peptide containing C-terminal EF hand domain sequences of *Drosophila* α-actinin does not bind Ca^{2+} [42], the native protein is unlikely to be Ca^{2+} regulated. Two subunits combine to form an anti-parallel homodimer in which the four spectrin repeats of the anti-parallel partners are aligned.[43,44] Since the configuration places actin binding domains at both ends of the molecule, α-actinin is presumed to participate in cross-linking actin filaments within the Z-band.[39,45] This assumption has been supported in part by compelling electron microscope images of actin filaments linked in vitro by transversely oriented α-actinin molecules that form ladder-like structures with thin filaments.[46] Cross linked actin filament pairs have been observed with opposite as well as common polarity.[47,48]

Chemical cross-linking studies and analysis of cryoelectron microscopy images of actin filaments decorated with the actin binding domain of α-actinin have identified residues 87-119 on domain 1 of actin as one of two α-actinin binding sites.[49] An alteration of this domain occurs in the *Drosophila* mutant *Act88F*^{E93K} which carries a mutation in the actin gene used predominately in the IFM and replaces an acidic amino acid with a basic one at residue 93.[50] It is likely that this alteration at the predicted α-actinin binding site on actin must significantly reduce the affinity between the two proteins, since IFM myofibrils from these mutants lack Z-band density, and immunoblots of SDS gels of glycerinated fibril preparations from three or four day old adult mutant flies have no detectable α-actinin.[50]

Drosophila α-actinin is encoded by a single gene *actn* mapped cytologically on the X chromosome at location 2C4-5.[41] Three isoforms result from alternative splicing of this gene, a nonmuscle (104kDa) and two muscle (104kDa and 107kDa) isoforms.[51-53] All three are identical except in the sequence of amino acids that join the actin binding domain to the first spectrin repeat.[51] This sequence is extended by an additional 22 amino acids in the larger (107kDa) isoform. Since three of the amino acids are proline, the extended sequence is probably not in an α-helical conformation, and the molecule is likely to be flexible in this region.

Northern and Western blot analysis suggests that the 107kDa isoform is expressed predominately in the supercontractile muscles of larvae and adult visceral tissues.[51,52]

The 104kD muscle isoform is used in adult somatic muscles (eg. leg and IFM).[51,52] In the Z-band of the honeybee IFM, α-actinin is hypothesized to form the connecting links, C1, C2, C3 and C5 that bind together anti-parallel actin filaments.[22] Since the center to center distance between thin filaments in the honeybee lattice varies between 170 and 240 Å and since the α-actinin molecule is between 300 to 400 Å long,[39] it is presumed that α-actinin does not have a simple transverse orientation in vivo as observed in α-actinin/thin filament ladders (350 Å spacing) prepared in vitro. If it is a major component of the connecting material, it may be bent or curved in the IFM Z-band and contribute both to the transverse and axial densities previously described.[22]

Drosophila α-actinin mutants have been analyzed and manipulated in an effort to sort out the role and requirements of α-actinin in Z-bands of different fiber types. Several characterized null mutants, lethal(1)Cb alleles, have inversions with breaks within the *actn* gene and fail to produce the protein.[54] All are lethal within two days of hatching, but surprisingly larvae are able to break through the egg case and move in the absence of α-actinin.[41,51,54] This suggests that other proteins must participate in forming stable cross-links between thin filaments at the Z-line to allow tension to be transmitted from one sarcomere to another.[41]

The genomes of these null mutants have been modified by introduction of recombinant transgenes that express altered forms of α-actinin so that the elements of domains essential for function can be identified. For example, when a transgene that encodes the adult (104kDa) α-actinin isoform is inserted into a null background, animals are viable, and larvae can move and develop into apparently normal adults.[53] The smaller adult protein is apparently interchangeable with the larger larval supercontractile muscle isoform (107kDa). This might not be anticipated given that the longer linking sequence between the actin binding domain and the first spectrin repeat of α-actinin in the larval isoform would be expected to provide necessary flexibility to accommodate the conformational changes observed in the perforated Z-bands.

While the adult (104kDa) isoform may be an adequate substitute for the larval isoform, the reverse is not the case. The mutant *fliA*^{3} has a single base substitution that prevents alternative splicing of a muscle specific exon and produces only the supercontractile (107kDa) muscle isoform. Adults are flightless and have leg weakness.[51] Levels of α-actinin are sharply reduced in the IFM, probably because the 107kDa isoform is not properly integrated within the highly structured Z-band lattice. Although Z-bands degenerate as the fly ages, myofibrils of newly eclosed flies look remarkably normal. This suggests that normal levels of α-actinin are not required for sarcomere assembly in the IFM.[51]

The failure of the larval isoform to substitute in the IFM is unlikely to be due simply to its larger size. α-actinin null flies with an α-actinin transgene, altered to include an additional (5th) spectrin repeat in the central rod region of the molecule, are not only viable and develop normally, but they can fly.[53] The additional spectrin repeat would be expected to increase the α-actinin length by 50-65Å, yet, unexpectedly, it neither causes disruption of the highly ordered Z-band nor alters its ability to support high frequency, oscillatory contractions. Perhaps other Z-band proteins provide the constraints that maintain the thin filament lattice spacings.[53] These findings suggest that although the length of the α-actinin molecule within the limits tested is not critical for performance, the appropriate spanning sequence between the actin binding domain and first spectrin repeat is. This region has proximity to the C-terminal domain of the antiparallel subunit and may influence its function.[44]

Kettin

Kettin is a high molecular weight protein that was identified by monoclonal antibodies raised against isolated IFM Z-disc preparations.[34-36] (see also chapter by Bullard, Leake, and Leonard in this volume) The protein was initially called Z(400/600) in *Drosophila* and *Apis*,[34,35] but since monoclonal antibodies directed against *Lethocerus* kettin and honeybee Z(400/600) recognize identical isoforms on immunoblots of *Drosophila* IFM myofibrils, (ref. 34 and unpublished observations) the antigens from these three insects are likely to be kettin family members. Kettin sequences have been found at the Z-band in all larval and adult tissues examined.[52]

Drosophila kettin in IFM is an alternatively spliced 540 kDa product of the *sallimus* (*sls*) gene,[55] previously called *D-titin*.[56] Kettin is an actin binding protein with 35 immunoglobulin domains separated by linker sequences.[57,58] The kettin/actin stoichiometry suggests that each Ig domain and one adjacent linker interacts with an actin monomer, and it is speculated that kettin may follow the genetic helix of actin.[59] Immunoelectron microscopy studies place the N-terminal domain of the molecule near the center of the Z-band where it is thought to overlap somewhat with kettin from an adjacent sarcomere.[59] The model predicts that actin is associated with 15 kettin Ig modules that span 45nm within the Z-band and with another 20 modules that extend 60 nm into the I-band where they might add structural support to the thin filament.[57,59,60] Since a bacterially expressed kettin Ig domain with flanking spacer sequences binds to α-actinin, it may be that kettin binds to both actin and α-actinin within the Z-band and helps to reinforce cross links between thin filaments.[59,61] Since kettin and α-actinin both bind noncompetitively to F-actin, they must have different binding sites on the thin filament.[59] In in vitro assays tropomyosin has been found to compete with both kettin and α-actinin for actin binding. However, immunoelectron microscopy evidence indicates that tropomyosin is restricted from the Z-band.[59]

Although the 540 kDa kettin protein is the major expressed isoform from the *sls* gene in *Drosophila* IFM, larger variants (700 kDa, 800 kDa, 1 MDa) have been found in this tissue.[62] Larger variants are also found in honeybee and *Lethocerus* IFM.[35,36] Domain targeting studies suggest that *sls* sequences are variably spliced to produce the larger isoforms.[62] The largest predicted transcript of the *sls* gene would encode a 1.9-2 MDa protein, but an antigen of this size has not been detected in either the IFM or other fiber types.[56,62]

Sequences in the *sls* gene downstream from the kettin sequence encode additional Ig domains, two PEVK domains and fibronectin type III domains.[56,63] Some of these domains may associate with myosin. An interesting study by Kulke et al has provided evidence that some or all of the *Drosophila* IFM *sls* gene variants associate with the thick filament and may contribute to muscle elasticity.[62] Sequences in the *sls* gene upstream from those encoding kettin are reported to have homology with the NH_2-terminal region of vertebrate titin,[56] but it is not known if these 5' sequences are also expressed in the IFM or if, and how, they might be integrated within the Z-band.

In wild type embryos the expressed *sls* gene product, identified with anti-kettin antibodies, is about 1 MDa.[56,64] The isoform appears 2-3 hours before other muscle structural proteins, and it is evident from studies with mutants that it is required for establishing and maintaining Z-band integrity and sarcomere organization.[58,63] Among kettin mutants that have been identified,[57,65] those with the most destructive phenotype, including ket^{14}, a null allele, are homozygous embryonic lethal.[58,63] Although some weak movements can be seen in these embryos, they fail to hatch.Thin and thick filaments are disorganized in the mutant muscles, and both kettin and α-actinin striation patterns normally observed by immunofluorescence microscopy in wild type Stage 17 embryos are undetectable in the severely affected individuals.[58,63]

Ket^{14} heterozygotes develop into adults but are haplo-insufficient. Although sarcomeric morphology appears normal in late pupae, after eclosion it becomes disrupted, and individuals are unable to fly.[58]

Projectin

Insect projectin (mini-titin) is a ~700-1200kD protein that forms all or a part of the connecting filaments that link thick filaments to the Z-band in insect fibrillar flight muscle (see refs. 24 and 66, for reviews and the chapter by Ayme-Southgate and Southgate in this volume). It was first labeled in IFM by antibodies that contaminated an anti-beetle paramyosin preparation.[67] Later it was identified by antibodies that were raised against a high molecular weight band on gels of isolated Z-discs purified from honeybee IFM myofibrils.[68] The polyclonal antibodies bound to the fine filaments projecting from the surfaces of isolated Z-discs and labeled the region between the edges of the Z-band and the lateral regions of the A-band [see also refs. 34, 36, 67, 69-72, (cf. refs. 73, 74)]. Subsequent studies with preparations of anti-projectin monoclonal antibodies (mAbs), prepared by using isolated Z-discs as immunogen, and specific polyclonal antibody preparations (see below) suggest that the protein extends into the Z-band rather than to its edges and that the polyclonal antibodies first prepared may have been directed against a proteolytic product of the parent protein.

In *Drosophila*, projectin is encoded by a single gene *bt* on chromosome 4, and it is variably spliced to produce different muscle specific isoforms.[75-78] It is related to the giant proteins twitchin and titin (connectin) that have repeating immunoglobulin type II and fibronectin type III domains, PEVK segments and a functional kinase domain near the C-terminus.[78-81] Isolated projectin molecules from locust flight muscle have a contour length of about 250Å and are long enough to extend from the Z-band to the terminal regions of the thick filaments.[69] Indirect immunofluorescence microscopy experiments with mAbs directed toward epitopes that are well separated in the protein confirm that a single molecule spans the distance across the short I band in the honeybee IFM (Fig. 7). Although the orientation of projectin in the IFM has not been established, its similarity to vertebrate titin in structure and function suggest that the N-terminus is likely to be toward the Z-band. This view is supported by preliminary immunoelectron microscopy studies showing that polyclonal antibodies, raised against N-terminal immunoglobulin domains of *Drosophila* projectin, bind within the Z-band (Fig. 8).

Since connecting filaments in honeybee IFM remain distinct and have the same lattice position in the I-band as thick filaments in the A-band,[20] they are assumed to follow a straight course into the Z-band where they form the hub of the C4 filaments within the Z-band lattice.[22,23] Projectin is likely to be the protein at this lattice position, but the identity of the C4 components is not known. In vertebrates titin's Z-spanning regions have Z-repeats,[82] sequences that can associate with the C-terminal regions of α-actinin,[83-86] but related sequences in projectin have not been identified.

In oblique sections through *Lethocerus* IFM, connecting filaments are less apparent than they are in the honeybee in the appropriate lattice positions in the I-band.[19,87] Longitudinal sections of *Lethocerus* show evidence that extensions of thick filaments branch laterally and associate with thin filaments outside the Z-band.[24] It is not certain whether such filament branches are those of kettin or projectin, or perhaps both. Kettin, and/or SLS protein, associate with both thick and thin filaments. Recent evidence suggests that projectin may also associate with thin, as well as thick filaments. Projectin has been shown to bind to actin in solid phase binding assays[72] and to influence the formation of actin paracrystals in vitro.[88] If projectin, like kettin, accompanies the thin filament into the Z-band in *Lethocerus*, it may be assembled there in a different manner from that predicted from three-dimensional reconstructions of honeybee Z-bands.

Projectin is expressed in all insect muscles, but in those outside the IFM it is limited to the A-band where it cannot function as a connecting filament.[34-36,52,71-73] (cf., ref. 69) Nonfibrillar muscles have longer I-bands than those in the IFM, and connecting filaments would likely be comprised of proteins considerably larger than projectin. Titin/connectin-like proteins of ~3 MDa have been detected in leg muscles of beetle, bumblebee and waterbug, and evidence suggests that they tether thick filaments to the Z-band.[71] Thus, in nonfibrillar insect muscles projectin and titin/connectin-like proteins coexist, and titin/connectin may play the role in these fibers that projectin plays in the IFM.

Zetalin

Four monoclonal antibodies raised against *Drosophila* isolated Z-disc preparations recognize a protein that migrates just behind myosin on 5% polyacrylamide SDS gels of *Drosophila* IFM myofibrils and is found within the Z-band matrix.[34] The protein was originally called Z(210) but was changed to zetalin (from the Greek 'zeta' or Z, and 'lin', short for line). Zetalin appears during development at pupal stage P8, when myofibrils begin to be assembled (Fig. 9). It is organized within the Z-band in the IFM and in a subset of jump muscle cells, but it is not

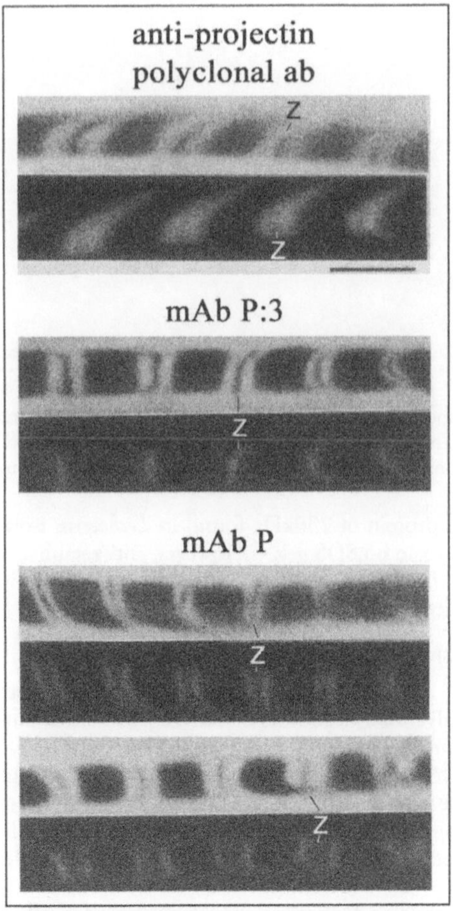

Figure 7. Localization of projectin epitopes in stretched honeybee myofibrils by indirect immuno-fluorescence microscopy. A polyclonal anti-honeybee projectin antibody labels the entire I-band in stretched sarcomeres. Anti-*Drosophila* projectin mAb P:3 labels the Z-line in stretched or unstretched (not shown) sarcomeres. Anti-honeybee projectin mAb:P labels the protein near the A-I junction, and the epitope moves away from the Z-band as the sarcomeres elongate. Phase (top) and fluorescence (bottom) images are presented for each antibody treated myofibril. Results demonstrate that a given projectin molecule in the IFM extends from the Z-band to the A-band.

detected in any other somatic or visceral fibers.[45,52] Of the 32 jump muscle fibers, the 28 that express zetalin are notably larger than the 4 anterior cells that lack the protein, and they use a different myosin isoform.[89] The large jump muscle fibers also differ, however, from the IFMs not only in sarcomere structure (e.g., 6:1 vs. 3:1 thin/thick filament ratios), but in contractile protein isoforms and mechanical properties.[90] Since zetalin is not widely expressed in adult muscles, it will be interesting to learn what common functional demands the IFM and large jump muscle cells share that require the use of this Z-band protein.

Recently, it was found that zetalin is less abundant than first assumed. The protein was shown to comigrate on 5% polyacrylamide SDS gels with a newly identified muscle protein, called myostrandin [a.k.a. A(225)], which can be resolved from zetalin on modified gels. Myostrandin is found in the A-band of the IFM and is an alternatively spliced product of the stretchin-MLCK gene.[91-94]

Figure 8. Ultrathin frozen longitudinal section of *Drosophila* IFM treated with antibody to the two N-terminal Ig domains of projectin. The primary antibody was localized by immunogold particles which accumulate at the Z-band. Magnification: 17,000X (results of A. Ayme-Southgate, A. Cammarato, S. Patel and J. Saide.)

An uncharacterized protein of 230kDa found in *Lethocerus* isolated Z-discs, like zetalin, migrates just behind myosin on SDS gels.[36] However, anti-zetalin antibodies do not recognize proteins in *Lethocerus*. Zetalin has not been detected in any other insect muscle or in vertebrates, and the gene encoding the protein has not yet been cloned.

Z(158) and Z(175)

Z(158) and Z(175) are immunologically similar proteins that are localized at the Z-line of honeybee IFM.[35] Polyclonal antibodies directed against either antigen cross-react with the other, and both proteins are recognized by a monoclonal antibody raised against honeybee isolated Z-discs. The 158kD and 175kD proteins have very similar peptide maps, and although the proteins are highly susceptible to proteolysis, measures to block protein degradation fail to eliminate the smaller isoform. Coomassie stained gels of honeybee isolated Z-discs indicate that the quantities of Z(158) and Z(175) are comparable to that of α-actinin in these preparations.[35]

Unlike α-actinin, kettin and projectin, which are found in all insect fiber types, Z(158) and Z(175) are expressed only in the IFM.[35] In this respect, these honeybee proteins are similar to zetalin which has a restricted distribution in the fly.[45] However, neither polyclonal nor monoclonal antibodies against Z(158/175) cross react with zetalin or any other *Drosophila* proteins.

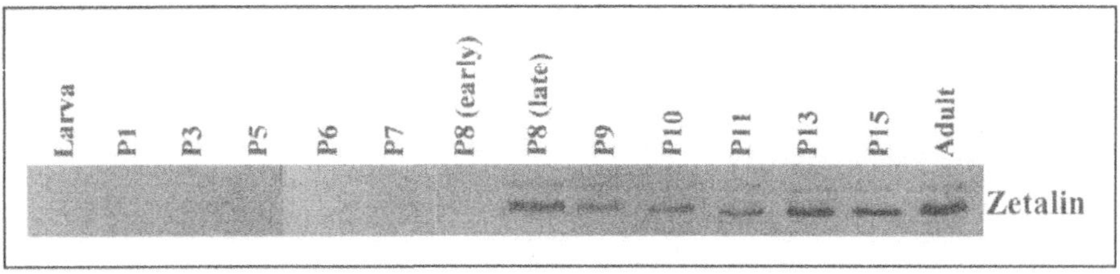

Figure 9. Immunoblot analysis of zetalin expression during *Drosophila* development. Third instar larvae and pupae, sampled at developmental stages indicated,[105] were processed for SDS-5% PAGE, blotted to nitrocellulose and probed with a monoclonal antibody (mAb 7G5) directed against zetalin. Two isoforms, ~225kDa and ~245kDa, are detected at pupal stage P8 (late) and accumulate through subsequent stages. Sample size of early and late P8 individuals was increased to improve temporal resolution of zetalin synthesis.

Capping Protein

Drosophila capping protein (CP) is an αβ heterodimer that binds to the barbed end of actin and is found at the Z-band in adult thoracic muscle.[95] It is related to the vertebrate Z-band protein, CapZ $_{(36/32)}$,[96] which not only caps thin filaments, but serves as a Z-band docking site for various signaling molecules.[97,98] Since procedures that interfere with the CapZ/actin interactions in cultures of chick myotubes delay thin filament organization into sarcomeres, [99] Cap Z is considered an important player in myofibrillogenesis.

In *Drosophila*, genes CG1050 and *cpb* encode the ~33kDa α and ~ 31kDa β chains, respectively. Several mutant *cpb* alleles have been produced, all of which are lethal in hemizygous individuals. Animals complete embryogenesis and hatch, but they die as 1st instar larvae.[95] Since CP regulates actin polymerization in many different cell types, it is not certain if muscle defects are the primary cause of the lethal phenotype. Of interest is that two mutant alleles were found to complement each other. In $cpb^{6.15}/cpb^{F19}$ heterozygotes the most obvious mutant phenotype is the abnormal appearances of the large bristles that require appropriate organization of actin filaments during morphogenesis. Interestingly, though levels of mutant CP β are half that of wild type, these individuals seem to walk, jump and fly without difficulty.[95] So far, no flies carrying a mutant *cpb* allele have been found to have muscle abnormalities (Kathryn G. Miller, personal communication). It may be that the capping function of CP Z can be substituted by other Z-band proteins.

MSP-300

MSP-300 is a ~300kDa dystrophin-like *Drosophila* protein, located in the Z-line of larval fibers and expressed in differentiating embryonic muscle.[100] In cultures of fusing myoblasts the protein is found at sites where actin filaments are connected to the sarcolemma, and it is concentrated at the leading edge of migrating myotubes. In embryos, during the early stages of muscle development, the protein is detected at regions of muscle-muscle and muscle-ectoderm attachment sites. Since the protein partially colocalizes with integrin, a transmembrane receptor for the extracellular matrix, and since integrin β-chain null mutations disturb the membrane targeting of MSP-300, the two proteins are likely to interact either directly or through other cytoplasmic integrin binding proteins. Although in differentiated larval muscle MSP-300 is predominately located at the Z-line,[100] it retains its membrane association, and may help to tether the Z-band of peripheral myofibrils to the sarcolemma. It is noteworthy that in cultured *Drosophila* myotubes integrin is located in the membrane in register with Z-lines.[101] MSP-300 has N-terminal actin binding domains and spectrin-like repeats, as do α-actinin and the giant vertebrate nesprins, other spectrin superfamily members that are Z-band proteins.[100,102,103] It is not known, however, if MSP-300 is integrated within the Z-band matrix or is restricted to its outer surface. Although MSP-300 is expressed in adult muscles,[100] there are no published reports of its subcellular distribution in these tissues.

Nonmuscle Myosin-II

Like MSP-300, nonmuscle myosin-II heavy chain, a product of the *zipper* (*zip*) gene, colocalizes with integrin during *Drosophila* embryonic myogenesis.[104] At stage 16 when muscle-specific myosin-II is diffusely distributed throughout the sarcoplasm, nonmusle myosin-II is concentrated with integrin at attachment sites of muscle termini where actin is linked to the membrane. At embryonic Stage 17 contractile proteins are organized into sarcomeres, and the process is dependent on nonmuscle myosin-II. In muscles of *zip* null mutants I-bands are not observed, and normal myofibril formation is arrested. The *zip* gene product is assumed to have a role in organizing arrays of thin filaments. This is further suggested by the finding that in 3rd instar larvae nonmuscle myosin-II is localized at the Z-line. It is speculated that the protein functions as an actin cross-linker and supports α-actinin in stabilizing the Z-band. How it might serve in this manner as a structural, rather than a contractile protein, in the presence of ATP is uncertain, however.[104]

The location of nonmuscle myosin-II at the Z-line of adult tissues has not been reported.

Concluding Remarks

This review has described the progress that has been made in characterizing Z-band associated proteins and in defining details of Z-band organization. One future challenge is to model components of the Z-band into its three-dimensional structure. Contour maps of the honeybee IFM Z-band are now available, and it is tempting to make predictions with information at hand about the location of specific proteins. It is likely, for example, that α-actinin forms at least part of the C1, C2, C3 and C5 densities that link anti-parallel thin filaments within the Z-band and contributes, with kettin, to the axial densities that form sleeves around actin filaments. CP Z may be located with C3 and C5 at filament ends. Projectin is positioned to form the hub of the C4 densities, and though Z(158/175) has not been characterized, it is reasonable to hypothesize that it binds to projectin and forms C4 densities. Z(158/175) is specific to the IFM, the only fiber type in which projectin is associated with the Z-band.

α-Actinin, kettin and, probably, CP Z are used in Z-bands of all muscles and are likely to serve similar functions. The variability of Z-band structure in different fibers is reflected, in part, by their use of isovariants of these common proteins. Expression of proteins that are specific to a given fiber underlie other unique structural and functional requirements of Z-bands. For example, *Drosophila* zetalin and honeybee Z(158/175) are restricted to the IFM, and, in the case of zetalin, to a subset of jump muscle fibers, as well. Since information is lacking on the association of MSP-300 and nonmuscle myosin II with Z-bands of adult tissues, it is not known if these proteins are limited to larval fibers.

It may be difficult to determine the arrangement of proteins in the more primitive Z-bands found in larval muscles and adult visceral and tubular somatic muscles, because there is no discernable framework into which proteins can be modeled. Nevertheless, there is active research addressing, in particular, the roles of proteins during development of larval muscles that should offer further insights into unstructured Z-bands.

Additional Z-band components are likely to be identified in the future. As further results of structural, biochemical, and genetic approaches converge, precise details of the associations of relevant domains of Z-band proteins will become more evident. Hopefully, it will be possible to understand how these multiple interactions result in assemblies of proteins that build these complex structures. It will be particularly interesting to learn how protein specialization in the Z-band of the IFM helps define the distinct functional properties of this remarkable tissue.

Note Added in Proof

After this chapter was written, studies that challenge the Z-band symmetry described by Cheng and Deatherage[22] were brought to my attention. The Z-band model of the latter authors predicts that each of the six actin filaments surrounding a thick filament in the A-band is rotated to present the same face to the thick filament at any given axial position in the sarcomere. Consequently, there would be regularly spaced rings of target areas for crossbridges around the thick filament, repeating axially every ~39 nm. Evidence that these target areas may, instead, be grouped as three sets staggered by 39/3 nm to mark a helix around each thick filament[106-108] is inconsistent with the thin filament arrangement modeled in the honeybee Z-band lattice, and it suggests that the symmetry of the Z-band may need to be reevaluated.

Acknowledgement

In memory of Dr. Benjamin Kaminer, former chairman of the Department of Physiology at Boston University School of Medicine, who provided kindness, support and encouragement for more than 25 years.

References

1. Candia Carnevali MD, de Equileo M, Valvassori R. Z-line morphology of funtionally diverse insect skeletal muscles. J Submicroscopic Cytol 1980; 12(3):427-446.
2. Candia Carnevali MD, Valvassori R. Z-line in insect muscles: Structural and functional diversities. Bolletino di Zoologia 1981; 48:1-9.
3. Candia Carnevali MD. The Z-line and supercontraction in the hydraulic muscular system of insect larvae. J Exp Zool 1978; 203:15-29.
4. Osborne MP. Supercontraction in the muscles of the blowfly larva: An ultrastructural study. Journal of Insect Physiology 1967; 13:1471-1482.
5. Goldstein MA, Burdette WJ. Striated visceral muscle of Drosophila melanogaster. J Morphol 1971; 134(3):315-334.
6. Rice MJ. Supercontracting and nonsupercontracting visceral muscles in the tsetse fly, Glossina austeni. J Insect Physiol 1970; 16(6):1109-1122.
7. Cook BJ, Pryor NW. Structural properties of the intrinsic muscles of the malpighian tubules of the female stable fly, Stomoxys calcitrans L. J Entomol Sci 1997; 32:138-147.
8. Hoyle G, McAlear JH. Mechanism of supercontraction in a striated muscle fiber. Science 1963; 141:712-713.
9. Hoyle G, McAlear JH, Selverston A. Mechanism of supercontraction in a striated muscle. J Cell Biol 1965; 26(2):621-640.
10. Hardie J. The tension/length relationship of an insect (Calliphora erythrocephala) supercontracting muscle. Experientia 1976; 32:714-716.
11. Hardie J, Hawes C. The three-dimensional structure of the Z disc in insect supercontracting muscles. Tissue Cell 1982; 14(2):309-317.
12. Hardie J. Z disc expansion during supercontraction: A passive mechanism? J Muscle Res Cell Motil 1980; 1(2):163-176.
13. Leyton RA, Ullrick WG. Z disc ultrastructure in scutal depressor fibers of the barnacle. Science 1970; 168(927):127-128.
14. Auber J. Distribution of the two kinds of myofilaments in insect muscles. Am Zoologist 1967; 7:451-456.
15. Squire J. Ultrastructure of diverse muscle types: Arthropod muscle. In: Squire J, ed. The Structural Basis of Muscular Contraction. New York: Plenum Press, 1981:381-414.
16. Auber J, Couteaux R. Ultrastructure de la strie Z dans des muscles de Diptères. J de Microscop 1963; 2:309-324.
17. Ashhurst DE. The fibrillar flight muscles of giant water-bugs: An electron-microscope study. J Cell Sci 1967; 2(3):435-444.
18. Ashhurst DE. The Z-line in insect flight muscle. J Mol Biol 1971; 55(2):283-285.
19. Ashhurst DE. The Z-line: Its structure and evidence for the presence of connecting filaments. In: Tregear RT, ed. Insect Flight Muscle. Amsterdam: Elsevier/North Holland Biomedical Press, 1977.
20. Saide JD, Ullrick WC. Fine structure of the honeybee Z-disc. J Mol Biol 1973; 79(2):329-337.
21. Trombitas K, Tigyi-Sebes A. The Z line of the flight muscle of honey-bee. Acta Biochim Biophys Acad Sci Hung 1975; 10(1-2):83-93.
22. Cheng NQ, Deatherage JF. Three-dimensional reconstruction of the Z disk of sectioned bee flight muscle. J Cell Biol 1989; 108(5):1761-1774.
23. Deatherage JF, Cheng NQ, Bullard B. Arrangement of filaments and cross-links in the bee flight muscle Z disk by image analysis of oblique sections. J Cell Biol 1989; 108(5):1775-1782.
24. Trombitás K. Connecting filaments: A historical prospective. In: Granzier HL, Pollack GH, eds. Elastic Filaments of the Cell. New York: Kluwer Academic/Plenum Publishers, 2000:481:1-23.
25. Garamvölgyi N. Inter-Z bridges in the flight muscle of the bee. J Ultrastruct Res 1965; 13(5):435-443.
26. Ernst E. Biophysics of the Striated Muscle 2nd ed. Budapest: Akadémiai Kiadó, Hungarian Academy of Sciences 1963.
27. Garamvölgyi N. Interfibrillar Z-bands in striated muscle. Acta Physiol Acad Sci Hung 1962; 22:235-241.
28. Trombitas K, Pollack GH. Visualization of the transverse cytoskeletal network in insect-flight muscle by scanning-electron microscopy. Cell Motil Cytoskeleton 1995; 32(3):226-232.
29. Guba F, Garamvölgyi N, Ernst E. On the electron microscopic structure of Z-lines. Intl Congr Elect Microsc 1958; 2:324-325.
30. Garamvölgyi N, Metzger-Torok G, Tigyi-Sebes A. The Z- and M-formations of straited muscle. Acta Physiol Acad Sci Hung 1962; 22:223-233.

31. Saide JD, Ullrick WC. Purification and properties of the isolated honeybee Z-disc. J Mol Biol 1974; 87(4):671-683.
32. Ferguson C, Lakey A, Hutchings A et al. Cytoskeletal proteins of insect muscle: Location of zeelins in Lethocerus flight and leg muscle. J Cell Sci 1994; 107(Pt 5):1115-1129.
33. Vigoreaux JO, Saide JD, Valgeirsdottir K et al. Flightin, a novel myofibrillar protein of Drosophila stretch-activated muscles. J Cell Biol 1993; 121(3):587-598.
34. Saide JD, Chin-Bow S, Hogan-Sheldon J et al. Characterization of components of Z-bands in the fibrillar flight muscle of Drosophila melanogaster. J Cell Biol 1989; 109(5):2157-2167.
35. Saide JD, Chin-Bow S, Hogan-Sheldon J et al. Z-band proteins in the flight muscle and leg muscle of the honeybee. J Muscle Res Cell Motil 1990; 11(2):125-136.
36. Lakey A, Ferguson C, Labeit S et al. Identification and localization of high molecular weight proteins in insect flight and leg muscle. EMBO J 1990; 9(11):3459-3467.
37. Ebashi S, Ebashi F. A new protein factor promoting contraction of actomyosin. Nature 1964; 203:645-646.
38. Masaki R, Endo M, Ebashi S. Localization of the 6S component of α-actinin at the Z-band. J Biochem 1967; 62:630-632.
39. Blanchard A, Ohanian V, Critchley D. The structure and function of alpha-actinin. J Muscle Res Cell Motil 1989; 10(4):280-289.
40. Royuela M, Astier C, Fraile B et al. Alpha-actinin in different invertebrate muscle cell types of Drosophila melanogaster, the earthworm Eisenia foetida, and the snail Helix aspersa. J Muscle Res Cell Motil 1999; 20(1):1-9.
41. Fyrberg E, Kelly M, Ball E et al. Molecular genetics of Drosophila alpha-actinin: Mutant alleles disrupt Z disc integrity and muscle insertions. J Cell Biol 1990; 110(6):1999-2011.
42. Dubreuil RR, Brandin E, Reisberg JH et al. Structure, calmodulin-binding, and calcium-binding properties of recombinant alpha spectrin polypeptides. J Biol Chem 1991; 266(11):7189-7193.
43. Djinovic-Carugo K, Young P, Gautel M et al. Structure of the alpha-actinin rod: Molecular basis for cross-linking of actin filaments. Cell 1999; 98(4):537-546.
44. Tang J, Taylor DW, Taylor KA. The three-dimensional structure of alpha-actinin obtained by cryoelectron microscopy suggests a model for Ca(2+)-dependent actin binding. J Mol Biol 2001; 310(4):845-858.
45. Vigoreaux JO. The muscle Z band: Lessons in stress management. J Muscle Res Cell Motil 1994; 15(3):237-255.
46. Podlubnaya ZA, Tskhovrebova LA, Zaalishtsbvili MM et al. Electron microscopic study of alpha-actinin. J Mol Biol 1975; 92(2):357-359.
47. Imamura M, Endo T, Kuroda M et al. Substructure and higher structure of chicken smooth muscle alpha-actinin molecule. J Biol Chem 1988; 263(16):7800-7805.
48. Meyer RK, Aebi U. Bundling of actin filaments by alpha-actinin depends on its molecular length. J Cell Biol 1990; 110(6):2013-2024.
49. Mimura N, Asano A. Further characterization of a conserved actin-binding 27-kDa fragment of actinogelin and alpha-actinins and mapping of their binding sites on the actin molecule by chemical cross-linking. J Biol Chem 1987; 262(10):4717-4723.
50. Sparrow J, Reedy M, Ball E et al. Functional and ultrastructural effects of a missense mutation in the indirect flight muscle-specific actin gene of Drosophila melanogaster. J Mol Biol 1991; 222(4):963-982.
51. Roulier EM, Fyrberg C, Fyrberg E. Perturbations of Drosophila alpha-actinin cause muscle paralysis, weakness, and atrophy but do not confer obvious nonmuscle phenotypes. J Cell Biol 1992; 116(4):911-922.
52. Vigoreaux JO, Saide JD, Pardue ML. Structurally different Drosophila striated muscles utilize distinct variants of Z-band-associated proteins. J Muscle Res Cell Motil 1991; 12(4):340-354.
53. Dubreuil RR, Wang P. Genetic analysis of the requirements for alpha-actinin function. J Muscle Res Cell Motil 2000; 21(7):705-713.
54. Fyrberg C, Ketchum A, Ball E et al. Characterization of lethal Drosophila melanogaster alpha-actinin mutants. Biochem Genet 1998; 36(9-10):299-310.
55. Bullard B, Linke WA, Leonard K. Varieties of elastic protein in invertebrate muscles. J Muscle Res Cell Motil 2002; 23(5-6):435-447.
56. Machado C, Andrew DJ. D-Titin: A giant protein with dual roles in chromosomes and muscles. J Cell Biol 2000; 151(3):639-652.
57. Kolmerer B, Clayton J, Benes V et al. Sequence and expression of the kettin gene in Drosophila melanogaster and Caenorhabditis elegans. J Mol Biol 2000; 296(2):435-448.

58. Hakeda S, Endo S, Saigo K. Requirements of Kettin, a giant muscle protein highly conserved in overall structure in evolution, for normal muscle function, viability, and flight activity of Drosophila. J Cell Biol 2000; 148(1):101-114.
59. van Straaten M, Goulding D, Kolmerer B et al. Association of kettin with actin in the Z-disc of insect flight muscle. J Mol Biol 1999; 285(4):1549-1562.
60. Bullard B, Goulding D, Ferguson C et al. Links in the chain: The contribution of kettin to the elasticity of insect muscles. Adv Exp Med Biol 2000; 481:207-218; discussion 219-220.
61. Lakey A, Labeit S, Gautel M et al. Kettin, a large modular protein in the Z-disc of insect muscles. EMBO J 1993; 12(7):2863-2871.
62. Kulke M, Neagoe C, Kolmerer B et al. Kettin, a major source of myofibrillar stiffness in Drosophila indirect flight muscle. J Cell Biol 2001; 154(5):1045-1057.
63. Zhang Y, Featherstone D, Davis W et al. Drosophila D-titin is required for myoblast fusion and skeletal muscle striation. J Cell Sci 2000; 113(Pt 17):3103-3115.
64. Machado C, Sunkel CE, Andrew DJ. Human autoantibodies reveal titin as a chromosomal protein. J Cell Biol 1998; 141(2):321-333.
65. Sliter TJ, Henrich VC, Tucker RL et al. The genetics of the Dras3-Roughened-ecdysoneless chromosomal region (62B3-4 to 62D3-4) in Drosophila melanogaster: Analysis of recessive lethal mutations. Genetics 1989; 123(2):327-336.
66. Ayme-Southgate A, Southgate R, McEliece MK. Drosophila projectin: A look at protein structure and sarcomeric assembly. Adv Exp Med Biol 2000; 481:251-262; discussion 262-264.
67. Bullard B, Hammond KS, Luke BM. The site of paramyosin in insect flight muscle and the presence of an unidentified protein between myosin filaments and Z-line. J Mol Biol 1977; 115(3):417-440.
68. Saide JD. Identification of a connecting filament protein in insect fibrillar flight muscle. J Mol Biol 1981; 153(3):661-679.
69. Nave R, Weber K. A myofibrillar protein of insect muscle related to vertebrate titin connects Z band and A band: Purification and molecular characterization of invertebrate mini-titin. J Cell Sci 1990; 95(Pt 4):535-544.
70. Vibert P, Edelstein SM, Castellani L et al. Mini-titins in striated and smooth molluscan muscles: Structure, location and immunological crossreactivity. J Muscle Res Cell Motil 1993; 14(6):598-607.
71. Ohtani Y, Maki S, Kimura S et al. Localization of connectin-like proteins in leg and flight muscles of insects. Tissue & Cell 1996; 28(1):1-8.
72. Weitkamp B, Jurk K, Beinbrech G. Projectin-thin filament interactions and modulation of the sensitivity of the actomyosin ATPase to calcium by projectin kinase. J Biol Chem 1998; 273(31):19802-19808.
73. Hu DH, Matsuno A, Terakado K et al. Projectin is an invertebrate connectin (titin): Isolation from crayfish claw muscle and localization in crayfish claw muscle and insect flight muscle. J Muscle Res Cell Motil 1990; 11(6):497-511.
74. Royuela M, Fraile B, De Miguel MP et al. Immunohistochemical study and western blotting analysis of titin-like proteins in the striated muscle of Drosophila melanogaster and in the striated and smooth muscle of the oligochaete Eisenia foetida. Microsc Res Tech 1996; 35(4):349-356.
75. Ayme-Southgate A, Vigoreaux J, Benian G et al. Drosophila has a twitchin/titin-related gene that appears to encode projectin. Proc Natl Acad Sci USA 1991; 88(18):7973-7977.
76. Fyrberg CC, Labeit S, Bullard B et al. Drosophila projectin: Relatedness to titin and twitchin and correlation with lethal(4) 102 CDa and bent-dominant mutants. Proc R Soc Lond B Biol Sci 1992; 249(1324):33-40.
77. Daley J, Southgate R, Ayme-Southgate A. Structure of the Drosophila projectin protein: Isoforms and implication for projectin filament assembly. J Mol Biol 1998; 279(1):201-210.
78. Southgate R, Ayme-Southgate A. Alternative splicing of an amino-terminal PEVK-like region generates multiple isoforms of Drosophila projectin. J Mol Biol 2001; 313(5):1035-1043.
79. Benian GM, Ayme-Southgate A, Tinley TL. The genetics and molecular biology of the titin/connectin-like proteins of invertebrates. Rev Physiol Biochem Pharmacol 1999; 138:235-268.
80. Maroto M, Vinos J, Marco R et al. Autophosphorylating protein kinase activity in titin-like arthropod projectin. J Mol Biol 1992; 224(2):287-291.
81. Fahrmann M, Fonk I, Beinbrech G. The kinase activity of the giant protein projectin of the flight muscle of Locusta migratoria. Insect Biochem Mol Biol 2002; 32(11):1401-1407.
82. Gautel M, Goulding D, Bullard B et al. The central Z-disk region of titin is assembled from a novel repeat in variable copy numbers. J Cell Sci 1996; 109(Pt 11):2747-2754.

83. Sorimachi H, Freiburg A, Kolmerer B et al. Tissue-specific expression and alpha-actinin binding properties of the Z-disc titin: Implications for the nature of vertebrate Z-discs. J Mol Biol 1997; 270(5):688-695.
84. Atkinson RA, Joseph C, Dal Piaz F et al. Binding of alpha-actinin to titin: Implications for Z-disk assembly. Biochemistry 2000; 39(18):5255-5264.
85. Young P, Gautel M. The interaction of titin and alpha-actinin is controlled by a phospholipid-regulated intramolecular pseudoligand mechanism. EMBO J 2000; 19(23):6331-6340.
86. Joseph C, Stier G, O'Brien R et al. A structural characterization of the interactions between titin Z-repeats and the alpha-actinin C-terminal domain. Biochemistry 2001; 40(16):4957-4965.
87. Ashhurst DE. Z-line of the flight muscle of belostomatid water bugs. J Mol Biol 1967; 27(2):385-389.
88. Podlubnaya ZA, Shpagina MD, Vikhlyantsev IM et al. Comparative electron microscopic study on projectin and titin binding to F-actin. Insect Biochem Mol Biol 2003; 33(8):789-793.
89. O'Donnell PT, Collier VL, Mogami K et al. Ultrastructural and molecular analyses of homozygous-viable Drosophila melanogaster muscle mutants indicate there is a complex pattern of myosin heavy-chain isoform distribution. Genes Dev 1989; 3(8):1233-1246.
90. Peckham M, Molloy JE, Sparrow JC et al. Physiological properties of the dorsal longitudinal flight muscle and the tergal depressor of the trochanter muscle of Drosophila melanogaster. J Muscle Res Cell Motil 1990; 11(3):203-215.
91. Champagne MB, Edwards KA, Erickson HP et al. Drosophila stretchin-MLCK is a novel member of the Titin/Myosin light chain kinase family. J Mol Biol 2000; 300(4):759-777.
92. Patel SR, Saide JD. A(225), a novel A-band protein of Drosophila indirect flight muscle. Biophys J 2001; 80:71a.
93. Patel SR, Saide JD. A(225), a Drosophila indirect flight muscle A-band protein, has a myosin dependent and a myosin independent isoform. Biophys. J 2003; 84:564a.
94. Patel SR. Characterization of myostrandin, a novel myofibrillar protein of Drosophila melanogaster muscle. Ph.D. thesis Boston University School of Medicine, 2003.
95. Hopmann R, Cooper JA, Miller KG. Actin organization, bristle morphology, and viability are affected by actin capping protein mutations in Drosophila. J Cell Biol 1996; 133(6):1293-1305.
96. Casella JF, Craig SW, Maack DJ et al. Cap Z(36/32), a barbed end actin-capping protein, is a component of the Z-line of skeletal muscle. J Cell Biol 1987; 105(1):371-379.
97. Ivanenkov VV, Dimlich RV, Jamieson Jr GA. Interaction of S100a0 protein with the actin capping protein, CapZ: Characterization of a putative S100a0 binding site in CapZ alpha-subunit. Biochem Biophys Res Commun 1996; 221(1):46-50.
98. Pyle WG, Hart MC, Cooper JA et al. Actin capping protein: An essential element in protein kinase signaling to the myofilaments. Circ Res 2002; 90(12):1299-1306.
99. Schafer DA, Hug C, Cooper JA. Inhibition of CapZ during myofibrillogenesis alters assembly of actin filaments. J Cell Biol 1995; 128(1-2):61-70.
100. Volk T. A new member of the spectrin superfamily may participate in the formation of embryonic muscle attachments in Drosophila. Development 1992; 116(3):721-730.
101. Volk T, Fessler LI, Fessler JH. A role for integrin in the formation of sarcomeric cytoarchitecture. Cell 1990; 63(3):525-536.
102. Rosenberg-Hasson Y, Renert-Pasca M, Volk T. A Drosophila dystrophin-related protein, MSP-300, is required for embryonic muscle morphogenesis. Mech Dev 1996; 60(1):83-94.
103. Zhang Q, Ragnauth C, Greener MJ et al. The nesprins are giant actin-binding proteins, orthologous to Drosophila melanogaster muscle protein MSP-300. Genomics 2002; 80(5):473-481.
104. Bloor JW, Kiehart DP. Zipper nonmuscle myosin-II functions downstream of PS2 integrin in Drosophila myogenesis and is necessary for myofibril formation. Dev Biol 2001; 239(2):215-228.
105. Bainbridge SP, Bownes M. Staging the metamorphosis of Drosophila melanogaster. J Embryol Exp Morphol 1981; 66:57-80.
106. Squire JM. Muscle filament lattices and stretch-activation: the match-mismatch model reassessed. J Muscle Res Cell Motil 1992; 13(2):183-189.
107. Reedy MK, Lucaveche C, Reedy MC et al. Experiments on rigor crossbridge action and filament sliding in insect flight muscle. Adv Exp Med Biol 1993; 332:33-44; discussion 44-36.
108. Edwards RJ, Reedy MK. The A-bee-Z problem of actin filament rotation in insect flight muscle. Biophys J 1994; 66:190a.

Chapter 13

Projectin, the Elastic Protein of the C-Filaments

Agnes Ayme-Southgate and Richard Southgate

The Original Experiments Linking Projectin to the IFMs C-Filaments and Their Proposed Role in Muscle Elasticity

In adult insects, the highly specialized indirect flight muscles (abbreviated as IFMs) are powerful muscles adapted for the rapid repeated contractions necessary for flight. These muscles are referred to as asynchronous muscles, because they undergo multiple rounds of contraction for each single nerve impulse, a property made possible by the stretch-activation mechanism.[1-5] The stretch-activation mechanism is explained as a "delayed increase in tension due to stretch" that activates the muscle and results in contraction. The IFMs are attached to the cuticle (exoskeleton) and because they are organized as two sets of nearly perpendicular muscles, their length oscillate in response to the stretch activation-contraction cycles. The stretch activation mechanism has been shown to be an intrinsic property of the myofibrillar apparatus,[1] and is made possible by several special physiological adaptations, such as a high resting stiffness. To explain some of the IFMs' properties, early models proposed the existence of an additional third filament system with elastic properties, which is usually referred to as the connecting or C-filament system. Electron microscopy studies of insect flight muscles revealed the presence of fine connections between the Z bands and the thick filaments.[6-11] In particular, electron microscopy of stretched myofibrils or purified Z disks of insect flight muscles have shown the presence of filaments extending or "projecting" from the Z band towards the myosin filaments and just overlapping the tip of the A band.[12,13] In honeybee IFMs, connecting filaments can be extended to well over ten times their normal rest length. When these muscles are stretched in rigor and then released, the recoil forces of the connecting filaments cause the sarcomere to shorten, leading to the crumpling of the thin filaments held in rigor.[9] The search for component(s) of the C-filament system led to the identification and characterization of the protein, projectin. Saide unequivocally demonstrated by antibody staining and biochemical analysis that the third connecting filament of honeybee flight muscles is composed of projectin.[12]

The Different Projectin-Related Proteins Found in Various Species: Arthropods and Nematodes

Projectin or mini-titin proteins were identified and characterized from several arthropod species, by principally two approaches: either through a biochemical isolation similar to the one used for the purification of vertebrate titin[14-18] or through the total analysis of myofibrillar and/or Z-line associated proteins.[12,13,19-21] Four different insect species were used for these studies (*Lethocerus, Apis, Drosophila* and *Locusta*), and in each case a high molecular weight polypeptide with an estimated size between 700 and 1200 kDa was characterized and named either projectin or mini-titin. Crayfish muscles were also used for the purification of an equivalent protein and

Nature's Versatile Engine: Insect Flight Muscle Inside and Out, edited by Jim Vigoreaux.

antibodies raised against the crayfish protein were also shown to cross-react with *Drosophila* muscles.[15,17,18] Most antibodies raised against projectin from one insect species show cross-reactivity to the equivalent protein in other insects. Some antibodies directed against titin were also able to recognize a few of these purified mini-titin proteins[15] and cross-reactivity between *Drosophila* projectin and antibodies to *C.elegans* twitchin was further demonstrated.[22] From the antibodies cross-reactivity and the biochemical profile of these arthropod proteins, it seems clear that mini-titins and projectins are equivalent proteins in the different species considered and are related to titin in vertebrates and twitchin in *C. elegans*.

The myofibrillar localization of these large proteins was determined by immunofluorescence studies on insect muscles, and in most reports the location of projectin is within the I-Z-I region of insect IFMs.[13,15,16,19,21] In electron microscopy imaging, the purified "mini-titin" fraction contains an elongated protein with a length between 0.2 and 0.3 μm and a width of 4 nm.[14,16] The length of one projectin molecule is, therefore, sufficient to cover the distance between the Z and A bands in IFMs, which ranges from 0.1 to 0.2 μm.[2]

The cumulative data, therefore, establishes the existence of a large elongated protein in the IFMs of several insect species, which belongs to the titin/twitchin protein family. Projectin is proposed to be the foremost, if not the sole, component of the elastic C-filament system, which is found between the Z band and the tip of the myosin thick filament. Some models further suggest that the projectin filaments play an important role in the stretch-activation mechanism, and in this hypothesis, projectin may behave as an elastic protein conferring high resting stiffness to the IFMs and/or capable of transferring stress to the thick filaments during stretching.

Complete Domain Structure of *Drosophila* Projectin

As shown in Figure 1 where the complete domain organization for projectin is presented, most of projectin is composed of two repeated motifs, called the Immunoglobulin C2 (Ig) and the Fibronectin III (FnIII) domains. Among muscle proteins, these two domains were first identified in the *C. elegans* protein, twitchin, and a consensus sequence was defined for both domains.[23] The largest encoded projectin protein contains 9120 amino acids with a molecular weight of approximately 1 MgDa. Projectin can be divided into five subregions: NH_2-terminal, core, intermediate, kinase and COOH-terminal regions.[24-28]

Overall, the Ig domains conform well to the consensus sequence derived originally for twitchin Ig domains, with an average homology of 61.5%. The FnIII domains share an average homology of 65% between themselves and with the original twitchin-derived consensus.[23]

The NH_2-terminal region is composed of two blocks of Ig domains separated by a large unique sequence identified as a PEVK-like region, but with no FnIII domains (see Fig. 1). The first block consists of eight Ig domains with short interspersed unique sequences.[28] These eight Ig domains conform well to the consensus sequence derived originally for the twitchin Ig domains, with a homology ranging from 50 to 72%. Besides the amino acids, which are part of

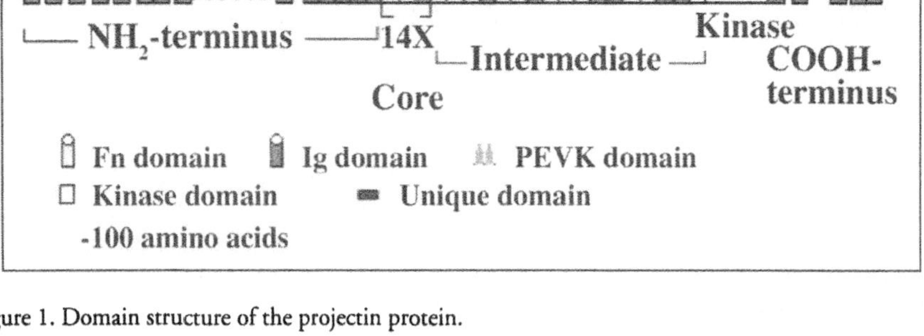

Figure 1. Domain structure of the projectin protein.

the Ig consensus, these eight Ig domains also display amino acid conservation at thirteen other positions, which are not conserved in the remainder of the projectin Ig domains.[28] These regions of higher conservation may reflect a relatively recent increase in the number of Ig domains by intragenic duplication within the NH_2-terminus of the projectin gene or an evolutionary pressure to conserve amino acids at these positions for folding and/or interaction requirements. In the second block found right after the PEVK-like region, the six Ig domains are more divergent from the Ig consensus,[27] but interestingly, individual Ig motifs in this part of projectin show a higher degree of similarity with the twitchin Ig motifs in the equivalent position. One projectin Ig motif also shows a higher degree of identity to some titin Ig motifs than to any twitchin Ig motif.[27] These patterns of similarity might be significant, if these motifs serve specific functions in assembly and anchoring of the projectin filament.

Within the NH_2-terminus, the largest unique sequence interspersed with the Ig domains was identified as a PEVK-like domain because of its high content in the amino acids proline (P), glutamic acid (E), valine (V) and lysine (K).[28] The average PEVK content is 46%, lower than the 75% reported for the PEVK domain of titin[29] (see below). Recently a repeat element has been described for the PEVK domain in titin, as well as in the *Drosophila* proteins D-titin and stretchin.[30-32] Unlike the above, the projectin PEVK-like domain displays a random distribution of the P, E, V and K amino acids with no clearly identifiable repeat structure. The proposed significance of the randomness and the lower PEVK content within the projectin PEVK-like domain is further discussed below.

The core region forms the bulk of the protein and is composed of the pattern [FnIII-FnIII-Ig] repeated 14 times. The same pattern is found in twitchin, whereas the core region of titin is composed of a different but probably functionally equivalent pattern of Ig and FnIII domains.[29,33,34] The intermediate region is also composed of Ig and FnIII domains but with no distinct pattern (see Fig. 1). A similar shift from the regular core pattern to the intermediate region is also found in vertebrate titin and *C. elegans* twitchin.

Projectin also contains a kinase domain located immediately after the intermediate region, which shows high homology to smooth muscle myosin light chain kinase.[25] Projectin has been shown to be a functional kinase protein capable of autophosphorylation (see below). The importance of the intermediate region for the correct positioning of the kinase domain with its substrate has been suggested in twitchin.[23,35]

The kinase domain is immediately followed at the COOH-terminus by five Ig domains interspersed with unique short sequences (see Fig. 1). The sequence of the unique regions is not conserved with equivalent regions within twitchin or titin, but their lengths are comparable.[27] It could be that the length, rather than the sequence, of these unique regions is important for correct positioning or assembly of projectin within the sarcomere.[27]

The global projectin domain structure, therefore, appears as a composite between twitchin and titin. The core, intermediate, and COOH-terminus are almost identical in their organization to the corresponding regions of *C. elegans* twitchin. The NH_2-terminus is more like a shortened version of titin's NH_2 terminus. The only important difference resides in the absence from the projectin sequence of the titin Z-repeats, which are involved in anchoring titin to the COOH-terminus of α-actinin within the Z band.[36]

Projectin Isoforms and Alternative Splicing

Early studies suggested that projectin is present as different isoforms. Projectin from various insect muscle types differs significantly in size as judged by mobility on SDS-polyacrylamide gels.[22] The isoform from asynchronous muscles (IFMs) is smaller than either of the two forms detected from various synchronous muscles. Partial proteolytic digests of the different isoforms from various muscle types yield similar but not identical patterns and it seems likely the various isoforms differ by the omission/inclusion of a few specific domains.[22] As projectin is a single copy gene, the most likely mechanism to generate several different isoforms is through alternative splicing events of the primary transcript. The coding region for the projectin gene covers 51.8 Kb with 46 exons and the longest projectin transcript is estimated to be 27.3 Kb[28] (see below).

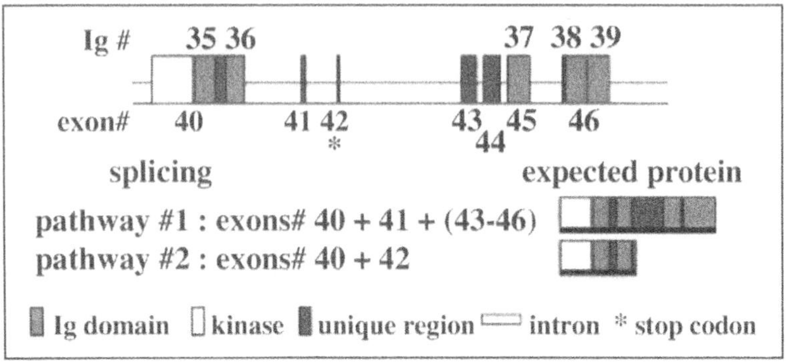

Figure 2. Alternative splicing at the 3'end of the projectin gene.

At this time, alternative splicing events have been described at both the 3' and 5' ends of the projectin gene. At the 3' end of the gene, two small exons are mutually exclusive (exons #41 and 42 in Fig. 2).

Use of exon #42 leads to the presence of an early termination codon that results in a protein shorter by 401 amino acids (or 40 kDa). The COOH-terminus of the shorter isoform only includes two Ig domains and one short unique sequence (Fig. 2, pathway #2). Use of the alternative exon #41 allows translation to continue, and the longer COOH-terminus includes five Ig motifs and all three unique regions[27] (Fig. 2, pathway #1). This alternative pathway, therefore, generates two isoforms with different COOH-termini (Fig. 2). The muscle-specificity (if any) of these two spliced forms still need to be confirmed but evidence from in situ hybridization points to the possibility that the longer form is synchronous muscle-specific (A. Ayme-Southgate, unpublished observations).

At the NH_2-terminus, the genomic region for the PEVK-like domain spans thirteen exons (#12 to 24, see Fig. 3). RT-PCR analysis using total adult RNA has indicated an extremely complex alternative splicing pattern for the PEVK-like region[28] identifying all the splicing possibilities indicated in Figure 3. Of the thirteen exons from the PEVK domain, only two are included in all the alternative splice forms characterized so far (exons #13 and 22). They are at this time considered as constitutive exons, whereas all the other exons are used alternatively.

As presented in Figure 3, over 70 different splice forms are theoretically possible, but it is still uncertain whether all the combinations between the three subregions do actually occur.[28]

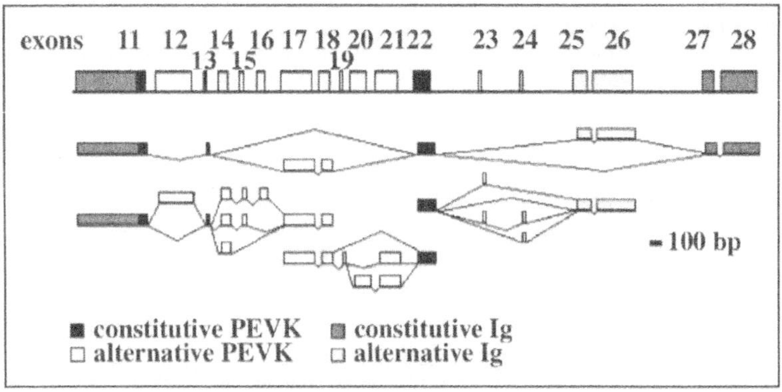

Figure 3. Alternative splicing within the PEVK-like domain.

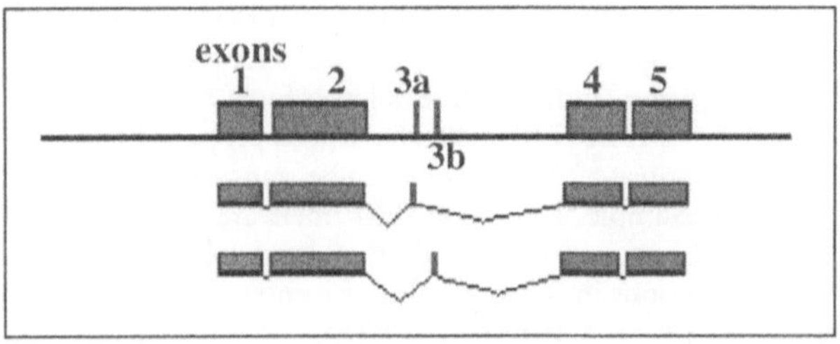

Figure 4. Alternative splicing within the NH_2-terminal region.

Due to this alternative splicing pattern, the length of the projectin's PEVK-like domain can range considerably from 100 to 624 amino acids. The PEVK content of the projectin PEVK-like region varies between 42 and 52% for the various exon combinations and lengths. A similar situation exists in titin with the PEVK domain ranging from 163 amino acids in cardiac muscle to 2174 amino acids in human soleus muscle.[37,38] The different lengths of the PEVK domain in titin have been correlated with the resting tension of the particular muscle type.[38-41]

Two other exons (#3a and #3b) within the NH_2-terminus of the gene are mutually exclusive (Fig. 4). RNA-PCR analysis revealed that exon #3a is used in projectin isoforms found in embryos and adults, whereas exon #3b is only amplified from IFM RNA preparations and may, presumably, be IFM-specific.[28] The amino acid sequence encoded by these two exons is a short unique sequence with no homology to any other known protein, including titin (#3a: PADEPKEKQID; #3b: THGKFSLLENEGQID).

An interesting exon/intron pattern is also present in the beginning of the core region, where four exons, each containing the pattern $[Fn\text{-}Fn\text{-}Ig]_n$, are present. Sequence and splicing pattern indicate that each of these exons or any combinations of them could be alternatively used.[27] This mechanism would generate proteins with varying length but probably not with very different overall protein structures. RNA-PCR analysis to test this idea is under progress. It is interesting to note that the central core domain of projectin is longer than the equivalent region in twitchin by four [Fn-Fn-Ig] patterns.

Possible Functions of the Different Projectin Domains Based on Analogies with Other Family Members

The Ig-like sequences in the twitchin/titin family are known to form Ig folds, as demonstrated for both one titin and one twitchin Ig domain.[42,43] Ig and FnIII domains, when present in extracellular and cell surface proteins, are known to be involved in recognition or adhesion properties.[44,45] Multiple lines of evidence support the notion that similar motifs in proteins of the twitchin/titin family also function in protein-protein interactions. For example,[46-48] two to seven Ig and FnIII domains of titin/connectin were expressed and purified from bacteria and shown to interact with the myosin rod and Myosin Binding Protein-C (abbreviated as MyBP-C). An Ig domain from the *Drosophila* protein, kettin, was also shown to bind to actin and α-actinin, but not to myosin.[49] MyBP-C was shown to have most of its myosin-binding activity localized to the COOH-terminal Ig domain.[50] MyBP-C has been found to also interact with the 11-domain super-repeat of the A-band titin/connectin but not with expressed portions from other regions of titin.[51] Based on this accumulated evidence of protein interactions mediated by Ig and FnIII domains, it is very probable that Ig and/or FnIII domains of projectin will also be implicated in such associations. Accordingly, initial studies show that a His-fusion protein representing part of projectin's core region interacts with myosin (AAS unpublished observations).

Multiple studies have indicated that the I band region of titin is a composite spring system. The Ig domains straighten at low stretch (without unfolding) followed by the unfolding of the

PEVK region at higher stretch.[39,40,52-56] Accordingly, at a given sarcomere length, shorter PEVK segments would lead to a high resting tension, whereas a longer extensible region would result in a lower fractional extension, and, therefore, a lower force.

The elastic properties of IFM C-filaments have been proposed to be a key determinant of the stretch-activation response.[2,3,9,11] For stress to be transmitted to the thick filaments, the fibers must be stiff; for example, *Lethocerus* IFM fibers are 28 times as stiff as rabbit psoas fibers.[57] This high stiffness is due to both the short I band and the elasticity of the C-filaments.[58,59] Contractions in IFMs are almost isometric and stretch activation develops with very small changes in sarcomere length[57] (approximately 2%). The extension of the C-filaments needed during stretch is, therefore, less than that required from the titin filaments in vertebrate muscles. Since projectin is believed to be the main protein component of the C-filaments (the other component, kettin,[60] is described in the chapter by B. Bullard et al), the high resting tension of the IFM fibers is likely due, at least in part, to projectin.[11,12,61] Using an antibody directed against the first two Ig domains of projectin, recent immuno-electron microscopy data indicates that the projectin molecule is orientated with its NH_2-terminus embedded within the Z-band (A. Ayme-Southgate and J. Saide, unpublished observation; see J. Saide's chapter in this volume for further details). This orientation would likely position the PEVK-like domain over the short I-band region. If the spring model described for titin's I-band holds true for projectin as well, we predict that the projectin isoform found within the IFMs' sarcomeres has a short PEVK-like region, consistent with the high resting tension of these muscles. The lower PEVK content of the PEVK-like domain of projectin could also be an adaptation to provide for lower extensibility of this region in projectin compared to titin's. A 40-50% PEVK content would have only limited potential for poly-Proline helix-coil structures,[30] but would be sufficient and adapted to the small amount of extension associated with stretch-activation, compared to the higher changes in sarcomere length which are characteristic of vertebrate muscles.

Kinase Activity and Phosphorylation

The projectin kinase domain is located towards the COOH-terminus at the same relative position as in twitchin and titin.[25] The catalytic core shares a high degree of similarity (61%) with twitchin kinase as well as smooth muscle myosin light chain kinase.[25] The region immediately COOH-terminal of the projectin kinase is well conserved with twitchin and, as shown for twitchin, could serve a regulatory function working as a kinase pseudosubstrate inhibitor.[62,63] Projectin is phosphorylated in vivo on serine residues[17] and both the IFM and synchronous isoforms of projectin are capable of autophosphorylation in vitro.[17,25] The substrate(s) for the projectin kinase are, however, still unknown, as well as the in vivo role, if any, of the autophosphorylation activity.

Projectin Mutant Alleles and Their Phenotypes

The projectin gene was mapped on the fourth chromosome[24] to polytene region 102C/D. The $bent^{Dominant}$ mutation (abbreviated as bt^D) was originally described as a deletion, probably because the bt^D chromosome failed to complement two complementation groups *l(4)2* and *l(4)23* on the fourth chromosome.[64,65] Genomic Southern analysis shows that the $bent^D$ chromosome does have a disruption in the projectin gene and one breakpoint of the bt^D rearrangement was mapped to either within or just NH_2-terminal to the projectin kinase domain.[25,26] The pattern of hybridization and data from in situ hybridization to polytene chromosomes suggest that the most probable rearrangement is either a large insertion or an inversion. The position of the potential second breakpoint is not known, but it presumably does affect the locus of the lost *l(4)23* complementation group. Flies heterozygous for bt^D have normal IFM ultrastructure and normal flight ability. However mechanics experiments on skinned fibers showed reduced stretch-activation and oscillatory work, consistent with projectin role as an elastic component of the C-filaments.[66,67]

In his mutational analysis of chromosome 4, Hochman described the *lethal(4)2* complementation group as a "highly mutable locus."[65] Three of the 35 alleles Hochman isolated in this complementation group are still available and all three alleles are homozygous recessive lethal, with the heterozygous condition normal and capable of flight. To date, there are no homozygous viable flightless mutants of projectin. The three *l(4)2* alleles fail to complement each other in any combination and, consistent with the original data, the bt^D chromosome fails to complement any of these three *l(4)2* alleles.[25] Accordingly, the three *l(4)2* alleles were renamed $bt^{l\text{-}a}$, $bt^{l\text{-}b}$ and $bt^{l\text{-}k}$. Alleles $bt^{l\text{-}a}$ and $bt^{l\text{-}k}$ show alterations in the projectin sequence large enough to produce detectable changes by total genomic Southern analysis.[25,26] The $bt^{l\text{-}a}$ allele is a small insertion in the core region of the projectin gene, causing premature termination[26] and the resulting polypeptide is severely truncated by ~50%. The $bt^{l\text{-}k}$ allele is either a large insertion or an inversion in the COOH-terminus of projectin, reducing the number of Ig motifs found at the COOH-terminus of the protein from five to two.[25]

The timing of death for all the alleles was determined and the analysis showed that the bt^D homozygotes die as late embryos, well after the formation of muscles. Muscle contractions can be seen in all embryos of this stock, even those that do not hatch, giving the impression that they were unable to emerge from the chorion. It is hypothesized that the weaker contractions do not produce the amount of necessary force to actually open the chorion.[25] Similar phenotypic analysis shows that $bt^{l\text{-}a}$ and $bt^{l\text{-}b}$ homozygotes die as late embryos. In contrast to the bt^D embryos, the $bt^{l\text{-}a}$ homozygotes never show any of the spontaneous contractions. This could be related to the severe truncation associated with the mutation, even though the mutant protein is expressed and stable (AAS, unpublished observations). The other allele, $bt^{l\text{-}k}$, survives the embryonic period and hatches. Death occurs during the larval stage, and interestingly some larvae die during first instar, whereas others survive up to early third instar (ref. 25 and unpublished observations). The effect of these mutated proteins on sarcomeric assembly is still under investigation.

Future Prospects

To assess the importance of projectin in the C-filaments for both flight physiology and myofibril assembly, we need to create a mutant fly which would be projectin-null only in the IFMs. Traditional mutagenesis schemes have not been successful at obtaining such a mutant allele, and the size of the projectin gene prevents the use of traditional P-element based transgenic approaches. A different approach might provide such a projectin-deficient stock through the combined use of RNA interference and control of transcription provided by the Gal4/UAS system to limit the formation of double-stranded RNA in the IFMs. This approach is under evaluation. With such projectin-null stocks, projectin's role as the elastic component of the C-filaments should be definitely established, as well as, the importance of projectin in sarcomere assembly and integrity.

References

1. Jewell BR, Ruegg C. Oscillatory contraction of insect fibrillar muscle after glycerol extraction. Proc R Soc Lond B 1966; 164:428-459.
2. Crossley AC. The morphology and development of the Drosophila muscular system. In: Ashburner M, Wright TRF, eds. The genetics and Biology of Drosophila, Vol 2B. London: Academic Press, 1978:499-559.
3. Pringle JWS. Stretch activation of muscle: Function and mechanism. Proc R Soc Lond B 1978; 201:107-130.
4. Wray JS. Filament geometry and the activation of insect flight muscle. Nature 1979; 280:325-326.
5. Squire JM. Muscle filament lattices and stretch activation: The match-mismatch model reassessed. J Muscle Res Cell Motil 1992; 13:183-189.
6. Auber J, Couteaux R. Ultrastructure de la strie Z dans des muscles de dipteres. J Microscopie 1963; 2:309-324.
7. Saide JD, Ullrick WC. Purification and properties of the isolated honeybee Z-Disc. J Mol Biol 1974; 87:671-683.

8. Ashhurst DE. The Z-line: Its structure and evidence for the presence of connecting filaments. In: Tregear RT, ed. Insect Flight Muscle: Proceedings of the Oxford Symposium. Amsterdam, North Holland: Elsevier, 1977:57-73.
9. Trombitas K, Tigyi-Sebe A. Fine Structure and mechanical properties of insect muscle gels to nitrocellulose sheets: Procedure and some applications. In: Tregear RT, ed. Insect flight Muscle: Proceedings of the Oxford Symposium. Amsterdam, North Holland: Elsevier, 1977:79-90.
10. Deatherage JF, Cheng N, Bullard B. Arrangement of filaments and cross-links in the bee flight muscle Z disk by image analysis of oblique sections. J Cell Biol 1989; 108:1775-1782.
11. Trombitas K. Connecting filaments: A Historical prospective. In: Pollack GH, Granzier H, eds. Proceedings: Elastic Filaments of the Cell. New York: Kluwer Academic/Plenum Publishers, 2000:1-23.
12. Saide JD. Identification of a connecting filament protein in insect fibrillar flight muscle. J Mol Biol 1981; 153:661-679.
13. Saide JD, Chin-Bow S, Hogan-Sheldon J et al. Characterization of components of Z-Bands in the fibrillar flight muscle of Drosophila melanogaster. J Cell Biol 1989; 109:2157-2167.
14. Hu DH, Kimura S, Maruyama K. Sodium dodecyl sulfate-gel electrophoresis of connectin-like high molecular weight proteins of various types of vertebrate and invertebrate muscles. J Biochem 1986; 99:485-1492.
15. Hu DH, Matsuno A, Terakado K et al. Projectin is an invertebrate connectin (Titin): Isolation from crayfish claw muscle and localization in crayfish claw muscle and insect flight muscle. J. Muscle Res Cell Motil 1990; 11:497-511.
16. Nave R, Weber K. A myofibrillar protein of insect muscle related to vertebrate Titin connects Z band and A band: Purification and Molecular Characterization of Invertebrate Mini-titin. J Cell Sci 1990; 95:535-544.
17. Maroto M, Vinos J, Marco R et al. Autophosphorylating protein kinase activity of Titin-like arthropod projectin. J Mol Biol 1992; 224:287-291.
18. Vibert P, Edelstein SM, Castellani L et al. Mini-Titins in striated and smooth molluscan muscles: Structure, location and immunological crossreactivity. J Muscle Res and Cell Motil 1993; 14:598-607.
19. Bullard B, Hammond KS, Luke BM. The site of paramyosin in insect flight muscle and the presence of an unindentified protein between myosin filaments and Z line. J Mol Biol 1977; 115:417-440.
20. Locker RH, Wild DJC. A comparative study of high molecular weight proteins in various types of muscle across the animal kingdom. J Biochem 1986; 99:1473-1484.
21. Lakey A, Ferguson C, Labeit S et al. Identification and localization of high molecular weight proteins in insect flight and leg muscles. EMBO J 1990; 9:3459-3467.
22. Vigoreaux JO, Saide JD, Pardue M-L. Structurally different Drosophila striated muscles utilize distinct variants of Z-Band associated proteins. J Muscle Res Cell Motil 1991; 12:340-354.
23. Benian GM, Kiff JE, Neckelmann N et al. Sequence of an unusually large protein implicated in regulation of myosin activity in C. elegans. Nature 1989; 342:45-50.
24. Ayme-Southgate A, Vigoreaux JO, Benian GM et al. Drosophila has a twitchin/titin-related gene that appears to encode projectin. Proc Natl Acad Sci USA 1991; 88:7973-7977.
25. Ayme-Southgate A, Southgate R, Saide J et al. Both synchronous and asynchronous muscle isoforms of projectin (the Drosophila bent locus product) contain functional kinase domains. J Cell Biol 1995; 128:393-403.
26. Fyrberg CC, Labeit S, Bullard B et al. Drosophila projectin: Relatedness to Titin and Twitchin and correlation with lethal (4) 102CDa and bent-dominant mutants. Proc R Soc Lond B 1992; 249:33-40.
27. Daley JK, Southgate R, Ayme-Southgate A. Structure of the Drosophila projectin protein: Isoforms and implication for projectin filament assembly. J Mol Biol 1998; 279:201-210.
28. Southgate R, Ayme-Southgate A. Alternative splicing of an amino-terminal PEVK-like region generates multiple isoforms of Drosophila projectin. J Mol Biol 2001; 313:1035-1043.
29. Labeit S, Kolmerer B. Titins: Giant proteins in charge of muscle ultrastructure and elasticity. Science 1995; 270:293-296.
30. Gutierrez-Cruz G, van Heerden A, Wang K. Modular motifs, structural folds and affinity profiles of PEVK segments of human fetal skeletal muscle Titin. J Biol Chem 2001; 276:7442-7449.
31. Machado C, Andrews D. D-TITIN: A giant protein with dual roles in chromosomes and muscles. J Cell Biol 2000; 151:639-651.
32. Champagne MB, Edwards KA, Erickson HP et al. Drosophila stretchin-MLCK is a novel member of the Titin/Myosin light chain kinase family. J Mol Biol 2000; 300:759-777.

33. Labeit S, Barlow DP, Gautel M et al. A regular pattern of two types of 100 residue motif in the sequence of Titin. Nature 1990; 345:273-276.
34. Labeit S, Gautel M, Lakey A et al. Towards a molecular understanding of Titin. EMBO J 1992; 11:1711-1716.
35. Benian GM, Ayme-Southgate A, Tinley TL. The genetics and molecular biology of the Titin/Connectin-like proteins of invertebrates. Rev Physiol Biochem Pharmacol 1999; 138:235-268.
36. Gautel M, Goulding D, Bullard B et al. The central Z-disk region of Titin is assembled from a novel repeat in variable copy numbers. J Cell Sci 1996; 109:2747-2754.
37. Kolmerer B, Witt CC, Freiburg A et al. The Titin cDNA sequence and partial genomic sequences: Insights into the molecular genetics, cell biology and physiology of the Titin filament system. Rev Physiol Biochem Pharm 1999; 138:21-55.
38. Freiburg A, Trombitás K, Hell W et al. Series of exon-skipping events in the elastic spring region of Titin as the structural basis for myofibrillar elastic diversity. Circ Res 2000; 86:1114-1121.
39. Gautel M, Lehtonen E, Pietruschka F. Assembly of the cardiac I-band region of Titin/connectin: Expression of the cardiac-specific regions and their structural relation to the elastic segments. J Muscle Res Cell Motil 1996; 17:4449-4461.
40. Linke WA, Ivemeyer M, Olivieri N et al. Towards a molecular understanding of the elasticity of Titin. J Mol Biol 1996; 261:62-71.
41. Cazorla O, Freiburg A, Helmes M et al. Differential expression of cardiac Titin isoforms and modulation of cellular stiffness. Circ Res 2000; 86:59-67.
42. Pfuhl M, Pastore A. Tertiary structure of an immunoglobulin-like domain from the giant muscle protein Titin: A new member of the I Set. Structure 1995; 3:391-401.
43. Fong S, Hamill S, Proctor M et al. Structure and stability of an immunoglobulin domain from Twitchin, a muscle protein of the nematode Caenorhabditis elegans. J Mol Biol 1996; 264:624-639.
44. Williams AF, Davis SJ, He Q et al. Structural diversity in domains of the immunoglobulin superfamily. Cold Spring Harbor Symp Quant Biol 1989; 54:637-647.
45. Bork P, Holm L, Sander C. The immunoglobulin fold: Structural classification, sequence patterns and common core. J Mol Biol 1994; 242:309-320.
46. Soteriou A, Gamage M, Trinick J. A Survey of interactions made by the giant protein Titin. J Cell Sci 1993; 104:119-123.
47. Gautel M. The super-repeats of Titin/connectin and their interactions: Glimpses at sarcomeric assembly. Adv Biophys 1996; 33:27-37.
48. Trinick J. Interactions of Titin/connectin with the thick filament. Adv Biophys 1996; 33:81-90.
49. Lakey A, Labeit S, Gautel M et al. Kettin, a large modular protein in the Z-Disc of insect muscles. EMBO J 1993; 12:2863-2871.
50. Okagaki T, Weber FE, Fischman DA et al. The major myosin-binding domain of skeletal muscle MyBP-C (C-Protein) resides in the COOH-terminal, immunoglobulin C2 motif. J Cell Biol 1993; 123:619-626.
51. Freiburg A, Gautel M. A molecular map of the interactions between Titin and myosin-binding protein C: Implications for sarcomeric assembly in familial hypertrophic cardiomyopathy. Eur J Biochem. 1996; 235:317-323.
52. Gautel M, Goulding D. A molecular map of Titin/connectin elasticity reveals two different mechanisms acting in series. FEBS Lett 1996; 385:11-14.
53. Linke WA, Ivemeyer M, Mundel P et al. Nature of PEVK-Titin elasticity in skeletal muscle. Proc Natl Acad Sci USA 1998; 95:8052-8057.
54. Linke WA, Rudy DE, Centner T et al. I-band Titin in cardiac muscle is a three-element molecular spring and is critical for maintaining thin filament structure. J Cell Biol 1999; 246:631-644.
55. Greaser ML, Wang SM, Berri M et al. Sequence and mechanical implications of cardiac PEVK. In: Pollack GH, Granzier H, eds. Proceedings: Elastic Filaments of the Cell. New York: Kluwer Academic/Plenum Publishers, 2000:53-63.
56. Granzier H, Helmes M, Cazorla O et al. Mechanical properties of Titin isoforms. In: Pollack GH, Granzier H, eds. Proceedings: Elastic Filaments of the Cell. New York: Kluwer Academic/Plenum Publishers, 2000:283-300.
57. Peckham M, White DCS. Mechanical properties of demembranated flight muscle fibers from a dragonfly. J Exp Biol 1991; 159:135-147.
58. White DCS. The elasticity of relaxed insect fibrillar flight muscle. J Physiol 1983; 343:31-57.
59. Granzier HL, Wang K. Passive tension and stiffness of vertebrate skeletal and insect flight muscles: The contribution of weak cross-bridges and elastic filaments. Biophys J 1993; 65:2141-2159.
60. Kulke M, Neagoe C, Kolmerer B et al. Kettin, a major source of myofibrillar stiffness in Drosophila indirect flight muscle. J Cell Biol 2001; 154:1045-1057.

61. Bullard B, Goulding D, Ferguson C et al. Links in the chain: the contribution of kettin to the elasticity of insect muscles. In: Pollack GH, Granzier H, eds. Proceedings: Elastic Filaments of the Cell. New York: Kluwer Academic/Plenum Publishers, 2000:207-220.
62. Hu S-H, Parker MW, Lei J et al. Intrasteric regulation of protein kinases: Insights from the crystal structure of Twitchin kinase. Nature 1994; 369:581-584.
63. Lei J, Tang X, Chambers T et al. The protein kinase domain of Twitchin has protein kinase activity and an autoinhibitory region. J Biol Chem 1994; 269:21078-21085.
64. Hochman B. Analysis of chromosome 4 in Drosophila melanogaster. II. Ethylmethanesulfonate induced lethals. Genetics 1971; 67:235-252.
65. Hochman B. Analysis of a whole chromosome in Drosophila. Cold Spr Harb Symp Quant Biol 1974; 38:581-589.
66. Moore JR, Vigoreaux JO, Maughan DW. The Drosophila projectin mutant, $bent^{D}$, has reduced stretch activation and altered flight muscle kinetics. J Musc Res Cell Motil 1999; 20:797-806.
67. Vigoreaux JO, Moore JR, Maughan DW. Role of the elastic protein projectin in stretch activation and work output of Drosophila flight muscles. In: Pollack GH, Granzier H, eds. Proceedings: Elastic Filaments of the Cell. New York: Kluwer Academic/Plenum Publishers, 2000:237-247.

CHAPTER 14

Some Functions of Proteins from the *Drosophila sallimus* (*sls*) Gene

Belinda Bullard, Mark C. Leake and Kevin Leonard

Abstract

Insect flight muscles contract at high frequencies and are activated by periodically stretching the muscles. For the stretch to have an effect, the muscles must be stiff. Two elastic proteins, projectin and kettin, are responsible for a large part of the muscle stiffness. Thin filaments containing actin emerge from Z-discs, which occur at periodic intervals along the myofibril, and thick filaments containing myosin interdigitate with the thin filaments. Both projectin and kettin form a mechanical link between the Z-discs and the ends of thick filaments. Kettin is made up of immunoglobulin-like (Ig) modules separated by linker sequences, and is associated with actin in the region of the Z-disc. The protein is an isoform derived from the *Drosophila sallimus* (*sls*) gene. Longer isoforms from the *sls* gene have additional, more extensible, sequence and these are found in non-flight muscles that are less stiff. Isoforms of the protein Sls have several different functions. Kettin causes thin filaments to align side-by-side in an anti-parallel fashion, which could nucleate Z-disc formation in developing myofibrils. Kettin is in the enlarged Z-discs close to the site of attachment of myofibrils to the cuticle, and may reinforce actin filaments in this region, giving the structure the required stiffness.

Sls appears early in development of the *Drosophila* embryo and is needed for fusion of myoblasts to form myotubes which will become muscle fibres. Sls is associated with the membrane at the site of myoblast fusion, together with other proteins (Duf and Rols) that are needed for fusion.

The elastic properties of single molecules of kettin have been measured using optical tweezers. The Ig domains unfold at relatively low stretching forces and refold at high forces. This suggests that kettin could be a folding-based spring, which may be relevant to its function in early muscle development, as well as in the adult myofibril.

Introduction

All striated muscles have large modular proteins that determine the elastic properties of the sarcomere. The *Drosophila* thorax contains muscles that vary widely in function, and the ultrastructure of the sarcomere is correspondingly varied. The structure of the indirect flight muscles (IFMs) is characteristic of asynchronous muscle. In these muscles, oscillatory contractions are produced by a delayed response to stretch, combined with the resonant properties of the thorax and wings. The high frequency (200 Hz) of oscillations in *Drosophila* IFM is possible because the muscles themselves are stiff; they have short I-bands and the sarcomere length changes very little during the contractile cycle. The relative inextensibility of IFM is due to connecting filaments extending from the Z-disc to the end of the thick filaments.[1] These filaments have two components: kettin and projectin, both of which contribute to the high resting tension.[2]

Nature's Versatile Engine: Insect Flight Muscle Inside and Out, edited by Jim Vigoreaux.

Other muscles in the *Drosophila* thorax have wider I-bands; their function would not require them to be unusually stiff and they have longer connecting filaments.

Kettin and larger isoforms from the *sls* gene have two distinct functions. The first is the regulation of muscle development in the early embryo and the second is in determining the elastic properties of different muscle types. We have recently reviewed the function of elastic proteins in the mechanics of IFM,[3] and here we will concentrate mainly on other functions of kettin and Sls. We also discuss some properties of single kettin molecules.

sls, the Gene

When a large modular protein containing immunoglobulin (Ig) domains is identified in the muscles of any animal, it is often called titin (or connectin). Invertebrate species have several proteins that have some similarity to vertebrate titin and a more specific nomenclature is needed. Vertebrate titin has a characteristic pattern of modules along the molecule, interspersed with domains of unique sequence: a region of tandem Igs, an elastic PEVK domain, more tandem Igs, and a region with a repeating pattern of fibronectin (Fn) and Ig domains, followed by a kinase domain and more Igs. We suggest that if a protein is called titin it should have these defining features, otherwise it should be given a different name.

The first gene coding for a modular elastic protein to be identified in *Drosophila* was the projectin gene.[4,5] Projectin has the pattern of domains characteristic of titin,[6] although the protein (1000 kDa) is about a third the size of titin. Kettin (from the German kette = chain) is a second modular protein in *Drosophila*[7-9] and it has been mapped to chromosome position 62C1-3. The complete cDNA has been sequenced and corresponds to a protein of 540 kDa, largely made up of Ig domains separated by linker sequences of 35 residues; at each end of the molecule there are tandem Ig domains and regions of unique sequence (Fig. 1). Andrew and colleagues[10] obtained partial cDNA sequences from a gene coding for a large protein (larger than kettin) with Ig and PEVK sequence, that is expressed during *Drosophila* embryogenesis; this they called *D-titin*. The gene is at the same chromosomal position as *kettin* and genetic mapping showed that known mutations in the *kettin* region, for example

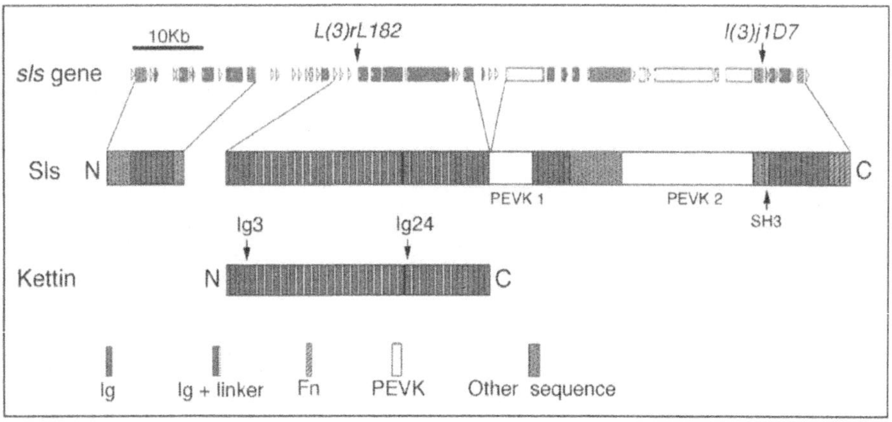

Figure 1. The *sls* gene and Sls protein. The map of open reading frames in the gene is taken from the sequence in FlyBase (http://flybase.bio.indiana.edu); the small triangles in the gene map are short ORFs. The positions of P-element insertions *l(3)rL182* and *l(3)jlD7* in the gene are shown. Domains in the largest protein that is predicted to be encoded by the gene are shown below the gene. The block of sequence at the N-terminus is expressed as an isoform, or spliced to different parts of the sequence downstream (Burkart et al, unpublished result). The domain structure of kettin is shown below the whole predicted Sls protein. Variable splicing pathways produce other isoforms. SH3 is a src homology domain. The positions of the epitopes in kettin (Ig3 and Ig24), reacting with the two antibodies used in stretching experiments are shown.

the P-element insertion *l(3)rL182,* failed to complement another P-element insertion, *l(3)j1D7,* in the region of unique sequence near the end of the gene (Fig. 1). This was evidence that the gene extended beyond *kettin.*[9,11,12] Since the *Drosophila* genome has been sequenced, the entire ~100 kb sequence of the gene is known. The gene is called *sallimus* (*sls*) (from the Finnish = fate) or *ket* in FlyBase and we have used *sls* as an alternative to *D-titin.*[3] There is no orthologous gene in vertebrates. The largest protein that could be encoded by the gene is predicted to be about 1.9 MDa; the modular structure is shown in Figure 1. The N-terminal region has tandem Ig domains and stretches of unique sequence, and this region is expressed as an isoform of Sls. Kettin is another isoform. Further downstream there are tandem Ig regions and two PEVK domains. At the C-terminus there are four Fn domains but the molecule does not have the super repeats of Ig and Fn domains found in titin, nor does it have a kinase domain.

Sls, the Protein

Details of splicing pathways in the *sls* gene have not yet been described. The predominant isoform of Sls in the adult *Drosophila* thorax (detected by SDS-PAGE) is 540 kDa kettin and there are minor amounts of 700 and 1000 kDa isoforms, both of which contain the entire kettin sequence. The flight muscle has mainly 540 and 700 kDa isoforms, while other thoracic muscles including leg muscles, also have the 1000 kDa isoform. All *Drosophila* muscles have 540 kDa kettin and muscles with long I-bands have larger isoforms in addition. Immuno-electron microscopy has shown that the N-terminus of kettin is in the Z-disc and the C-terminal region of the molecule is some way out from the Z-disc.[13] Kettin is associated with actin, and the stoichiometry, determined by binding assays with kettin from the giant waterbug *Lethocerus,* suggests each Ig domain binds to an actin monomer; the linker sequence between Ig domains would space the domains to coincide with helically arranged actin monomers. The cartoon in Figure 2a illustrates the second of the alternative models suggested by van Straaten et al.[13] Recent immunolabelling results, using antibodies to different regions of kettin, favour a model in which kettin crosses the whole width of the Z-disc and follows the long-pitch helix of actin. Labelling with antibody to Ig24 is shown in Figure 2b. Tandem Ig domains in the N-and C-terminal regions of the molecule probably would not bind to actin because there are no linker sequences. The effect of kettin binding to actin is to reinforce the thin filament near the Z-disc. Kettin inhibits the binding of tropomyosin to actin, and the region of the thin filament close to the Z-disc that is occupied by kettin does not have tropomyosin.[13]

There is no protein in *Drosophila* with the domain structure of nebulin in vertebrate skeletal muscle. Kettin has some features in common with nebulin. Both proteins are anchored in the Z-disc, and regularly spaced modules bind to actin monomers; however, kettin, unlike nebulin, could not determine the length of thin filaments because it is not long enough. Sls, like nebulin, has a src homology (SH3) domain near the C-terminus (Fig. 1). Nebulin is oriented with the C-terminus in the Z-disc so that this domain is close to the Z-disc, whereas in Sls, which has the opposite polarity, the SH3 domain is separated from the Z-disc by almost the whole length of the molecule. The vertebrate nebulin SH3 domain binds the signalling molecule, myopalladin, which is essential for myofibril assembly,[14] and the SH3 domain in Sls may also have a signalling function.

In IFM, the region of the thin filament with bound kettin crosses the short I-band; and the C-terminus of kettin is attached to the end of the thick filament, either directly or through association with projectin. This link is estimated to be responsible for 70% of the passive stiffness of IFM.[2] In other thoracic muscles, the longer I-bands are spanned by Sls isoforms that include varying lengths of sequence downstream of kettin, including the two elastic PEVK domains.[15] It is likely that the Fn domains at the C-terminus of the gene are spliced onto the several longer isoforms of Sls, and that these domains link the ends of the molecules to the thick filaments.

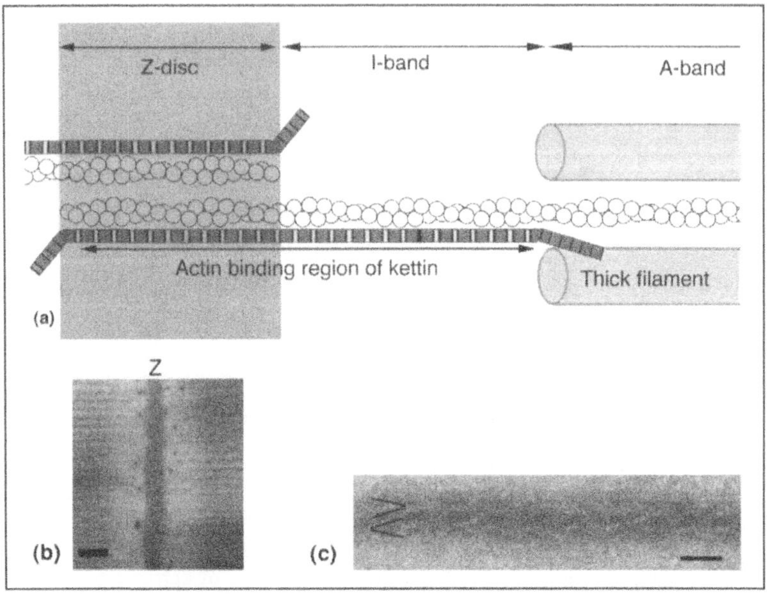

Figure 2. a) A cartoon showing the layout of kettin in the IFM sarcomere near the Z-disc. Each Ig domain in the Ig-linker region of kettin binds to an actin monomer. The model shows kettin crossing the Z-disc and shows that a kettin molecule following the long-pitch actin helix would extend over about 5 half-repeats of the actin helix. Ig domains at the ends of kettin are not separated by linker sequences and are not expected to bind to actin. Regions of unique sequence may be extensible. The C-terminal region is associated with the thick filament. The model is based on the positions of Ig3, Ig16, Ig24 and Ig35 determined by labelling cryosections with antibodies. b) A cryosection of the Z-disc region of a sarcomere labelled with antibody to Ig 24 and protein A-gold. c) Negatively stained paired actin filaments decorated with myosin S1 showing anti-parallel association of filaments. Arrowheads show the polarity of the S1 on actin. Scale bars are 100 nm for (b) and 40 nm for (c).

Kettin may be responsible for the formation of the Z-disc at an early stage in the development of the sarcomere. Thin filaments from neighbouring sarcomeres interdigitate in the Z-disc and adjacent filaments with an anti-parallel orientation are crosslinked by α-actinin. When the 540 kDa isoform of kettin is added to isolated actin filaments, the filaments associate with each other laterally; the orientation of the filaments is shown by decorating them with myosin S1. The arrowheads formed by S1 face in different directions in the filaments of a pair, which suggests that kettin molecules in one filament interact with oppositely oriented ones in another filament[13] (Fig. 2c). Thus actin filaments with associated kettin molecules could form a lattice of interdigitating filaments of alternating polarity in the Z-disc. The requirement for interacting kettin molecules to be anti-parallel would prevent cross-linking of thin filaments in the I-band, where neighbouring filaments have the same orientation.

The IFMs in *Drosophila* are attached to the cuticle by a modified Z-disc at the end of the myofibril. The terminal Z-disc is linked to an epithelial tendon cell and bundles of microtubules in the tendon cell join spike-like tonofibrillae that are inserted into the cuticle. The terminal Z-disc is modified into a wide region of high density.[16] The whole terminal Z-disc is labelled by antibody to kettin, suggesting the structure is composed of actin filaments reinforced by kettin (Fig. 3). No regular lattice has been observed in this terminal Z-disc, even in high quality electron micrographs.[16] This may be because all the thin filaments are likely to have the same polarity and formation of a regular lattice by cross-linking oppositely oriented filaments with α-actinin would not be possible. α-Actinin also binds to kettin[7] and both Sls

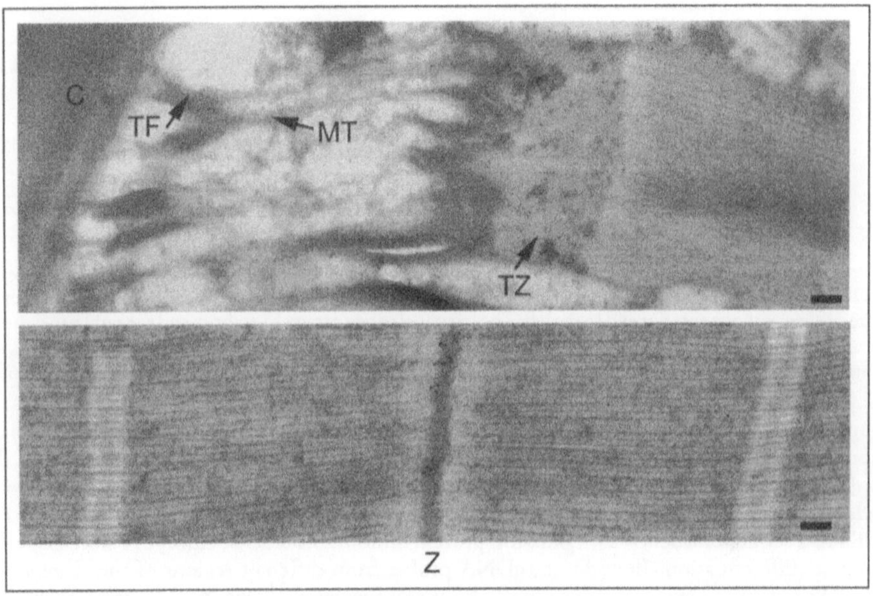

Figure 3. The position of kettin in the terminal Z-disc of *Drosophila* flight muscle. Cryosections are labelled with monoclonal antibody to an epitope in the kettin sequence and protein A-gold. The top panel shows the end of a myofibril and the connection to the cuticle. A modified terminal Z-disc is linked to the cuticle by bundles of microtubules that are attached to tonofibrillae inserted into the cuticle. Kettin is present across the whole of the terminal Z-disc. The lower panel shows kettin in a normal Z-disc in a myofibril, labelled with the same antibody. TZ, terminal Z-disc; MT, microtubule bundle; TF = tonofibrillae; C = cuticle; Z = Z-disc. Scale bar 100 nm.

and α-actinin are essential for the formation of the terminal Z-disc. The structure is particularly sensitive to mutations in these genes. Homozygous *sls* mutations are lethal at stages that vary from embryonic to pupal, but heterozygotes develop into viable adults.[8,9,11,12] Hakeda et al[8] describe a null mutant (*ket*^{14}) with a deletion in the N-terminal region of the kettin sequence. Electron micrographs of flight muscle in adult flies that are heterozygous for this mutation have much reduced terminal Z-discs, although Z-discs in the rest of the sarcomere appear normal, at least before the flight muscle contracts at eclosion. Similarly, electron micrographs of heterozygotes of the α-actinin null allele *l(1)2Cb* show terminal Z-discs that are reduced in size, resulting in disrupted muscle insertions, while the structure of the rest of the myofibril is unaffected.[17] Homozygotes of the *fliA* series with point mutations in the α-actinin gene also have reduced or absent terminal Z-discs, but in this case, the rest of the myofibril is abnormal too.

The requirement that IFM is resistant to stretch means that the connection of the myofibrils to the cuticle must be fairly rigid. The large terminal Z-disc in which actin is reinforced with kettin may provide the correct compliance. Kettin molecules are likely to be arranged end-to-end along actin in the structure; it is also possible that there are longer isoforms of Sls in this part of the sarcomere and that these include PEVK sequence to give just the right stiffness to the connection between muscle fibre and cuticle.

Sls in the *Drosophila* Embryo

Sls appears at an early stage of embryogenesis. The mRNA and protein are detected in the embryo at early stage 11 (5:20 to 7:20 h after egg laying at 25°C), mainly in the gut region. Later during stage 11, the mRNA and protein are seen in the precursors of somatic and visceral

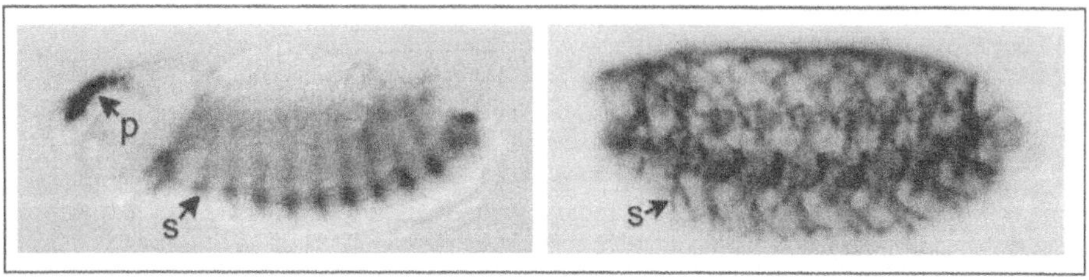

Figure 4. Expression of Sls RNA in *Drosophila* embryos. RNA was detected by in situ hybridization with cDNA coding for part of the sequence in the PEVK 2 domain of Sls. Whole mount embryos are seen in lateral view with the anterior end to the left: The left panel shows a stage 13 embryo and the right panel shows a stage 16 embryo. *Sls* transcripts are seen in pharyngeal and somatic muscles. p = pharyngeal muscle, s = somatic muscle. The figure is modified from Machado C, Andrew DJ. J Cell Biol 2000; 151(3):639-652.

muscles and in the head.[8,9,11,12] Sls is detected several hours before myoblast fusion, which starts at stage 12.[18] It is not clear which isoforms of Sls are expressed in the embryo. In all the studies quoted, antibody to the kettin region of the molecule was used to detect the protein in whole mount embryos. However, Machado et al[10] found that antibodies both to kettin and to PEVK-2 sequence labelled a megadalton protein on immunoblots of embryos before and after muscle differentiation. In addition, cDNA probes from different regions of the *sls* gene detected the same expression pattern throughout embryogenesis[11] (Fig. 4). Therefore, the whole 1.9 MDa protein is probably expressed in the embryo, though there may be other isoforms as well. Myofibrils are formed at stage 16[19] and by this stage, Sls RNA and protein are concentrated at muscle attachment sites. This is seen clearly at the ends of body wall muscles where they are attached to the epidermis.[8,11,12]

Sls and Myoblast Fusion

The importance of Sls in myoblast fusion is shown by the effect of mutants on the fusion process. Unfused myoblasts are seen in stage 16 embryos of homozygous *sls* mutants stained with antibody to kettin; body wall muscles are disorganized and attachment sites are absent, even though many of the mutants express normal levels of protein.[8,11,12] Some idea of the function of Sls in myoblast fusion can be gained by examining the relationship of *sls* to other genes involved in the fusion process. The events leading to myoblast fusion are characterised by the effects of a series of mutants.[20] The body wall muscles of the *Drosophila* embryo are derived from two types of myoblasts originating from the somatic mesoderm. Founder cells determine the characteristics of particular muscles, and these combine with fusion competent myoblasts (FCMs) to form multinucleated myotubes of the type specified by the founder cells.[21] Most genes involved in the fusion process, including *sls*, are expressed in both types of myoblast, but a few are expressed either in founder cells or in FCMs. Two proteins associated with the membrane are present only in founder cells, dumbfounded (Duf), or only in FCMs, sticks and stones (Sns).[22,23] Duf attracts FCMs to the founder cells before fusion. Both Duf and Sns are predicted to be transmembrane proteins with extracellular Ig domains and the two may interact during fusion.[24,25] Changes both at the cell surface and inside the cell, are necessary for successful fusion and rolling pebbles (Rols), another protein specific to founder cells, is thought to link processes occurring in the two regions.[26] The interest of Rols here is that it is needed for migration of Sls to the site of membrane fusion between muscle precursors and FCMs.[26] Rols accumulation at the membrane coincides with Duf, and in *duf* mutants, Rols is not membrane-associated and Sls remain cytosolic. However, Sls is also concentrated in the membrane of FCMs at the site of cell contact with myotubes, even though FCMs do not express Rols; some other unknown mechanism is thought to produce this localisation in FCMs.[26] In general, the expression of genes needed for fusion is transitory and stops once fusion is complete and myotubes have their full complement of nuclei. The localization of Sls once Rols expression has decreased is not known; Sls may no longer be associated with the myotube membrane at sites of fusion and could take up its second function in the mature muscle sarcomere. Again, an antibody specific to the kettin region of Sls was the only one used to localize the protein in fusing myoblasts. Sls isoforms vary in size from 200 kDa to 1.9 MDa and not all isoforms have the kettin sequence. The use of antibodies recognizing different regions of the molecule would show which isoforms are important in the fusion process.

The early appearance of *sls* (*kettin*) RNA was detected in screens for genes that are expressed in the mesoderm.[27] The effect of mutations that reduce or increase the amount of mesoderm, on gene expression were compared up to stage 12. There was a clear difference in the effect of these mutations on the expression of *sls* and *bent* (the projectin gene). In mutants with less mesoderm, *sls* expression is reduced at stage 11-12 compared with expression in wild type embryos, and in those with more mesoderm, *sls* expression is increased from stage 9-11; but *bent* is hardly affected because very little is expressed at these early stages. Thus Sls has a function in the embryo that projectin does not, although both are large modular proteins with Ig domains that have some functions in common in the adult fly. A similar screen using mutant embryos in which either founder cells or FCMs were enriched, showed there is significantly more *sls* expression in FCMs than in founder cells.[28] It is possible that several different isoforms of Sls that will be needed in the final muscle are produced in FCMs.

Kettin as a Spring

The kettin sequence in Sls is highly conserved in different invertebrates, probably because the molecule is associated with actin in the same way in all muscles.[3] As mentioned above, the predominant isoform in *Drosophila* IFM is 540 kDa. Since the function of kettin in determining the stiffness of muscle fibres is rather different from the function of vertebrate titin, it is of interest to compare the elastic properties of the two molecules. The properties of single molecules of 540 kDa kettin isolated from *Lethocerus*,[13] have been determined by mechanical stretch experiments using optical tweezers. Antibodies to two regions of the sequence that are 19 complete Ig-linker modules apart (Fig. 1) were bound to microspheres held in independent optical traps (Fig. 5a,b).[29] Molecules of kettin were held between the microspheres by attachment to the antibodies, and were stretched and released at different rates by changing the separation of the microspheres; force was measured during stretches and releases (Fig. 5c). The force-extension traces showed hysteresis due to step-like events during both stretch and release (Fig. 5d). The calculated change in contour length, using a freely-jointed chain model[30] corresponding to these steps was about 30 nm, which is consistent with unfolding and refolding of single Ig domains. The range of lengths of the stretched region predicted from the sequence[8,9] was larger than that modelled from polymer fits to the curves. This suggested that the inter-Ig linkers act as more condensed structures than a simple string of independent residues, with the long-range order of each extending over at least two neighbouring residues.

Kinetic analysis of the forces at which the steps occurred suggested the spontaneous rate of Ig domain unfolding was high ($\sim 5 \times 10^{-3}\ s^{-1}$). This is two orders of magnitude higher than the spontaneous rate of unfolding of the more stable Ig domains of titin; although it is comparable to the rate of unfolding of some less stable titin Igs that are thought to unfold under physiological conditions.[31] Kettin Ig domains can refold at significantly higher forces than found for titin Ig domains at equivalent rates of loading; this is borne out by the fact that refolding events occur at forces in excess of ~10 pN (Fig. 5d). In the IFM myofibril, most of the kettin sequence is bound to actin, which would mean Ig domains would not be subjected to the high forces experienced by titin. However, the C-terminal tandem Ig domains extend towards the myosin filaments and would be subjected to periodic stretching forces during oscillatory contraction. These Igs may act as shock absorbers; a potentially damaging stretch could result in limited unfolding of Ig domains, and refolding at relatively high forces would

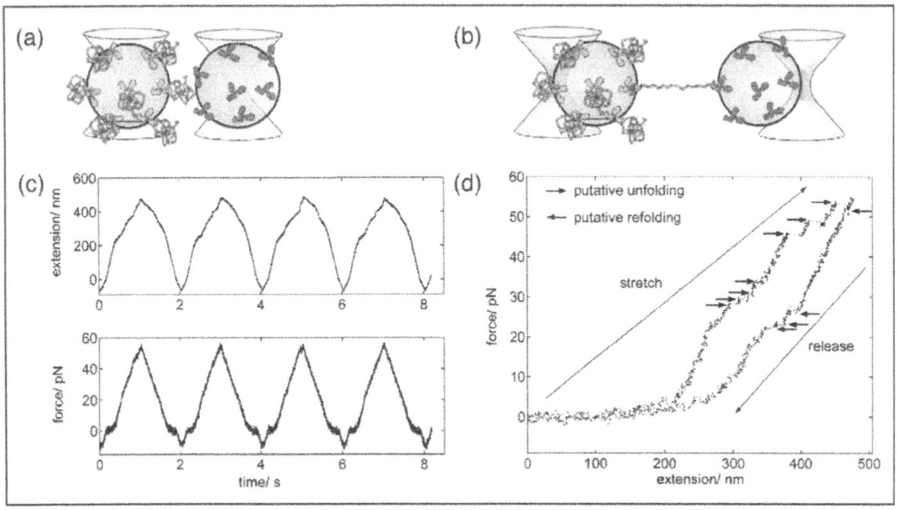

Figure 5. Unfolding of Ig domains during extension of kettin. a) Schematic representation of double laser optical tweezers with two microspheres trapped at the cross over of the laser beam. Single kettin molecules are bound between the microspheres by two antibodies (Fig. 1). b) Beads are subjected to cycles of stretch and release by movement of the right hand laser trap. The bead on the left is held stationary by a feedback system controlling the position of the laser trap. As domains unfold, the elastic force drops and the bead on the right moves towards the centre of its trap. c) Typical data for extension (upper trace) and force (lower trace). Some step events can be made out by eye. d) Example of a stretch-release cycle showing hysteresis. Arrows mark the position of positive (stress-relaxation) and negative (recovery) steps attributed to unfolding and refolding of Ig domains. Changes of slope in the lower part of the curves were not detected as steps. Ig domains unfold at moderate forces and refold at significantly higher forces than observed for vertebrate titin. The figure is modified from Leake MC, Wilson D, Bullard B et al. FEBS Lett 2003; 535(1-3):55-60.

not require the sarcomere to be slack before normal operation could be resumed. The ease with which kettin Ig domains refold suggests that kettin may act as an efficient folding-based spring. This may be relevant to the function of Sls in embryogenesis (see below).

Concluding Remarks

Sls is an elastic molecule that would vary in mechanical properties depending on which modules a particular isoform contained. This means the protein can have more than one function. The *sls* gene has a conserved region coding for kettin. Kettin binds strongly to actin and reinforces the attachment of actin to the Z-disc. In IFM, the C-terminus binds to the thick filament and this contributes to the high stiffness of IFM needed for oscillatory contraction. Alternative splicing of kettin to different ORFs downstream would produce isoforms differing in size and extensibility, which could cater for the differing requirements for the elasticity of different muscle types. When the modular composition and position of minor isoforms of Sls in the IFM sarcomere has been determined, it will be possible to make a more accurate prediction of the contribution of the sum of the isoforms to the passive elasticity of the muscle. The relative ease of unfolding and refolding of kettin Igs is not typical of Ig domains. Measurement of the elastic properties of recombinant peptides from different regions of Sls, using optical tweezers or atomic force microscopy, will also be necessary before the contribution of isoforms to passive tension can be determined; as has been shown for vertebrate titin.[31,32]

The need for Sls in myoblast fusion is unexpected because it is not clear what the function of the defined number of actin binding modules would be, or how they might attach Sls to the region of fusion between myoblast membranes. Nor is it known what the function of long elastic PEVK regions might be in fusing myoblasts. Recently, it has been shown that mechanical stresses are sensed by cells in the developing *Drosophila* embryo.[33,34] As the cells in the embryo change shape and migrate, forces are produced by movement relative to the extracellular matrix. These forces affect the patterning of the embryo by changing gene expression. Sls, as an elastic molecule, is expected to respond to stress. A possible function of Sls in embryogenesis might be as an intermediary between a mechanical change and signalling to the nucleus. The kettin region of Sls may function as a folding based spring during embryogenesis, before it has any function in Z-disc nucleation. More is known about the properties of Sls than any other protein in the cascade involved in fusion. Most are known only as genes. Sls might be the protein to start with in studying the interaction of proteins making up the complex at the site of membrane fusion.

References

1. White DC. The elasticity of relaxed insect fibrillar flight muscle. J Physiol 1983; 343:31-57.
2. Kulke M, Neagoe C, Kolmerer B et al. Kettin, a major source of myofibrillar stiffness in Drosophila indirect flight muscle. J Cell Biol 2001; 154:1045-1057.
3. Bullard B, Linke WA, Leonard K. Varieties of elastic protein in invertebrate muscles. J Muscle Res Cell Motil 2002; 23:435-447.
4. Ayme-Southgate A, Vigoreaux J, Benian G et al. Drosophila has a twitchin/titin-related gene that appears to encode projectin. Proc Natl Acad Sci USA 1991; 88:7973-7977.
5. Fyrberg CC, Labeit S, Bullard B et al. Drosophila projectin: Relatedness to titin and twitchin and correlation with lethal(4)102 CDa and bent-dominant mutants. Proc R Soc Lond B Biol Sci 1992; 249:33-40.
6. Southgate R, Ayme-Southgate A. Alternative splicing of an amino-terminal PEVK-like region generates multiple isoforms of Drosophila projectin. J Mol Biol 2001; 313(5):1035-1043.
7. Lakey A, Labeit S, Gautel M et al. Kettin, a large modular protein in the Z-disc of insect muscles. EMBO J 1993; 12:2863-2871.
8. Hakeda S, Endo S, Saigo K. Requirements of Kettin, a giant muscle protein highly conserved in overall structure in evolution, for normal muscle function, viability, and flight activity of Drosophila. J Cell Biol 2000; 148:101-114.
9. Kolmerer B, Clayton J, Benes V et al. Sequence and expression of the kettin gene in Drosophila melanogaster and Caenorhabditis elegans. J Mol Biol 2000; 296(2):435-448.
10. Machado C, Sunkel CE, Andrew DJ. Human autoantibodies reveal titin as a chromosomal protein. J Cell Biol 1998; 141:321-333.
11. Machado C, Andrew DJ. D-Titin: A giant protein with dual roles in chromosomes and muscles. J Cell Biol 2000; 151:639-652.
12. Zhang Y, Featherstone D, Davis W et al. Drosophila D-titin is required for myoblast fusion and skeletal muscle striation. J Cell Sci 2000; 113 (Pt 17):3103-3115.
13. van Straaten M, Goulding D, Kolmerer B et al. Association of kettin with actin in the Z-disc of insect flight muscle. J Mol Biol 1999; 285:1549-1562.
14. Bang ML, Mudry RE, McElhinny AS et al. Myopalladin, a novel 145-kilodalton sarcomeric protein with multiple roles in Z-disc and I-band protein assemblies. J Cell Biol 2001; 153:413-427.
15. Bullard B, Hååg P, Brendel S et al. Mapping kettin and D-titin in the invertebrate sarcomere. J Mus Res Cell Motil 2002; 22:602.
16. Reedy MC, Beall C. Ultrastructure of developing flight muscle in Drosophila. II. Formation of the myotendon junction. Dev Biol 1993; 160(2):466-479.
17. Fyrberg E, Kelly M, Ball E et al. Molecular genetics of Drosophila alpha-actinin: Mutant alleles disrupt Z disc integrity and muscle insertions. J Cell Biol. 1990; 110(6):1999-2011.
18. Hartenstein V. Atlas of Drosophila development. In: Bate M, Martinez Arias A, eds. The Development of Drosophila melanogaster Vol II. Plainview, New York: Cold Spring Harbor Laboratory Press, 1993.
19. Bate M. The mesoderm and its derivatives. In: Bate M, Martinez Arias A, eds. The Development of Drosophila melanogaster. Vol II. Plainview, New York: Cold Spring Harbor Laboratory Press, 1993:1013-1090.
20. Taylor MV. Muscle differentiation: How two cells become one. Curr Biol 2002; 12(6):R224-228.
21. Bate M. The embryonic development of larval muscles in Drosophila. Development 1990; 110:791-804.

22. Ruiz-Gomez M, Coutts N, Price A et al. Drosophila dumbfounded: A myoblast attractant essential for fusion. Cell 2000; 102(2):189-198.
23. Bour BA, Chakravarti M, West JM et al. Drosophila SNS, a member of the immunoglobulin superfamily that is essential for myoblast fusion. Genes Dev 2000; 14(12):1498-1511.
24. Frasch M, Leptin M. Mergers and acquisitions: Unequal partnerships in Drosophila myoblast fusion. Cell 2000; 102(2):127-129.
25. Dworak HA, Charles MA, Pellerano LB et al. Characterization of Drosophila hibris, a gene related to human nephrin. Development 2001; 128(21):4265-4276.
26. Menon SD, Chia W. Drosophila Rolling pebbles: A multidomain protein required for myoblast fusion that recruits D-Titin in response to the myoblast attractant Dumbfounded. Dev Cell 2001; 1(5):691-703.
27. Furlong EE, Andersen EC, Null B et al. Patterns of gene expression during Drosophila mesoderm development. Science 2001; 293(5535):1629-1633.
28. Artero R, Furlong EE, Beckett K et al. Notch and Ras signaling pathway effector genes expressed in fusion competent and founder cells during Drosophila myogenesis. Development 2003; 130(25):6257-6272.
29. Leake MC, Wilson D, Bullard B et al. The elasticity of single kettin molecules using a two-bead laser-tweezers assay. FEBS Lett 2003; 535(1-3):55-60.
30. Smith SB, Cui Y, Bustamante C. Overstretching B-DNA: The elastic response of individual double-stranded and single-stranded DNA molecules. Science 1996; 271(5250):795-799.
31. Li H, Linke WA, Oberhauser AF et al. Reverse engineering of the giant muscle protein titin. Nature 2002; 418(6901):998-1002.
32. Watanabe K, Mühle-Goll C, Kellermayer MS et al. Different molecular mechanics displayed by titin's constitutively and differentially expressed tandem Ig segments. J Struct Biol 2002; 137(1-2):248-258.
33. Farge E. Mechanical induction of Twist in the Drosophila foregut/stomodeal primordium. Curr Biol 2003; 13(16):1365-1377.
34. Scott IC, Stainier DY. Developmental biology: Twisting the body into shape. Nature 2003; 425(6957):461-463.

Section III

Towards a Systems Level Analysis of Muscle

Chapter 15

Sustained High Power Performance:
Possible Strategies for Integrating Energy Supply and Demand in Flight Muscle

Vivek Vishnudas and Jim O. Vigoreaux

Abstract

The high power output necessary for insect flight has resulted in the evolution of muscles with large and abundant myofibrils, the so called 'myofibrillar' muscles. In principle, this modification should come with a trade-off as the broader diameter of the myofibril would slow ATP/ADP flux and potentially constrain muscle speed (myosin ATPase). However asynchronous flight muscle exhibits no such trade-off as it simultaneously displays speed, power, and endurance. Insect flight muscle appears to lack the components for a phosphagen shuttle system that would provide temporal and spatial buffering of nucleotides. The reliance on a phosphagen shuttle is partly alleviated by the proximity of mitochondria to myofibrils. We present a model for how IFM meets its operational demands by minimizing nucleotide diffusion and facilitating the import and export of nucleotides to the myofibril.

Introduction

Insect flight commands a very high rate of fuel utilization and is considered the most metabolically demanding form of animal locomotion.[1] While different insects have come to rely on various substances for fuel, (e.g., trehalose, proline or diacylglycerols), flight is an intrinsically aerobic process. The higher energetic cost is explained, in large part, by the higher frequency of contraction in flight muscle than in nonflight muscle.[1] Respiratory rates have been shown to increase as much as 50-100 fold in some insect species when going from rest to flight, compared to a 14-fold increase in birds and a 5-fold increase in vertebrate skeletal muscles from rest to maximum work.[2-4] While both synchronous and asynchronous flight muscles have high oxygen utilization rates, asynchronous muscles are, as a rule, more oxidative than synchronous muscles[2] and some insects with asynchronous muscles have been reported to achieve the highest mass-specific metabolic rates of any animal.[5] Insect flight muscles exhibit a very high ATP turnover rate, over 1900 μmoles ATP g^{-1} min^{-1} during sustained flight, the highest recorded in the animal kingdom.[6] This rate is more than three times higher than the rate measured for hummingbird and more than sixty times higher than human leg muscle during sustained running.[7] The primary consumer of ATP in asynchronous flight muscles is myofibrillar myosin since these muscles have very limited sarcoplasmic reticulum (SR; and hence, Ca^{2+} pumps)[8] and, unlike synchronous muscles, expend little energy in Ca^{2+} cycling. Asynchronous muscles possess a myosin isoform with higher Ca^{2+}-ATPase activity than nonflight muscle or rabbit skeletal muscle[9] (see chapters by Sparrow and Geeves and Swank and Maughan) and also possess more myosin per cross-sectional myofibril area than synchronous muscle (thick to thin

Nature's Versatile Engine: Insect Flight Muscle Inside and Out, edited by Jim Vigoreaux.

filament ratio of 1:3 vs 1:6).[10] It has been suggested that nearly 100% of the ATP generated by the mitochondria is used up by myosin for contractile activity.[11]

This chapter examines the adaptations in insect flight muscle that allows it to cope with the extreme energetic demands of flight (sustained and fast contractions) and discusses the structured arrangement of organelles, glycolytic enzymes, and other proteins involved in energy metabolism while emphasizing the similarities and differences with other muscle types.

Meeting the Energetic Demands of Flight

In comparison to vertebrate skeletal muscle, the activity of most enzymes involved in fuel metabolism (e.g., lipases, citric acid cycle) are much higher in insect flight muscle.[12] Most flight muscles utilize carbohydrates as their primary fuel and mitochondrial ATP synthesis starting with the complete oxidation of sugars (trehalose and glycogen) appears to meet most of the ATP requirements of the working insect muscles.[12] Fatty acid and amino acid oxidation also generate acetyl CoA and citric acid cycle intermediates that contribute to oxidative phosphorylation in some insect species that take up long duration flights.[13] Examination of the flight muscle ultrastructure provides some insight into how this tissue copes with its high energetic demand. Two of the most prominent features are the extensive tracheolar system and the abundance of giant mitochondria (Fig. 1). A high density network of tracheoles in the flight muscles is consistent with high respiratory rates during flight. The deeply penetrating tracheoles reduce intracellular diffusion distances and provide for direct infusion of gaseous oxygen to the mitochondria.[1] It also has been suggested that a tracheal-based O_2 delivery system helps achieve an efficient rest to work transition in insect flight muscles, making them independent of a phosphagen based energy reserve system.[7]

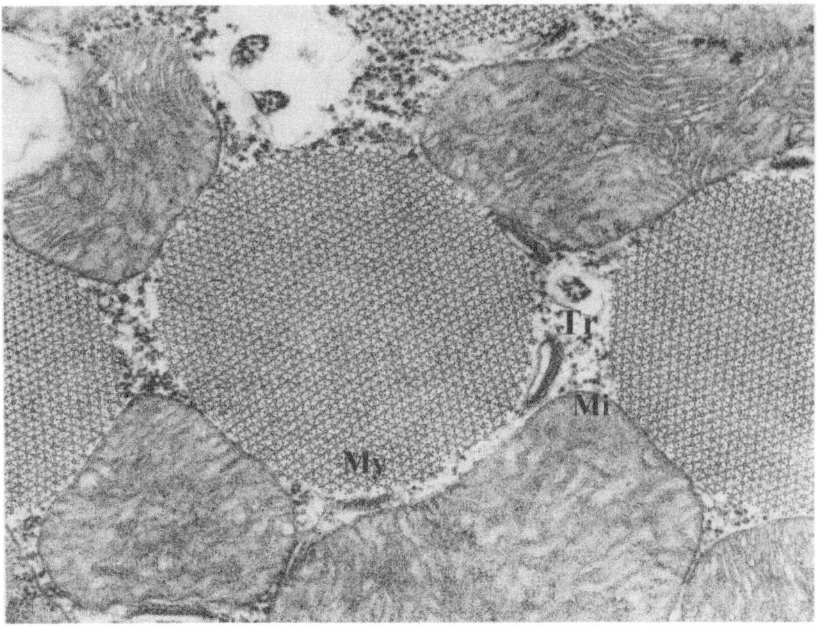

Figure 1. Cross sectional view of *Drosophila* indirect flight muscle by electron microscopy. Observe the close proximity of mitochondria to the myofibril that reduces diffusion distance for adenine nucleotide. The even distribution of tracheoles around myofibrils allows rapid uptake of oxygen. My, myofibril; Mi, mitochondria; Tr, tracheole. (50,000X mag).

Mitochondria occupy ~24% (synchronous) to 37% (asynchronous) of the cellular space in flight muscle. The ratio of mitochondria to myofibril is 1:3 (v:v) in synchronous muscle but twice that (1:1.5) in asynchronous muscle.[8] Mitochondria are not only large and abundant, they are also more densely packed with cristae (inner mitochondrial membrane).[11] Another significant design modification is that mitochondria are strategically distributed to appose myofibrils, in many cases completely enveloping individual myofibrils (Fig. 1). This proximity between the primary energy producer (mitochondria) and primary energy consumer (myofibril) minimizes the diffusion distance for nucleotides (ADP and ATP), possibly downplaying the need for a spatial phosphagen-based buffering system (see below).

The Role of Glycolytic Enzymes

Given the highly oxidative nature of insect flight muscle, one would expect that ATP generated directly by the glycolytic pathway makes but a small contribution to the energy supply during flight. Activity of various glycolytic enzymes, however, are generally higher in flight muscle than in nonflight muscle[14] leading to the production of pyruvate and α-glycerophosphate but no lactate.[14,15] As in vertebrate muscles, glycolytic enzymes in insect flight muscle have a high affinity for the myofibril and may exist as part of a complex that assist in the local generation of ATP.

In *Drosophila* indirect flight muscles (IFM), the significance of three-dimensional organization of enzymes in muscle function was first examined in mutants for the enzyme glycerol-3- phosphate dehydrogenase (GPDH).[16] A high concentration of GPDH was found in the M line and reduced levels in the Z band, though the association of GPDH with the latter appears to be more stable than with the former.[17] The IFM-specific isoform of GPDH, GPDH-1, is different from other isoforms (GPDH-2 and GPDH-3) in having the C-terminal tripeptide glutamine-asparagine-leucine. Transgenic GPDH-1 null flies expressing GPDH-3 in their IFM show an extramyofibrillar distribution, demonstrating that the presence of the C-terminal tripeptide is necessary for the proper localization of GPDH to the myofibril. More importantly, other glycolytic enzymes (aldolase and glyceraldehyde-3-phosphate dehydrogenase, GAPDH) failed to colocalize in the sarcomere of GPDH null mutants as well as in the sarcomere of transgenic lines expressing GPDH-3 isoform.[16] These observations are consistent with a possible role of GPDH as an adaptor protein that facilitates the binding of other enzymes to the M-line.

The myofibrillar localization of glycolytic enzymes has important functional implications.[16] GPDH-3 transgenic flies are flight impaired despite normal levels of GPDH activity in their IFM.[17] It thus seems likely that glycolytic enzymes exist as complexes involved in maintaining compartmentalized metabolite pools important for flight. Furthermore, levels of glycolytic intermediates (e.g., fructose-6-phosphate) remain unchanged during flight, consistent with the view that metabolic enzymes form multiprotein complexes.[18] The colocalization of glycolytic enzymes in the sarcomere is likely a feature of many muscles, including vertebrate skeletal[19] and cardiac muscles.[20] While it remains unclear whether the enzymes simply colocalize or form part of a large multienzyme complex, their close proximity to the major site of ATP consumption increases the efficiency of cellular metabolism.[21]

One important difference between insect flight muscle and vertebrate skeletal muscle is the regulation of phosphofructokinase (PFK), one of the two key regulatory enzymes (the other being hexokinase) of glycolysis. Unlike vertebrate PFK, insect flight muscle (locust) PFK is not down regulated by H^+ or citrate. Instead, fructose 2,6-biphosphate acts as a potent activator of PFK by acting synergistically with AMP.[22] The near constant pH in flight muscle together with the large fractional increase in [AMP] during flight would ensure high activity of PFK and a constant glycolytic flux. Insects also appear to lack the hexokinase isozyme that binds mitochondrial porins.[23] Hence the integration of glycolysis and oxidative phosphorylation is likely to be different in insects and mammals.

The Role of the Phosphagen System in Insect Flight Muscle

In vertebrate muscle, mobilization of the high-energy phosphagen, phosphocreatine (PCr), is the primary means for meeting the energy demand of the contracting muscle.[24]

$$PCr + ADP + H^{+} \rightarrow creatine + ATP \quad (1)$$

A mitochondrial creatine kinase (CK) works in concert with a cytosolic CK to provide intracellular energy transport and to maintain [ATP] constant in the working muscle.[25,26] As such, the CK reaction plays a central role in integrating ATP supply from oxidative phosphorylation to ATP demand from contractile activity.[26] In insects, however, the phosphagen system is likely to contribute minimally to ATP supply during sustained contractions. Under resting conditions, the level of phosphoarginine (PArg) (the insect phosphagen) in flight muscle is only 10 to 15 % that of PCr in skeletal muscles. Arginine kinase (AK) activity is surprisingly low, opposite of what one would expect for a very fast muscle based on vertebrate studies which have shown that CK activity keeps pace with myosin ATPase (i.e., higher CK activity in fast twitch muscle than in slow twitch muscle).[7] In asynchronous IFM of blowfly, [PArg] drops precipitously during the first few seconds after initiation of flight but remain constant from there on.[18] The concentration of ATP followed a similar trend.[18] In locust (synchronous flight muscle), PArg levels have been shown to remain constant at the onset of flight.[1] This suggests that, unlike the CK reaction in vertebrate skeletal muscle, the AK reaction plays a minimal role in providing metabolic capacitance during sustained flight.[3,24] This is supported by studies in which the levels of PArg were found to be more than two times lower in flight muscle than in nonflight (tibia) muscle. Interestingly, asynchronous flight muscle has lower levels of PArg than synchronous flight muscle, and the phosphagen levels are inversely proportional to the wing beat frequency.[25]

The role of the AK reaction in providing spatial buffering (i.e., facilitated diffusion) of ATP/ADP is also in doubt. In *Drosophila* IFM, both the mitochondrial and the M-line bound isoforms of AK have been reported to be absent[27] and mitochondria lack AK activity.[25] The presence of this buffering system may not be critical in the IFM given the high mitochondrial density and short diffusion distance between mitochondria and myofibrils, a fact borne out by computer simulations.[24] As mentioned earlier, the pH in flight muscle (locust) has been shown to remain constant under physiological conditions.[22] Because a drop in pH is one way of increasing phosphagen kinase activity (equation 1), the above observation suggest that the insect phosphagen system is not exploited to the same extent as the vertebrate phosphagen system.

While IFM may have solved the problem of inter-organellar energy transport through morphological design, the large diameter of IFM myofibrils (and correspondingly more ATP-consuming myosins) makes the problem of intra-myofibrillar radial diffusion a critical one. For example, myofibrils from asynchronous muscles (beetle) are more than three times wider than myofibrils from synchronous muscles (locust) ($3.7\ \mu m^2$ vs $0.82\ \mu m^2$).[8] The question arises as to whether there is an active mechanism to prevent the formation of nucleotide gradients (that may impinge on the function of the myosin ATPase) and to expedite the exodus of ADP from the diffusionally restricted myofibrillar space. In the following section we discuss recent evidence that suggest the IFM myofibril may have coopted the mitochondria of its most efficient mechanism of nucleotide transport, an ADP/ATP translocase.

Nucleotide Transport—The Challenge for Asynchronous Muscles

The large increase in ATP turnover rate that occurs during the transition from rest to flight, coupled with the sustained demand during extended flight, suggest that passive diffusion of nucleotides may not adequately meet the energetic quota. A comparison of mean diffusion lengths (i.e., the average distances traversed after the lifetimes[28]) of ATP and ADP versus other metabolites (PCr and Cr) in vertebrates confirms that ADP is the most diffusion- limited of all metabolites. The mean diffusion lengths of PCr (57 μm) and Cr (37 μm) were observed to be significantly higher than those of ATP and ADP (22 and 1.8 μm, respectively).[28] The net flux (J) of metabolites (PCr, Cr, ATP and ADP) by direct diffusion[29] has been estimated in rat skeletal muscle to be proportional to the product of the free diffusion coefficient (D_f) and concentration gradient (dC/dr).[30]

$$J \sim D_f * (dC/dr) \tag{2}$$

Assuming equal concentration gradients, it has been shown that JPCr ≈ 5 JATP and JCr >> JADP. The direct implication of these findings is that while forward flux of ATP (to myofibrils) can be sustained by direct diffusion, the return of ADP to the mitochondria needs a facilitatory process, in this case creatine kinase-mediated diffusion of creatine.[30] It has been estimated that direct diffusion of ADP and ATP would compromise metabolic capacity for diffusion lengths greater than 2 μm.[30] The limited diffusion potential of ATP and ADP and the absence of a phosphagen shuttle may explain the need for close proximity between mitochondria and myofibrils in the flight muscle. Moreover since ATP synthesis is completely oxidative, ADP, a key regulator of oxidative phosphorylation, must be effectively transported to the mitochondria. Most of the ADP emerges from the myofibril given the limited role of anaerobic glycolysis and SR Ca^{2+}ATPase in IFM. A direct shuttling of this myofibrillar ADP would effectively couple ATP supply with contractile demand.

A more significant problem in insect flight muscle is the constraint on radial diffusion of ADP imposed by the wide myofibrils. The absence of an M-line bound phosphagen kinase, which in vertebrate muscle rephosphorylates actomyosin generated ADP to ATP, leaves diffusion as the only mechanism for removing ADP from the myofibril. Simulations of cardiac muscle myofibrils suggest that the diffusion coefficient of ADP along the radial axis is lower than the diffusion coefficient along the longitudinal axis.[31]

An emerging theme in muscle research is the direct communication among separate subcellular compartments.[7] One recent study showed that a SR ankyrin isoform interacts with obscurin, a myofibrillar protein, thus providing a direct protein link between the two organelles.[32] Other studies have presented evidence for the existence of direct interactions between mitochondria and the main energy consuming organelles, SR and myofibrils (see Kaasik et al[33] and references therein). Though the nature or composition of these 'direct channels', or the mechanisms by which they operate to ensure efficient organelle cross-talk, have yet to be defined, they are seen as important regulators of intracellular compartmentation of adenine nucleotides.[34] Some have proposed the existence of intracellular energetic units (ICEUs), structural and functional complexes between myofibrils, mitochondria, and SR in cardiac cells that create barriers restricting the diffusion of adenine nucleotides.[35] ICEUs have not been investigated in insect flight muscles but the characteristic distribution of mitochondria closely abutting myofibrils makes ICEUs a distinct possibility.

A recent study using mass spectrometry identified adenine nucleotide translocase (ANT), a product of the *stress sensitive B* (*sesB*) locus, as one of the components of the *Drosophila* IFM myofibril.[36] This protein is found primarily in the inner membrane of the mitochondria where it catalyzes the exchange of cytosolic ADP with ATP generated by oxidative phosphorylation. Its presence as a myofibrillar component has not been confirmed by other techniques. It would be of interest to examine if a nucleotide transporter-like protein is a component of structural connections between the mitochondria and myofibrils where it can operate by enhancing an ADP gradient out of the myofibril, thereby overcoming potential diffusional limitations, and/or by enhancing an ATP gradient into the myofibril. Such a mechanism would help maintain an exact balance between ATP consumption and supply, a critical issue given that the transition from rest to flight increases ATP turnover several hundredfold. In the synchronous flight muscles of the locust, for example, all of the available energy-rich phosphates (ATP + PArg) would support only 1 second of flight.[22]

Another possibility is that a translocase-like protein is associated with the thick filaments where it may assist in nucleotide exchange near the motor domain. Accumulation of ADP near

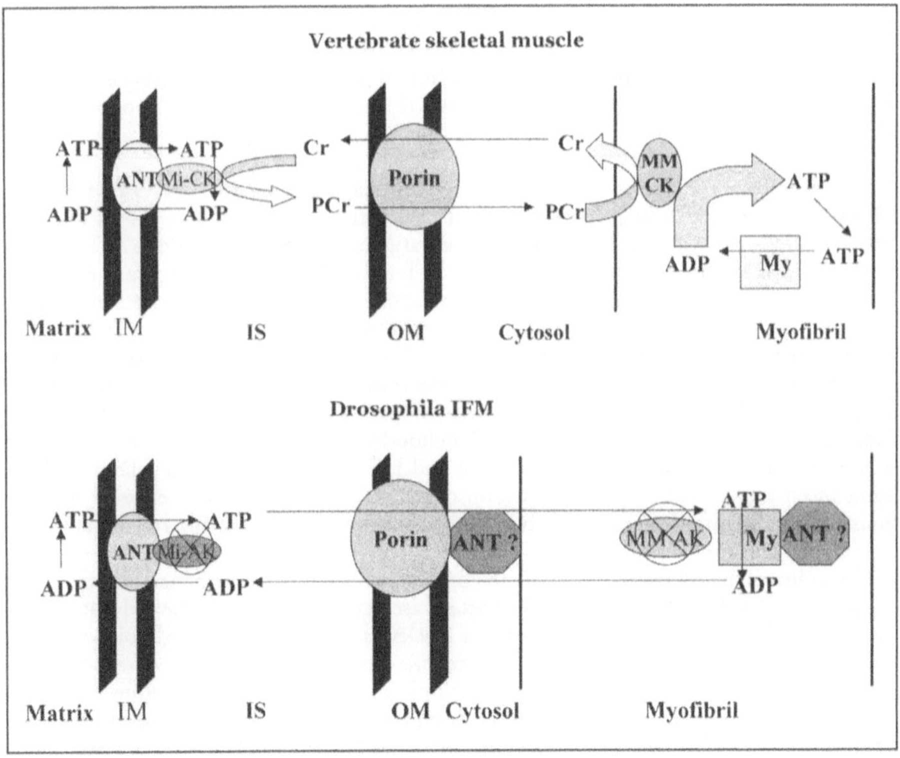

Figure 2. Top) Schematic representation of the creatine kinase shuttle in vertebrate skeletal muscle. The coupled activity of mitochondrial and cytosolic (M-line bound) creatine kinase provides transport of energy nucleotides between mitochondria and myofibril, supplying ATP for contractile activity. The narrow diameter of the myofibril allows diffusion of PCr for in situ regeneration of ATP. IM, Inner mitochondrial membrane; IS, Intermembrane space; OM, Outer mitochondrial membrane; ANT, Adenine nucleotide transporter; MiCK-Mitochondrial creatine kinase; MMCK, Cytosolic creatine kinase; My, Myosin; Cr, Creatine; PCr, Phosphocreatine. Bottom) Schematic representation of a possible adaptation in insect flight muscles for nucleotide exchange between mitochondria and myofibril. Insect flight muscles do not depend on phosphagen and lack both M-line bound (MM AK) and mitochondrial Arginine kinase (Mi AK). Instead, the close proximity of mitochondria and myofibril may allow for direct exchange of nucleotides, perhaps aided by a nucleotide translocase-like protein (ANT?).

the motor domain has been shown to slow down myosin ATPase and crossbridge cycling rate in vertebrate cardiac muscle.[33,37] Given the large increase in [ADP] during flight (up to 400-fold in locust)[22] muscle performance would stand to benefit from a mechanism that actively removes ADP from the vicinity of the myosin motor (Fig. 2). A scenario similar to this has been proposed for *Drosophila* dynamin. A recent study identified an allele of *sesB* (*orangi*) as an enhancer of temperature sensitive paralysis mutants of *shibire*, the gene that encodes dynamin.[38] Because the function of dynamin relies on a source of GTP (from ATP via nucleoside diphosphate kinase), recovery of a *sesB* enhancer points to the importance of spatial and temporal trafficking of nucleotides in sustaining processes that readily consume ATP.

Other mitochondrial proteins are known to fulfill functions outside this organelle. Cytochrome c released from mitochondria in response to apoptotic stimuli is responsible for caspase activation and subsequent cell death.[39] The β chain of ATP synthase has been shown recently to function as a hepatic cell membrane receptor for plasma apolipoprotein A-1.[40] Similarly, the outer mitochondrial membrane voltage-dependent anion channel also is found in the cell membrane.[41] Hexokinase is known to perform different functions depending on its subcellular distribution.[42] The characterization of a myofibril-associated nucleotide translocase would be another example of nature co-opting its available repertoire of tools to address new problems.

Sustainable Muscle Performance

The abundance of large, double-cristae mitochondria in IFM no doubt plays a major role in dictating the endurance of insect flight. Studies in vertebrate muscle have shown that force often declines well before ATP is depleted. Thus fatigue, and not a limitation to ATP synthesis and supply, is mainly responsible for decline in force.[43] IFM is fatigue resistant due to the absence of lactate dehydrogenase (i.e., no lactate buildup) and the presence of the glycerol-3-phoshate shuttle responsible for the reoxidation of NADH.[44]

In vertebrate muscle, glycolytic flux has been shown to make a significant contribution to ATP synthesis during aerobic work. It has been proposed that this flux sets the limits to oxidative phosphorylation.[43] Glycolytic intermediates are acidic (protogenic) and their accumulation increases intracellular [H^+]. Mitochondrial oxidative phosphorylation responds to two important intracellular signals, ADP and H^+. ADP is a positive regulator while H^+ ion concentration is an inhibitory signal for oxidative phosphorylation. Hence for sustained muscle performance, it is essential that the levels of some glycolytic intermediates (mostly acidic) are kept low.

Most insect flight muscles studied so far appear well adapted for sustained oxidative phosphorylation. The absence of lactate dehydrogenase activity confers another advantage since catalysis by this enzyme generates H^+, and the low levels of citrate and other glycolytic products help maintain a nearly constant pH.[18,22] Exercise cause a drop in pH in vertebrate muscle which inhibits the rise in [ADP], potentially limiting oxidative phosphorylation.[43] Thus the insect system appears fine tuned to hold glycolytic flux at an optimum level to maintain constant pH, allow [ADP] to rise, and maximize oxidative phosphorylation. The colocalization of glycolytic enzymes in the myofibril could serve the goal of maintaining low concentration of glycolytic intermediates. One prediction of this hypothesis is that increases in glycolytic flux, through genetic, physiologic, or pharmacologic interventions, should attenuate muscle and flight performance. Recent findings showing that energy metabolism and physiological performance are quantitative traits subject to complex epistatic and pleiotropic effects[45] illustrate the need for reconciling studies of enzyme activity with broad based 'metabolome' approaches in order to uncover the fundamental biochemical nature of physiological performance.

Concluding Remarks

The emergence of flight is widely accredited as one of the major, if not the main, events that led to the evolutionary success of insects. Many studies that have examined the ultrastructural, mechanical, and biochemical properties of the IFM have identified key differences between this muscle and other muscles, differences that undoubtedly account for the extreme specialization of this muscle type. Here we argue that a mechanism for minimizing nucleotide diffusion dependence, a metabolic adaptation, has played a major role in the evolution of this muscle. The provision for a channeled exchange of nucleotides between myofibril and mitochondria, perhaps mediated by a myofibril associated nucleotide translocase-like protein, would have been an important adjustment of the IFM in its quest for speed, power and endurance. Circumventing diffusional limitations of catalytic products is increasingly viewed as a key adaptation in muscle.[7]

The dependence of IFM on aerobic metabolism necessitates an efficient mechanism for coupling ATP demand with ATP supply. The high power demands of flight have driven the evolution of large myofibrils and a reliance on stretch activation, with a decrease in the need for SR and for extensive Ca^{2+} cycling, at a substantial energy savings, energy that can now be allocated for contractile force. Given that the energetic demand arises almost exclusively from myosin, a direct coupling of the myofibril to the mitochondria would provide an advantage over a cytosolic phosphagen shuttle. The enhancement of a mitochondria-myofibril interaction, physical or functional, would appear to be a step taken in the right direction towards better synchronization between energy supply and demand pathways. A system designed to provide a direct conduit between the myofibril and the mitochondria would maintain [ADP] high in the vicinity of the mitochondria, eliciting maximal oxidative capacity and sustained power. Defining the molecular basis of this coordinated activity remains an important challenge in understanding the adaptations of flight muscle to fulfill the demands of aerial locomotion.

Acknowledgements

We thank David Maughan for comments on the manuscript. This publication was made possible by support from NSF MCB-0090768 and MCB-0315865, and by the Vermont Genetics Network through the NIH Grant Number 1 P20 RR16462 from the BRIN program of the NCRR.

References

1. Harrison JF, Roberts SP. Flight respiration and energetics. Annu Rev Physiol 2000; 62:179-205.
2. Kammer AE, Heinrich B. Insect flight metabolism. In: Treherne JE, Berridge MJ, Wigglesworth VB, eds. Advances in Insect Physiology, Vol. 13. London: Academic Press, 1978:133-228.
3. Sacktor B. Utilization of fuels by muscle. In: Candy DJ, Kilby BA, eds. Insect Biochemistry and Function. London: Chapman and Hall, 1975:1-81.
4. Beenakkers AMT, Van der Horst DJ, Van Marrewijk WJA. Biochemical processes directed to flight muscle metabolism. In: Kerkut GA, Gilbert LI, eds. Comprehensive Insect Physiology, Biochemistry and Pharmacology. Oxford: Pergamon Press, 1985:10:451-486.
5. Lindstedt SL, Hokanson JF, Wells DJ et al. Running energetics in the pronghorn antelope. Nature 1991; 353(6346):748-750.
6. Casey TM, Ellington CP, Gabriel JM. Allometric scaling of muscle performance and metabolism: Insects. Adv Biosci 1992; 84:152-162.
7. Hochachka PW. Muscles as molecular and metabolic machines. Boca Raton: CRC Press, 1994.
8. Josephson RK, Malamud JG, Stokes DR. Asynchronous muscle: A primer. J Exp Biol 2000; 203(Pt 18):2713-2722.
9. Swank DM, Bartoo ML, Knowles AF et al. Alternative exon-encoded regions of Drosophila myosin heavy chain modulate ATPase rates and actin sliding velocity. J Biol Chem 2001; 276(18):15117-15124.
10. Squire JM. Muscle: Design, Diversity, and Disease. Menlo Park: Benjamin/Cummings Publishing Co., 1986.
11. Lindstedt SL, McGlothlin T, Percy E et al. Task-specific design of skeletal muscle: Balancing muscle structural composition. Comp Biochem Physiol B Biochem Mol Biol 1998; 120(1):35-40.
12. Crabtree B, Newsholme EA. Comparative aspects of fuel utilization and metabolism by muscle. In: Usherwood PNR, ed. Insect Muscle. London: Academic Press, 1975.
13. O'Brien DM, Suarez RK. Fuel use in hawkmoth (Amphion floridensis) flight muscle: Enzyme activities and flux rates. J Exp Zool 2001; 290(2):108-114.
14. Crabtree B, Newsholme EA. The activities of phosphorylase, hexokinase, phosphofructokinase, lactate dehydrogenase and the glycerol 3-phosphate dehydrogenases in muscles from vertebrates and invertebrates. Biochem J 1972; 126(1):49-58.
15. Sacktor B. Biochemical adaptations for flight in the insect. Biochem Soc Symp 1976; (41):111-131.
16. Wojtas K, Slepecky N, von Kalm L et al. Flight muscle function in Drosophila requires colocalization of glycolytic enzymes. Mol Biol Cell 1997; 8(9):1665-1675.
17. Sullivan DT, MacIntyre R, Fuda N et al. Analysis of glycolytic enzyme colocalization in Drosophila flight muscle. J Exp Biol 2003; 206(Pt 12):2031-2038.
18. Sacktor B, Hurlbut EC. Regulation of metabolism in working muscle in vivo. II. Concentrations of adenine nucleotides, arginine phosphate, and inorganic phosphate in insect flight muscle during flight. J Biol Chem 1966; 241(3):632-634.
19. Pette D. Cytosolic organization of carbohydrate-metabolism enzymes in cross-striated muscle. Biochem Soc Trans 1978; 6(1):9-11.
20. Lange S, Auerbach D, McLoughlin P et al. Subcellular targeting of metabolic enzymes to titin in heart muscle may be mediated by DRAL/FHL-2. J Cell Sci 2002; 115(Pt 24):4925-4936.

21. Kraft T, Hornemann T, Stolz M et al. Coupling of creatine kinase to glycolytic enzymes at the sarcomeric I-band of skeletal muscle: A biochemical study in situ. J Muscle Res Cell Motil 2000; 21(7):691-703.
22. Wegener G. Flying insects: Model systems in exercise physiology. Experientia 1996; 52(5):404-412.
23. Suarez RK. Shaken and stirred: Muscle structure and metabolism. J Exp Biol 2003; 206(Pt 12):2021-2029.
24. Sweeney HL. The importance of the creatine kinase reaction: The concept of metabolic capacitance. Med Sci Sports Exerc 1994; 26(1):30-36.
25. Ellington WR. Evolution and physiological roles of phosphagen systems. Annu Rev Physiol 2001; 63:289-325.
26. Kushmerick MJ, Conley KE. Energetics of muscle contraction: The whole is less than the sum of its parts. Biochem Soc Trans 2002; 30(2):227-231.
27. Wyss M, Maughan DM, Wallimann T. Reevaluation of the structure and physiological function of guanidino kinases in fruitfly (Drosophila), sea urchin (Psammechinus miliaris) and man. Biochem J 1995; 309:255-261.
28. Yoshizaki K, Watari H, Radda GK. Role of phosphocreatine in energy transport in skeletal muscle of bullfrog studied by 31P-NMR. Biochim Biophys Acta 1990; 1051(2):144-150.
29. Jacobus WE. Theoretical support for the heart phosphocreatine energy transport shuttle based on the intracellular diffusion limited mobility of ADP. Biochem Biophys Res Commun 1985; 133(3):1035-1041.
30. de Graaf RA, van Kranenburg A, Nicolay K. In vivo (31)P-NMR diffusion spectroscopy of ATP and phosphocreatine in rat skeletal muscle. Biophys J 2000; 78(4):1657-1664.
31. Kongas O, van Beek JH. Diffusion barriers for ADP in the cardiac cell. Mol Biol Rep 2002; 29(1-2):141-144.
32. Bagnato P, Barone V, Giacomello E et al. Binding of an ankyrin-1 isoform to obscurin suggests a molecular link between the sarcoplasmic reticulum and myofibrils in striated muscles. J Cell Biol 2003; 160(2):245-253.
33. Kaasik A, Veksler V, Boehm E et al. Energetic crosstalk between organelles: Architectural integration of energy production and utilization. Circ Res 2001; 89(2):153-159.
34. Andrienko T, Kuznetsov AV, Kaambre T et al. Metabolic consequences of functional complexes of mitochondria, myofibrils and sarcoplasmic reticulum in muscle cells. J Exp Biol 2003; 206(Pt 12):2059-2072.
35. Vendelin M, Eimre M, Seppet E et al. Intracellular diffusion of adenosine phosphates is locally restricted in cardiac muscle. Mol Cell Biochem 2004; 256-257(1-2):229-241.
36. Ashman K, Houthaeve T, Clayton J et al. The application of robotics and mass spectrometry to the characterisation of the Drosophila melanogaster indirect flight muscle proteome. Letters Peptide Science 1997; 4:57-65.
37. Ventura-Clapier R, Veksler V, Hoerter JA. Myofibrillar creatine kinase and cardiac contraction. Mol Cell Biochem 1994; 133-134:125-144.
38. Rikhy R, Ramaswami M, Krishnan KS. A temperaturesensitive allele of Drosophila sesB reveals acute functions for the mitochondrial adenine nucleotide translocase in synaptic transmission and dynamin regulation. Genetics 2003; 165(3):1243-1253.
39. Jiang X, Wang X. Cytochrome C-Mediated Apoptosis. Annu Rev Biochem 2004; 73:87-106.
40. Martinez LO, Jacquet S, Esteve JP et al. Ectopic beta-chain of ATP synthase is an apolipoprotein A-I receptor in hepatic HDL endocytosis. Nature 2003; 421(6918):75-79.
41. Buettner R, Papoutsoglou G, Scemes E et al. Evidence for secretory pathway localization of a voltage-dependent anion channel isoform. Proc Natl Acad Sci USA 2000; 97(7):3201-3206.
42. Frommer WB, Schulze WX, Lalonde S. Plant science. Hexokinase, Jack-of-all-trades. Science 2003; 300(5617):261-263.
43. Conley KE, Kemper WF, Crowther GJ. Limits to sustainable muscle performance: Interaction between glycolysis and oxidative phosphorylation. J Exp Biol 2001; 204(Pt 18):3189-3194.
44. Klowden MJ. Physiological systems in insects. San Diego: Academic Press, 2002.
45. Montooth KL, Marden JH, Clark AG. Mapping determinants of variation in energy metabolism, respiration and flight in Drosophila. Genetics 2003; 165(2):623-635.

Chapter 16

X-Ray Diffraction of Indirect Flight Muscle from *Drosophila* in Vivo

Thomas C. Irving

Abstract

The indirect flight muscle (IFM) of the fruit fly, *Drosophila,* represents a powerful model system for integrated structure and function studies because of the ease of genetically manipulating this organism. Recent advances in synchrotron technology have allowed collection of high quality two dimensional X-ray fiber diffraction patterns from the IFM of living fruit flies both at rest and during tethered flight. Based on many decades of X-ray and electron microscopic studies of vertebrate muscle and IFM from the waterbug, *Lethocerus,* there now exists a framework for interpreting changes in the X-ray diffraction patterns in terms of structural changes at the myofilament level. These developments allow testing of hypotheses concerning muscle function in a truly in vivo system.

Introduction

Small-angle X-ray diffraction has told us much about what we know about the molecular events involved in muscle contraction. X-ray diffraction is the only technique that can provide molecular level information in tissue under hydrated, physiological conditions at the physiologically relevant milli-second time scale. Some of the key findings due to X-ray diffraction are the relative constancy of length of the thick and thin filaments,[1,2] the movement of crossbridges from their resting configuration at the onset of activation and prior to tension development,[2-4] and the change in crossbridge configuration[5-9] during mechanical quick-release protocols designed to synchronize their action.[10] Thus the results of X-ray diffraction experiments have been very important in the formulation of our current picture of crossbridge action in active muscle. Some of the outstanding questions remaining that X-ray diffraction could potentially help to address are: (1) the configuration of myosin heads (crossbridges) in various intermediate states of the crossbridge cycle, (2) the length of the power stroke as a function of load and (3) the molecular mechanisms behind length dependent activation/de-activation. Because acto-myosin motors are ubiquitous in the cell, the knowledge gained from studies aimed at addressing these questions may have implications well beyond muscle physiology.

Over the last decade or so, the indirect flight muscle (IFM) of *Drosophila melanogaster* has been emerging as an attractive model system for combined genetic, mechanical, behavioral and structural studies. Much of its value as a model system stems from the availability of the fly genome[11] and advanced tools such as P-element transformation[12] leading to the exceptional ease of genetic manipulation of this system.[13] Particularly advantageous is the fact that many muscle proteins exist in IFM specific isoforms resulting from unique IFM specific genes (e.g., actin)[14] or differential splicing pathways (e.g., myosin heavy chain)[14] so that one

Nature's Versatile Engine: Insect Flight Muscle Inside and Out, edited by Jim Vigoreaux.

can genetically engineer flies with severe IFM phenotypes and still have viable flies to study. Mechanical and behavioral studies of mutant flies have been shown to be powerful strategies to probe specific aspects of muscle function.[15-17] There have been, however, relatively few structural studies of this system.

The IFM from the large waterbug, *Lethocerus sp.* has long been appreciated as a particularly well-ordered system for structural studies. This has resulted in a substantial literature of X-ray diffraction[18-26] and electron microscopic studies.[27-35] While electron microscopic studies yield images that can be relatively easily interpreted, the samples, whether in conventional thin section or in various frozen hydrated forms, require extensive processing increasing the danger of artifacts. A disadvantage of *Lethocerus* IFM is that all the structural studies (that we are aware of) have been on so-called glycerinated (de-membranated) fibers isolated from the thoraces of these insects. While these preparations allow great freedom in manipulating the ionic milieu of the myofilaments while permitting simultaneous structural and mechanical studies, they are not the same as a truly physiological, living system. Furthermore, since *Lethocerus* specimens must be collected from the wild, there is no possibility of genetically manipulating the IFM in this organism at this time.

The possibility of obtaining X-ray diffraction patterns from live Dipterans, which includes *Drosophila* , was demonstrated about 4 decades ago.[18,36] These early patterns taken on the large blowfly, *Calliphora sp.*, although not of high quality by current standards, showed indications of the high degree of structural order of the better studied *Lethocerus* system. This interesting "preparation", however, had not been exploited again until quite recently when, using then newly developed synchrotron-based X-ray technology, it was demonstrated[37] that it was possible to obtain high quality static and time resolved X-ray diffraction patterns from living flies under resting conditions and during tethered flight. In this chapter, I will review the methods used to obtain X-ray patterns from living *Drosophila* under static and time-resolved conditions, the kinds of information one can obtain from such patterns, and indicate some future directions.

Methods for Obtaining X-Ray Diffraction Patterns from *Drosophila*

Preparation of Specimens

The strength of the X-ray diffraction signal represents a balance between the diffraction from a given thickness of sample and the absorption of the diffracted X-rays by the same sample. As a rule of thumb, the ideal thickness for a muscle specimen would be that required to reduce the transmitted X-ray intensity to $1/e$ of its incident value. This corresponds to about 3 mm for the 12 keV X-rays used in the diffraction experiments, considerably thicker than a typical fly's thorax (~1 mm). Considerable gains can be achieved by making the fly as large as possible, for example by growing under un-crowded conditions, to increase the thickness of the muscle fibers intercepted by the X-ray beam. Prior to mounting, flies need to be anesthetized, preferably by cooling to 4°C using a thermoelectrically cooled stage. Flies recover from cold within a few minutes but an hour is typically used to be safe. CO_2 gas is also commonly used to anesthetize flies and recovery times are similar. Such flies, however, do not always fly or fly as long after recovery. For flies used for static shots the top of the head, the neck and thoracic cuticle adjacent to the neck are either glued to a stainless steel pin or a fine tungsten wire. Flies used for time resolved patterns are tethered by gluing a single pin to the upper region of the thorax allowing for free movement of the head and wings during flight. Flies so tethered will spontaneously flap their wings for many minutes.

X-Ray Instrumentation

The fly IFM system, either the intact fly or isolated fibers, presents a particularly challenging diffraction problem. The near crystalline lattice made up of the myofilaments in individual myofibrils consists of a number of large ordered assemblies, with similar long spacing repeats (40-200 nm). The reciprocal relationship between the spacings in repeating structure and scattering angle implies that the scattering angles will be small, generally less than 1 degree. This, and the coexistence of several long spacings in the ordered structures, implies that high angular resolution (~200 nm order to order, 50 nm first order) in diffraction patterns is required to adequately resolve features in the X-ray diffraction patterns. Even highly crystalline muscle specimens such as IFM are much more weakly diffracting than, say a typical protein crystal, so in order to obtain interpretable patterns on a physiologically useful timescale of seconds (or less), exceptionally strong sources of X-rays are required. In order to resolve the weak signals in the presence of high backgrounds, very well collimated X-ray beams and sensitive detectors with good spatial resolution are required. These technical requirements are satisfied by modern synchrotron beamlines at so-called third generation synchrotron facilities that are optimized to provide X-ray light from devices called undulators. The experiments described here were all carried out using the small angle instrument at the BioCAT undulator-based beam line 18-ID [38,39] at the Advanced Photon Source (APS), Argonne National Laboratory.

The overall experimental arrangement is shown in Figure 1. The APS undulator "A" as installed in the BioCAT beamline 18-ID can deliver a very high flux (~10^{13} photons/s) to the sample with angular divergences of <15 micro-radian vertical and <25 micro-radian horizontal full width half maximum (FWHM) and a source size that is typically 30 micrometers vertical by ~650 micrometers horizontal FWHM (specifications as of fall 2003). These source properties allow the use of highly demagnifying X-ray optics with independent horizontal and vertical focusing so that one can optimally match the beam dimensions at the sample and detector for a particular experiment. In the case of the intact fly preparation, this allows focusing horizontally at the sample (<200 micrometers FWHM) and vertically at the detector <40 micrometers FWHM). The beam size at the sample under these conditions is then ~200 micrometers square. Since the fibers comprising the dorsal longitudinal muscles (DLM's) are about 150 microns in diameter this optical arrangement allows selection of these muscles to the exclusion of other nearby muscle systems. In order to collect diffraction patterns it is convenient to use an X-ray beam energy of 12 keV (wavelength = 0.103 nm), and a specimen-to-detector distance in the range of 1.5-2 m. All paths traversed by the X-ray beam are, to the extent possible, enclosed in vacuum chambers with mica or Kapton™ windows to prevent absorption and scattering by air.

Another important requirement is to have a good match between the properties of the beam and those of the detector. Detectors must be efficient in order to reduce radiation damage. These properties are satisfied by the best of modern CCD (Charge-coupled Device)-based X-ray detectors. Diffraction patterns have been recorded[37] using a 6 x 6 cm active area, 1024 x1024 pixel CCD detector[40] or a 5.5 cm x 8.8 cm, 1028 x 1798 pixel CCD detector.[41] The latter detector is particularly advantageous because of its high spatial resolution (~65 microns FWHM for the point spread function), which is well matched to the BioCAT focused beam size and its high sensitivity (near single photon counting capability) because of the absence of a de-magnifying fiber optic taper. The commercial MAR 165 CCD detector (MAR*USA*, Evanston IL, 2k x 2k 80 micron pixels 2.7:1 taper 165 mm circular active area) was also recently tested and shown to yield satisfactory performance for this application. Analog to digital converters in such detectors digitize the signal to 16 bits for each pixel. True dynamic range in such detectors range from 6000-30000 X-ray photons/pixel/read depending on the overall gain and noise characteristics of the system.

Another important consideration is control of X-ray dose. The relationship between diffracted intensity and radiation damage in biological materials is complex[42] and very difficult to model, especially in living systems such as the intact fly. Increasing exposure time will not necessarily yield more diffracted intensity since increased dose will increase the incoherent background from disordered (i.e., radiation damaged) material at the expense of the undamaged muscle that diffracts coherently. Since living tissues have radiation damage repair systems (such as those involving the free radical scavenging enzymes superoxide dismutase (SOD) and catalase) it can sometimes be advantageous to spread the dose over a longer time to allow time for the organism to recover. Another mechanism of damage is heating of the specimen

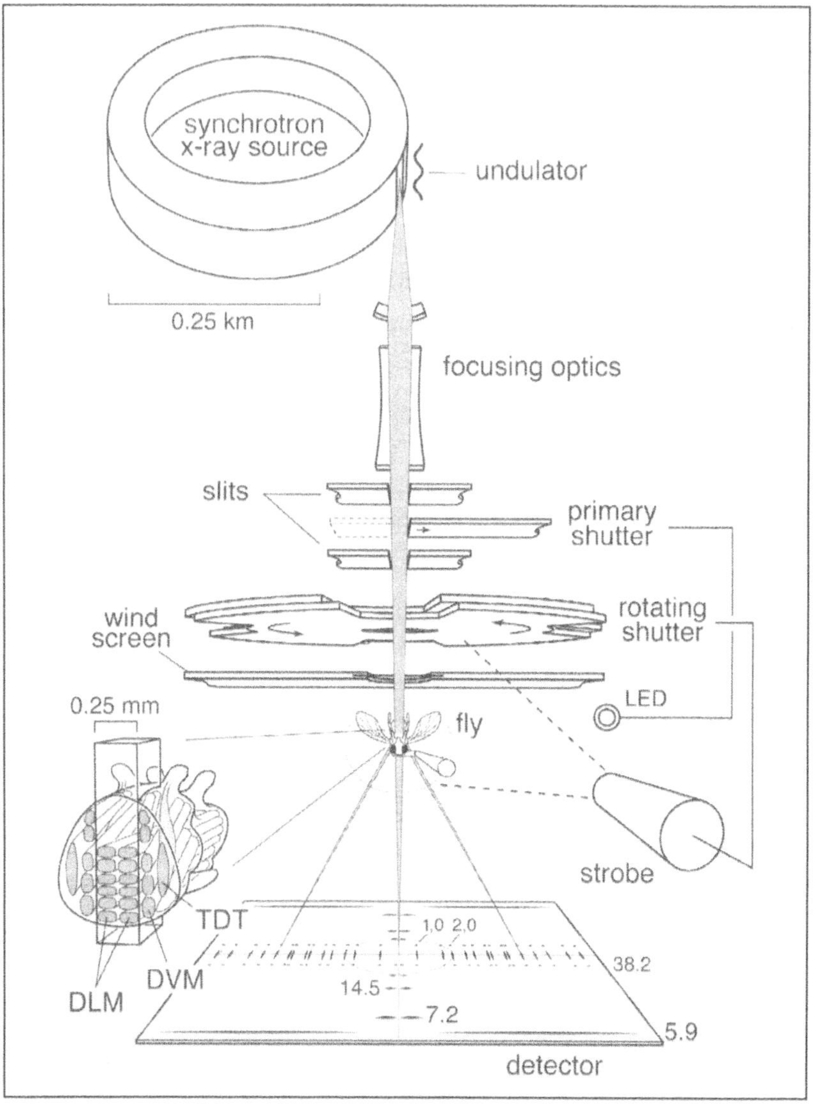

Figure 1. Overall experimental set-up and preparation. The BioCAT beamline 18-ID takes X-ray irradiation from an undulator in the Advanced Photon Source storage ring. X-ray optics focus the beam horizontally at the fly and vertically at the detector. The incident beam intensity may be modulated by aluminum filters of various thicknesses (maximum flux, ~2.0×10^{13} photons/s; wavelength, 0.103 nm). The camera length was 1.5-2 m (fly-to-detector distance); the spot size at the CCD-based detector was ~40×200 μm (full width, half maximum); the spot size at the sample was ~200×200 μm. The support wire glued to the fly could be manipulated by a motorized stage to orient the dorsal longitudinal muscle (DLM) fibers in the X-ray beam path. The inset shows details of the thoracic musculature (DLM, DVM and TDT), with the orientation of the X-ray beam indicated. Also shown in the figure is the shuttering arrangement used in the experiments of Irving and Maughan (2000). During the wing beat, the X-ray aperture was modulated by a rotating shutter, coupled to a stroboscope. Shutter opening was synchronized to wing beat frequency by manually adjusting the strobe frequency and shutter disc rotation until the wing image 'froze' at a desired position. For additional details, see text.

by the absorbed X-ray beam. In extreme cases this can cause heat induced shortening of the IFM causing the wings to be raised. This occurs in 1-0.5 s exposure to an incident dose of 1 x 10^{13} photons/s or an absorbed dose of about $1x10^{-2}$ Gy. Spreading the dose over a longer time will also minimize specimen heating. In practice, exposure time is determined empirically. In the experiments we have done, duration of exposure and beam intensity was adjusted appropriately in the range of 10^{11}-10^{13} photons/s using aluminum attenuators to obtain adequate counting statistics in the X-ray pattern with minimal radiation damage. Exposures for static specimens are typically 1 s with an incident intensity of around 10^{12} photons/s in order to record the equator and the stronger features on the meridian of the X-ray patterns.

Static Patterns from Live Flies

After recovery from anesthesia, the fly is mounted with the long axis of its thorax perpendicular to the X-ray beam (head up) such that the 0.2 mm wide collimated X-ray beam intersects the two parallel sets of DLM fibers located on both sides of the median plane of the thorax (Fig. 1). It is important to position the beam relatively low in the thorax in order to miss the fused thoracic-abdominal ganglion that contains the neuronal circuits that drive the flight muscles. This will reduce radiation damage to the nervous system and reduce X-ray background from other structures. The resulting X-ray diffraction pattern will be the summed pattern from 12 well-aligned DLM fibers. Flight activity can be induced by blowing gently on the fly, using a long flexible tube that extends from the experimental enclosure to the control area of the beamline.

Time Resolved Studies from Living Flies

The first time- resolved experiments[37] employed a rotating shutter to isolate two phases of the wing beat cycle (Fig. 1). This consisted of a chopper-type shutter mounted on a DC motor that had an opening time of 1/6th of a complete cycle. The shutter consisted of a sandwich of two 1/8th inch-thick aluminum discs with 4 equally spaced slots, the width of which could be adjusted by rotating the discs with respect to one another. The center of the temporal aperture was set such that the wings were either at the top of the up-stroke or at the bottom of the down-stroke, i.e., at times that included the maximum and minimum muscle length, respectively. Wing position was monitored using a video camera and stroboscope, which illuminated the fly every fourth wing beat. Shutter frequency was coupled electronically to strobe frequency, which was adjusted by means of a potentiometer to find a frequency at which the wings appeared to 'freeze' in either the up or the down position and the primary shutter (upstream of the fast shutter, set for 1 s opening time) was opened by the operator to allow illumination of the specimen. The average wing beat frequency was ~200 Hz, so the effective time resolution of the exposure was ~0.8 ms [i.e., $(1/6 \times 200\ s^{-1})^{-1}$].

A difficulty with these experiments was that the wing beat frequency of the flies varied continuously and was seldom stable for more than a few seconds. This variation in wing beat frequency precluded attempts to systematically vary the phase of aperture opening through a full contraction-extension cycle (as has been done with glycerinated IFM from *Lethocerus*[43]) thus these experiments were restricted to collecting patterns from only the fully extended DLMs (wings raised) and the fully contracted patterns (wings lowered). Even under these conditions, there was likely to be signal contamination due to phase fluctuations that could be as much as ~15%. Also, because the temporal aperture was a significant portion of the wing beat cycle (1/6th, or 27% by area), muscle lengths other than the extremes were sampled, thereby adding to the signal contamination.[37]

In order to obtain higher time resolution from a more stable system we subsequently adopted a different approach using the so-called "Flight Arena" described by M. Dickinson and colleagues.[44] This "tethered flight simulator" consists of an optical wingbeat analyzer capable of tracking the movements of the animal's wings and a panoramic electrical display capable of

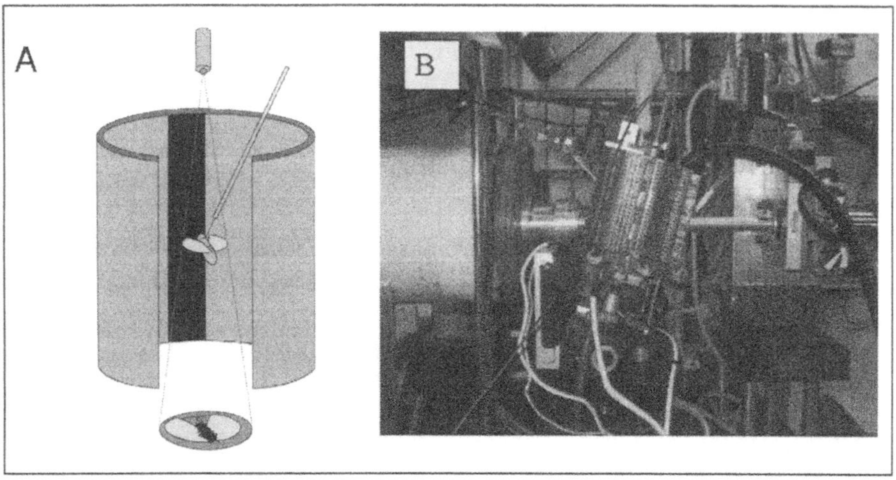

Figure 2. A) Flight arena schematic showing position of fly with respect to the surrounding LED display. Light from an infrared diode at the top impinges on the fly and casts shadows of the animal's wings on the two optical sensors of the wing beat analyzer, which track the back and forth motion of the wings within the stroke plane. The raw analog signals of these optical sensors connect to dedicated online circuitry to determine the stroke amplitude of both wings as well as the instantaneous wingbeat frequency. The LED display panels can be used for closed loop control of the wingbeat frequency. For details see text. B) The flight arena in the X-ray apparatus. The X-rays come from the right in the figure and enter the 2m long evacuated flight tube that fills the space between the sample and the detector (see Fig. 1).

presenting the animal with dynamic visual stimuli. Figure 2A shows a schematic diagram of this arrangement and Figure 2B shows the flight arena as installed in the X-ray diffraction apparatus. Using this apparatus, animals will fly from between 30-120 minutes until they run out of energy stores. When placed in the arena, an infrared diode mounted at the top of the arena casts shadows of the animal's wings on the two optical sensors of the wing beat analyzer, which track the back and forth motion of the wings within the stroke plane. From the raw analog signals of these optical sensors, dedicated online circuitry determines stroke amplitude of both wings in each stroke as well as the instantaneous wingbeat frequency.

The visual display surrounding that animal consists of a cylindrical array of 1584 green light emitting diodes. Programmed images (bitmaps) are scanned onto the display at 1 KHz, at least 5 times faster than the animal's intrinsic flicker fusion rate. In open loop mode, the animal is presented with a moving pattern and its responses are measured with the wingbeat analyzer. In closed loop mode, the behavior of the animal (measured by the wingbeat analyzer) is used to control the motion of the visual display. In a typical open loop configuration, termed *fixation*, the rotational velocity of a 30 deg vertical stripe about the vertical axis (the yaw axis of the animal) is controlled by the difference in left-right stroke amplitude. In this paradigm, the animal will actively adjust its stroke amplitude to maintain the stripe in the frontal region of its visual field. The wingbeat analyzer provides a trigger pulse that may be used to synchronize external devices with any precise phase of the stroke cycle. In our case, the trigger signal corresponded to a precise point in the wingbeat cycle just before the DLM's completed shortening. A simple program running on a laptop computer was used to monitor the trigger pulse signal via the parallel port, calculate the average wingbeat frequency from the running average of 10 wingbeat periods, and trigger a shutter (opening time 350 μ/sec), after a delay set to a preset fraction of the wingbeat period, once for every 10 wing beats. This arrangement has allowed collection of X-ray diffraction patterns at rest and at 8 precisely-determined equally spaced points in the ~5 ms wing beat cycle (Dickinson et al., in preparation).

Diffraction Patterns from *Drosophila* IFM

Origin of Diffraction Features in the X-Ray Pattern

When properly aligned, the DLM's of living flies give rich two-dimensional X-ray fiber diffraction patterns (Fig. 3A). The strongest feature is the so-called equator with 16-20 closely spaced sharp reflections (Fig. 3B). This part of the pattern arises from the hexagonally packed thick and thin filaments in the A-band of the IFM (~56 nm inter-thick filament spacing). Parallel to the equator are various layer lines. These arise from the helical substructure of the

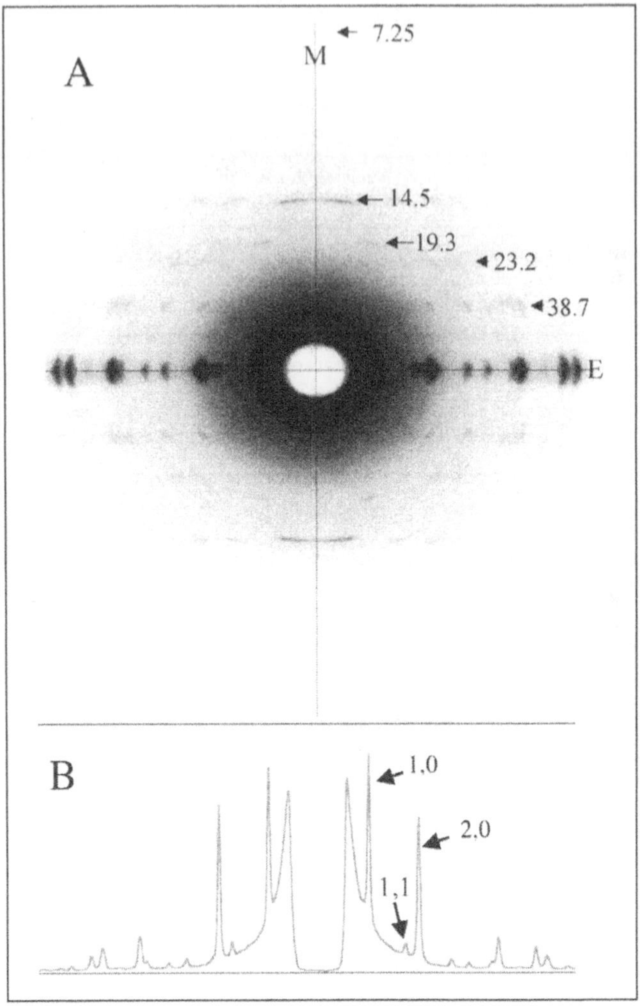

Figure 3. A) Two-dimensional X-ray diffraction pattern from a living fly. The strongest feature is the so-called equator (E) with numerous closely spaced sharp reflections shown in one-dimensional projection as indicated in (B). This part of the pattern arises from the hexagonally packed thick and thin filaments in the A-band of the IFM (~56 nm inter-thick filament spacing). Parallel to the equator are various layer lines indexing as orders of 232 nm i.e., they are found at an axial position of n/232 nm where n is an integer. The 38.7 nm, (n = 6), 1/23.2 nm (n = 10), 1/19.3 nm (n = 12), 1/14.5 nm (n = 16) and 1/7.25 nm (n = 32) layer lines are as indicated.

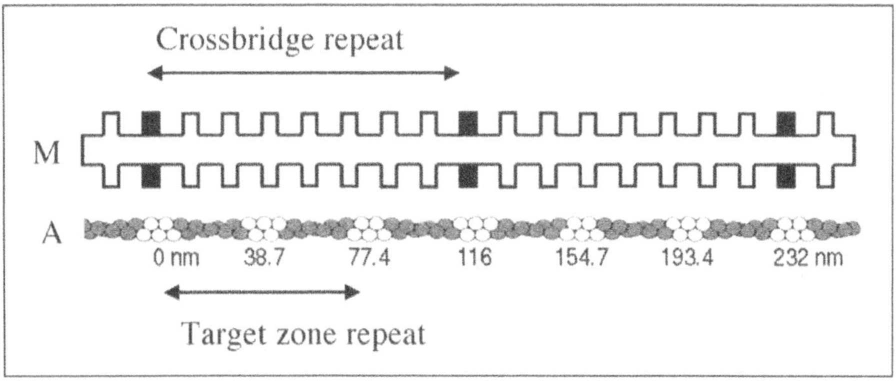

Figure 4. The origin of the 232 nm repeat. The actin containing thin filament (A) is a two-stranded helical structure composed of G-actin monomers. So called "target zones" where the myosin heads may bind are indicated as alternating 2- and 3-monomer patterns located axially every 38.7 nm. The rotational symmetry produces an identical actin target zone repeat every 77.4 nm. The four-stranded thick filament (M) cross-bridge structure repeats every 116 nm, so that the combined repeats come into register every 232 nm. (After Fig. 4, in Tregear et al. Biophys J 2004; 86:3009-19.)

actin containing thin filaments and myosin containing thick filaments and the various accessory proteins associated with them (helix repeats of 5-39 nm, pitches of 5-160 nm). The fundamental long spacing repeat is 232 nm so that the layer lines will appear in the diffraction pattern at an axial position of n/232 nm where n is an integer.

The 232 nm repeat can be thought of as the frequency of the "beating" period of the fundamental repeat distances of the thick filament with that of the thin filament. The actin containing thin filament can be visualized as a two-stranded helical structure composed of the G-actin monomers. So called "target zones" where the myosin heads may bind are located axially every 38.7 nm. The rotational symmetry produces an identical actin target zone repeat every 77.4 nm. The four-stranded thick filament cross-bridge structure repeats every 116 nm,[26] so that the combined repeats come into register every 232 nm (Fig. 4).

The most prominent layer line is at 1/38.7 nm, (n = 6) which has contributions from both the 2-stranded actin containing thin filaments and the four-stranded myosin containing thick filaments. Other prominent layer lines in Figure 3 are seen at 1/23.2 nm (n = 10), 1/19.3 nm (n = 12), 1/14.5 nm (n = 16) and 1/7.25 nm (n = 32). It is sometimes possible to see the actin based layer line at 1/5.95 nm (n = 39). Because of the relatively high degree of crystallinity of IFM (compared to mammalian or amphibian muscle), the layer lines are "sampled", i.e., intensity is only observed at the hexagonal lattice positions. So corresponding to each reflection on the equator will be a "row line" parallel to the so-called "meridian" or midline of the pattern.

The observed diffraction pattern arises from the periodic array of the thick and the thin filaments in the A-bands of the sarcomeres. Since the sarcomeres are connected end to end by way of the Z-band into the myofibrils, each myofibril represents a single crystallite. The myofibrils are all aligned axially with the muscle fiber but have random orientations azimuthally. Thus the observed diffraction pattern consists of the cylindrical average of the patterns from each myofibril. This is convenient in the sense that it is unnecessary to rotate the specimen, as it is in conventional crystallography, to obtain all the diffraction features. There is, however, considerable loss of information because of the cylindrical averaging that makes conventional crystallographic analysis difficult. Because of the rotational averaging, there is superposition of individual X-ray reflections in the observed diffraction spots making it difficult to assign intensities to each distinct X-ray reflection unambiguously, and the low resolution of the diffraction information means that standard crystallographic techniques for solving the so-called "phase problem" (see below) and deduce a "structure" are impractical.

Another practical difficulty is obtaining accurate intensity measurements in the presence of the diffuse background. In any intact muscle the coherent diffraction from the myofilaments is superimposed on a strong background composed of incoherent scattering from all other components of the muscle, including nerves, mitochondria, and soluble proteins. This will make intensity measurements from the inner peaks less accurate since some model must be assumed for the background (typically a 2D Gaussian or an exponential decay model) and this introduces error when measuring the peak intensity. There are also several global background fitting procedures built into the program XFIX, part of the CCP13 suite (http://www.ccp13.ac.uk) of fiber diffraction software. While any one of these procedures may be quite useful to remove background in selected regions of the pattern, it is not possible generally to reliably remove the background globally so they should be used with caution.

Even at the low resolution of typical muscle fiber patterns, however, the observed intensities and positions in the low-angle pattern from muscle is quite sensitive to the number, structure and lattice disposition of crossbridge linkages between myosin and actin.[46] There have been a number of attempts to model the arrangement of crossbridges starting with the known atomic level structures of myosin subfragment-1 and the known spacings and symmetries of their arrangement around the thick filament. Models may also attempt to incorporate the thin filaments using model structures based on the atomic coordinates of G-actin, tropomyosin and troponin.[45] This type of analysis is very time consuming and it has only been done with strongly diffracting specimens such as *Lethocerus* IFM,[26] fin muscle from bony fish[46] and skeletal muscle from rabbit[47] and, so far, under only static conditions such as relaxation or rigor. It is not clear, at present, whether it will be possible to perform such analysis on patterns from living *Drosophila* IFM. Fortunately, there is enough known about the relationship of certain structural components to diffraction intensity to allow hypothesis testing on the basis of relatively few, strong parts of the pattern without the need for large-scale molecular simulation. The ways in which this may be done are outlined in the sections that follow.

Interpretation of the Equatorial Diffraction Pattern

The Equatorial Pattern and Lattice Spacing

A detailed description of fiber diffraction theory is beyond the scope of this article. For an excellent background as it applies to muscle see chapter 2 of Squire.[48] The intensities in the observed diffraction pattern are related to the Fourier transform of the electron density distribution in the original object. A crystalline object can be described mathematically as the convolution of the repeating unit, called the crystallographic unit cell, and a "lattice " describing the arrangement of the unit cells in two or three dimensional space. Convolution theorem dictates then that the intensity in the diffraction pattern will be related to the product of the Fourier transform of the density distribution within the unit cell (which will be a complicated continuous function) and the Fourier transform of the lattice which will be just another discontinuous arrangement of points in Fourier, also called reciprocal space. Thus, because of the crystalline symmetry, intensity is not continuously distributed but confined to discrete diffraction spots. Each diffraction spot comes from the energy deposited in the detector by absorbed X-ray photons. Photons have wave-like properties in that they can be described by an amplitude and a phase with respect to the origin in reciprocal space. The wave amplitude is proportional to the square root of the observed intensity in a given spot. In principle, if one obtains the intensity of all the diffraction spots in the pattern and can arrive at the phases of each of these reflections, one can calculate the electron density distribution (a "structure") in the myofibrils in either two or three dimensions by Fourier reconstruction using standard crystallographic techniques. There is no general way to arrive at the phase information from the X-ray pattern alone. In conventional crystallography various techniques may be used to derive the phases. As described above, these techniques are not generally applicable to the muscle case but that model structures may be used to simulate the X-ray diffraction pattern and, if they provide a good fit to the data, provide the best available estimate of the real structure.

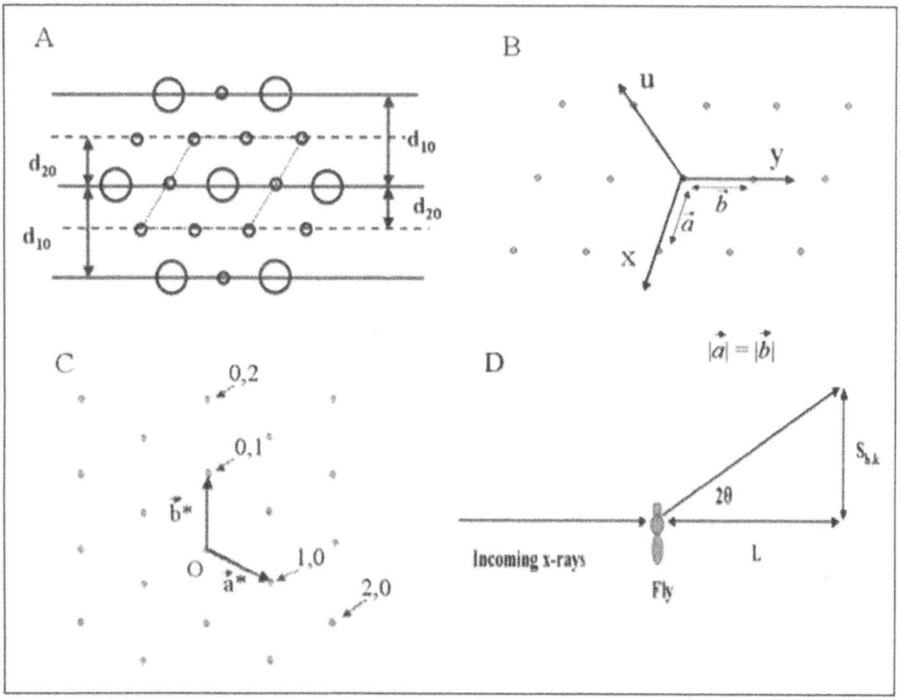

Figure 5. Origin of the equatorial diffraction pattern from Drosophila IFM. A) shows schematically a cross section through the A-band of a myofibril. The large circles represent myosin containing thick filaments, the small circles actin containing thin filaments. Also shown are the crystallographic 1,0 and 2,0 Bragg planes with the distance between them being d_{10} and d_{20} respectively. The trapezoidal region bounded by the dotted line is the crystallographic unit cell. The crystal structure can be generated by convoluting the unit cell with the hexagonal lattice of points shown in (B). Note that it requires 3 axes, x,y and u to completely specify a hexagonal lattice. The Fourier transform of the hexagonal lattice in real space is the reciprocal lattice shown in (C). Note that the lattice is rotated 90 with respect to the real space lattice. The reciprocal lattice is defined by the lattice vectors a* and b* which are proportional to 1/|a| and 1/|b| respectively where a and b are the real space lattice vectors shown in 4b. Reciprocal lattice points are identified by their Miller indices h and k which can be thought of as coordinates in reciprocal lattice space. The 1,0 (i.e., h = 1,k = 0), 2,0, 0,1 and 0,2 lattice points are indicated. Each lattice point will correspond to an X-ray reflection that could potentially be observable in the diffraction pattern. D) shows the experimental geometry. The distance $S_{h,k}$ in the diffraction pattern can be related to $d_{h,k}$ using Bragg's law and knowing the X-ray wavelength and L.

Here we apply these concepts to the analysis of the equatorial pattern from *Drosophila* IFM. Figure 5A shows a schematic diagram of the regular two-dimensional lattice comprising the thick and thin filaments in insect IFM. Note the trapezoidal area bounded by dotted lines. This is the so-called crystallographic unit cell and contains one thick filament and three thin filaments (4 half filaments on the sides + 2 one-sixth filaments + 2 one-third filaments at the corners). The lattice function in this case (Fig. 5B) consists of a simple hexagonal arrangement of points. The convolution process places an entire until cell centered (i.e., the position of the thick filament in this instance) at each one of these lattice points to produce the real crystalline array.

For every real hexagonal lattice (with characteristic spacing a) in "real space" one can construct an imaginary lattice (in "reciprocal space") with characteristic spacing a* where a* µ1/a (Fig. 5C). Individual points in this "reciprocal lattice" can be defined by their so-called Miller indices, h and k, each of which corresponds to a possible X-ray reflection that may be observed in the diffraction pattern. The distance between the origin of the reciprocal lattice (marked O in Fig. 5C) and a reciprocal lattice position defines a vector (of magnitude $\sqrt{(h^2 + k^2 + hk)}/d_{10}$, that is inversely proportional to the perpendicular distance between imaginary lattice planes that can be drawn in the crystallographic unit cell (basic repeating unit) of the myofibrils (Fig. 5A). The lattice planes and lattice vectors for the two sets of Miller indices h = 1 and k = 0, and h = 2 and k = 0, are shown in Figure 5A, c respectively. These correspond to the two strongest reflections observed on the equator. As many as 20 diffraction orders have been observed in some patterns (out to h = 5, k = 3), the strongest reflections being 1,0; 1,1; 2,0; 2,1; 3,0; 3,1; 3,2 and 4,1. The rotational averaging implies that there are contributions to the intensity of an observed reflection from three distinct hexagonal reciprocal lattice points. In Figure 5C, note that the distance from the origin to the 1,0 reciprocal lattice point is the same as that for h = 0, k = 1 reflection and the h = 1, k = -1, so the intensity of the observed "1,0" equatorial spot will be the sum of the three reflections. Likewise the observed 2,0 diffraction spot has contributions from the h = 0 k = 2 and the h = 2 and k = -2 reflections.

Figure 5D shows a diagram of the experimental geometry showing the sample, the detector and the diffraction angle 2θ. The relationship between the distance of a given diffraction spot ($S_{h,k}$) from the center of the pattern and the distance between the corresponding lattice plane ($d_{h,k}$) in the muscle unit cell is given by Bragg's law: $n\lambda = 2d\sin\theta$. For the geometry given in Figure 5D, $\tan 2\theta = S/L$, where λ is the wavelength of the incident X-ray photons (0.103 nm) and L is the distance from the sample to the detector and n is an integer. For small angles, the Bragg relationship can be reduced to $d_{hk} = \lambda L/S_{hk}$. The innermost, strongest reflection from the 1,0 lattice plane (which joins the thick filaments) is called d_{10} (n = 1 in Bragg's law) and it is the perpendicular distance between the d_{10} planes that is meant by lattice spacing. The third reflection, which is the next strongest, from the 2,0 lattice planes is called d_{20} (n = 2 in Bragg's law) and it is half the perpendicular distance between the d_{10} planes. The d_{10} lattice spacing can be converted to the inter-thick filament spacing by multiplying d_{10} by $2/\sqrt{3}$. Thus the equatorial pattern is a convenient and accurate way to measure the inter-filament spacing in muscle.

Estimating the Lattice Spacing

For a quick estimate of the separation of the 1,0 and 2,0 equatorial reflections, one can use the distance measuring tools in either the FIT2D (http://www.esrf.fr/computing/scientific/FIT2D) or the ImageJ programs (http://rsb.info.nih.gov/ij/) which allow you to visualize the X-ray pattern on the screen of a personal computer and manually estimate the center of mass of the desired reflections. This method can quickly give the value of S_{10}, the 1,0 reflection spacing, to 1% accuracy, especially if two to three measurements, ideally by different people, are averaged. The d_{10} lattice spacings are then calculated using the reduced version of Bragg's Law given above.

The manual method just described is the one of choice in disordered systems such as skinned cardiac muscle, but for the highly crystalline *Drosophila* IFM lattice much higher accuracies are achievable. We routinely see 15 reflections (out to the 6,0 reflection) on the equator, sometimes as many as 20 (out to the 5,3 reflection). The distances from the center of the pattern to each of these outer reflections ($S_{h,k}$) are related to the 1,0 distance, S_{10}, by $S_{h,k} = S_{10}\sqrt{(h^2 + k^2 + hk)}$ where h and k are the Miller indices of each reflection (see above). This selection rule allows construction of a nonlinear least squares fitting procedure[49] to model the one dimensional projection (obtained using FIT2D or ImageJ) of the intensities along the equator as a series of Gaussian-shaped peaks superimposed on a background assumed to be the sum of two exponentials. The peak positions are constrained to obey the selection rule or the hexagonal lattice. Using this procedure, it is routine to estimate d_{10} to accuracies of better than 0.05 nm or 0.1 %.

Hypothesis Testing Based on the Lattice Spacing

An old idea in muscle physiology is that muscle observes constant lattice volume ($= 2d_{10}^2/\sqrt{3}$ x sarcomere length) whereby changes in muscle fiber length are compensated by changes in lattice spacing. Constant volume was first demonstrated in relaxed, membrane-intact, skeletal muscle by Matsubara and Elliott[50] and appears to be a consistent feature of relaxed muscle.

There have been, however, a number of observations showing that constant volume is not always observed in contracting muscle (see Millman[51] for review). The notion that constant lattice volume can be used to explain muscle behavior under contracting conditions still persists, however.

Irving and Maughan[37] performed a critical test of the hypothesis by observing the lattice spacing when the wings were up (DLM's extended) and wings were down (DLMs shortened). Under these conditions the DLM's would change length by about 3.5%,[52] which, according to the constant volume hypothesis, would predict changes in lattice spacing of about 1.7%. The observed changes were less than 0.25%, ~4-7 times less than predicted. Thus the DLM's of living *Drosophila* do not maintain constant lattice volume during oscillatory contraction. Another observation was that there was a small but significant lattice shrinkage when the IFM was activated. On the basis of these observations, it was proposed[37] that a net radial force generated as crossbridges attach appears to bring the thick filaments to an 'equilibrium' inter-thick filament spacing (of ~56 nm) that is maintained for several thousand wingbeat cycles. Since under contracting conditions the most important radial forces at physiological lattice spacings will be due to the crossbridges,[51] the attractive and repulsive components of the crossbridge force must be in balance.

Use of the I_{20}/I_{10} Intensity Ratio

The fitting procedure described above yields the relative intensities of the equatorial reflections as the areas of their respective Gaussian models, which are free parameters in the fits. As stated above, the strongest of these reflections are the first and third i.e., the 1,0 and 2,0 reflections respectively. The ratio of the intensities, I_{20}/I_{10}, by analogy to the I_{11}/I_{10} ratio in vertebrate muscle, has a straightforward interpretation. In the filament lattice of IFM, there is only one 1,0 plane entirely in the unit cell and it is populated by one thick filament and one thin filament. There will be two 2,0 planes in the unit cell one entirely within the unit cell and two planes shared with the adjacent unit cells (see Fig. 5A). One plane is coincident with the 1,0 plane so will contain one thick filament and two half thin filaments. The two shared planes join the thin filaments in the planes located half way in between the 1,0 planes and each contain one 1/2 thin filament, one 1/3 filament and one 1/6 thin filament. Thus, there will be a total of one thick filament and three thin filaments contributing to the intensity of the 2,0 reflection. If mass, in the form of crossbridges, moves from the region of the thick filament to that of the thin filament, there will be a loss of electron density on the 1,0 and a gain on the 2,0 plane. This will increase the intensity of the 2,0 at the expense of the 1,0 equatorial reflection. Thus the $I_{2,0}/I_{1,0}$ intensity ratio can be used as a measure of mass shifts that accompany changes in physiological state or in comparing mutant and wild type strains of flies.

This feature has been exploited[37] to compare two lines of transformed flies *Mlc2*^{S66,67A}, a flightless mutant in which two phosphorylated serine residues (66 and 67) of the myosin regulatory light chain (RLC) are replaced with unphosphorylatable alanines, and a flight-capable control line *Mlc2*$^{+}$, in which wild type myosin RLC protein is expressed in a MLC2 null background.[16,17] The results showed that the I_{20}/I_{10} equatorial intensity ratio of the mutant fly was 35% less than that of wild type. The reduced mass associated with the thin filament (implied by the reduced I_{20}/I_{10} ratio) supported the hypothesis that myosin heads that lack phosphorylated RLC remain close to the thick filament backbone. On this basis, it was speculated[37] that the mutant flies are flightless because too few myosin heads are positioned for productive actomyosin interactions.

It is also possible, in principle, to use the observed equatorial intensities with a set of phases for each reflection derived by some means to calculate maps of the electron density distribution (an "electron density map") in the unit cell using inverse Fourier transformation. These correspond to the projection of the mass in the A-band onto a plane perpendicular to the long axis of the fiber. These have been used in vertebrate muscle[49,53] to estimate changes in filament dimensions and calculate shifts of mass from the region of the thick filaments to that of the thin filament (presumably due to the movement of crossbridges). There have not been attempts to phase insect IFM to date. This problem may be addressed using a combination of intensity modeling,[49,54,55] shrinking/ swelling experiments[49,56] and comparison to electron micrographs.[57,48]

Peak Widths

The width of the Gaussian representing a given diffraction peak $\sigma_{h,k}$ can be expressed as $\sqrt{(\sigma_c^2 + \sigma_d^2 S_{hk} + \sigma_s^2 S_{hk}^2)}$ where $S_{hk} = \sqrt{(h^2 + k^2 + hk)}$.[49,53] σ_c is the known width of the X-ray beam, σ_d is related to the amount of heterogeneity in inter-filament spacing among the myofibrils, and σ_s is related to the amount of paracrystalline (liquid-like) disorder of the myofilaments in the hexagonal lattice.[53] Both σ_d and σ_s may be used as free parameters of the fits. In wild-type flies with well-ordered muscles, both these parameters are very small (<<1) but can increase significantly in the case of the IFM of some mutant flies. Since these parameters can be estimated to high precision using the fitting program, they can be used to characterize disorder in the lattices of IFM from transgenic flies.

Contributions from Other Muscles

In flies where the thorax is slightly off-axis with the X-ray beam, one can observe distinct equatorial patterns from the dorsal–ventral muscle (DVM) and tergal depressor of the trochanter (TDT) (see Fig. 3 of ref. 37). The DVM and TDT patterns are at increasing angles to the DLM pattern. The TDT can be identified by its relatively strong 1,1 equatorial reflection which arises from the different filament arrangement in these leg muscles. This capacity for simultaneous observation of two and three different muscle systems has been used[37] as an additional test of the constant volume hypothesis. Since the DLM and DVM are antagonistic muscles, one set will lengthen while the other shortens, and vice versa. Going from rest to activation, the lattice spacings of the DVM moved in the same direction and by about the same amount as those of the DLM. It was concluded that the reduced lattice spacing in working DLM is most likely related to an inwardly directed radial component of the crossbridge force upon activation rather than to changes in cell length. Changes in the lattice spacing of the TDT, showed no consistent relationship to changes in the DLM or DVM spacings.

Interpretation of the Near-Meridional 14.5 and 7.2 nm Reflections

The axial displacement of one myosin head to the next one in a given helical strand in all thick filaments is 14.5 nm reflecting the packing of the constituent myosin molecules in the backbone of the filament. This periodic density gives rise to a strong reflection on the meridian in most striated muscles. Reflections are also observed at spacings of 14.5/n nm where n is an integer. The most widely studied are the 14.5 nm (n = 1) reflection and the 7.2 nm (n = 2) reflection. A considerable body of work on vertebrate muscle now exists[59-61] relating changes in the intensity of the 14.5 nm reflection to changes in the axial orientation of the myosin heads as they change physiological state or go through their crossbridge cycle. More recently, studies of the 7.2 nm reflection (also in vertebrate muscle) indicate that most of the intensity of this reflection comes from backbone structures.[62,63] Thus, insofar as these results can be extrapolated to the IFM system, changes in the 14.5 nm spacing and intensity can be related to axial motions of the myosin heads and changes in the spacing and intensity of the 7.2 nm reflection to changes in the thick filament backbone.

The pattern in Figure 3 shows that unlike *Lethocerus* in which the 14.5 nm layer is strong on the meridian, in *Drosophila* the 14.5 nm meridional reflection is split, with reduced intensity on the meridian itself. Likewise, the 7.2 nm reflection is partially split (Fig. 3), with the first true meridional reflection appearing at 4.83 nm (not shown). Elliott[36] also observed this splitting in his study of fixed and embedded *Calliphora* IFM. Irving and Maughan[37] suggested that unlike *Lethocerus* where all thick filaments are in axial register, adjacent thick filaments in flies are staggered by one third of the 14.5 nm spacing. The diffracting structures are probably groups of myosin heads making up 5-6 nm-thick 'crowns" spaced by 14.5 nm along the filament.[24] Thus filament stagger would cause the density of the crowns on one filament to spread into the axial projection of the crown density on adjacent filaments, resulting in an apparent increase in the frequency of these crowns with the observed consequences on the diffraction pattern. Since the intensity of the 14.5 and 7.2 nm reflection presumably still originates from the axial density distribution of the myosin heads and thick filament backbone respectively, changes in the spacing and intensity of these reflections can still be interpreted as in other muscle systems.

Peak positions and intensities of the 14.5 and 7.2 nm near-meridional reflections can be obtained from 1-dimensional projections along the meridian using the multiple 1-D fitting routines in FIT2D. Any change in spacing in the 14.5 nm reflection can be interpreted in terms of changes in the position of the center of mass of crossbridges. Intensity changes can be interpreted as changes in the angle of the crossbridges with respect to the long axis of the filament. The intensity will be strongest when the crossbridges are perpendicular to the axis of the filament and weakest when they are at an oblique angle. Any changes in the spacing of the 7.2 nm reflection when the filaments are bearing tension can potentially be used to estimate the elasticity of the thick filament backbone. Changes in the intensity of the 7.2 nm reflection do not yet have a simple interpretation.

Recent work appears to bear out this interpretation. Time-resolved studies using the flight arena revealed that the 14.5 nm reflection do undergo large changes in intensity during the wingbeat cycle indicating large angular changes of the myosin heads with respect to the long axis of the thick filament (Dickinson et al, in preparation). Small changes in the position of the 14.5 nm reflection indicate that there are also small axial displacements in the catalytic domain of the myosin heads during the cycle. Changes in the 7.2 nm reflection spacing and intensity are much smaller and support the idea that they arise from different structures in the thick filament. The details of, and implications of these findings, will be discussed elsewhere.

There is an additional difference between *Lethocerus* and *Drosophila* IFM.[37] The integrated intensity of the 14.5 nm reflection in resting *Drosophila* is low and becomes higher during activation (Dickinson et al, in preparation) in marked contrast to that observed in glycerinated *Lethocerus* DLM, where the intensity of the 14.5 nm reflection decreases by ~50% upon activation by stretch.[25] In *Lethocerus*, the loss of intensity is assumed to be due to myosin heads losing their thick-filament based symmetry when they bind to 'target sites' on the thin filaments. It has been reported[64] that the myosin heads in *Drosophila* seemed relatively disordered as compared to *Lethocerus* and this may be the explanation for the reduced 14.5 nm intensity in the resting state.

Interpretation of the Intensities on the First Row Line

Tregear and colleagues[25] in their analysis of X-ray patterns from stretch-activated, glycerinated (skinned) IFM from *Lethocerus* showed that stretch activation was associated with a loss of intensity on the first row line spot (aligned with the 1,0 equatorial reflection) on the 38.7 nm layer line and a gain of intensity on the first row line spot on the 19.3 nm layer line. This reciprocal change in intensity is most likely due to labeling by the attached crossbridges of so-called target zones which are actin binding sites on the thin filament located half-way between troponin subunits. Binding of the heads half way between the troponin repeats strengthens the 19.3 nm reflection and weakens the 38.7 because of destructive interference of X-rays from the two repeating structures. Figure 3, which shows a pattern from contracting IFM in *Drosophila*, shows a strong 19.3 nm first row line spot (as indicated by the arrow) with the 38.7 nm reflection being absent. In similar patterns from resting muscle the reverse is true, consistent with previous observations.[25] We have recently made time resolved measurements of the intensities of these first row line reflection in order to detect target zone binding in various parts of the wingbeat cycle. The results of these experiments will be presented and discussed elsewhere.

Concluding Remarks

In conclusion, with the *Drosophila* system it is possible to obtain detailed, two-dimensional X-ray diffraction patterns from working muscle in a living organism. Diffraction features in these patterns can be and have been used for testing fundamental hypotheses concerning crossbridge function in vivo. We expect that the ability to carry out future X-ray studies in transgenic flies with defined alterations of muscle proteins, many of which have homologous function in humans, will be a powerful tool to address more specific and insightful questions that could not be answered any other way.

Note Added in Proof

The work referenced as "Dickinson et al, in preparation" is published as: Dickinson M, Farman G, Frye M et al. Molecular dynamics of cyclically contracting insect flight muscle in vivo. Nature 2005; 433:330-333

Acknowledgements

Use of the Advanced Photon Source was supported by the U.S. Department of Energy, Basic Energy Sciences, Office of Energy Research, under Contract No. W-31-109-ENG-38. BioCAT is a U.S. National Institutes of Health-supported Research Center RR08630.

References

1. Elliott GF, Lowey J, Millman BM. Low angle diffraction studies of living striated muscle during contraction. J Mol Biol 1967; 25:33-35.
2. Huxley HE, Brown W. The low angle X-ray diagram of vertebrate striated muscle and its behaviour during contraction and rigor. J Mol Biol 1967; 30:383-434.
3. Haselgrove JC, Huxley HE. X-ray evidence for radial crossbridge movement and for the sliding filament model in actively contracting skeletal muscle. J Mol Biol 1973; 77:549-568.
4. Huxley HE, Haselgrove JC. The structural basis of contraction in muscle and its study by rapid X-ray diffraction methods. In International Boehringer Mannheim Symposia "Myocardial failure", Springer: Berlin 1976; 4-15.
5. Huxley HE, Simmons RM, Faruqi AR et al. Millisecond time-resolved changes in X-ray reflections from contracting muscle during rapid mechanical transients, recorded using Synchrotron Radiation. Proc Natl Acad Sci USA 1981; 78:2297-2301.
6. Huxley HE, Simmons RM, Faruqi AR et al. Changes in the X-ray reflections from contracting muscle during rapid mechanical transients and their structural implications. J Mol Biol 1983; 169:469-506.
7. Irving M, Lombardi V, Piazzesi G et al. Myosin head movements are synchronous with the elementary force-generating process in muscle. Nature 1992; 357:156-8.
8. Dobbie I, Linari M, Piazzesi G et al. Elastic bending and active tilting of myosin heads during muscle contraction. Nature 1998; 396:383-7.
9. Piazzesi G, Reconditi M, Linari M et al. Mechanism of force generation by myosin heads in skeletal muscle. Nature 2002; 415:659-62.
10. Huxley AF, Simmons RM. Proposed mechanism of force generation in striated muscle. Nature 1971; 173:971-973.
11. Celniker SE, Rubin GM. The Drosophila melanogaster genome. Annu Rev Genomics Hum Genet 2003; 4:89-117.
12. Rubin GM, Spradling AC. Genetic transformation of Drosophila with transposable element vectors. Science 1982; 218:348-353.
13. Maughan DW, Vigoreaux JO. An integrated view of insect flight muscle: Genes, motor molecules and motion. News Physiol Sci 1999; 14:87-92.
14. Bernstein SI, O'Donnell PT, Cripps RM. Molecular genetic analysis of muscle development, structure and function in Drosophila. Int Rev Cytol 1993; 143:63-152.
15. Peckham M, Molloy JE, Sparrow JC et al. Physiological properties of the dorsal longitudinal flight muscle and the tergal depressor of the trochanter muscle of Drosophila melanogaster. J Muscle Res Cell Motil 1990; 11:203-15.
16. Tohtong RH, Yamashita H, Graham M et al. Impairment of muscle function caused by mutations of phosphorylation sites in myosin regulatory light chain. Nature 1995; 374:650-653.

17. Dickinson MH, Hyatt CJ, Lehman FO et al. Phosphorylation-dependent power output of transgenic flies: An integrated study. Biophys J 1997; 73:3122-3134.
18. Worthington CR. X-ray diffraction studies on the large-scale molecular structure of insect muscle. J Mol Biol 1961; 3:618-633.
19. Miller A, Tregear RT. Evidence concerning crossbridge attachment during muscle contraction. Nature 1970; 226:1060-1061.
20. Armitage PM, Tregear RT, Miller A. Effect of activation by calcium on the X-ray diffraction pattern from insect flight muscle. J Mol Biol 1975; 92:39-53.
21. Holmes KC, Tregear RT, Barrington-Leigh J. Interpretation of the low angle X-ray diffraction from insect muscle in rigor. Proc R Soc Lond [Biol.] 1980; 207:13-33.
22. Goody RS, Reedy MC, Hofmann W et al. Binding of myosin subfragment 1 to glycerinated insect flight muscle in the rigor state. Biophys J 1985; 47:151-169.
23. Rapp G, Güth K, Maèda Y et al. Time-resolved X-ray diffraction studies on stretch activated insect flight muscle. J Muscle Res Cell Motil 1991; 12:208-215.
24. Reedy MK, Lucaveche C, Naber N et al. Insect crossbridges, relaxed by spin-labeled nucleotide, show well-ordered 90 degrees state by X-ray diffraction and electron microscopy, but spectra of electron paramagnetic resonance probes report disorder. J Mol Biol 1992; 227:678-697.
25. Tregear RT, Edwards RJ, Irving TC et al. Stretch-activation of insect flight muscle changes the low-angle actin based diffraction pattern. Biophys J 1998; 74:1439-1451.
26. Al-Khayat HA, Hudson L, Reedy MK et al. Myosin head configuration in relaxed insect flight muscle: X-ray modeled resting crossbridges in a prepowerstroke state are poised for actin binding. Biophys J 2003; 85:1063-1079.
27. Reedy MK, Reedy MC. Rigor crossbridge structure in tilted single filament layers and flared-X formations from insect flight muscle. Biophys J 1985; 47:151-69.
28. Taylor KA, Reedy MC, Cordova L et al. Three-dimensional image reconstruction of insect flight muscle. I. The rigor myac layer. J Cell Biol 1989a; 109:1085-102.
29. Taylor KA, Reedy MC, Cordova L et al. Three-dimensional image reconstruction of insect flight muscle. II. The rigor actin layer. J Cell Biol 1989b; 109:1103-1123.
30. Taylor KA, Reedy MC, Reedy MK et al. Crossbridges in the complete unit cell of rigor insect flight muscle imaged by three-dimensional reconstruction from oblique sections. J Mol Biol 1993; 233:86-108.
31. Taylor KA, Schmitz H, Reedy MC et al. Tomographic 3D reconstruction of quick-frozen, Ca^{2+}-activated contracting insect flight muscle. Cell 1999; 99(4):421-431.
32. Winkler H, Reedy MC, Reedy MK et al. 3-D structure of nucleotide-bearing crossbridges in situ: Oblique section reconstruction of insect flight muscle in AMPPNP at 23°C. J Mol Biol 1996; 264:302-22.
33. Schmitz H, Lucaveche C, Reedy MK et al. Oblique section 3-D reconstruction of relaxed insect flight muscle reveals the cross-bridge lattice in helical registration. Biophys J 1994; 67:1620-33.
34. Schmitz H, Reedy MC, Reedy MK et al. Electron tomography of insect flight muscle in rigor and AMPPNP at 23°C. J Mol Biol 1996; 264:279-301.
35. Schmitz H, Reedy MC, Reedy MK et al. Tomographic three-dimensional reconstruction of insect flight muscle partially relaxed by AMPPNP and ethylene glycol. J Cell Biol 1997; 139:695-707.
36. Elliott GF. X-ray diffraction from insect flight muscle. J Mol Biol 1965; 13:956-958.
37. Irving TC, Maughan DW. In vivo X-ray diffraction of indirect flight muscle from Drosophila melanogaster. Biophys J 2000; 78:2511-2515.
38. Irving TC, Fischetti R, Rosenbaum G et al. Fiber diffraction using the bioCAT undulator beamline at the advanced photon source. Nucl Instrum Meth (A) 2000; 448:250-254.
39. Irving TC, Fischetti RF. Fibre diffraction using the bioCAT facility at the advanced photon source. Fiber Diffraction Review 2001; 9:58-62.
40. Naday I, Westbrook EM, Westbrook ML et al. Characterization and data collection on a direct-coupled CCD X-ray detector. Nucl Instrum Methods (A) 1994; 348:635-640.
41. Phillips WC, Stewart A, Stanton M et al. High-sensitivity CCD-based X-ray detector. J Synchrotron Rad 2002; 9:36–43.
42. Helliwell JR. Macromolecular crystallography with synchrotron radiation. Cambridge University Press, 2003.
43. Tregear R, Miller A. Evidence of crossbridge movement during contraction of insect flight muscle. Nature 1969; 222:1185-1186.
44. Lehmann FO, Dickinson MH. The changes in power requirements and muscle efficiency during elevated force production in the fruit fly, Drosophila melanogaster. J Exp Biol 1997; 200:1133-1143.
45. Squire JM, Harford JJ, al-Khayat HA. Molecular movements in contracting muscle: Towards "muscle—the movie". Biophys Chem 1994; 50:87-96.

46. Hudson L, Harford JJ, Denny RC et al. Myosin head configuration in relaxed fish muscle: Resting state myosin heads must swing axially by up to 150 A or turn upside down to reach rigor. J Mol Biol 1997; 273:440-55.
47. Gu J, Xu S, Yu LC. A model of cross-bridge attachment to actin in the A*M*ATP state based on X-ray diffraction from permeabilized rabbit psoas muscle. Biophys J 2002; 82:2123-33.
48. Squire JM. The structural basis of muscular contraction. New York: Plenum 1981.
49. Irving TC, Millman BM. Changes in thick filament structure during compression of the filament lattice in vertebrate striated muscle. J Muscle Res Cell Motility 1989; 10:385-396.
50. Matsubara I, Elliott GF. X-ray diffraction studies on skinned single fibres of frog skeletal muscle. J Mol Biol 1972; 72:657-69.
51. Millman BM. The filament lattice of striated muscle. Physiol Rev 1998; 78:359-391.
52. Chan WP, Dickinson MH. In vivo length oscillations of indirect flight muscles in the fruit fly Drosophila virilis. J Exp Biol 1996; 199:2767-74.
53 Yu LC, Steven AC, Naylor GR et al. Distribution of mass in relaxed frog skeletal muscle and its redistribution upon activation. Biophys J 1985; 47(3):311-21.
54. Yu LC. Analysis of equatorial X-ray diffraction patterns from skeletal muscle. Biophys J 1989; 55(3):433-40.
55. Malinchik S, Yu LC. Analysis of equatorial X-ray diffraction patterns from muscle fibers: Factors that affect the intensities. Biophys J 1995; 68:2023-31.
56. Worthington CR, McIntosh TJ. Direct determination of the electron density profile of nerve myelin. Nature New Biol 1973; 245(143):97-9.
57. Trus BL, Steven AC, McDowall AW et al. Interactions between actin and myosin filaments in skeletal muscle visualized in frozen-hydrated thin sections. Biophys J 1989; 55:713-24.
58. Hawkins CJ, Bennett PM. Evaluation of freeze substitution in rabbit skeletal muscle. Comparison of electron microscopy to X-ray diffraction. J Muscle Res Cell Motil 1995; 16:303-18.
59. Huxley HE, Simmons RM, Faruqi AR et al. Changes in the X-ray reflections from contracting muscle during rapid mechanical transients and their structural implications. J Mol Biol 1983; 169:469-506.
60. Irving M, Lombardi V, Piazzesi G et al. Myosin head movements are synchronous with the elementary force-generating process in muscle. Nature 1992; 357:156-8.
61. Irving M, Piazzesi G, Lucii L et al. Conformation of the myosin motor during force generation in skeletal muscle. Nat Struct Biol 2000; 7:482-5.
62. Huxley HE, Stewart A, Irving TC. Spacing changes in the actin and myosin filaments during activation and their implications. Adv Exp Med Biol 1998; 453:281-7.
63. Huxley HE, Reconditi M, Stewart A et al. What the higher order meridional reflections tell us. Biophys J 2003; 84:139a.
64. Menetret JF, Schröder RR, Hoffman W. Cryo-electron microscopic studies of relaxed striated muscle thick filaments. J Muscle Res Cell Motil 1990; 11:1-11.

CHAPTER 17

Functional and Ecological Effects of Isoform Variation in Insect Flight Muscle

James H. Marden

Abstract

Nearly all of the known structural molecules in insect flight muscles exist as multiple isoforms. Both post-transcriptional and post-translational mechanisms are responsible for this variability. Among these mechanisms, alternative splicing is noteworthy for the ability to create a large number of combinatorial arrangements of alternative exons. For example there are over 1K possible distinct combinations of the characterized isoforms of troponin-T and projectin, which are just two of the many alternatively spliced proteins in insect muscle. The potential number of combinatorial possibilities for larger suites of insect muscle proteins is exponentially larger, i.e., numbers that far exceed the total number of coding genes in the insect genome. Thus, isoform variation is a potent source of variation in insect flight muscle and other tissue types, and the control of alternative splicing and other mechanisms that generate protein isoforms is a likely target of natural selection. Presently we know very little about the realized extent of this potential to generate protein variation, and no studies have yet examined constraints such as coordinate regulation of isoform expression of multiple protein species within insect flight muscles.

Functional studies of naturally occurring isoform variation, including the effects of phosphorylation of myosin light chain and alternative splicing of troponin-T, have revealed quantitative effects on muscle contractility and flight performance. In the case of troponin-T in a dragonfly, the isoform mixture affects the calcium sensitivity of muscle activation, power output of intact flight muscle, and wingbeat kinematics. High muscle power output is positively related to success in territoriality and mating, but increases the energetic cost of flight. Thus, alternative splicing of troponin-T in this species appears to be an important mechanism for adjusting between energetically expensive high-performance flight or less costly low-performance flight. Dragonflies show a strong positive relationship between fat reserves and muscle power output, which suggests that there is a signaling pathway that allows an alternative splicing mechanism to adjust muscle energy expenditure rates in accordance with levels of energy reserves. Infection of dragonflies by protozoan gut parasites is associated with a disappearance of the ability to match muscle power output to energy reserves, i.e., it appears that parasites disrupt this signaling pathway. This example shows that isoform variation in insect flight muscle has interesting effects not only on muscle contractile physiology, but is involved with many aspects of whole organism physiology and ecology. The taxonomic and mechanistic diversity within insect flight muscles provides a rich source of material for future studies that seek to understand the functional, ecological and evolutionary context of molecular diversity.

Nature's Versatile Engine: Insect Flight Muscle Inside and Out, edited by Jim Vigoreaux.

Introduction

This chapter presents an overview and catalog of isoform diversity within insect flight muscles, and summarizes studies that have examined the quantitative effect of naturally occurring isoform variation on muscle contraction and flight performance. Isoforms are defined here as molecules that exist as multiple types because of modifications during or after transcription from a single gene. Small variations in the composition of a protein in insect flight muscle can affect both ultrastructure and mechanics, sometimes independently.[1] Thus, achieving a full understanding of the structure and function of insect flight muscles requires knowledge about proteins that exist as multiple isoforms.

An overview of isoform variation also provides an important general perspective because it addresses a central question raised by genome sequencing: how could complex multicellular organisms evolve with as little as a 2-3 fold increase in the number of coding genes compared to unicellular eukaryotes? Protein isoform diversity has been proposed as one of the key mechanisms that has allowed a huge increase in organismal complexity with only a modest inflation in gene number.[2-4] By surveying the array of molecular variants known to exist within one tissue type, along with the known functional effects and the as yet poorly documented combinatorial possibilities, we can begin to appreciate the capacity of a relative handful of protein-coding genes to give rise to an exponentially greater array of functionally variable protein combinations.

The final part of this chapter extends this theme by exploring in a more focused manner the biology of isoform diversity in one protein in the flight muscles of dragonflies. By combining mechanistic and ecological perspectives, this work exemplifies the emerging sub discipline of functional ecological and evolutionary genomics.[5] Also, because it is based on a species other than *Drosophila melanogaster*, it provides additional taxonomic breadth and an indication of the diversity of flight muscle physiology within insects, the most taxonomically diverse group of organisms.

Nature's Versatile Engine

The diversity of flying insects necessitates a highly adaptable flight motor. Flight-capable insects range in body mass over about five orders of magnitude (< 1mg to ca. 80 g), with wide variation in wingbeat frequency and contractile mechanics. Certain tiny wasps and flies have wingbeat frequencies of 500-1000 Hz, whereas large butterflies and moths use wingbeat frequencies as low as 10-20 Hz. Carpenter bees[6,7] are active at midday in hot deserts with muscle temperatures up to 48°C, whereas small winter-active geometrid moths[8] can fly with muscle temperatures as low as –3°C. These large differences in contraction frequency and operating temperature have resulted in the evolution of considerable diversity in the ultrastructure and contractile physiology of insect flight muscles.[9-12]

Insect flight muscles can also undergo phenotypic variation over time within an individual during adult maturation and/or in response to environmental variation. An extreme example is that of bark beetles (*Ips pini*; Scolytidae), which undergo degeneration of their flight muscles during mating and brood rearing within trees, but can within a few days regenerate their flight muscles and regain the ability to fly.[13] Other insects undergo fairly predictable changes in flight muscle size and ultrastructure during the course of adult maturation. The flight muscles of Tsetse flies are not fully mature until after they have consumed a number of blood meals,[14] undergoing a total increase in mass of about 75%. Nearly all taxa of dragonflies (Odonata) undergo substantial growth of their flight muscles during adult maturation,[15] with dragonflies of the genus *Libellula* showing as much as a doubling or tripling of muscle size.[16]

There are important molecular changes during adult maturation of insect flight muscles,[17-19] and presently there is fairly little knowledge of how molecular changes are associated with variation in muscle size and usage (but see ref. 20). The extreme taxonomic and functional diversity of insects, along with the ecological importance of variation in flight performance versus energetic costs (a major tradeoff for any flying animal) suggests that this is a rich field for further exploration,[21] especially for biologists seeking to relate molecular mechanisms underlying phenotypic plasticity with quantitative functional assays in an ecological context. The remainder of this chapter presents an overview of progress that has been made to date along these lines.

The Underlying Genetics: An Underinflated Genome and a Hyperinflated Transcriptome and Proteome

As argued above, the taxonomic and ecological radiation of insects has required tremendous versatility and adaptability of their flight motor, including the ability of individual insects to make adjustments in their muscle size and contractility. As will be shown below, phenotypic adjustments in muscle protein composition generally involve changes in transcripts and proteins rather than changes in expression within gene families. This realization is part of a general paradigm shift away from the "one gene - one protein" model, which has for many decades been a core element of the central dogma of molecular biology. Movement away from this paradigm has been gaining considerable momentum during the last few years in not just muscle physiology, but all aspects of biology. A primary driver has been the revelation that gene number is not hugely different between simple and complex organisms. The first complete eukaryotic genome sequence was for yeast, which revealed a total of 6K coding genes. At that time, estimates of gene number in more complex eukaryotes varied widely, with 100K being the most widely accepted projection for the number of coding genes in humans.[22] A common underlying assumption was that evolution of more complex and versatile creatures must have involved a great proliferation of gene number. We now know that complex multicellular eukaryotes have only about 14-30K genes, which is but a 2-5 fold enumeration of the genome size of yeast. This is a rather startling finding, for it reveals that features such as multiple tissue types, nervous systems, behavior, complex life histories, and the ability to make quantitative phenotypic adjustments in the functionality of tissues such as flight muscles, evolved with only a modest inflation of gene number. As this realization has come about, it has also become clear that complex eukaryotes rely heavily on mechanisms that create protein diversity at the transcriptional, post-transcriptional, and post-translational stages. For example, it is presently estimated that at least 40% of human genes are alternatively spliced,[23] whereas there are less than a dozen known alternatively spliced genes in yeast.

Isoform diversity appears to be especially common in nerve[24,25] and muscle,[26] where the ability to vary the molecular composition of ion channels and contractile filaments allows for fine-tuning of the kinetics of electrical or mechanical outputs and therefore specialization of function. A muscle containing a certain protein isoform may have mechanical properties quite different from a neighboring muscle containing a different isoform. Accordingly, it is common in insects to find one particular isoform expressed only in flight muscle and a different isoform in leg muscle. There are also examples in which a single insect flight muscle contains different protein isoforms or isoform mixtures over time or in different ecological settings. Thus, both the functional specificity between muscles, and the ability to adjust to ontogenetic or ecological conditions within muscles appears to be based in large part on the ability to vary protein isoform composition.

Creation of multiple protein isoforms from a single gene occurs in insect flight muscles by a variety of mechanisms, including alternative start codons (e.g., PAR domain protein 1),[27] alternative splicing (e.g., troponin-T),[28,29] and post-translational modification of proteins by phosphorylation (e.g., flightin).[19] Table 1 shows a catalog of proteins that are known to exist as multiple isoforms within insect flight muscles, or have an isoform expressed in flight muscle that is different from isoforms expressed in other muscles. Known functional effects are noted. This list includes components of the thick filaments (myosin heavy chain and myosin light chain), thin filaments (tropomyosin and all of the troponins), structural proteins that affect stiffness, elasticity, and filament anchoring (kettin, projectin, alpha actinin), metabolic enzymes (GPDH), regulatory enzymes (Mlck and Pdp 1), and ion channels (Slowpoke, BSC1).

Table 1. Catalog of insect flight muscle proteins that exist as multiple isoforms

Molecule	Insect	Variant	Effects	Refs.
Myosin heavy chain	*Drosophila*	IFM and TDT specific expression of alt. exon 3, 9, 11, 15. At least 14 isoforms known	Filament sliding velocity; ATPase rates; ultrastructure; subtle effects on flight ability	32,38-41, 53-58
Myosin regulatory light chain	*Drosophila*	Multiple phospho-variants that change with adult age in IFM	19-28% decrease in mechanical power output of null-mutant flies transformed with unphosphorylatable MLC construct	17,33-35
Actin	*Drosophila*	IFM expresses only ACT88F, one of six actin genes	Ultrastructure; flightless when expressing endogenous human actin that differs by 15aa or *Drosophila* construct that differs by 18aa. Individual aa replacements affect in vitro motility of myosin.	42,59-63
Tropomyosin I	*Drosophila*	IFM and TDT express only one isoform	Transformants expressing non-flight muscle isoform regain ability to jump and fly	30,64,65
Troponin H	*Drosophila* *Lethocerus*	Two different size isoforms expressed in IFM	Functional effects unknown	66-68
Myosin alkali light chain	*Drosophila*	IFM accumulates unique alt. spliced isoform	Functional effects unknown	69
Glyceraldehyde -3-phosphate dehydrogenase	*Drosophila*	One splice variant in IFM	IFM specific isoform localizes GPDH at M-lines and Z-discs and co-localizes GAPDH and aldolase. Mutants expressing non-IFM isoform lose localization of all three of these enzymes and are flightless.	31
Alpha actinin	*Drosophila*	Two muscle isoforms	Loss of contractility in muscles that do not express the protein, although ultrastructure is normal	70
Troponin C	*Drosophila; Lethocerus; Anopheles*	One major and one minor isoform in IFM	One isoform has a single calcium binding site whereas the other has two calcium binding sites. Initial activation of IFM is hypothesized to occur at isoforms with 2 Ca^{++} binding sites, with stretch activation regulated by the isoform with one Ca^{++} binding site.	43,71

continued on next page

Table 1. Continued

Molecule	Insect	Variant	Effects	Refs.
Troponin I	*Drosophila*	Alt. splicing creates ten isoforms that show muscle specific qualtitative and quantitative patterns of expression.	Mutation that prevents one of the four alternative versions of exon 6 prevents development of the flight muscles that normally express large amounts of this isoform.	43
Troponin T	*Drosophila, Apis, Libellula*	One isoform in Drosophila IFM and TDT; mixture of multiple splice variants in dragonfly FM. Evidence of phosphovariants in Drosophila.	Relative abundance of certain isoforms is correlated with calcium sensitivity of fiber activation, force and power output of intact muscle, and wingbeat kinematics.	28,29, 44,72
Myosin light chain kinase	*Drosophila*	Three isoforms formed by alt. splicing	Isoforms differ in regulatory domain; effects on muscle function unknown	73,74
Projectin	*Drosophila*	Complex pattern of alt. splicing in this giant protein; in one functional domain, 14 of 19 exons are alt. spliced	May affect muscle elasticity and passive stiffness, plays a role in stretch activation	36,75,76
Kettin (titin)	*Drosophila; Lethocerus*	At least two isoforms probably created by alt. splicing	Functional effect unknown; may affect muscle elasticity and passive stiffness	77-79
Flightin	*Drosophila*	Phosphorylation at multiple sites creates 11 protein variants	Alterations in phosphorylation affect dynamic viscous modulus.	19,80-83
Stretchin-Mlck	*Drosophila*	Internal promoters and poly-A sites enable expression of 7 distinct transcripts	Functional effect unknown	74
BSC1 (putative sodium channel)	*Blattella*	Two alt. splice variants are present in muscle	Functional effect unknown	84
Slowpoke (calcium activated potasium channel)	*Drosophila*	Large array of transcripts from tissue specific promoters and alt. splicing. One isoform restores flight ability in null mutants.	Presumably affects channel conductance	85-87
PAR domain protein 1	*Drosophila*	Trascription factor that controls expression of Tropomyosin I. Alt. start codons and alt. splicing create an array of Pdp1 transcripts	Functional effect unknown	27

Abbreviations: IFM= indirect flight muscle; TDT= tergal depressor of the trochanter; alt. = alternative; aa = amino acids.

Actin and Troponin C are included in Table 1 for the sake of completeness, although these proteins have multiple isoforms that are encoded by a gene family rather than the one-gene, many-protein pattern of expression that is the operational definition of isoform variation used in this chapter. Altogether, Table 1 includes almost all of the myofibrillar proteins known to exist in *Drosophila* indirect flight muscle.[1]

Functional Effects of Isoform Variation

Some of the proteins listed in Table 1 have an isoform that is expressed only in flight muscle. Experiments in which genetic manipulations have caused a nonflight muscle isoform to be expressed in the flight muscle of a *Drosophila* null mutant have shown effects ranging from no readily observable phenotypic change (tropomyosin[30]) to flightlessness (GPDH[31]) or intermediate effects (MHC[32]). This mixed bag of results prevents the general conclusion that flight muscle specific isoform expression is essential for proper function, although the majority of cases seem to indicate that this is true.

The Importance of Quantitative Measures

Functional variation caused by isoform switching can be subtle and detectable only by quantitative rather than qualitative methods. For example, restoration of flight ability in *Drosophila* by replacement of IFM-specific tropomyosin with another tropomyosin[30] is a qualitative measure that does not address how the level of flight ability might be affected. The following example illustrates how quantitative changes may be detectable only by using sophisticated measurement techniques.

In *Drosophila* flight muscles, myosin light chain kinase and other phosphorylases appear to become active during the first few hours following adult emergence, since only dephosphorylated MLC is present in late pupae and phosphorylated MLC accumulates in the hours following adult emergence.[17,33] MLC phosphorylation increases the ATPase activity of purified *D. melanogaster* myosin.[18,33] These observations, along with the similar time course of MLC phosphorylation and flight acquisition in newly emerged adults, suggest that MLC phosphorylation upregulates muscle contractility.

Genetic manipulations have been used to characterize the in vivo functional effects of variability in MLC phosphorylation.[34,35] In these experiments, flightless heterozygotes of homozygous-lethal MLC null mutants were rescued to normal muscle ultrastructure and flight ability by P-element transformation with the wild-type allele. Site-directed mutagenesis was subsequently used to create cDNA constructs in which 2 serine residues, the sites of MLC2 phosphorylation by myosin light chain kinase (MLCK), were replaced by unphosphorylable alanines. These constructs were transposed into MLC2 null mutants, resulting in lines of flies in which the only full-length, functional copy of MLC2 lacked either one or both of the sites that can be phosphorylated by MLCK. The resulting flies were examined for flight muscle ultrastructure, skinned fiber mechanical characteristics, aerodynamic power output and metabolic power input during tethered flight.[34,35] Muscles from the flies transformed with MLC2 lacking one or both MLCK phosphorylation sites showed no apparent changes in myofibrillar ultrastructure during rest, maximal activation, or rigor, nor did they show significantly altered calcium sensitivity, cross-bridge kinetics, or maximum steady-state isometric tension. Mutant muscles did show mechanical features indicative of a reduced recruitment of force-producing cross-bridges during stretch activation. Mechanical power output of mutant lines during tethered flight was reduced by 19-28% compared to wild type transformants, along with a similar decrease in metabolic power input, with no change in efficiency. Mutant flies could generally produce sufficient vertical net aerodynamic force to support their body weight, but significantly less than the 1.35 force/weight ratio produced by wild-type rescued and unmanipulated control flies.

This example shows quite clearly that subtle but important functional effects can result from changes as small as the addition of one or two phosphates on a single protein.

Alternative Splicing and the Generation of Combinatorial Complexity

Phenomena such as tissue-type specificity of isoform expression and functional effects of protein phosphorylation are reasonably familiar to most biologists. What is considerably less familiar, and completely missing from undergraduate textbooks in cellular and molecular biology, is the fact that a single tissue can express myriad forms of alternatively spliced transcripts from a single gene.

Figure 1 shows the splicing patterns of two genes that each encode a diversity of transcripts and proteins within insect muscle. The troponin T gene (TnT) in dragonflies encodes seven distinct transcripts,[28] including all but one of the eight possible combinations of a cassette of three alternatively spliced exons. The relative abundance of these transcripts within an individual dragonfly matches, at least qualitatively, the relative abundance of different TnT protein variants from flight muscle on 2-d gels,[29] which indicates that these transcripts are translated and the protein is incorporated into muscle. A more elaborate example is the projectin gene in *Drosophila* (for more detail, see chapter by Ayme-Southgate and Southgate), which has been shown to encode at least 16 distinct transcripts[36] (and probably many more, since the basic pattern shown in Figure 1B allows at least 144 combinations of the 13 alternatively spliced exons). These projectin transcripts were charaterized from total adult RNA, so it remains to be determined which isoforms are expressed in flight muscle. Suppose for the sake of illustration that these two genes undergo independently regulated alternative splicing in a single insect flight muscle. In that case, there could be 7 x 144 = 1,008 combinations of distinct proteins, a total that approaches 10% of the number of coding genes in an insect genome. Expansion of these combinatorial possibilities by inclusion of other alternatively spliced genes in muscle, or the plethora of post-translational modifications such as phosphorylation[19] would quickly expand this number of protein combinations to very large numbers.

Although there is presently very little knowledge of the extent to which organisms actually use this potential to generate different protein combinations, and no studies have yet examined constraints such as coordinate regulation of isoform expression of multiple protein species in insect flight muscles, it is clear that the capacity for generating different combinations of proteins is extremely high. It is also interesting to note that natural selection might commonly affect loci that control alternative splicing, thereby causing shifts in the relative abundance of isoforms of alternatively spliced proteins. Such a response to selection could result in large changes in function despite little or no genetic change at the loci of the relevant structural proteins or enzymes. A rough example of this is the recent finding that the chromosomal locations of quantitative trait loci significantly associated with the activity of a number of glycolytic enzymes in *Drosophila* (glycogen synthase, hexokinase, phosphoglucomutase, trehalase) is different from the chromosomal locations of those enzymes.[37] There is no indication as yet that this particular example involves isoform variation, but it serves to illustrate the point that genetic variation underlying functional differences does not necessarily reside in genes that encode the proteins that carry out a particular function.

Functional Consequences of Naturally Occurring Isoform Variation

Most of the work on functional effects of insect flight muscle isoforms has focused on genetic manipulations that cause expression of nonIFM isoforms in flight muscles,[30-32,38-42] or mutations that cause a failure to express the wild type IFM isoform or isoform mixture.[43] Such studies are excellent tools for understanding the molecular basis of muscle development and contraction, but they reveal little about naturally occurring variation because they create phenotypes that do not exist in nature. There has been relatively little work aimed at determining the functional consequences of naturally occurring variation in IFM isoform content. The example of MLC2 phosphorylation[17,18,33-35] discussed above comes close to doing this, but does not squarely hit the mark because variation in MLC2 phosphorylation is only known to occur during early adult maturation, so that flight-capable wild type flies

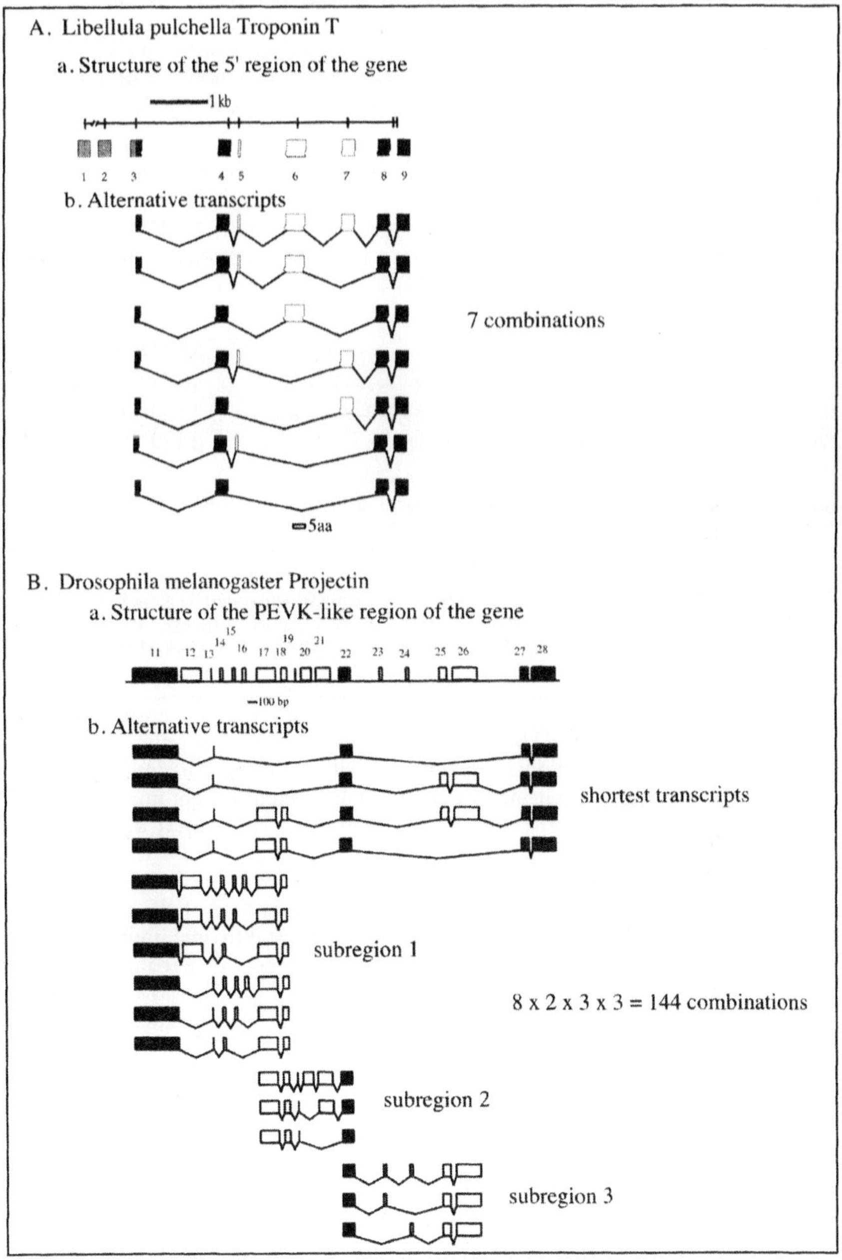

Figure 1. Gene structure and alternative splicing pattern of two insect muscle genes. Panel A shows the 5' end of the troponin-T gene from the dragonfly *Libellula pulchella*.[28,29,45] Alternative exons are white, constitutive exons are black, and exons within the 5' UTR are grey. Panel B shows the PEVK-like region of the *Drosophila melanogaster* projectin gene (adapted from Southgate R, Ayme-Soutgate A. Alternative splicing of an amino-terminal PEVK-like region generates multiple isoforms of Drosophila projectin. J Mol Biol 2001; 313:1035-1043). Alternative exons are white; constitutive exons are black. The estimate of the number of unique transcripts is for all possible combinations of the two alternate expression patterns of exons 25-26 and the exon combinations in characterized subregions 1, 2, and 3.

Table 2. Fragment sizes (number of nucleotides) and mean relative abundances (% of total TnT transcripts within flight muscles from an individual dragonfly) of the PCR products obtained from amplification of the 5′ alternatively spliced region of* L. pulchella *TnT cDNA

cDNA Fragment Size	Deduced Amino Acid Sequence of Variable Region	Mean Relative Abundance (Range; SD)
243	MSDEEEYSEEEEEV...	4.5 (0 – 10.4; 3.8)
246	MSDEEEYSEEEEEV......K..	9.9 (0 – 42.0; 10.4)
258	MSDEEEYSEEEEEV...RPRGK......	53.4 (35.9 – 82.1; 14.7)
261	MSDEEEYSEEEEEV......K..................................RPRGK......	26.8 (0 – 47.4; 12.9)
267	MSDEEEYSEEEEEV...............KEPEKKTE.........................	4.7 (0 – 13.3; 4.0)
270	MSDEEEYSEEEEEV......K......KEPEKKTE.........................	0.3 (0 – 2.7; 0.8)
285	MSDEEEYSEEEEEV......K......KEPEKKTE......RPRGK......	not in flight muscle
exon	3/4 5 6 7	

Sample size is 21 adult dragonflies, inlcuding both males and females, ranging from newly emerged to sexually mature. Also shown are the deduced amino acid sequences from characterized cDNA's that correspond to the 5′ variable region of these fragments. Exons are separated by gaps. Constitutive exon 8 (not shown) begins after the 3′-most base shown for each sequence. The exon numbering scheme differs from what we published previously[28] due to our discovery in genomic sequence of an additional exon in the 5′ UTR.

are presumably invariant in their MLC2 phosphorylation state. The only studies that have specifically addressed the functional effects of naturally occurring variation in flight muscle isoforms examine alternative splicing of troponin-T in dragonfly flight muscles.[28,29] Here I present an overview of that work and place it in an ecological context that broadens the ability to appreciate functional significance.

As shown in Figure 1, the troponin-T gene in the dragonfly *Libellula pulchella* contains three alternative exons near the 5' end of the coding region. One of the alternative exons contains only three nucleotides that encode a single amino acid (lysine), thus demonstrating that alternative splicing can provide the finest possible level of control over amino acid content of a protein. Interestingly, *Drosophila* troponin-T also has three alternative exons in the 5' coding region, including a micro-exon that encodes a single lysine residue.[44] In *Drosophila* however, flight muscles contain only one splice variant and the microexon is expressed only in adult hypodermic and visceral muscles. From this it appears that there is wide variation among insects in their patterns of expression of troponin-T splice variants.

We have characterized six distinct transcripts of *L. pulchella* troponin-T from either cDNA clones or PCR products from flight muscles, along with a seventh cDNA that occurs in leg and body wall muscles but is absent from flight muscles.[28,29,45] A PCR fragment that corresponds with the size of the eighth possible combination of the three alternative exons is sometimes detectable as a rare transcript but has not been captured in any of our sequenced subclones.

To quantify the relative abundance of the different splice forms of TnT, we used PCR primers for constitutive regions flanking the cassette of three alternatively spliced exons. One of these primers carried a fluorescent tag, which allowed the PCR product to be fractionated and quantified according to the relative abundance of each fragment size.[28,29,45] The primer pair used in this experiment generated TnT transcript fragments that were 243, 246, 258, 261, 267, 270, and 285 nucleotides in length (Table 2). This array of fragment sizes agrees precisely with sizes predicted from sequence data of the seven known splice variants.

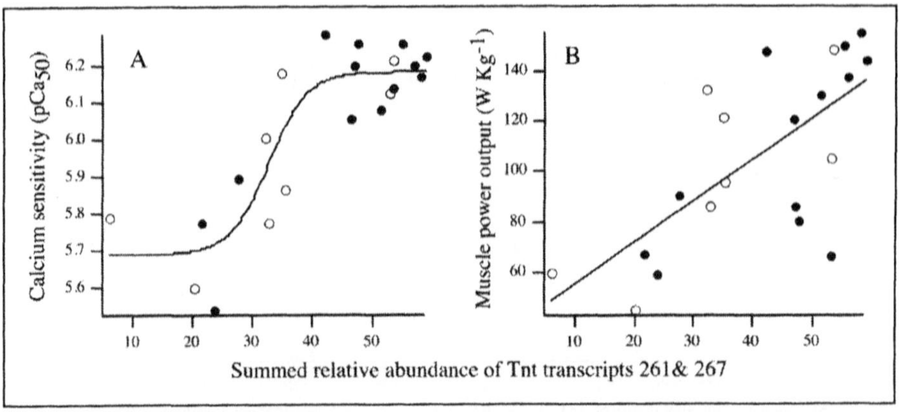

Figure 2. Relationship between the abundance of two troponin T transcripts and A) the calcium sensitivity of skinned fibers and B) power output of the intact basalar muscle during workloop contractions. Data are from reference 28. Open symbols are females; closed symbols are males.

To determine how variation in the alternative splicing ofTnT affects muscle function, we combined measurements of the relative abundance of TnT splice variants within an individual dragonfly with assays of muscle contractile performance. One important caveat in these experiments is that in order to isolate a sufficient amount of RNA, all of the flight muscles from one half of the thorax were homogenized prior to generating cDNA. Thus, when comparing the relative abundance of different troponin-T transcripts to the contractile performance of a single muscle or to the flight performance of the dragonfly, we made the simplifying assumption that within an individual the major flight muscles are fairly homogeneous in their isoform composition.

At the cellular level, there was a strong correlation between the summed relative abundance of transcripts 261 and 267 and the sensitivity of skinned fibers to activation by calcium (Fig. 2A). There was nearly as strong a correlation between the summed relative abundance of transcripts 261 and 267 and the power output of the intact basalar muscle, which drives the downstroke of the forewing leading edge (Fig. 2B). (Note that these results are consistent with the way alternatively spliced forms of troponin-t affect the calcium sensitivity and other contractile properties of human cardiac muscle).[46] From high-speed video recordings of free-flying dragonflies, we showed that there is a significant correlation between the relative abundance of transcripts 261 and 267 and wingstroke amplitude and frequency (Fig. 3A), the main kinematic variables that insects use to adjust aerodynamic force and power output. Finally, dragonflies with greater wingbeat amplitude and frequency were shown to have significantly higher rates of flight metabolism (Fig. 3B).

Male dragonflies engage in vigorous and sometimes highly escalated flight contests to establish and defend territories, and to acquire and defend mates. Thus, it is not surprising to find that both territorial and mating success have a significant positive relationship with muscle power output[47] (Fig. 4).

If having high muscle power output is strongly related to territorial and mating success, what is the purpose and utility of dragonflies being able to vary the contractility and power output of their flight motor? One answer comes from ontogenetic studies, where we have shown that contractility increases steadily during the course of adult maturation.[29] *L. pulchella* dragonflies approximately double in body mass between adult emergence and sexual maturity, and mating and territoriality occur only when dragonflies are fully mature. Thus, it appears that muscle and flight performance are up-regulated only at maturity when intense aerial battles are used by males to establish and defend territories and acquire mates, and by females to evade

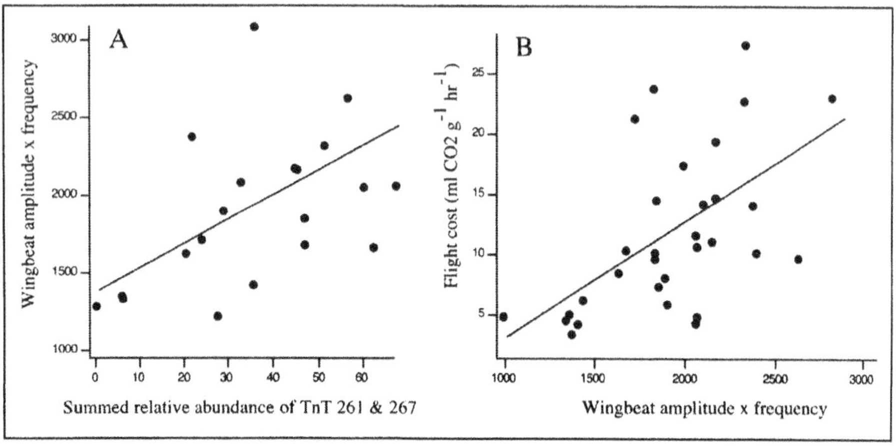

Figure 3. Relationship between the abundance of two troponin t transcripts and A) the product of wingbeat amplitude and frequency during free flight, and B) effect of the product of wingbeat amplitude and frequency on the energetic cost of flight. Each data point represents one individual dragonfly. Data are from reference 29.

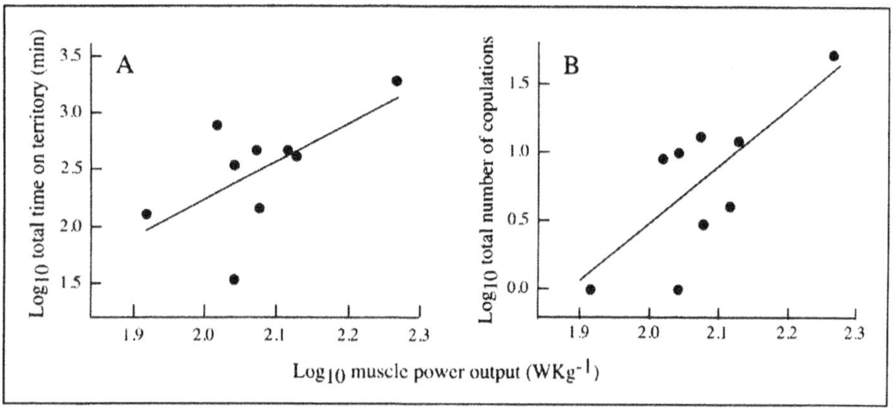

Figure 4. Relationship between muscle power output of male *L. pulchella* dragonflies and their lifetime territorial (A) and mating (B) success. Data are from reference 47.

unwanted copulation attempts by males. Reduced power output by immature adults should reduce their energetic cost of flight, which may be critically important given that approximately two-thirds of newly emerged *L. pulchella* adults lose body mass and disappear from the population (i.e., they appear to starve).[48] Seen in this context, it appears that dragonflies use isoform variation of troponin T to adjust the tradeoff between muscle performance and the energetic cost of flight (Fig. 5).

Not all *L. pulchella* fully upregulate their muscle performance at maturity, as some have a relatively low muscle power output even after they have been mature for a number of weeks (i.e., the low power output data points in Fig. 4). This brings us back to the question of why this species, even at sexual maturity when flight performance is critical for territorial and mating success, has evolved the ability to down-regulate muscle contractility and flight performance. We have recently obtained what appears to be a good answer to this question, for we

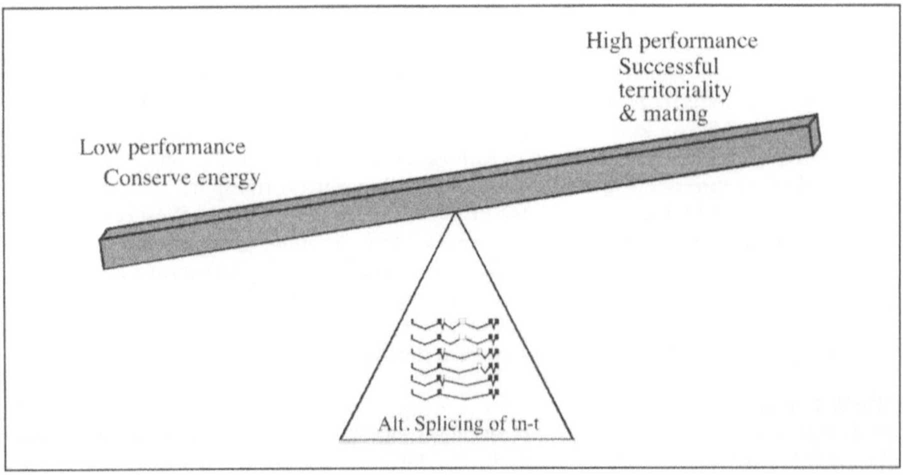

Figure 5. Alternative splicing of troponin-T allows dragonflies to achieve either low performance, low cost flight that helps conserve energy, or high performance flight that increases territorial and mating success.

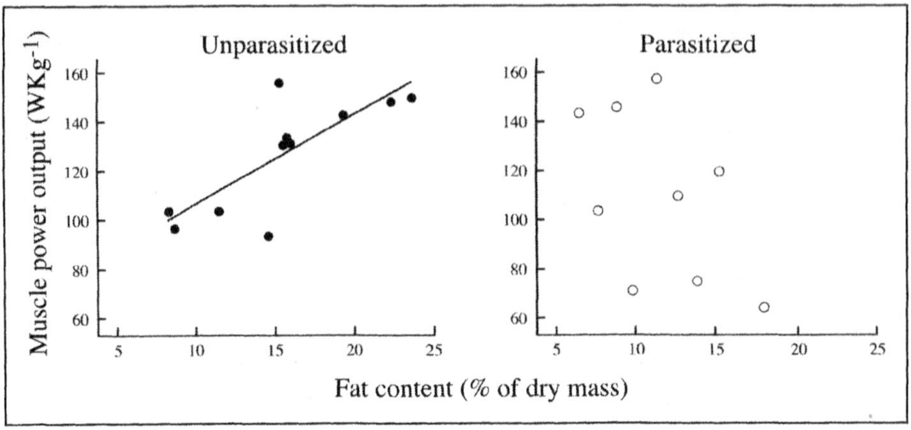

Figure 6. Relationship between total body lipid content and the power output of dragonfly flight muscles. Healthy dragonflies show a strong positive relationship between fat reserves and muscle power output, but this relationship is absent in individuals infected with gregarine gut parasites. Data are from reference 47; all individuals were sexually mature males.

have found that nutritional status is an important factor determining the variation in muscle performance among mature males.[47] Males that are not infected with gregarine gut parasites (protozoans) show a strong positive correlation between total body lipid content and muscle power output, whereas infected individuals show no such relationship (Fig. 6).

The ability of healthy dragonflies to adjust their muscle contractile performance may allow their maximum energy consumption rate to match the rate at which energy can be mobilized from storage pools. Variation among individuals in levels of stored fat may be caused by short term differences in territorial effort and foraging success (which are affected strongly by weather), but there are also likely to be long-term differences in foraging ability that keep some individuals relatively energy-poor throughout their adult lives.

In *L. pulchella* dragonflies infected with gregarine gut parasites, the apparent loss of the ability to match muscle contractility to the size of the energy storage pool suggests that gregarines may cause physiological changes that affect signaling pathways and energy homeostasis. Indeed, we have found that dragonflies with gregarine trophozoites in their midgut have chronically activated p38 map kinase in their flight muscles (Schilder and Marden, unpub. data). In vertebrates, this molecule is involved in pathways that control a wide variety of cellular functions and gene expression,[49] including insulin signaling, glucose transport, and the function of fat storage cells.[50,51] Changes in p38 MAPK signaling are also known to affect certain components of the molecular machinery that controls alternative splicing.[52] Although we have only begun to scratch the surface of this intriguing interaction between gut parasites and the ability of dragonflies to adjust their performance and energetic costs, what we have found so far in this single species suggests that there is likely to be a wealth of interesting biology involved in the physiology and ecology of isoform variation in the flight muscles.

References

1. Vigoreaux JO. Genetics of the Drosophila flight muscle myofibril: A window into the biology of complex systems. Bioessays 2001; 23(11):1047-1063.
2. Herbert A, Rich A. RNA processing and the evolution of eukaryotes. Nat Genet 1999; 21(3):265-269.
3. Ewing B, Green P. Analysis of expressed sequence tags indicates 35,000 human genes. Nat Genet 2000; 25(2):232-234.
4. Maniatis T, Tasic B. Alternative pre-mRNA splicing and proteome expansion in metazoans. Nature 2002; 418(6894):236-243.
5. Feder ME, Mitchell-Olds T. Evolutionary and ecological functional genomics. Nat Rev Genet 2003; 4(8):651-657.
6. Chappell MA. Temperature regulation of carpenter bees (Xylocopa californica) foraging in the Colorado Desert of southern California. Physiological Zoology 1982; 55:267-280.
7. Baird JM. A field study of thermoregulation in the carpenter bee Xylocopa virginica virginica (Hymenoptera: Anthophoridae). Physiological Zoology 1986; 59:157-167.
8. Heinrich B, Mommsen TP. Flight of winter moths near O°C. Science 1985; 228:177-179.
9. Josephson RK, Young D. A synchronous insect muscle with an operating frequency greater than 500 Hz. J Exp Biol 1985; 118:185-208.
10. Josephson RK, Young D. Fiber ultrastructure and contraction kinetics in insect fast muscles. American Zoologist 1987; 27:991-1000.
11. Josephson RK. Contraction dynamics of flight and stridulatory muscles of tettigonid insects. J Exp Biol 1984; 108:77-96.
12. Marden J. Evolutionary adaptation of contractile performance in muscle of ectothermic winter-flying moths. J Exp Biol 1995; 198(Pt 10):2087-2094.
13. Robertson IC. Flight muscle changes in male pine engraver beetles during reproduction: The effects of body size, mating status and breeding failure. Phsiological Entomology 1998; 23:75-80.
14. Bursell E. The behaviour of tsetse flies (Glossina swynnertoni Austen) in relation to problems of sampling. Proceedings of the Royal Entomological Society of London Series A 1961; 36:9-20.
15. Anholt BR, Marden JH, Jenkins DM. Patterns of mass gain in adult odonates. Canadian Journal of Zoology 1991; 69:1156-1163.
16. Marden JH. Bodybuilding dragonflies: Costs and benefits of maximizing flight muscle. Physiological Zoology 1989; 62:505-521.
17. Takahashi S, Takano-Ohmuro H, Maruyama K. Regulation of Drosophila myosin ATPase activity by phosphorylation of myosin light chains—I. Wild-type fly. Comp Biochem Physiol B 1990; 95(1):179-181.
18. Takahashi S, Takano-Ohmuro H, Maruyama K et al. Regulation of Drosophila myosin ATPase activity by phosphorylation of myosin light chains—II. Flightless mfd- fly. Comp Biochem Physiol B 1990; 95(1):183-185.
19. Vigoreaux JO, Perry LM. Multiple isoelectric variants of flightin in Drosophila stretch-activated muscles are generated by temporally regulated phosphorylations. J Muscle Res Cell Motil 1994; 15(6):607-616.
20. Fitzhugh G, Marden J. Maturational changes in troponin T expression, Ca2+-sensitivity and twitch contraction kinetics in dragonfly flight muscle. J Exp Biol 1997; 200(Pt 10):1473-1482.

21. Marden JH. Variability in the size, composition, and function of insect flight muscles. Annu Rev Physiol 2000; 62:157-178.
22. Lee C. The incredible shrinking human genome. Trends Genet 2001; 17(4):187-188.
23. Modrek B, Resch A, Grasso C et al. Genome-wide detection of alternative splicing in expressed sequences of human genes. Nucleic Acids Res 2001; 29(13):2850-2859.
24. Grabowski PJ, Black DL. Alternative RNA splicing in the nervous system. Prog Neurobiol 2001; 65(3):289-308.
25. Black DL, Grabowski PJ. Alternative pre-mRNA splicing and neuronal function. Prog Mol Subcell Biol 2003; 31:187-216.
26. Goldspink G. Gene expression in skeletal muscle. Biochem Soc Trans 2002; 30(2):285-290.
27. Reddy KL, Wohlwill A, Dzitoeva S et al. The Drosophila PAR domain protein 1 (Pdp1) gene encodes multiple differentially expressed mRNAs and proteins through the use of multiple enhancers and promoters. Dev Biol 2000; 224(2):401-414.
28. Marden JH, Fitzhugh GH, Girgenrath M et al. Alternative splicing, muscle contraction and intraspecific variation: Associations between troponin T transcripts, Ca(2+) sensitivity and the force and power output of dragonfly flight muscles during oscillatory contraction. J Exp Biol 2001; 204(Pt 20):3457-3470.
29. Marden JH, Fitzhugh GH, Wolf MR et al. Alternative splicing, muscle calcium sensitivity, and the modulation of dragonfly flight performance. Proc Natl Acad Sci USA 1999; 96(26):15304-15309.
30. Miller RC, Schaaf R, Maughan DW et al. A nonflight muscle isoform of Drosophila tropomyosin rescues an indirect flight muscle tropomyosin mutant. J Muscle Res Cell Motil 1993; 14(1):85-98.
31. Wojtas K, Slepecky N, von Kalm L et al. Flight muscle function in Drosophila requires colocalization of glycolytic enzymes. Mol Biol Cell 1997; 8(9):1665-1675.
32. Swank DM, Knowles AF, Suggs JA et al. The myosin converter domain modulates muscle performance. Nat Cell Biol 2002; 4(4):312-316.
33. Takahashi S, Maruyama K. Activity changes in myosin ATPase during metamorphosis of fruitfly. Zoological Science 1987; 4:833-838.
34. Tohtong R, Yamashita H, Graham M et al. Impairment of muscle function caused by mutations of phosphorylation sites in myosin regulatory light chain. Nature 1995; 374(6523):650-653.
35. Dickinson MH, Hyatt CJ, Lehmann FO et al. Phosphorylation-dependent power output of transgenic flies: An integrated study. Biophys J 1997; 73(6):3122-3134.
36. Southgate R, Ayme-Southgate A. Alternative splicing of an amino-terminal PEVK-like region generates multiple isoforms of Drosophila projectin. J Mol Biol 2001; 313(5):1035-1043.
37. Montooth KL, Marden J, Clark AG. Mapping determinants of variation in energy metabolism, respiration and flight in Drosophila. Genetics 2003; 165:623-635.
38. Swank DM, Bartoo ML, Knowles AF et al. Alternative exon-encoded regions of Drosophila myosin heavy chain modulate ATPase rates and actin sliding velocity. J Biol Chem 2001; 276(18):15117-15124.
39. Swank DM, Wells L, Kronert WA et al. Determining structure/function relationships for sarcomeric myosin heavy chain by genetic and transgenic manipulation of Drosophila. Microsc Res Tech 2000; 50(6):430-442.
40. Swank DM, Knowles AF, Kronert WA et al. Variable N-terminal regions of muscle myosin heavy chain modulate ATPase rate and actin sliding velocity. J Biol Chem 2003; 278(19):17475-17482.
41. Bernstein SI, Milligan RA. Fine tuning a molecular motor: The location of alternative domains in the Drosophila myosin head. J Mol Biol 1997; 271(1):1-6.
42. Fyrberg EA, Fyrberg CC, Biggs JR et al. Functional nonequivalence of Drosophila actin isoforms. Biochem Genet 1998; 36(7-8):271-287.
43. Barbas JA, Galceran J, Torroja L et al. Abnormal muscle development in the heldup3 mutant of Drosophila melanogaster is caused by a splicing defect affecting selected troponin I isoforms. Mol Cell Biol 1993; 13(3):1433-1439.
44. Benoist P, Mas JA, Marco R et al. Differential muscle-type expression of the Drosophila troponin T gene. A 3-base pair microexon is involved in visceral and adult hypodermic muscle specification. J Biol Chem 1998; 273(13):7538-7546.
45. Wolf MR. Molecular and functional characterization of troponin T in Libellula pulchella and Periplanta americana. University Park: Biology. Pennsylvania State University, 1999.
46. Gomes AV, Guzman G, Zhao J et al. Cardiac troponin T isoforms affect the Ca2+ sensitivity and inhibition of force development. Insights into the role of troponin T isoforms in the heart. J Biol Chem 2002; 277(38):35341-35349.
47. Marden JH, Cobb JR. Territorial and mating success of dragonflies that vary in muscle power output and presence of gregarine gut parasites. Animal Behaviour 2004; In press.

48. Marden JH, Rowan B. Growth, differential survival, and shifting sex ratio of free-living Libellula pulchella (Odonata: Libellulidae) dragonflies during adult maturation. Annals of the Entomological Society of America 2000; 93:452-458.
49. Cowan KJ, Storey KB. Mitogen-activated protein kinases: New signaling pathways functioning in cellular responses to environmental stress. J Exp Biol 2003; 206(Pt 7):1107-1115.
50. Fujishiro M, Gotoh Y, Katagiri H et al. Three mitogen-activated protein kinases inhibit insulin signaling by different mechanisms in 3T3-L1 adipocytes. Mol Endocrinol 2003; 17(3):487-497.
51. Carlson CJ, Koterski S, Sciotti RJ et al. Enhanced basal activation of mitogen-activated protein kinases in adipocytes from type 2 diabetes: Potential role of p38 in the downregulation of GLUT4 expression. Diabetes 2003; 52(3):634-641.
52. van der Houven van Oordt W, Diaz-Meco MT, Lozano J et al. The MKK(3/6)-p38-signaling cascade alters the subcellular distribution of hnRNP A1 and modulates alternative splicing regulation. J Cell Biol 2000; 149(2):307-316.
53. Bernstein SI, Hansen CJ, Becker KD et al. Alternative RNA splicing generates transcripts encoding a thorax-specific isoform of Drosophila melanogaster myosin heavy chain. Mol Cell Biol 1986; 6(7):2511-2519.
54. Standiford DM, Sun WT, Davis MB et al. Positive and negative intronic regulatory elements control muscle-specific alternative exon splicing of Drosophila myosin heavy chain transcripts. Genetics 2001; 157(1):259-271.
55. Wells L, Edwards KA, Bernstein SI. Myosin heavy chain isoforms regulate muscle function but not myofibril assembly. EMBO J 1996; 15(17):4454-4459.
56. Kronert WA, Edwards KA, Roche ES et al. Muscle-specific accumulation of Drosophila myosin heavy chains: A splicing mutation in an alternative exon results in an isoform substitution. EMBO J 1991; 10(9):2479-2488.
57. Hastings GA, Emerson Jr CP. Myosin functional domains encoded by alternative exons are expressed in specific thoracic muscles of Drosophila. J Cell Biol 1991; 114(2):263-276.
58. O'Donnell PT, Collier VL, Mogami K et al. Ultrastructural and molecular analyses of homozygous-viable Drosophila melanogaster muscle mutants indicate there is a complex pattern of myosin heavy-chain isoform distribution. Genes Dev 1989; 3(8):1233-1246.
59. Lovato TL, Meadows SM, Baker PW et al. Characterization of muscle actin genes in Drosophila virilis reveals significant molecular complexity in skeletal muscle types. Insect Mol Biol 2001; 10(4):333-340.
60. Brault V, Sauder U, Reedy MC et al. Differential epitope tagging of actin in transformed Drosophila produces distinct effects on myofibril assembly and function of the indirect flight muscle. Mol Biol Cell 1999; 10(1):135-149.
61. An HS, Mogami K. Isolation of 88F actin mutants of Drosophila melanogaster and possible alterations in the mutant actin structures. J Mol Biol 1996; 260(4):492-505.
62. Anson M, Drummond DR, Geeves MA et al. Actomyosin kinetics and in vitro motility of wild-type Drosophila actin and the effects of two mutations in the Act88F gene. Biophys J 1995; 68(5):1991-2003.
63. Drummond DR, Peckham M, Sparrow JC et al. Alteration in crossbridge kinetics caused by mutations in actin. Nature 1990; 348(6300):440-442.
64. Tansey T, Mikus MD, Dumoulin M et al. Transformation and rescue of a flightless Drosophila tropomyosin mutant. EMBO J 1987; 6(5):1375-1385.
65. Basi GS, Boardman M, Storti RV. Alternative splicing of a Drosophila tropomyosin gene generates muscle tropomyosin isoforms with different carboxy-terminal ends. Mol Cell Biol 1984; 4(12):2828-2836.
66. Karlik CC, Fyrberg EA. Two Drosophila melanogaster tropomyosin genes: Structural and functional aspects. Mol Cell Biol 1986; 6(6):1965-1973.
67. Cripps RM, Sparrow JC. Polymorphism in a Drosophila indirect flight muscle-specific tropomyosin isozyme does not affect flight ability. Biochem Genet 1992; 30(3-4):159-168.
68. Bullard B, Leonard K, Larkins A etal. Troponin of asynchronous flight muscle. J Mol Biol 1988; 204(3):621-637.
69. Falkenthal S, Graham M, Wilkinson J. The indirect flight muscle of Drosophila accumulates a unique myosin alkali light chain isoform. Dev Biol 1987; 121(1):263-272.
70. Roulier EM, Fyrberg C, Fyrberg E. Perturbations of Drosophila alpha-actinin cause muscle paralysis, weakness, and atrophy but do not confer obvious nonmuscle phenotypes. J Cell Biol 1992; 116(4):911-922.
71. Qiu F, Lakey A, Agianian B et al. Troponin C in different insect muscle types: Identification of two isoforms in Lethocerus, Drosophila and Anopheles that are specific to asynchronous flight muscle in the adult insect. Biochem J 2003; 371(Pt 3):811-821.

72. Domingo A, Gonzalez-Jurado J, Maroto M et al. Troponin-T is a calcium-binding protein in insect muscle: In vivo phosphorylation, muscle-specific isoforms and developmental profile in Drosophila melanogaster. J Muscle Res Cell Motil 1998; 19(4):393-403.
73. Kojima S, Mishima M, Mabuchi I et al. A single Drosophila melanogaster myosin light chain kinase gene produces multiple isoforms whose activities are differently regulated. Genes Cells 1996; 1(9):855-871.
74. Champagne MB, Edwards KA, Erickson HP et al. Drosophila stretchin-MLCK is a novel member of the Titin/Myosin light chain kinase family. J Mol Biol 2000; 300(4):759-777.
75. Moore JR, Vigoreaux JO, Maughan DW. The Drosophila projectin mutant, bentD, has reduced stretch activation and altered indirect flight muscle kinetics. J Muscle Res Cell Motil 1999; 20(8):797-806.
76. Vigoreaux JO, Moore JR, Maughan DW. Role of the elastic protein projectin in stretch activation and work output of Drosophila flight muscles. Adv Exp Med Biol 2000; 481:237-247; discussion 247-250.
77. Kolmerer B, Clayton J, Benes V et al. Sequence and expression of the kettin gene in Drosophila melanogaster and Caenorhabditis elegans. J Mol Biol 2000; 296(2):435-448.
78. Kulke M, Neagoe C, Kolmerer B et al. Kettin, a major source of myofibrillar stiffness in Drosophila indirect flight muscle. J Cell Biol 2001; 154(5):1045-1057.
79. Hakeda S, Endo S, Saigo K. Requirements of Kettin, a giant muscle protein highly conserved in overall structure in evolution, for normal muscle function, viability, and flight activity of Drosophila. J Cell Biol 2000; 148(1):101-114.
80. Vigoreaux JO, Hernandez C, Moore J et al. A genetic deficiency that spans the flightin gene of Drosophila melanogaster affects the ultrastructure and function of the flight muscles. J Exp Biol 1998; 201(Pt 13):2033-2044.
81. Vigoreaux JO. Alterations in flightin phosphorylation in Drosophila flight muscles are associated with myofibrillar defects engendered by actin and myosin heavy-chain mutant alleles. Biochem Genet 1994; 32(7-8):301-314.
82. Vigoreaux JO, Saide JD, Valgeirsdottir K et al. Flightin, a novel myofibrillar protein of Drosophila stretch-activated muscles. J Cell Biol 1993; 121(3):587-598.
83. Henkin JA, Maughan DW, Vigoreaux JO. Mutations that affect flightin expression in Drosophila alter the viscoelastic properties of flight muscle fibers. Am J Physiol Cell Physiol 2004; 286:C65-C72.
84. Liu Z, Chung I, Dong K. Alternative splicing of the BSC1 gene generates tissue-specific isoforms in the German cockroach. Insect Biochem Mol Biol 2001; 31(6-7):703-713.
85. Atkinson NS, Brenner R, Chang W et al. Molecular separation of two behavioral phenotypes by a mutation affecting the promoters of a Ca-activated K channel. J Neurosci 2000; 20(8):2988-2993.
86. Atkinson NS, Brenner R, Bohm RA et al. Behavioral and electrophysiological analysis of Ca-activated K-channel transgenes in Drosophila. Ann N Y Acad Sci 1998; 860:296-305.
87. Brenner R, Yu JY, Srinivasan K et al. Complementation of physiological and behavioral defects by a slowpoke Ca(2+) -activated K(+) channel transgene. J Neurochem 2000; 75(3):1310-1319.

CHAPTER 18

Muscle Systems Design and Integration

Fritz-Olaf Lehmann

Abstract

The recent advances in experimental technology allows us to assess the mechanical power output and function of the *Drosophila* flight muscle within the context of the flying animal. In an intact animal, production and control of aerodynamic forces during flight depend on several factors including the maximum power output of the musculature and the interplay in neural activation between the two functionally, physiologically and anatomically distinct classes of flight muscles: the asynchronous power muscles and the synchronous flight control muscles. Although the maximum mechanical power output and the efficiency of the locomotory musculature can be estimated from in vitro biophysical experiments, the values determined from such experiments in *Drosophila* are substantially lower than the maxima that must occur in the flying animal. As a consequence, the systems-level perspective on power production is a necessary bridge in any attempt to link the function and performance of flight musculature with its specific role for wing motion and flight force control in the behaving animal.[1,2] Moreover, the cost of locomotion in flying insects is rarely constant but varies as the animal changes speed and direction. Ultimately the muscles of the insect must compensate for these changing requirements by varying the amount of muscle power that they produce. This chapter considers mainly the mechanisms by which mechanical power output of the asynchronous flight muscles is regulated to match the changing requirements during flight control behaviours and summarizes vital muscle parameters including muscle efficiency, measured in intact fruit flies.

Power Requirements for Flight

In a flying insect, the mechanical power generated by the muscular-skeletal flight system drives the wings up and down and must match the power requirements for flight (Fig. 1).[3-5] In a behaving animal, these energetic requirements can be derived by measuring the motion of the insect body and the wings, and the resultant flight force that the animal is producing. Total power requirements for flapping flight may be divided into 4 major power terms: *parasite*, *profile*, *inertial*, and *induced* power.[6-8] Parasite power is the rate of work required to overcome the fluid drag on the body as the animal moves through the air. For insects that are hovering and traveling with low speed, parasite power is comparatively small and can be ignored in many instances. At high forward speed, parasite power increases rapidly because it depends on the cube of forward velocity. At high forward velocity parasite cost is the predominant factor of flight costs and distinctly forms the 'classic *U*- or *J*-shaped curve' of total power expenditures in flying insects, birds and bats.[1,9,10] In contrast, profile power is the cost associated with the work required to overcome the drag on the beating wings and is increasing with the cube of the product between stroke amplitude and stroke frequency. Drag, in turn, linearly depends on the drag coefficient that changes with Reynolds number for wing motion. In small insects such as *Drosophila* Reynolds number is relatively low (approximately 90-230) indicating that viscous

Nature's Versatile Engine: Insect Flight Muscle Inside and Out, edited by Jim Vigoreaux.

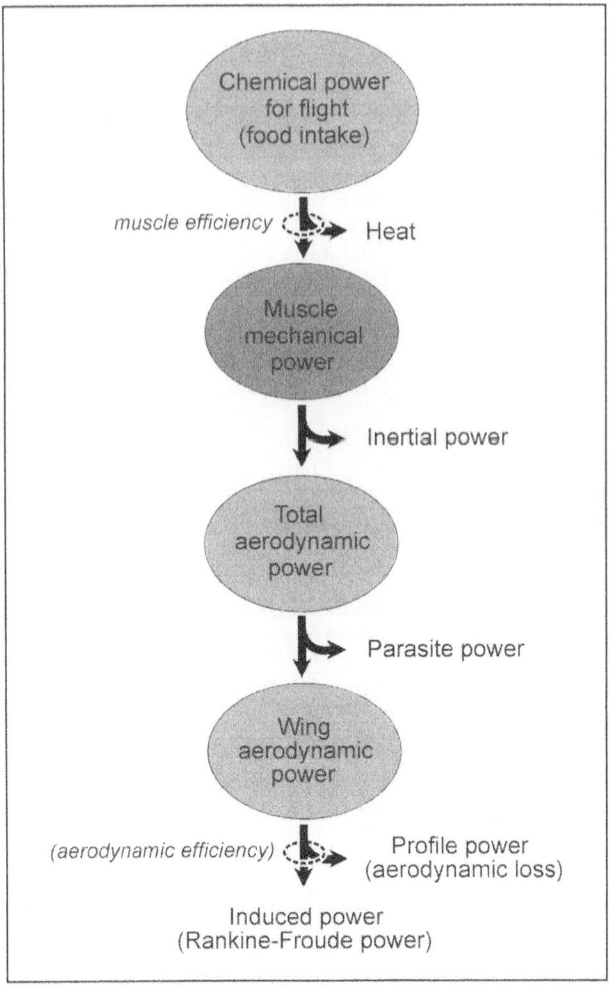

Figure 1. Pathway of power expenditures in flapping flight of insects. Chemical energy due to food intake is converted into muscle mechanical power. In most insects, the efficacy (muscle efficiency) of this conversion process is low ranging from 3.0 to 17%.[3-6,8,38-40] In endothermic insects metabolic heat production is used to facilitate the mechanical power output of the flight muscles at low ambient temperatures.[41,42] In freely flying *Drosophila* a minimum ambient temperature of approximately 15 °C is required to produce flight forces that are equal to body weight.[43] Inertial power is the cost to accelerate the wings at the beginning of each half stroke. Parasite power, the cost to overcome the drag on the body of the animal, is negligible at low forward speed and hovering. Profile power is the cost associated with aerodynamic drag that the beating wings face when moving through the air. Induced power is the cost to accelerate a fluid momentum downwards and thus representing the cost of lift production. The Rankine-Froude estimate of induced power represents the minimum power requirements for lift production.[8] Aerodynamic efficiency is the ratio between Rankine-Froude power and the sum of induced and profile power (aerodynamic loss). Modified from Casey.[6]

forces of the air are relatively large and profile cost may dominate. Larger insects that fly at higher Reynolds number benefit from a relative reduction in profile power, and thus muscle mechanical power output can be reduced. In many cases conventional drag coefficient estimates based on Reynolds number ($C_D = 7Re^{-0.5}$)[8] underestimate profile power because lift

***Table 1. Flight parameters of* Drosophila melanogaster**‡

	Minimum	Hovering	Maximum
Stroke amplitude (deg)	148 ± 9	162 ± 8	169 ± 7
Stroke frequency (Hz)	190 ± 18	209 ± 15	212 ± 12
Wing velocity (m s^{-1})	1.38 ± 0.14	1.66 ± 0.14	1.76 ± 0.14
Flight force (μN)	4.4 ± 2.0	10.3 ± 1.2	13.6 ± 1.5
Muscle strain amplitude (%)	2.7 – 4.6†	2.7 – 4.6†	2.7 – 4.6†
Mean C_L	0.97 ± 0.40	1.59 ± 0.20	1.88 ± 0.29
Mean C_D	0.50 ± 0.21	1.02 ± 0.13	1.74 ± 0.27
P^*_{MR} (W kg^{-1})	519 ± 123	664 ± 100	727 ± 119
P^*_{ind} (W kg^{-1})	7.1 ± 5.3	21.4 ± 1.0	32.4 ± 6.1
P^*_{pro} (W kg^{-1})	22.0 ± 6.1	69.9 ± 12	139 ± 26
P^*_{aero} (W kg^{-1})	29.1 ± 11.4	91.3 ± 13	171 ± 32
P^*_{acc} (W kg^{-1})	43.5 ± 13.5	67.4 ± 13.5	77.0 ± 16.5
η_M in vivo (%)	5.6 ± 1.2§	13.8 ± 1.8	23.5 ± 2.6

‡Flight parameters were measured during extremes of force production that fell within the top 1% (maximum muscle power output) or bottom 1% (minimum muscle power output) of flight force or within 1% of body weight (muscle power output during hovering flight). Oscillations in mean muscle strain (strain amplitude) were derived from the length changes of the thoracic exoskeleton in flying *Drosophila virilis* during an entire contraction-relaxation cycle.[37] Mean drag coefficient (C_D) of the moving wings was derived from mean lift coefficient (C_L) during translational motion of a robotic fruit fly wing moving at Reynolds number of 134 that is typical for *Drosophila* flight.[12] Flight specific power is given in units of W kg^{-1} flight muscle mass that amounts to approximately 30% of total body mass in the fruit fly. P^*_{MR}, metabolic power (total flight costs) derived from the release of carbon dioxide during flight. In diptera the energy conversion factor is 21.4 J ml^{-1} CO_2; P^*_{ind}, induced power; P^*_{pro}, wing profile power calculated using the drag coefficient of the robotic wing; P^*_{aero}, aerodynamic power; P^*_{acc}, inertial power; η_M, efficiency of the asynchronous flight muscle. Mean body weight of *Drosophila melanogaster* females is 1.05 ±0.13 mg. Data are shown as means ±S.D. *N* = 27 flies. † = no flight force measurements were conducted in these experiments. § = assuming 100% elastic energy storage.

production in *Drosophila* and other insects is greatly enhanced by unsteady aerodynamic effects resulting in up to 3-4 times higher profile power cost.[11,12] In *Drosophila*, profile power rises steeply with increasing locomotor performance and exceeds inertial power expenditures at forces that are approximately equal and above body weight (Table 1, Fig. 2A).

Inertial power is needed to accelerate the wings and the surrounding fluid (virtual wing mass) from rest to maximum velocity within the first half of each half stroke cycle. Inertial power depends on several factors including wing mass, mass distribution given by the wing shape and length, wing velocity and the wing's acceleration profile throughout the stroke cycle. The most conservative model of motion is one in which the wing movements are described by a pure sinusoidal function. Under these conditions inertial cost is approximately 20% higher compared to the most liberal model of wing motion in which the wings have a short period of high acceleration followed by a long region of constant velocity kinematics (triangular function). In the case of varying flight forces in an intact flying fruit fly, inertial cost can be estimated from the product of wing velocity squared and stroke frequency. Induced power is the cost to generate a downward jet of air that keeps the insect airborne. This term is equal to the total flight force produced by the beating wings multiplied by the mean velocity of the wake that the flying animal leaves behind. Producing flight forces at relatively low wing velocities are energetically favorable and may lower the energetic expenditures of the insect's flight muscles.

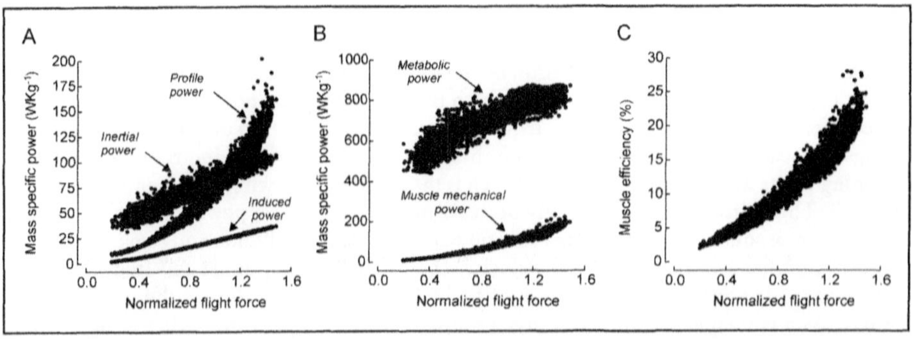

Figure 2. A) Flight power requirements in *Drosophila* increases with increasing flight force production. Induced power is the smallest cost in total power balance. Profile power estimates based on drag coefficients measured in a robotic fruit fly wing are smaller than inertial power requirements at low flight force but higher when the fly produces flight forces that exceed the body weight of the animal. B) Aerodynamic power requirements that are estimated from wing kinematics and flight force measurements, match the mechanical power output of the asynchronous flight muscles. The changes in metabolic power are derived from measurements of carbon dioxide release of the fly during flight in a respirometric flight chamber. C) Muscle efficiency plotted as the ratio between muscle mechanical power output and metabolic power as shown in (B). Due to different slope and offset of both data, muscle efficiency is steeply increasing with flight force. Data are recorded from a single fruit fly flying in a virtual reality flight arena.

Power Reduction

Elastic Energy Recycling

The total flight cost in an insect is not a simple linear relationship between profile, inertial and induced power but rather depends on the complex interaction between the costs that are associated with the motion of the fluid around the wing (profile and induced power) and the cost that is due to the motion of the wing *sensu stricto* (inertial power). Inertial energy at mid half stroke has two potential fates.[13] In one case, the kinetic energy stored in the wing motion serves as a energy source while the wing is slowing down in the second half of each half stroke. This energy in turn may be used to overcome profile drag on the wing and to generate downward lift. Alternatively, the wing could store its kinetic energy elastically in the skeletal system of the flight motor, for instance in the highly elastic protein resilin that is spread throughout the wing and the thoracic cuticle,[14,15] or in the elastic components of the asynchronous flight muscle. If the kinetic energy freed in the deceleration phase of each half stroke is lower than the combined profile and induced power requirements (aerodynamic power) no elastic energy recycling is required to minimize the mechanical power output of the flight muscle (Fig. 2B). Elastic energy recycling is beneficial for the flying insect only when inertial cost exceeds aerodynamic power requirements. In *Drosophila* this situation is given at flight forces below approximately 90% of the animal's body weight (Fig. 2A). In contrast, due to the high wing drag that is expected during elevated flight force production (flight forces that are equal and above body weight), calculations show that in the fruit fly inertial power is smaller than aerodynamic power. In this case, the elastic components in the entire flight system serve as springs that damp mechanical stress on both the skeletal structures and the myofilaments of the flight muscle rather than providing a mechanism to lower mechanical power output of the asynchronous flight muscles.

Behavioral Strategies

Flying insects may reduce their total energetic costs by increasing the efficiency with which chemical energy is turned into aerodynamic flight forces. Potentially, this can be achieved by at least three distinct mechanisms. First, the insect could increase the efficiency of chemical energy conversion into mechanical work by the filaments of the flight muscle or increase the efficiency of chemical energy production by the mitochondria (muscle efficiency). Second, the insect could increase the efficiency with which muscle mechanical power is turned into flight forces (aerodynamic efficiency). This requires an increase in the lift to drag ratio of the beating wings that can be achieved by a decrease in angle of attack during wing translation and by flapping wings with a high aspect ratio, which is the ratio between wing length and mean depth (chord). Third, insects may potentially decrease muscle mechanical power output at constant flight force production by changing the ratio between stroke amplitude and stroke frequency. In conventional aerodynamics, wing velocity (the product of amplitude and frequency) predominantly determines the magnitude of flight force production during the up and down stroke. As a consequence, an insect may produce the same flight force either at high stroke frequency and low stroke amplitude or by a low frequency and high amplitude.

In comparison, flight cost is minimal when the animal accelerate a large amount of air at low speed because in this case the kinetic energy (= fluid mass x wake velocity squared) of the moving fluid is small. Since the volume of fluid that the wings accelerate downwards directly depends on stroke amplitude given at the angle between the dorsal and ventral excursion during wing flapping, the smallest muscle power output is required when the insect maximizes stroke amplitude while minimizing stroke frequency. This relationship implies that nervous activation of the muscle fibers by the thoracic ganglion, and thus muscle contraction dynamics, might be constrained primarily by the energetic cost to keep the animal airborne. However, measurements in a virtual flight simulator show that a tethered flying *Drosophila* produces maximum muscle mechanical power even at a flight force roughly equal to the animal's own body weight (hovering condition) that lies approximately 30% below the fly's maximum flight force. The additional power input at hovering flight conditions offers the fruit fly a broader range of different combinations between stroke amplitude and frequency for force production that should in turn enhance the animal's ability to control its flight force in free flight (Fig. 3B). Analytical models based on *Drosophila* kinematics show that the additional energetic expenses around hovering conditions due to high stroke frequency are relatively small and solely amount to approximately 10% of total induced power.[16]

Power Constraints on Steering Capacity

Since the power output sustained by the flight muscles may directly constrain wing kinematics in the fruit fly at elevated power requirements for flight, it potentially lowers the ability of the insect to modulate wing kinematics at elevated aerodynamic performance and may thus limit flight maneuverability. High aerial maneuverability of an insect may be useful in a large variety of behavioural contexts including predator avoidance, prey catching, mating success, and male-male competition.[17,18] Behavioral observation in the European beewolf *Philanthus triangulum* suggest a close correlation between flight maneuverability and mating success. The males of this species establish small territories near female nests and defend this territory against other males in air combats that require fast changes in wing motion and power requirements.[19] Figure 3B shows that maximum mechanical power output of the asynchronous muscles constraints the kinematic envelope of *Drosophila* up to a unique combination of amplitude and frequency at maximum force production when stroke amplitude has reached its mechanical limit near 180 degs. Previous results have shown that fruit flies mainly control forces and moments by changing stroke amplitude of the two wings.[20,21] A collapse in kinematic envelope during peak force production should thus attenuate greatly the maneuverability and stability of animals in free flight because stroke amplitude can not be modulated without a reduction in flight force production. The attenuation in steering performance can be demonstrated in tethered flies flying in a flight simulator, in which the animal is stabilizing actively the angular velocity of a visual pattern (black bar) displayed in the arena by controlling the bilateral difference between left and right stroke amplitude. When flight force increases, the animal's steering capacity decreases indicating that the ability of the fly to control yaw moments around its

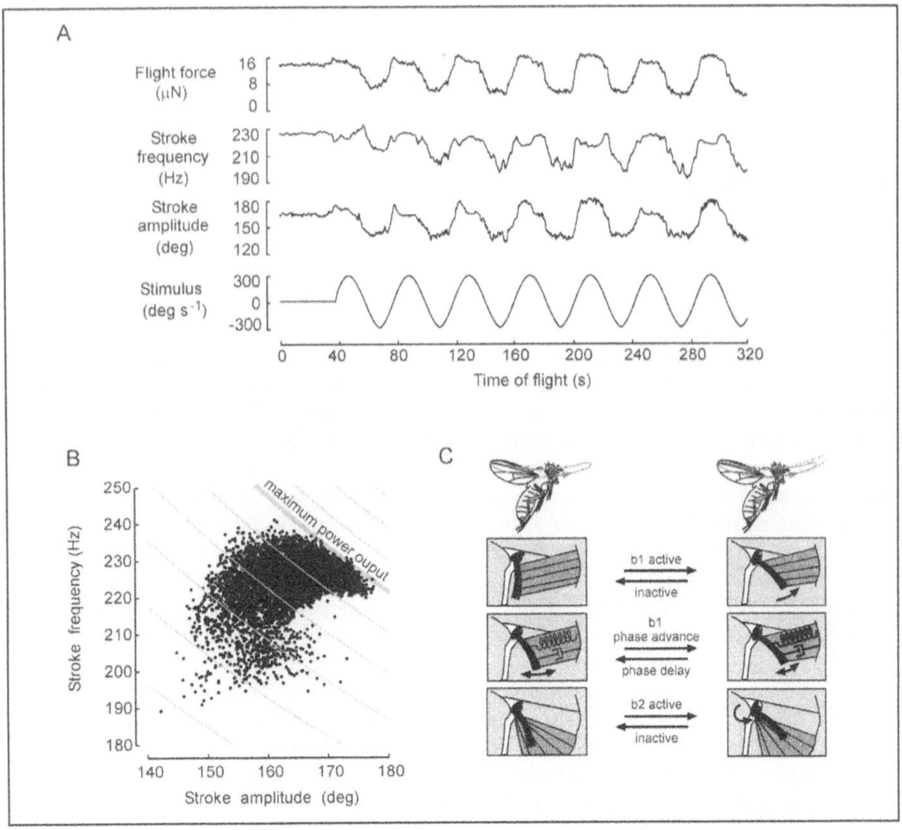

Figure 3. A) Flight sequence of a tethered fruit fly during vertical oscillation of a visual lift stimulus (diagonal stripes) displayed in a virtual reality flight arena. In the attempt to stabilize the retinal slip of the visual pattern on the compound eye, the tethered flying *Drosophila* varies its production of total flight forces.[5] This force modulation is accomplished by changes in stroke amplitude and stroke frequency due to alterations of spike activity in both the power muscles and at least two different sets of flight control muscles (basalar and axillar muscles). The rectification of stroke frequency when the fly generates highest forces indicates that the animal is producing maximum muscle mechanical power. B) In vivo working range of the *Drosophila* flight motor. Flight kinematics is constraint by power availability and the morphological limit of stroke amplitude. The values are plotted as a function of instantaneous stroke amplitude during a flight sequence of a single fly such as shown in (A). The hyperbolic lines represent isolines at which flight muscle power is constant. Flight forces that are equal to the body weight of the animal are produced at a stroke amplitude approximately between 160 and 165 degs. The variance of possible amplitude and frequency combinations is decreasing from the left-hand side (low forces) to the right-hand side (high forces). At maximum flight force *Drosophila* is constraint to a unique combination between amplitude and frequency due to maximum power output of mean 171 W kg^{-1} muscle tissue. The restriction of wing kinematics at highest force lowers steering capacity and thus potentially attenuates maneuverability and stability in freely flying fruit flies. C) Model of muscular mechanism of stroke amplitude control in the blow fly *Calliphora vicina*. The basalar 1 and 2 flight control muscles (b1 and b2) control the position of the basalar sclerite (black) that reconfigures the wing hinge by a long tendon. During flight b1 typically fires one spike in each stroke cycle to maintain a tonic tension on the basalar sclerite. The timing of b1 spike initiation phase within the stroke cycle alters the position of the basalare by either increasing muscle tension (spike phase advanced) or decreasing muscle tension (spike phase delayed). The mechanical muscle properties are indicated as a dashpot and a spring. Higher muscle tension correlates with an increase in stroke amplitude when a higher flight force is required as shown in (A). Activation of b2 produces similar changes in the basalare motion. However, in contrast to b1, the b2 is typically inactive during flight and only becomes active during turns. Figure 3C is redrawn partly from Tu and Dickinson.[26]

vertical body axis and flight direction is impaired.[22] Although the constraint of mechanical power output on wing kinematics might be small in an unloaded animal, the limited power output of the flight muscles may significantly lower maneuverability and stability when the insect is carrying prey or additional loads, leaving the animal susceptible to a higher risk of predation.[23]

Balancing Power and Control

Kinematic Changes

In many insect species, flight is not solely a mode of transport but also a means of prey capture, mating display and territorial maintenance—behaviors that can demand rapid an elaborate maneuvers. For a moth darting away from an attacking bat or a mosquito trying to escape from an attacking dragonfly, the ability to rapidly *change* power output may represent as important a selectional criterion for the insect flight motor as the absolute level of power production.[24,25] In contrast to insects with synchronous flight muscles, electrophysiological evidence indicates that in diptera small control muscles adjust the amplitude, angle of attack, rotational timing at the stroke reversals, and wing trajectory during the up and down stroke (for a more detailed description of control muscle function see Fig. 4). In comparison to the 'big and dumb' asynchronous flight muscles, the 14 pairs of flight control muscles offer the fly's nervous system a means of rapidly controlling wing kinematics and power output of the flight motor. There are two pairs of control muscles that are responsible for alteration in stroke frequency by changing the stiffness of the resonating thoracic box: the pleurosternal muscles one and two. In contrast, stroke amplitude is modulated by at least the first and second basalar muscle (b1 and b2) and the first control muscle of the pterale (I1). Direct electrical activation of b1 and b2 in a flying fly results in an increase in stroke amplitude during flight whereas activation of the I1 control muscle induces a collapse of stroke amplitude.[26-28] When no flight power is required for wing motion such as during cleaning behaviour and courtship song production, some control muscles are powerful enough to move the wing to the appropriate position for cleaning or to vibrate the wings with a low stroke amplitude.[29,30] Control muscles, however, are too small to generate enough power to accommodate directly the changes in wing kinematics occurring in a flying animal.

Negative Work of Control Muscles

In order to function for steering in a flying insect, it is not necessary that flight control muscles generate a large amount of muscle power. Since most control muscles in diptera insert directly onto the sclerites of the wing hinge, they function by reconfiguring the motion of the wing hinge within the up and down stroke. These modifications in the wing gear mechanisms thus may alter the transmission of muscle power from the asynchronous flight muscles to the moving wings. This biomechanical arrangement consisting of fast responding muscle actuators and mechanical gear components allows the fly to conduct modifications of wing motion (such as stroke amplitude, Fig. 3A and C) within a few stroke cycles although the neural activation of the power muscles could not instantaneously support the associated changes in power requirements. Nevertheless, the nervous system of *Drosophila* is not simply switching the power muscles on and off during flight, but rather changing the frequency of muscle spikes in accordance with the mean power requirements of the entire flight system (Fig. 5).

The reconfiguration of the wing hinge is not necessarily requiring that control muscles are doing positive work during contraction. In the one case in which a work-loop analysis was performed on a steering muscle (b1 muscle), the high strain frequency ranging from 100 to 200 Hz and the typical low content of contractile filaments makes the b1 muscle fiber incapable of generating positive work during flight. In other words, regardless of its

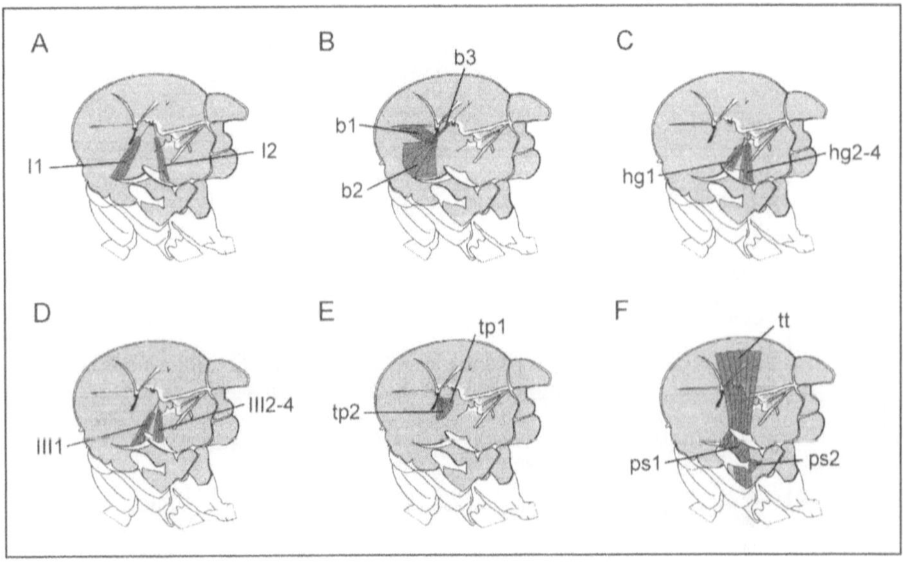

Figure 4. Morphology and function of the 14 flight control muscles in flies. A) The muscles of the first pterale: I1 and I2. Activity of I1 and I2 is correlated with a decrease in stroke amplitude during flight. B) The muscles of the basalare: b1, b2 and b3. Muscle b1 typically fires a single action potential in every stroke cycle whereas b2 is only active during turning maneuvers in (tethered) flying animals. Both muscles increase stroke amplitude by rotating the basalar apodeme forward. Spikes of control muscles occur in a narrow phase band of the wing beat cycle that determines their efficacy on wing kinematics. In *Drosophila* electrical stimulation of b2 at different phase bands indicates that the phasic effect of muscle contraction amounts to a approximately 10% modulation of the tonic effect on stroke amplitude. The b2 also controls the rotational speed of wing rotation at the end of each stroke cycle. The b3 is counter balancing force generation of b1 and b2. See (Fig. 3C) for more details. C) The muscles of the posterior notal wing process: hg1 and hg2-4. During turning behavior the hg3 is active on the outer side of the turn and inactive on the contralateral side. D) The muscles of the third pterale: III1 and III2-4. The III1 pulls the wing back when active but also increases stroke amplitude during flight turns. The III2-4 are likely to act similar to III1. E) The indirect tergopleural muscles: tp1 and tp2. These muscles are likely to control angle of attack during wing stroke. F) The indirect control muscles: tergotrochanter (tt) and pleurosternal muscles ps1 and ps2. The tergotrochanter muscle inserts on the inner side of the scutum and connects via a long apodeme to the trochanter of the mesothoracic leg. It is activated during take-off and straightens the middle leg to elicit a 'jump' start of the fly. In *Calliphora* and the house fly *Musca* stroke frequency increases when spike frequency of ps muscle increases supposedly due to alterations in tension between sclerites within the wing hinge. Results are taken from various authors.[26,27,44-52] Figures are redrawn from Dickinson and Tu.[53]

nervous activation, the b1 muscle is stretched throughout the stroke due to the relative changes in distance between the basalar sclerite of the scutum and the wing hinge (Fig. 3C). The power required for lengthening the muscle in turn comes from the power muscles. As a consequence, the b1 muscle does not produce positive work on the wing hinge; it functions more as an active spring than a force producing element.[31] This concept in control of muscle power and wing kinematics allows the animal to exert a rather constant tension on the complex wing hinge throughout the entire stroke cycle which is likely to facilitate the control performance of the mechanical thoracic resonator. Moreover, through their influence on stroke parameters such as amplitude and frequency, the nervous activity in flight control muscles is critical in regulating the power output of the much larger asynchronous muscle fibers.

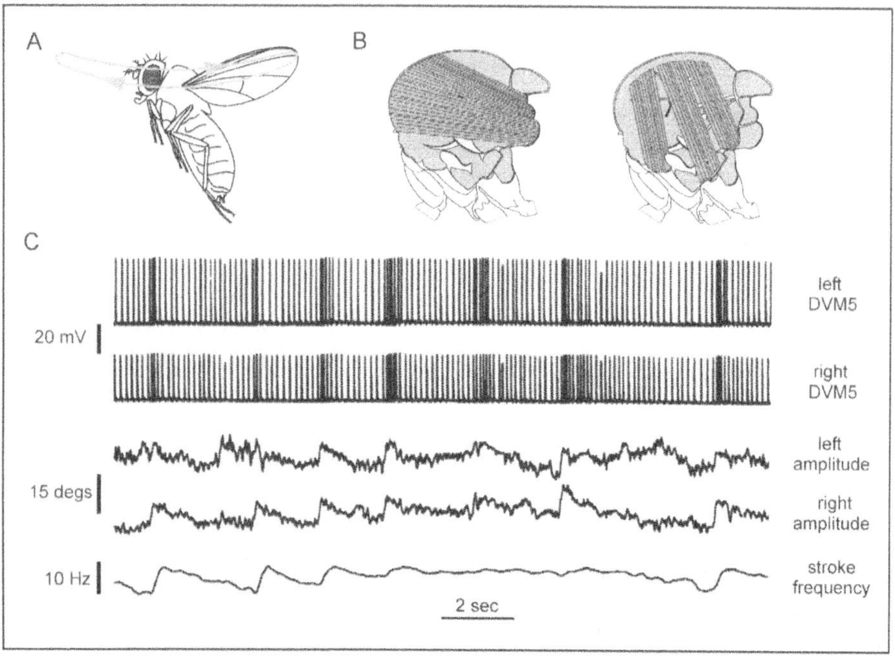

Figure 5. A) Hovering posture of a freely flying fruit fly. The main stroke plane is oriented approximately horizontally indicating that the lift vector produced by the beating wings is oriented perpendicular to the stroke plane. There are 4 stroke phases: 2 translational phases (up and down stroke) and 2 rotational phases at the end of each half stroke in which the wings are quickly rotated around the longitudinal wing axis. B) Generalized morphology of the indirect power muscles in flies. The 6 dorsal longitudinal muscles fibers (DLM, left), and 7 dorso-ventral fibers (DVM, right) are seen from the medial section of the right side. Indirect power muscles do not insert directly on or near the wing hinge but are attached throughout the thoracic cuticle. Their contractions generate distortions of the thoracic exoskeleton which an elaborate hinge transforms into the up and down wing motion. C) Neural drive of the asynchronous muscle increases during flight maneuvers of *Drosophila*. The top two traces show intracellular recordings from a bilateral pair of DVM. The firing rate of these muscles correlate with changes in stroke amplitude and stroke frequency during flight. According to conventional aerodynamic theory a simultaneous increase in both amplitude and frequency results in an increase in flight force production and vice versa. Wing kinematic responses were elicited using a moving visual stimulus in front of the fly. The increase in spike firing rate is assumed to be linked to an increase in free intracellular calcium concentration that in turn facilitates mechanical power output of flight muscle fibers due to an increase in recruitment/activation of myofibrils during stretch activation.

Changes in Muscle Efficiency in Vivo

The changing power requirements for flight demands alter the mechanical power requirements that face the asynchronous flight muscles to rapidly commit additional mechanical power during flight maneuvers. In the fruit fly, spike frequency in the indirect flight muscles increases when the wings undergo larger stroke angles due to an activation of basalar control muscles which is correlated with an increase in flight force production.[32] This increase in neural input might raise the release of calcium from internal stores, although fibrillar muscle contains only a small fraction of sarcoplasmic reticulum. An alternative explanation that is consistent with the presence of calcium currents in the asynchronous fibers is that calcium enters the muscle from outside through voltage gated calcium channels.[33,34] Although the steady-state calcium activation curve is steep,[35] the calcium influx from a burst in spikes of the DLM and DVM could function to recruit cross bridges into force production. As a consequence, the fuel consumption of the muscle fibers increases. The benefit of nervous muscle activation for providing a higher level of mechanical power, however, depends on muscle efficiency. Although typically assumed constant, muscle efficiency might also vary during flight force production, helping to assuage the increased metabolic demands resulting from elevated force production.

Muscle efficiency can be estimated by deriving at least two of the three factors: total metabolic rate, metabolic heat production and muscle mechanical power output. Because it is difficult to record exact measures of heat production in small insects, biologists derive muscle efficiency from the ratio between simultaneous measurements of carbon dioxide release and the mechanical work done by the muscle fibers. Strain-stress measurements in vitro of isolated muscles fibers are done in a muscle rig in which the isolated fibers undergo cyclic oscillations at a rate that would occur during flight in the intact animal. However, in many instances the mechanical power output estimated from these experiments is significantly lower than those predicted in the behaving animal.[36] Alternatively, muscle power can be derived from the power requirements for flight by simultaneously measuring stroke amplitude, stroke frequency and total flight force. To modulate wing kinematics and thus power requirements in flight, fruit flies are flown in a virtual flight simulator and stimulated with visual patterns moving vertically around the fly.[5] *Drosophila* responds robustly to this moving pattern by modulating the production of flight forces and thus power requirements in flight (Fig. 3A). These experiments are roughly analogous to electronically changing the weight of the animal and allow modulation of muscle power output in the intact animal. Muscle efficiency in *Drosophila* is lowest (5.6%) when the animal produces small flight forces and rises to a maximum (23.5%) during maximum force production (Table 1, Fig. 2C). The increase in muscle efficiency reflects the steep increase in profile power requirements with rising flight force, while the animal produces carbon dioxide at moderate rates even at elevated locomotor activity. The different slopes in metabolic rate and aerodynamic power requirements for flight help the fruit fly satisfy the increased energetic demands resulting from elevated flight force production when carrying loads and when performing flight maneuvers.

Concluding Remarks

Like all flying insects, the fruit fly *Drosophila* must regulate the amount of power that is generated by the flight musculature to match the changing demands of the locomotory output. The control of power output results from two separate effects: the changes in the asynchronous muscle's own neural drive as well as the activation of smaller, less powerful control muscles. Through their action on the stiffness of the thoracic box and the wing hinge, control muscles can alter the strain rate of power muscles by changing stroke amplitude and the frequency of the mechanical resonator, thereby changing both the power output and eventually flight force. Although the flight motor of flies differs from muscle systems design in other insects with synchronous power muscles, it reveals the trade-off between power generation and control in flapping flight which may potentially limit flight performance and thus aerial maneuverability in many flying insects.

References

1. Ellington CP. Limitations on animal flight performance. J Exp Biol 1991; 160:71-91.
2. Josephson RK. Contraction dynamics and power output of skeletal muscle. Ann Rev Physiol 1993; 55:527-546.
3. Josephson RK, Stevenson RD. The efficiency of a flight muscle from the locust, Schistocerca americana. Journal of Physiology 1991; 442:413-429.
4. Stevenson RD, Josephson RK. Effects of operating frequency and temperature on mechanical power output from moth flight muscle. J Exp Biol 1990; 149:61-78.

5. Lehmann F-O, Dickinson MH. The changes in power requirements and muscle efficiency during elevated force production in the fruit fly, Drosophila melanogaster. J Exp Biol 1997; 200:1133-1143.
6. Casey TM. A comparison of mechanical and energetic estimates of flight cost for hovering sphinx moths. J Exp Biol 1981; 91:117-129.
7. Tucker VA. Bird metabolism during flight: Evaluation of theory. J Exp Biol 1973; 58:689-709.
8. Ellington CP. The aerodynamics of insect flight. VI. Lift and power requirements. Phil Trans R Soc Lond B 1984; 305:145-181.
9. Pennycuick CJ. Power requirements for horizontal flight in the pigeon Columba livia. J Exp Biol 1968; 49:527-555.
10. Pennycuick CJ. Mechanics of flight. In: Farner DS, King JR, eds. Avian Biology. London: Academic Press, 1975:1-75.
11. Lehmann F-O. The constraints of body size on aerodynamics and energetics in flying fruit flies: An integrative view. Zoology 2002; 105:287-295.
12. Dickinson MH, Lehmann F-O, Sane S. Wing rotation and the aerodynamic basis of insect flight. Science 1999; 284:1954-1960.
13. Dickinson MH, Lighton JRB. Muscle efficiency and elastic storage in the flight motor of Drosophila. Science 1995; 268:87-89.
14. Gorb SN. Serial elastic elements in the damselfly wing: Mobile vein joints contain resilin. Naturwissenschaften 1999; 552-555.
15. Ardell DH, Andersen SO. Tentative identification of a resilin gene in Drosophila melanogaster. Insect Biochemistry and Molecular Biology 2001; 31:965-970.
16. Lehmann F-O. The efficiency of aerodynamic force production in Drosophila. Comparative Biochemistry & Physiology. Part A, Molecular & Integrative Physiology 2001; 131:77-88.
17. Roeder KD, Treat AE. The detection and evasion of bats by moths. Am Sci 1961; 49:135-148.
18. Currie RW. The biology and behavior of drones. Bee World 1987; 68:129-143.
19. Strohm E, Lechner K. Male size does not affect territorial behaviour and life history traits in a sphecid wasp. Anim. Behav 2000; 59:183-191.
20. Götz KG, Wehrhan C. Optomotor control of the force of flight in Drosophila and Musca I. Homology of wingbeat-inhibiting movement detectors. Biol Cybernetics 1984; 51:129-134.
21. Götz KG. Bewegungssehen and Flugsteuerung bei der Fliege Drosophila. In: Nachtigall W, ed. BIONA- report 2. Stuttgart: Fischer, 1983:21-34.
22. Lehmann F-O, Dickinson MH. The production of elevated flight force compromises flight stability in the fruit fly Drosophila. J Exp Biol 2001; 204:627-635.
23. Marden JH. Bodybuilding dragonflies: Costs and benefits of maximizing flight muscle. Physiol Zool 1989; 62:505-521.
24. Marden JH, Waage JK. Escalated damselfly territorial contests are energetic wars of attrition. Anim Behav 1990; 39:954-959.
25. Marden JH, Rollins RA. Assessment of energy reserves by damselflies engaged in aerial contests for mating territories. Anim Behav 1994; 48:1023-1030.
26. Tu MS, Dickinson MH. The control of wing kinematics by two steering muscles of the blowfly, Calliphora vicina. J Comp Physiol A 1996; 178:813-830.
27. Lehmann F-O, Götz KG. Activation phase ensures kinematic efficacy in flight-steering muscles of Drosophila melanogaster. J Comp Physiol 1996; 179:311-322.
28. Lehmann F-O. Aerodynamische, kinematische und electrophysiologische Aspekte der Flugkrafterzeugung und Flugkraftsteuerung bei der Taufliege Drosophila melanogaster. Max-Planck-Institute for biological Cybernetics. Tübingen: Thesis, University of Tübingen, 1994:131.
29. Ewing AW. The neuromuscular basis of courtship song in Drosophila: The role of direct and axillary wing muscles. J Comp Physiol 1979; 130:87-93.
30. Bennet-Clark HC, Ewing AW. The wing mechanism involved in the courtship of Drosophila. J Exp Biol 1968; 49:117-128.
31. Tu MS, Dickinson MH. Modulation of negative work output from a steering muscle of the blowfly Calliphora vicina. J Exp Biol 1994; 192:207-224.
32. Heide G, Spüler M, Götz KG et al. Neural control of asynchronous flight muscles in flies during induced flight manoeuvers. In: Wendler G, ed. Insect Locomotion. Berlin: Paul Parey, 1985:215-222.
33. Patlak J. The ionic basis for the action potential in the flight muscle of the fly, Sarcophaga bullata. J Comp Physiol 1977; 107:1-11.
34. Salkoff L, Wyman R. Ionic currents in Drosophila flight muscle. J Physiol 1983; 337:687-709.
35. Peckham M, Molloy JE, Sparrow JC et al. Physiological properties of the dorsal longitudinal flight muscle and the tergal depressor of the trochanter muscle of Drosophila melanogaster. J Muscle Res Cell Motil 1990; 11:203-215.

36. Tohtong R, Yamashita H, Graham M et al. Impairment of muscle function caused by mutations of phosphorylation sites in myosin regulatory light chain. Nature 1995; 374:650-653.
37. Chan WP, Dickinson MH. In vivo length oscillations of indirect flight muscles in the fruit fly Drosophila virilis. J Exp Biol 1996; 199:2767-2774.
38. Casey TM, Ellington CP. Energetics of insect flight. In: Wieser W, Gnaiger E, eds. Energy Transformations in Cells and Organisms. Stuttgart: Thieme Verlag, 1989:200-210.
39. Dickinson MH, Hyatt CJ, Lehmann F-O et al. Phosphorylation-dependent power output of transgenic flies: An integrated study. Biophys J 1997; 7:3122-3134.
40. Wakeling JM, Ellington CP. Dragonfly flight III. Lift and power requirements. J Exp Biol 1997; 200:583-600.
41. Josephson RK. Temperature and the mechanical performance of insect muscle. In: Heinrich B, ed. Insect thermoregulation. New York: John Wiley & Sons, 1981:19-44.
42. Heinrich B. Temperature regulation of the sphinx moth, Manduca sexta. I. Flight energetics and body temperature during free and tethered flight. J Exp Biol 1971; 43:141-152.
43. Lehmann F-O. Ambient temperature affects free-flight performance in the fruit fly Drosophila melanogaster. J Comp Physiol B 1999; 169:165-171.
44. Nachtigall W, Wilson DM. Neuro-muscular control of dipteran flight. J Exp Biol 1967; 47:77-97.
45. Heide G. Die Funktion der nicht-fibrillären Flugmuskeln bei der Schmeißfliege Calliphora. Teil I: Lage, Insertionsstellen und Innervierungsmuster der Muskeln. Zool Jb Physiol 1971; 76:87-98.
46. Heide G. Die Funktion der nicht-fibrillären Flugmuskeln bei der Schmeißfliege Calliphora. Teil II: Muskuläre Mechanismen der Flugssteuerung und ihre nervöse Kontrolle. Zool Jb Physiol 1971; 76:99-137.
47. Heide G, Götz KG. Optomotor control of course and altitude in Drosophila is achieved by at least three pairs of flight steering muscles. J Exp Biol 1996; 199:1711-1726.
48. Egelhaaf M. Visual afferences to flight steering muscles controlling optomotor responses of the fly. J Comp Physiol A 1989; 165:719-730.
49. Kutsch W, Hug W. Dipteran flight motor pattern: Invariabilities and changes during postlarval development. J Neurobiol 1981; 12:1-14.
50. Wisser A. Mechanisms of wing rotating regulation in Calliphora erythrocephala (Insecta, Diptera). Zoomorphol 1987; 106:261-268.
51. Wisser A, Nachtigall W. Functional-morphological investigations on the flight muscles and their insertion points in the blowfly Calliphora erythrocephala (Insecta, Diptera). Zoomorphol 1984; 104:188-195.
52. Pfau HK. Funktion einiger direkter tonischer Flügelmuskeln von Calliphoar erythrocephala Miegen. Verh Dtsch Zool Ges 1977; 70:275.
53. Dickinson MH, Tu MS. The function of Dipteran flight muscle. Comp Biochem Physiol A 1997; 116A:223-238.

Chapter 19

Molecular Assays for Acto-Myosin Interactions

John C. Sparrow and Michael A. Geeves

Abstract

The indirect flight muscles of insects are highly specialised to produce power for flight. Asynchronous flight muscle contraction is largely isometric (3-4% shortening in vivo) and can occur at high oscillatory frequencies. Contraction kinetics are the property of the myofibrils and differences in contraction speed between insects must result from evolution of the acto-myosin system. With actin sequences so closely conserved the kinetic differences are due to sequence changes in myosin. So detailed kinetic study of insect flight muscle myosins from different insects are required to understand how these myosins support the often rapid wingbeat frequencies and produce the power required for flight. In addition, our ability to understand the relationship between the myosin amino acid sequence and its kinetics by using molecular genetic approaches with transgenic *Drosophila* relies on having the ability to search for differences in the acto-myosin cross-bridge kinetics. In this chapter we identify the technical developments that have allowed these goals to be achieved, the further applications that can be made from them, the results from kinetic measurements and the issues raised from considering these kinetics and their relevance to the evolution of flight muscle function.[1]

Introduction

Insects show a very wide range of wingbeat frequencies in vivo – from about 30Hz in the largest (e.g., *Lethocerus)* to over 200Hz in *Drosophila,* but spanning three orders of magnitude if one considers all flying insects! At least in asynchronous flight muscle the wingbeat frequencies and contraction speeds correlate with the mechanical kinetics of skinned fibre preparations.[1] So contraction kinetics are a property of the fibres/myofibrils themselves. Contraction is produced by the interactions of actin and myosin, so the differing physiological parameters of insect flight muscle are likely due, in large part at least, to changes in the mechanochemical properties of the acto-myosin cross-bridge cycle. In vertebrates, including mammals, there are strong clear correlations between muscle fibre shortening velocity and myosin isoform expressed in the fibre.

Biochemical kinetics studies of insect flight muscle myosins have so far been restricted to two species of the water-bug, *Lethocerus cordofanus, Lethocerus maximus,* the dung beetle *Heliocropis japetus*[2,3] and the fruitfly, *Drosophila melanogaster,*[4-6] though data[7] exist comparing purified myosins and myosin S1 subfragments from *Lethocerus indicus, Lethocerus griseus, Drosophila melanogaster, Melontha melontha* (coleopteran maybug), *Vespula vulgaris* (hymenopteran, common wasp), *Bombus terrestris* (hymenopteran, land bee), *Calliphora vomitoria* (dipteran, blue bottle), *Eristalis tenax* and *Episyrphus balteatus* (dipteran, hoverflies).

Nature's Versatile Engine: Insect Flight Muscle Inside and Out, edited by Jim Vigoreaux.

A major problem for the biochemical study of insect myosins is that, with the exception of particularly large insects such as *Lethocerus*, the amount of starting material for purification of myosin and actin is very small, especially from a model organism such as *Drosophila*. Recent molecular genetic approaches to muscle and acto-myosin function in *Drosophila* has generated interest and importance in adapting, or adopting, existing techniques to analyse the kinetics of acto-myosin from relatively small amounts of dissected flight muscles. To date all studies of insect flight muscle myosin have started with dissected flight muscles[4-6] (see also Miller and Bernstein, this book).

Myosin Purification and Preparation of the S1 Fragment

Muscle myosin IIs, in all animals, are hexamers (see Fig. 1, Miller and Bernstein, this book) with MWs of ~500 kDa. Myosin is insoluble at physiological salt conditions (where it exists as filaments) due to the propensity of the C-terminal parts of the myosin heavy chain to form coiled-coils and to dimerise via side-by-side associations to form the thick filaments. Myosin's insolubility makes it quite difficult to work with biochemically, especially under conditions which reflect its function in vivo—where it and F-actin, are present at high concentrations and salt concentrations of ~90 mM. For studies of myosin a large suite of well-established solution biochemistry techniques have been developed over the last 30-40 years for investigations of vertebrate and invertebrate, largely molluscan, myosins. Muscle myosins can be solubilised in high salt concentrations,[8] usually in excess of 500mM KCl and maintained in solution above 300mM KCl. Insect myosins require rather more extreme conditions for their extraction from IFM[2,4,5,6] than those from vertebrate muscles. All protocols use 1M KCl with 10mM sodium pyrophosphate at pH7.0. The IFM must be dissected since different insect muscles express different myosin isoforms. The dorso-longitudinal muscles (DLM) of *Drosophila* are more easily removed than the dorso-ventral muscles (DVM). An adult *Drosophila* weighs about 1 mg -0.9 mg for males, 1.1mg for females—and the DLM account for approximately 10% of the flies' mass so each set of muscles has a wet mass

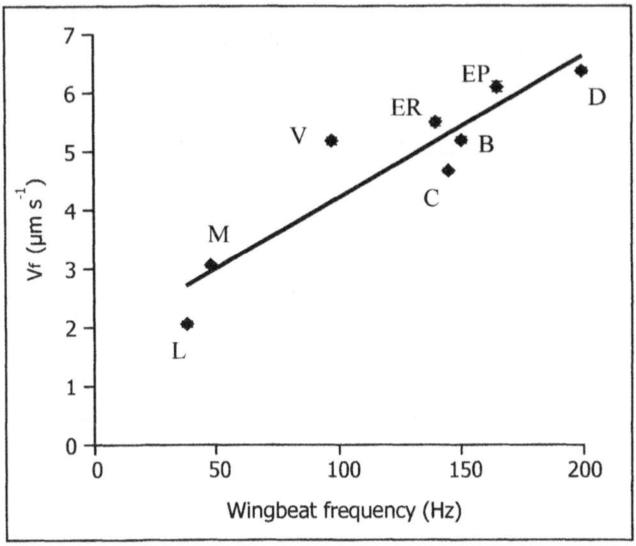

Figure 1. Correlation between wingbeat frequency and the in vitro motility velocities (V_f; mean ±s.d.) of F-actin filaments driven by myosins isolated from the indirect flight muscles of various insects. L, *Lethocerus indicus*; M, *Melontha melontha*; V, *Vespula vulgaris*; C, *Calliphora vomitoria*; ER, *Eristalis tenax*; B, *Bombus terrestris*; EP, *Episyrphus balteatus*; D, *Drosophila melanogaster*. (Redrawn from Iliffe).[7]

of 100 μg. In the case of *Drosophila* current protocols yield about 2 – 3 μg of myosin per fly, a significant fraction of all the IFM myosin. The myosin retains its enzymatic activities (see below) and the light chains remain fully associated with the extracted myosin. The insect myosins remain soluble at lower salt concentrations and can be stored in 500 mM KCl, though purified insect flight muscle myosins lose activity fairly quickly.[4,8]

Whole myosin can be used to quantitate the biochemical properties of the molecule, either in solution (in high salt) or in 'solid-state' assays where the myosins are surface bound (see below). Protocols to isolate soluble, enzymatically active fragments of vertebrate myosins were developed many years ago (see Miller and Bernstein, this volume). The two major soluble fragments used are subfragment 1 (S1), consisting of single myosin motor domains and 'heavy meromyosin' (HMM), being double-headed molecules in which the larger part of the myosin coiled-coil domain is removed. So far only enzymatically active S1 has been successfully recovered from insect flight muscles; *Lethocerus*,[3] *Drosophila*[6] and *M. melontha, V. vulgaris, B. terrestris, C. vomitoria, Er. tenax* and *Ep. balteatus*.[7]

Protocols to obtain S1 from vertebrate myosins vary to cope with differences in proteolytic sensitivities of different myosins. IFM S1 was first prepared from *Lethocerus* flight muscles[3] using a chymotryptic digestion of rigor myofibrils developed by Cooke.[10] This protects the S1 from supernumerary proteolytic cleavages that inactivate it. This has proved a successful approach to recover stable preparations of S1 with ATPase and F-actin binding activity from *Drosophila*[6,7] and other small insects.[7] Detailed protocols are described in White et al.[3] and Silva et al.[6] Slight modifications are required to optimise the conditions for different insects.[7] S1 yields from *Lethocerus* myofibrils were about 30% of total myosin heads available in the myofibrils;[3] the yield from 125 dissected *Drosophila* IFM was approximately 35 μg – 0.28 μg/fly.[6] Recently greater yields, 1.25 μg/fly, have been reported.[11] Gel separations of S1 show that the myosin essential light chains are still bound,[3,6] but the regulatory light chains have been lost. This is usual following chymotryptic digestion of vertebrate myosins.

Purification of Flight Muscle Actin

Insect flight muscle actin was first prepared from *Lethocerus* and *Heliocropis*[2] using the standard methods for extracting vertebrate muscle actin. Similar approaches were used[12] to obtain ACT88F isoforms (the IFM-specific actin; see Sparrow, this book) from either dissected *Drosophila* IFM or by ion-exchange chromatography of actin extracts obtained from an acetone-dried powder of whole flies.

Although the biochemical kinetics of flight muscle actin interacting with flight muscle myosins have been studied,[2,4] actin from vertebrate sources, usually rabbit skeletal muscle can be used. From the few comparative studies that have been done, it appears that only very minor differences in acto-myosin kinetics can be ascribed to the source of the actin; myosin is the major influence on acto-myosin kinetics.

Assays of Myosin and Acto-Myosin

Apart from two earlier studies on *Lethocerus*,[2,3] little work was done on insect muscle myosin enzymatics until recently.[4-7,11] All the assays used for insect myosins and acto-myosins were originally developed for studies of the vertebrate proteins.

Steady-State ATPase Assays

Steady-state solution assays of myosin, S1, or acto-myosin ATPase activities in which the proteins are incubated with ATP, and the rates of product (ADP and inorganic phosphate, Pi) release are measured, give indications of the overall cycling rate. Different assay techniques have been used on insect myosins, S1 and acto-myosins. Detection of myosin ATPase activity by phosphate release is readily performed using colorimetric assays e.g., the malachite green assay[13] has been used successfully to measure *Drosophila* myosin ATPase (Sparrow, unpublished). A more sensitive radio-assay of Pi release that has been successfully developed for

Drosophila myosin ATPase measurements, uses ammonium molybdate to trap the^{32}Pi released from ATP-γ-^{32}P into a complex which is extracted into an organic phase of isobutanol-benzene, and then quantified by Cerenkov counting.[4] HPLC separation of nucleotide reaction products has been used to determine the ATP/ADP ratio[3,6] either by absorbance at 260nm or, more sensitively, by the use of ^{14}C-ATP.[3] Most recently a linked enzymatic assay system that monitors ADP production via the absorbance changes of NAD/NADH using the pyruvate kinase/lactate dehydrogenase coupled assay has also been exploited.[11]

Stop-Flow Approaches

Steady-state ATPase assays are readily performed, even on the amounts of myosin available from small insects, but only report on the overall cycling. Stop-flow is a powerful tool for determining the rapid kinetic transitions between intermediate acto-myosin states in the cross-bridge cycle. This technique, in which small aliquots of proteins and their reactants are rapidly mixed and the reactions followed optically, can provide kinetic information on the millisecond timescale. However, it usually requires relatively large volumes of protein. Recent developments that reduce the reactant volumes required[14] have allowed accurate determination of kinetic parameters of acto-myosin and myosin/nucleotide interactions with microgram quantities of protein, using fluorescence of a pyrene label covalently attached to actin to report myosin or S1 binding to actin.

Pyrene labelling requires substantial amounts of actin and cannot be used to label the quantities of insect flight muscle actin usually available. However, by using pyrene-labelled rabbit actin with purified, unlabelled *Drosophila* actin in competitive binding experiments for rabbit S1, kinetic differences between wild-type and mutant (ACT88F^{E93K}) actins have been determined.[12] Attempts to measure *Drosophila* myosin binding to rabbit actin by acto-myosin light scattering in stopped flow assays showed that the signals were very poor at the concentrations (> 100 nM) accessible.[6] Previous use of pyrene-labelled actin allowed actin and rabbit S1 interactions to be measured at concentrations as low as 30 nM, but measurements could not be made with *Drosophila* myosin apparently because the pyrene signal change is much smaller for *Drosophila* compared to mammalian myosins; signals were also weak with *Drosophila* S1 (only 1.5% vs. 20% for rabbit S1) but were usable.[6]

Flash Photolysis/Light Scattering

Recent technical developments using flash photolysis allows acto-myosin kinetics to be performed on very small amounts (microgram and sub-microgram) of the proteins.[15-17] The approach is to use laser flash photolysis to release caged-ATP within small (20 μl) samples of proteins. A white light source monitors the amount of ATP released by the changed absorbance at 405 nm and the acto-myosin association/dissociation by light-scattering. The advantages are that rapid changes are easily and reliably detected, but also, that repeated caged-ATP releases allow many experimental repetitions to be performed on the same sample. This technique has been used to study acto-myosin kinetics on the relatively small amounts of myosin available from wild type *Drosophila* IFM[6] and for different native or chimeric *Drosophila* myosin isoforms expressed transgenically in the IFM.[11]

In Vitro Motility Assays

Myosins bound onto glass or nitrocellulose-coated glass substrates produce ATP-dependent movement of fluorescently-labelled (rhodamine-phalloidin) F-actin.[18] For detailed protocols see Kron et al.[18] F-actin velocities can be measured by tracking of filaments using video-recorded data (see Homsher et al[20] for a critical assessment of this analysis). Under optimal conditions the velocities are dependent on the type of myosin (largely independent of the F-actin source) and correlate with the contraction speeds of the muscles from which the myosin was purified. This technique requires only small amounts of proteins, especially F-actin, and was initially used to study *Drosophila* IFM-specific actin.[12,21,22]

Compared to vertebrate myosins, movement generated by insect myosins was more difficult to achieve, though reliable measurements are now routinely obtained with *Drosophila* myosin.[4,5,23] Part of the problem is the relative lability of the *Drosophila* flight muscle myosin.[4] In addition, the best F-actin movement (greatest number of filaments moving smoothly and continuously) occurs using 0.5 mg ml^{-1} of myosin, 10-fold higher than typically used with rabbit HMM[19] and at the high end of optimal values (0.1–0.5 mg ml^{-1}) reported for myosins from other muscle types;[24] movement on *Drosophila* myosin is not achieved below 0.3 mg ml^{-1}.[4] F-actin diffuses off the *Drosophila* myosin-coated surface after ATP addition unless methylcellulose (0.4%) is present and motility is optimal when the ionic strength is about 25 mM (no added KCl). Higher ionic strengths decrease the number of filaments moving and some filaments leave the myosin surface, even in the presence of methylcellulose. The mean F-actin velocity on *Drosophila* IFM myosin, 6.5 µm s^{-1} [4] is faster than that found with rabbit myosin, 4-5 µm s,$^{-1}$ [24] even with *Drosophila* IFM F-actin.[12] *Drosophila* IFM myosin exhibits rather different properties than rabbit skeletal myosin in this assay. Overall smooth, optimal F-actin velocities on other insect flight muscle myosins requires high myosin concentrations, methylcellulose and low ionic concentrations.[7] This raises the issue of whether the flight muscle myosins have a lower affinity for F-actin (see discussion below).

This in vitro motility assay is a powerful tool for studying insect myosins but has limitations when comparing the F-actin in vitro velocities with shortening velocities measured from skinned fibres. The major limitations are: (a) that in vitro velocities are measured without an applied load, except for the viscous drag experienced by the filaments, and (b) that the in vitro system is disordered i.e., the myosin (or myosin subfragments) are randomly oriented whereas the myofibrillar lattice in striated muscle maintains myosin and actin in close proximity. This is especially true in asynchronous insect flight muscles, which are remarkable for their high degree of order including the same axial spacing of the F-actin/thin filament repeat and the myosin heads in the thick filaments. This high order gives effective concentrations of actin and myosin binding sites, which determine the probability of effective collisions, much higher than those achievable in vitro.

Comparisons of in vitro motility velocities on myosins prepared from the insects with different wingbeat frequencies show (Fig. 1) a good correlation.[7]

Single Molecule Optical Trap Measurements

The in vitro motility assay does not provide detailed kinetic or mechanical measurements. More recently single acto-myosin cross-bridge mechanical experiments have been carried out using optical tweezers transducers in the 'three-bead' configuration.[25-28] Two latex beads are independently trapped just below the focus of a laser beam by the forces produced by the pressure of light refracted through the beads. The forces are sufficient to reduce bead movement against thermal motion and of similar magnitude to those produced by single myosin molecules. The two beads, attached to the ends of a single actin filament, hold the filament close to a glass bead coated with a low density of myosin, or more usually, HMM molecules. Interactions between single myosin heads and the actin filament are detected by focussing an image of one bead onto an optical detector. They are seen as short periods of reduced noise in the position signal, most frequently displaced from the mean position of the signal. Three measurements can be made; the step size, the force produced and, from the duration of the attachments at different ATP concentrations, the second order rate constant for the detachment of rigor heads by ATP binding.

The *Drosophila* flight muscle myosin working-strokes interacting with either rabbit or *Drosophila* IFM actin, were 3.91 ± 2.36 nm[4] using the 'Molloy' analysis.[26] These values compare to those of 4.20 ± 0.93nm for rabbit myosin and S1 (4.73 nm papain S1 and 3.81nm chymotryptic S1[29]) under the same conditions, but rather smaller than values for rabbit HMM (5.5 nm).[27]

Major Conclusions Relating to the Enzymatic Properties of Insect Flight Muscle Acto-Myosin

Clearly decent biochemical kinetic parameters can be measured from the very limited amounts of protein, myosin, S1 and actin that can be obtained from dissected insect flight muscles.

The in vitro motility studies show that higher myosin concentration and methylcellulose are required to produce smooth F-actin movement compared to vertebrate myosins.[4] This is consistent with the K_m for actin of *Lethocerus* S1, measured from actin-activated ATPase studies, which is 6-15 times greater than vertebrate fast muscle[3] and consistent with earlier observations on *Lethocerus* myosin.[2] Thus, it seems likely that IFM myosins may generally have low actin affinities.

Rate Limiting Step

What is the rate-limiting step of the IFM cross-bridge cycle? The meaning of the question is different for the oscillatory IFM than for most muscle since the cross-bridges are operating for only a very short time during each contraction cycle. There is no equivalent to the steady state in an enzyme reaction. In the following we have attempted to address this question using data available from studies on *Drosophila*. The wings of *Drosophila* beat at ~200 Hz so each complete wingbeat lasts for only 5 ms. This time includes both the shortening and recovery stroke of each muscle and so the time during which the muscle shortens is at most 2.5 ms. The typical length change in each contraction was estimated as 3.5 % at a sarcomere length of 3.3 µm. Thus each half sarcomere shortens by ~60 nm in 2.5 ms and then has 2.5 ms to recover before the next shortening event.

The fraction of cross-bridges attached at any one moment during shortening of *Drosophila* IFM is unknown. X-ray diffraction studies of *Lethocerus* IFM[30] during active contraction have shown that at least 30% of all the myosin heads have moved away from their axially ordered position on the myosin helix and that 7-23% of the myosin heads are stereo-specifically attached to the actin helix of the thin filament at the peak of tension. Interestingly, a number of mechanical studies and X-ray diffraction studies on vertebrate fibres suggest figures as low as 5% for the fraction of heads bound at any one time during shortening. In the following calculations for *Drosophila* we have used two estimates for the fraction of heads bound—5% and 20%. Certainly given the higher wingbeat and faster fibre kinetics of *Drosophila* fibres[1,31] the fraction of heads cycling per wingbeat and bound at any time may be fewer than in *Lethocerus*; steady state fibre ATPase (unstretched) is fairly constant across insects with a wide range of wingbeat frequencies.[1] For convenience we have rounded the estimated *Drosophila* cross-bridge step to 5 nm (the data are not significantly different from 5 nm—mean less than one standard deviation from this value).[4] Combining the estimate of 5% [20%] heads and a 5 nm cross-bridge throw then the fraction of cross-bridges contributing during each contraction cycle is 60/5 x 5 = 60% of the total [240%]. At 60% each myosin cross-bridge would contribute only every second stroke (recovery time is thus 5 ms). With 20% of heads attached the estimates leads to the figure in brackets, 240%. This would mean the heads are contributing 2.4 times per cycle, so the system is closer to a true steady state and each head turns over 2.4 ATP molecules per cycle. If the fraction attached is less than 5 % (our lower estimate) the fraction contributing per stroke is even less and the recovery time even longer. The fraction of myosin heads bound may be a factor of 2-fold lower if a cross-bridge comprises only one of the pair of heads on each myosin.

The lifetime of a cross-bridge can be estimated as follows. The filaments are sliding at a speed of 60 nm in 2.5 ms or 24 nm/ms. If the cross-bridge remains attached for the working stroke (5 nm) then the lifetime of the attached cross-bridge can be no more than 5/24 = 0.208ms without imposing a load on the shortening filament. We can calculate the lifetime of the unstrained A.M.D and A.M states from the kinetics measured in solution and compare these to the lifetime of the cross-bridge calculated above. The A.M state binds ATP and

Box 1. Detachment rate and ATP concentration

The full description of the ATP concentration dependence of A.M detachment is:

$k_{detach} = [ATP]K_1k_{+2}/(1 + K_1[ATP])$ which is hyperbolic in [ATP]

(k_{+2} is the maximal value of k_{detach} at infinite [ATP] and $1/K_1$ is the concentration of ATP required for k_{detach} to be half maximal)

Thus k_{detach} must be $4000s^{-1}$ if the detachment of AM is not to be rate limiting for the velocity.

This can only happen if k_{+2} (the maximal value of k_{detach}) $> 4000s^{-1}$. We can estimate the possible size of K_1 and k_{+2} for ATP dissociation given that we know $K_1k_{+2} = 0.8\ \mu M^{-1}s^{-1}$.

If

$k_{+2} = 4000$, then $K_1 = 205M^{-1}$ and $1/K_1 = 4.9$ mM, $k_{detach} = 4000$ then [ATP] >30 mM

$k_{+2} = 5000$, then $K_1 = 164M^{-1}$ and $1/K_1 = 6.1$ mM, $k_{detach} = 4000$ then [ATP]=24 mM

$k_{+2} = 6000$, then $K_1 = 136M^{-1}$ and $1/K_1 = 7.3$ mM, $k_{detach} = 4000$ then [ATP] =14.9 mM

$k_{+2} = 8000$, then $K_1 = 102M^{-1}$ and $1/K_1 = 9.8$ mM, $k_{detach} = 4000$ then [ATP] =9.8 mM

Thus the value of k_{+2} would need to be >> than $3000s^{-1}$ for k_{detach} to be $> 3000s^{-1}$ and for such a rate to be achieved at accessible concentrations of ATP.

detaches at a rate controlled by the ATP concentration times the apparent 2nd order rate constant of ATP binding ($0.82\ \mu M^{-1}s^{-1}$).[6] (see Box 1 for a more complete description). The A.M state lifetime is then 1/([ATP] x 0.82)s. If the ATP concentration in the flight muscle is > 6 mM, the AM state lifetime will be < 0.2 ms, the expected life time of the cross-bridge calculated above. The A.M.D state lifetime is controlled by the rate constant of ADP release (k_{-AD}). This has not been measured, but the equilibrium dissociation constant for ADP binding (K_{AD}; see Fig. 2) has been estimated as 400 μM.[6,11] If the ADP binding rate constant (k_{+AD}) is diffusion controlled, then it is expected to be $\geq 10^7M^{-1}s^{-1}$ (the collision frequency for a small molecule and a protein is estimated as $10^9 - 10^{10}M^{-1}s^{-1}$ and a target zone of 1% gives $10^7 - 10^8M^{-1}s^{-1}$).[32] The ADP dissociation rate constant can then be estimated from $k_{-AD} = K_{AD}.k_{+AD}$ and $k_{-AD} = 4000\ s^{-1}$. The lifetime of the A.M.D state is therefore 1/4000 = 0.25 ms.

Thus the lifetime of both AM.ADP (defined by the rate of ADP release) and AM (defined by the rate of ATP binding) are of the correct order to define the lifetimes of the attached cross-bridges. The data so far,[7] from comparisons of myosins from insects with different wingbeat frequencies, indicates that the rate of ATP binding changes, but with little change in ADP affinity. This suggests that the A.M state lifetime may define the maximum shortening speed of IFM across the insects.

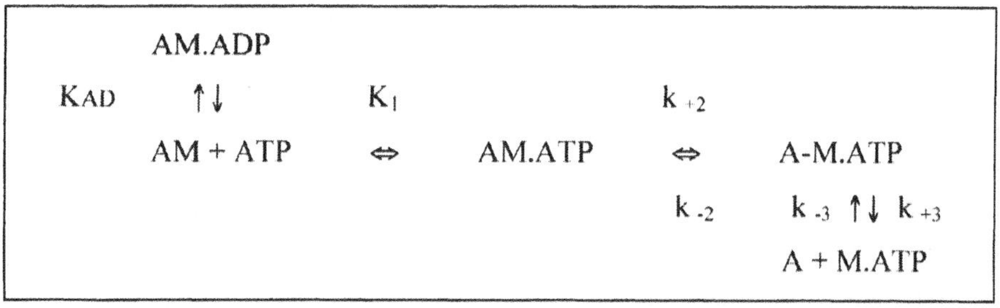

Figure 2. Part of acto-myosin crossbridge cycle to show ADP release and ATP binding steps and the nomenclature of the rate constants referred to in the text (redrawn from Kurzawa-Goertz et al).[33]

A further issue relating to the ADP affinity, which is weak compared to mammalian fast muscle myosin,[33] is that the K_{AD} value remains high for insect muscle. Possibly ADP affinity remains high to prevent the effects of ADP build up in fatigue (but see chapter by Vishnudas and Vigoreaux for an alternative explanation). The ATP concentration must also remain high and the requirement, to maintain the ATP/ADP ratio, is undoubtedly linked to the observation that in the IFM 50% of the fibre volume is occupied by mitochondria.

Major Questions about Insect Flight Muscle Acto-Myosin Kinetics That Remain

The major question to arise from the ability to obtain kinetic measurements of insect acto-myosin is how the kinetics have evolved to suit the physiological needs of insects of different sizes, and therefore different wingbeat frequencies.

A major issue remains the rate-limiting step. We have suggested where this may be. Detailed kinetic analysis of the effects of ATP and ADP concentration on contraction velocity (fibres), motility and acto-myosin kinetics in solution are required, but many of the reactions steps remain difficult to access. However, estimates of some of these kinetics can also be obtained from fibre mechanics experiments (see chapter by Maughan and Swank).

Drosophila as an insect, and as a model organism for the molecular genetics of muscle function and human disease, has a major role to play in furthering our understanding of both muscle function and specifically the myosin molecule. This is exemplified by the elegant use of transgenic flies to exploit both the alternative spliced exons of the myosin heavy chain transcript and the recovery of sufficient proteins for kinetic analysis. To fully use insect flight muscles to achieve these aims we need to develop further the biochemical and biophysical technologies for studying the contractile proteins. The challenge is that for many insects only very small amounts of protein are available. The progress made to date, largely within the last decade suggests that much more can be achieved, both with *Drosophila* flight muscle myosin, and with those from the enormous variety of insect species with different requirements for contraction velocity and power development. There is a vast natural experiment in insect flight muscles that can enrich and enhance our understanding of how muscle and muscle protein functions have evolved.

References

1. Molloy JE, Kyrtatas V, Sparrow JC et al. Kinetics of flight muscles from insects with different wingbeat frequencies. Nature 1987; 328:449-451.
2. Bullard B, Dabrowska R, Winkelman L. The contractile and regulatory proteins of insect flight muscle. Biochem J 1973; 135:277-286.
3. White DCS, Zimmerman RW, Trentham DR. The ATPase kinetics of insect fibrillar flight muscle myosin subfragment-1. J Muscle Res Cell Motil 1986; 7:179-192.
4. Swank DM, Bartoo ML, Knowles AF et al. Alternative exon-encoded regions of Drosophila myosin heavy chain modulate ATPase rates and actin sliding velocity. J Biol Chem 2001; 276:15117-15124.
5. Swank DM, Knowles AF, Kronert WA et al. Variable N-terminal regions of muscle myosin heavy chain modulate ATPase rate and actin sliding velocity. J Biol Chem 2003; 278:17475-17482.
6. Silva R, Sparrow JC, Geeves MA. Isolation and kinetic characterisation of myosin and myosin S1 from the Drosophila indirect flight muscles. J Muscle Res Cell Motil 2003; 24:489-498.
7. Iliffe CA. The kinetics and mechanics of insect flight muscle. PhD thesis University of York 2002.
8. Margossian SS, Lowey S. Preparation of myosin and its subfragments from rabbit skeletal muscle. Methods Enzymol 1982; 85:55-71.
9. Tanikawa M, Ueyama A, Maruyama K. Instability of insect myosin ATPase activity and its protection. Comp Biochem Physiol 1987; 86B:63-65.
10. Cooke R. A new method of preparing myosin subfragment-1. Biochem Biophys Res Commun 1972; 49:1021-1028.
11. Miller BM, Nyitrai M, Bernstein SI et al. Kinetic analysis of Drosophila muscle myosin isoforms suggests a novel mode of mechanochemical coupling. J Biol Chem 2003; 278:50293-50300.
12. Razzaq A, Schmitz S, Veigel C et al. Actin residue E93 is identified as an amino acid affecting myosin binding. J Biol Chem 1999; 274:28321-28328.

13. Kodama T, Fukui K, Kometani K. Initial phosphate burst in ATP hydrolysis by myosin and subfragment-1 as studied by modified malachite green method for determination of inorganic phosphate. J Biochem (Tokyo) 1986; 99:1465-1472.
14. Kurzawa SE, Geeves MA. A novel stopped-flow method for measuring the affinity of actin for myosin head fragments using microgram quantities of protein. J Muscle Res Cell Motil 1996; 17:669-676.
15. Weiss S, Chizov I, Geeves MA. A flash photolysis fluorescent/light scattering apparatus for use with sub microgram quantities of muscle proteins. J Muscle Res Cell Motil 2000; 21:423-432.
16. Weiss S, Rossi R, Pellegrino MA et al. Differing ADP release rates from myosin heavy chain isoforms define the shortening velocity of skeletal muscle fibers. J Biol Chem 2001; 276:45902-45908.
17. Clark RJ, Nyitrai M, Webb MR et al. Probing nucleotide dissociation from myosin in vitro using microgram quantities of myosin. J Muscle Res Cell Motil 2003; 24:315-321.
18. Kron S, Spudich JA. Fluorescent actin filaments move on myosin fixed to a glass surface. Proc Nat Acad Sci (USA) 1986; 83:6272-6276.
19. Kron S, Toyoshima YY, Uyeda TQP et al. Assays for actin sliding movement over myosin coated surfaces. Methods Enzymol 1991; 85:57-60.
20. Homsher E, Wang F, Sellers JR. Factors affecting movement of F-actin filaments propelled by skeletal heavy meromyosin. Am J Physiol 1992; 262:c714-c723.
21. Bing W, Razzaq R, Sparrow JC et al. Tropomyosin and troponin regulation of wild-type and E93K mutant actin filaments from Drosophila flight muscle: Charge reversal on actin changes actin-tropomyosin from ON to OFF state. J Biol Chem 1998; 273:15016-15021.
22. Schmitz S, Clayton J, Nongthomba U et al. Drosophila ACT88F indirect flight muscle-specific actin is not N-terminally acetylated: A mutation in N-terminal processing affects actin function. J Mol Biol 2000; 295:201-210.
23. Palmiter Littlefield K, Swank DM, Sanchez BM et al. The converter domain modulates kinetic properties of Drosophila myosin. Am J Physiol Cell Physiol 2003; 284:c1031-c1038.
24. Sellers JR, Cuda G, Wang F et al. Myosin-specific adaptations of the motility assay. Methods Cell Biol 1993; 39:23-49.
25. Finer JT, Simmons RM, Spudich JA. Single myosin molecule mechanics: Piconewton forces and nanometre steps. Nature 1994; 368:113-119.
26. Molloy JE, Burns, JE, Sparrow JC et al. Single molecule mechanics of heavy meromyosin and S1 interacting with rabbit or Drosophila actins using optical tweezers. Biophys J 1995; 68(4 Suppl):298s-305s.
27. Molloy JE, Burns JE, Tregear RT et al. Movement and force produced by a single myosin head. Nature 1995; 378:209-212.
28. Veigel C, Bartoo ML, White DC et al. The stiffness of rabbit skeletal acto-myosin cross-bridges determined with an optical tweezers transducer. Biophys J 1998; 75:1424-1438.
29. Molloy JE, Kendrick-Jones J, Veigel C et al. An unexpectedly large working stroke from chymotryptic fragments of myosin II. FEBS Lett 2000; 480:293-297.
30. Tregear RT, Edwards RJ, Irving TC et al. X-ray diffraction indicates that active cross-bridges bind to actin target zones in insect flight muscle. Biophys J 1998; 74:1439-1451.
31. Peckham M, Molloy JE, Sparrow JC et al. Physiological properties of the dorsal longitudinal flight muscle and the tergal depressor of the trochanter muscle of Drosophila melanogaster. J Muscle Res Cell Motil 1990; 11:203-215.
32. Gutfreund H. Kinetics for the life sciences. Cambridge: Cambridge University Press, 1995.
33. Kurzawa-Goertz SE, Perrecault-Micale CL, Trybus KM et al. Loop 1 can modulate ADP affinity, ATPase activity and motility of different scallop myosins. Transient kinetic analysis of S1 isoforms. Biochemistry 1998; 37:7517-7525.

Chapter 20

Insect Flight Muscle Chemomechanics

David Maughan and Douglas Swank

Abstract

The biochemical and mechanical basis of insect flight has captivated the interest of biologists for decades. This chapter presents a brief review of the approaches used and results obtained by investigators intent on understanding the chemomechanical basis of contraction in insect muscle. We are much closer now than we have ever been to understanding the details of the contractile mechanism. This has been in large measure due to the great expansion of tools available to us, and the increasing number of investigators interested in the problem. We start with an overview of the physical methods used to investigate the mechanical properties of insect flight muscle (the biochemical and single molecule methods are covered in the chapter by Sparrow/Geeves, this volume). The physical methods are largely based on analyzing the response in muscle force to perturbations in muscle length. Next, we present a contemporary view of the contractile mechanism, based on these methods. We discuss the role of myosin in relation to its interaction with actin and other proteins of the myofilament lattice, focusing on those factors that determine muscle kinetics. One of the distinguishing characteristics of insect musculature is the extremely wide range of contractile speeds displayed by a diverse set of muscle fiber types. The chemomechanics of *Lethocerus* and *Drosophila* flight muscle are discussed and compared, highlighting recent experiments designed to elucidate the role of specific structural regions of the *Drosophila* myosin heavy chain in setting muscle fiber kinetics. We conclude the chapter with a brief discussion of the highly specialized structural proteins ancillary to myosin and actin that enable the high power output required for flight.

Methods of Measuring Mechanical Properties of Insect Muscle

The primary methods used to evaluate the mechanical properties of the insect indirect flight muscle (IFM) include (1) the evaluation of force responses to step-wise changes in length, pioneered by Jewell and Ruegg in the mid-sixties;[1] (2) large amplitude sinusoidal length perturbation analysis, expressed as work loops, pioneered by Machin and Pringle in the late fifties;[2] and (3) small amplitude sinusoidal length perturbation analysis, pioneered by Machin and Pringle in the early sixties.[3] For historical overviews of the early studies based on these methods, the reader is referred to superb reviews by Squire[4] and Tregear[5] published in the early eighties.

Step Analysis

Step analysis has been used extensively to study muscle properties since the pioneering work of A.V. Hill.[6] In insects, step-wise changes in length are the best means to demonstrate the classic stretch activation and shortening deactivation phenomenon[7] that underlies the ability of the flight muscle to generate oscillatory work and power (see chapters by Moore and Reedy,

Nature's Versatile Engine: Insect Flight Muscle Inside and Out, edited by Jim Vigoreaux.

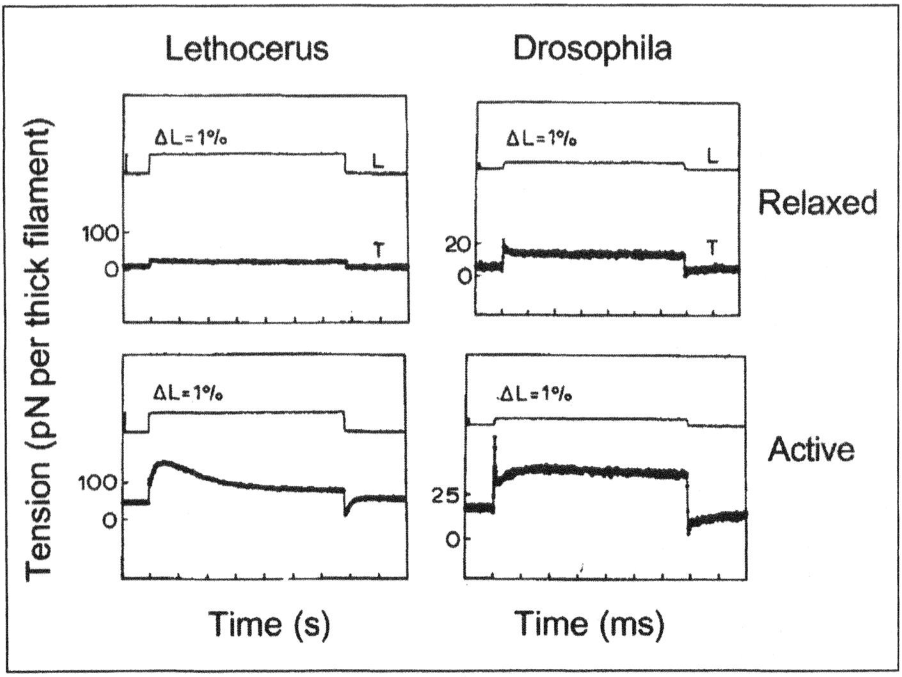

Figure 1. Force transients in response to step-wise changes in length in mechanically skinned indirect flight muscle of *Lethocerus indicus* (left) and *Drosophila melanogaster* (right), at rest (pCa 6.5, upper trace) and at maximal Ca^{2+} activation (pCa 4.5, lower trace). Note different time scales. 15°C, 10 mM [P_i]. Reprinted with permission from Peckham M, Molloy JE, Sparrow JC, et al. J Mus Res Cell Motil 1990; 11:203-215.

this volume). Figure 1 displays examples from mechanically skinned *Lethocerus* and *Drosophila* IFM,[8] which focus on the force response to step-wise stretches. In Ca^{2+}-activated IFM (lower trace), the response consists of: (1) an increase in force concurrent with stretch, (2) an abrupt fall in force after the stretch is complete, (3) a delayed redevelopment of force (i.e., the classic 'stretch activation' response), with an exponential time constant that varies widely, depending on species, and finally (4) a very slow fall in force. In relaxed IFM (Fig. 1 upper trace) the sharp increase in force that is associated with a step-wise stretch is followed by a rapid, then slower, decline in force referred to as stress-relaxation. Stress relaxation is also evident in Ca^{2+}-activated IFM as a fast initial decline and slow final decline in force that brackets the stretch activation response. The time course of stress relaxation, which resembles that of viscoelastic (rubber-like) materials undergoing rearrangement of polymeric subunits,[9] can be approximated by the expression $A(t^{-k})$ for $t > 0$. Stress relaxation originates in the viscoelasticity of the filaments that connect the Z-disc to the thick filaments, as well as in the viscoelasticity of the Z-disc, thick filaments, and thin filaments themselves.[10-12] A high frequency passive stiffness component that is removed by gelsolin, an actin depolymerizing agent, probably originates in weak-binding cross-bridges.[11]

For very small amplitudes, generally <0.125% muscle length, the response to a step decrease in length (i.e., a release) mirrors that of a step increase (stretch); that is, the response is 'linear',[13,14] and can therefore be compared to small amplitude sinusoidal length perturbation analysis (described below). An example of a linear response from glycerol-extracted IFM from *Lethocerus maximus* is shown in Figure 2A (bottom trace), where the force transient during release has the same time constant (τ) as the force transient during stretch. This response is in contrast to the asymmetrical, nonlinear, force transients seen at larger step amplitudes (Fig. 1

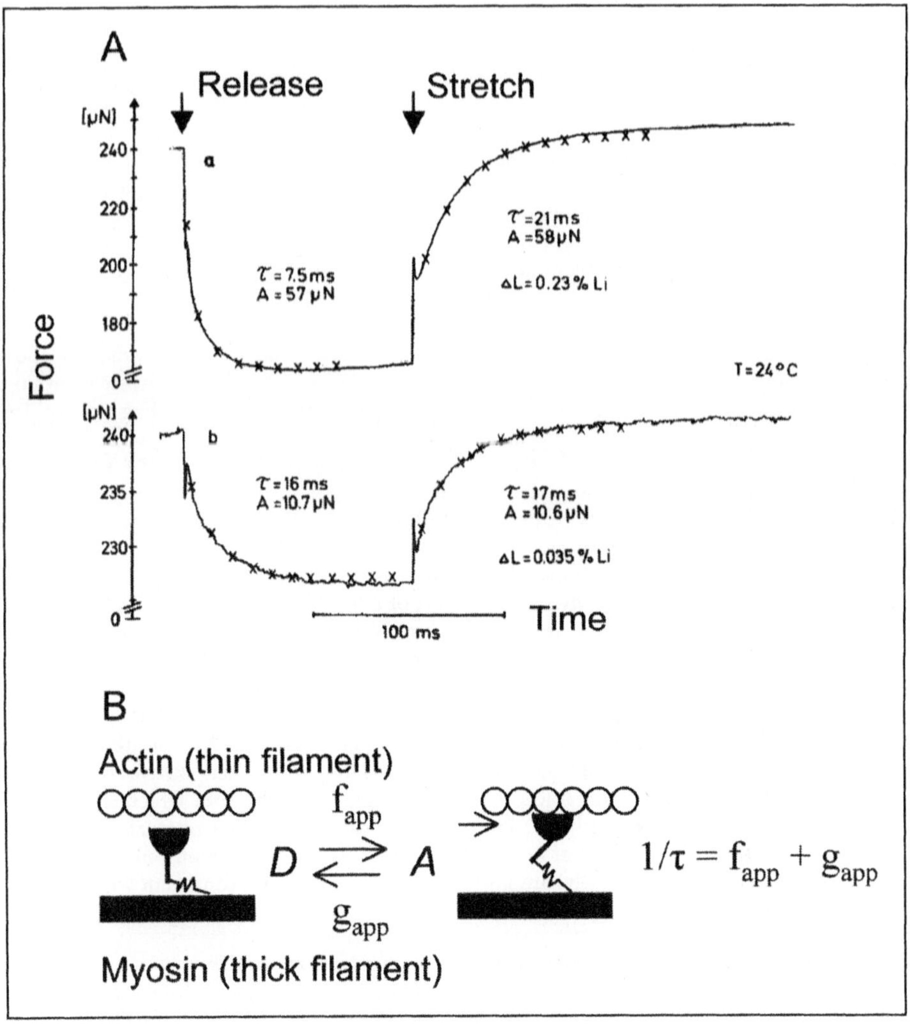

Figure 2. A) Transients at maximal Ca^{2+} activation (pCa 4.5) in glycertol-extracted indirect flight muscle fibers of *Lethocerus maximus*. First part of trace is a response to a step-wise release, the second part to a stretch. Note similar time constants (τ) for low release-stretch amplitudes (linear response, lower trace) versus dissimilar time constants for higher release-stretch amplitudes (nonlinear response, upper trace). 24°C; no added P_i. B) Two-state cross-bridge scheme. The symbol D represents the fraction of detached myosin heads; A, the fraction of attached heads. f_{app} and g_{app} are the apparent rate constants of attachment and detachment, respectively. Reprinted with permission from Guth K, Kuhn HJ, Tsuchiya T, et al. Biophys Struct Mech 1981; 7:139-169.

and Fig. 2A, upper trace). The drop in force in response to a large amplitude step reduction in length in Ca^{2+}-activated muscle is referred to as shortening deactivation.[15,16]

Step analysis is probably most useful in investigating the nonlinear properties of muscle fibers, in particular, the strain dependency of the actomyosin cross-bridge rate constants. A.F. Huxley, R. Simmons and others have studied the nonlinear properties in detail in vertebrate skeletal muscle.[17-19] In insect flight muscle, Abbott and Steiger[20] were the first to employ step analysis, followed by Guth and colleagues.[21] Like vertebrate muscle, step analysis in insect flight muscle was used to show that the delayed force response to large releases and restretches can be related to simple two-state chemomechanical models.[22] The exponential rate constant at which force increases after a release-restretch pulse can be equated to $f_{app} + g_{app}$; that is, the sum of the apparent attachment and detachment rate constants, respectively, of work-producing cross-bridges that cycle between a detached nonforce-generating state and an attached force-generating state[23,24] (Fig. 2B). Since the pioneering two-state model of Huxley,[25] mechanical studies have showed that a more complex multiple state model is required in order to account for the rather complex, rapid tension responses of muscle to sudden changes of load or length.[19,23,26] Nevertheless, two-state models are still useful as simple descriptors of the apparent rates of cross-bridge attachment and detachment in insect flight muscle. Guth and colleagues[22] even used the technique in *Lethocerus indicus* IFM to investigate the strain dependency of actomyosin ATPase activity. They found values of $f_{app} + g_{app}$ derived from the frequency dependency of ATPase activity (22.9 s^{-1} + 10.6 s^{-1} at 20°C) that were comparable to those obtained from the time course of stretch activation (28.6 s^{-1} + 2 s^{-1}, 20°C), consistent with the notion that ATP hydrolysis and mechanical work are tightly coupled in the cycling cross-bridge.

Large Amplitude Sinusoidal Length Perturbation Analysis (Work Loops)

Large amplitude perturbations generally refer to changes in muscle fiber length that produce filament axial displacements greater than the unitary step displacement of the cross-bridge, e.g., ~7.3 nm in *Drosophila melanogaster*.[27] Work loops produced by large amplitude displacements are very useful for measuring work and power output under conditions that replicate the actual muscle length changes in vivo, e.g., changes of 2-5% in length of the IFM in *Drosophila virilis*,[28] or ~34-85 nm per half sarcomere. The process by which stretch activation and shortening deactivation leads to maintained oscillatory contraction can be easily visualized by this simple but powerful technique (see chapter by Josephson, this volume). In pioneering work on the fruit fly *Drosophila melanogaster*, Peckham and colleagues[8] compared work loops generated by isolated, glycerinated IFM of this animal to those generated by IFM from the water bug *Lethocerus indicus* (reproduced in Fig. 3). Both preparations generate net positive work, but the maximum work per cycle is much less in *Drosophila* than in *Lethocerus*, which shows broad open loops, as observed previously.[29,30] However, because the frequency of maximum work of *Drosophila* is about ten times that of *Lethocerus*, maximal values of power output (~2.3 vs. ~1.4 fW per thick filament, respectively) are similar in both preparations, consistent with the flight muscle's primary role as power producer.

Small Amplitude Sinusoidal Length Perturbation Analysis

Recently, small amplitude sinusoidal length perturbation analysis has been a method of choice to investigate kinetic differences between muscle types and to deduce details of the chemomechanical events underlying kinetic differences.[31-34] While details of the sinusoidal length perturbation approach are explained elsewhere,[14,35] a brief summary of the approach and its relationship to step-wise length perturbation analysis is illustrated in Figure 4.

The characteristic force response to a step-wise change in length is given in the left panel (boxes) of Figure 4. The corresponding force response to a sinusoidal change in length of the same amplitude is given in the middle panel (top). Typically, perturbations of <0.25 % peak-to-peak length changes are applied over a range of frequencies, and the phase and amplitude relations between the applied length change, and resulting force change measured. Transient responses to both step and sinusoidal length changes occur because the rate constant of at least one reaction step is strain sensitive.[14,17,31] The resulting instability in the cross-bridge cycle causes a redistribution of cross-bridge states that produce the force transients.

Nyquist plots, an example of which is given in the lower right panel of Figure 4, are often used to display the complex moduli, i.e., indices of muscle viscoelasticity that are independent of muscle length and cross-sectional area.[14] Complex moduli are calculated by dividing the change in fiber tension (ΔF per cross-sectional area) by the fractional change in fiber length (ΔL per length). The Nyquist plot displays the complex modulus as the vector sum of the

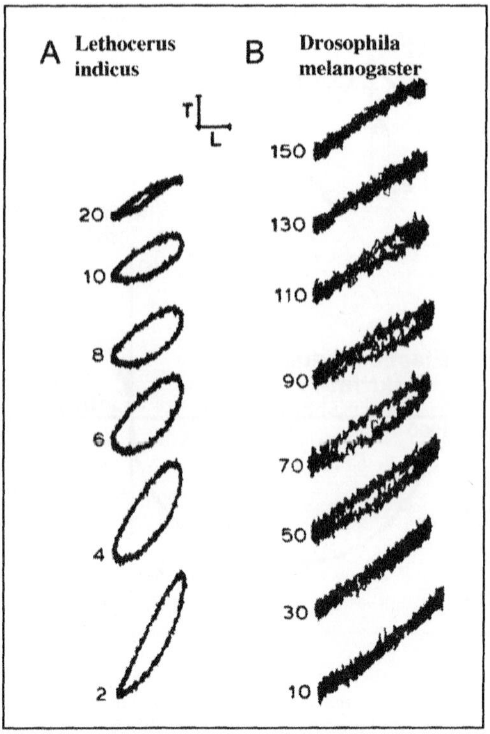

Figure 3. Work loops, i.e., plots of tension (T) versus length (L), at the stated frequencies of oscillation (Hz) in Ca^{2+}-activated, mechanically skinned *Lethocerus indicus* (panel A) and *Drosophila melanogaster* (panel B) indirect flight muscles. Calibration: T, 30 pN per thick filament; L, 0.2%. Stretch and release cycles carried out at frequencies at which oscillatory power is maximal in intact muscles create a loop in the stress-strain plot (traversed counterclockwise), where tension (stress) is greater during shortening than during lengthening. 15°C, 10 mM [P_i]. Reprinted with permission from Peckham M, Molloy JE, Sparrow JC, et al. J Mus Res Cell Motil 1990; 11:203-215.

viscous modulus (force response out-of-phase with the length change) and the elastic modulus (force response in-phase with the length change). The length of the vector (arrow) represents the magnitude of the complex modulus. The angle (θ) measured from the X-axis counterclockwise represents the phase advance of force with respect to length.

There are a variety of ways in which one can extract components of the Nyquist plots, to which physical-chemical correlates can be ascribed.[31,35] One approach is illustrated in Figure 4, where three components have been extracted using a curve-fitting model.[36] Similar models have been applied to mouse[37] and human[35] cardiac muscle. The Nyquist plot is deconvolved into a nonexponential component *A* (straight line), roughly equivalent to stress-relaxation $A(t^{-k})$ in the time domain (for t >0), and two exponential processes *B* and *C* (semi-circles), equivalent to $-B(e^{-2\pi b t})$ and $C(e^{-2\pi c t})$ in the time domain (left panel, Fig. 4). Thus the characteristic frequencies of *B* and *C* (closed and open squares, respectively) are related to the exponential time constants by $\tau_3^{-1} = 2\pi b$ and $\tau_2^{-1} = 2\pi c$. Process *B* corresponds to 'phase 3', process *C* corresponds to 'phase 2' and the high-frequency component of *A* corresponds to 'phase 1' of the tension response to a step change in length according to the Huxley-Simmons nomenclature.[17,31,38]

Processes *B* and *C* have been identified with transitions between cross-bridge states on the basis of their high Q_{10} (temperature coefficient) of the rate constants and marked dependency of both rate and amplitude on [MgATP], [MgADP], and [Pi].[31,35] Component *A*, on the other

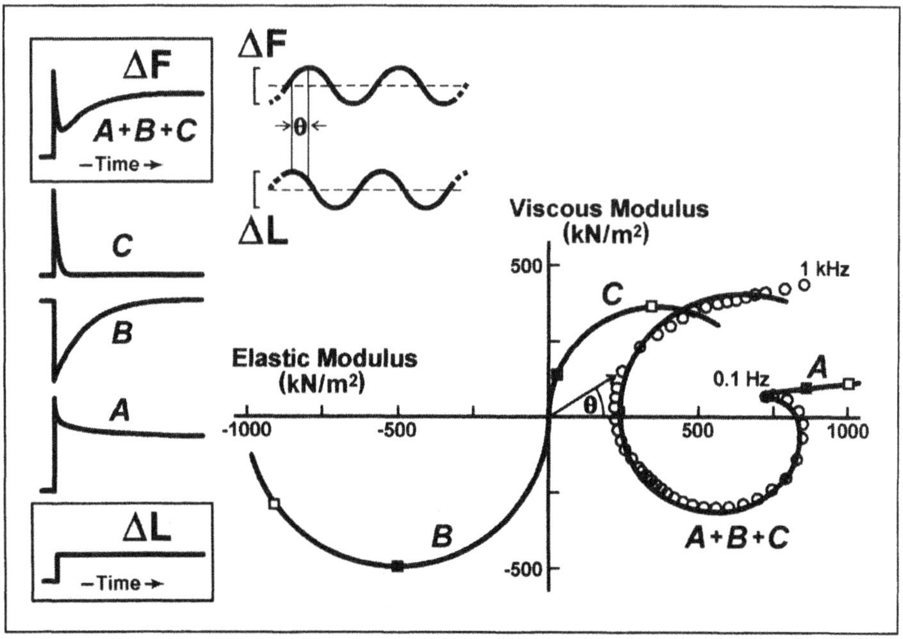

Figure 4. Small amplitude step-wise (left) and sinusoidal (right) length perturbation analysis, shown schematically. The force response to step-wise stretch can be deconvolved into three separate processes, *A*, *B* and *C*. Process *B* and *C* are exponential while process *A* is nonexponential. Process *A* contributes primarily to the time course of the initial phase of the rapid decay and the final slow decline in force. Process *B* contributes primarily to the time course of redeveloped force, and process *C* contributes primarily to the later phase of the rapid decay preceding force redevelopment. The force response to sinusoidal length perturbations can be deconvolved similarly. The complex modulus obtained from sinusoidal analysis shown at right can be expressed as $y(f) = A\,(i2\pi f/\alpha)^k - B\, if/(b + if) + C\, if/(c + if)$, which is the inverse Fourier transformation of the step response $F(t) = A(t^{-k}) - B(e^{-2\pi bt}) + C(e^{-2\pi ct})$ shown at left ($t > 0$; $i = \sqrt{-1}$, $\alpha = 1$ Hz, and k = a unitless exponent). Coefficients A, B and C are the magnitudes of *A*, *B*, and *C*, respectively (in units of mN per mm^2 fiber cross-sectional area). The characteristic frequencies of *B* and *C* are *b* and *c* (in units of $s,^{-1}$ or Hz). Exemplar data (open circles) is given for a Ca^{2+}-activated (pCa 4.5) skinned IFM (*D. melanogaster*) in the Nyquist plot. The curve fit returned constants A = 657 kN $m,^{-2}$ B = 978 kN $m,^{-2}$ C = 723 kN $m,^{-2}$ k = 0.07, b = 53 $s,^{-1}$ c = 537 $s.^{-1}$ The closed squares denote values at the characteristic frequency *b*; the open squares, values at *c*. The open circles denote values obtained from a frequency sweep. Under linear conditions (small amplitude length perturbations) the time constants of exponential processes in the time domain can be directly related, at least theoretically, to the apparent rate constants of exponential processes in the frequency domain.[14,36]

hand, has been identified with passive viscoelastic elements,[36,37] with a rate constant that has a relatively low Q_{10} (like that of rubber). A detailed analysis of component *A* has revealed both a fixed (A_{nonCa}) component associated with structural elements of the myofibril outside the actomyosin cross-bridge, and a Ca^{2+}-dependent (A_{Ca}) component that emerges when cross-bridges form and is thus associated with the structural "backbone" of the cross-bridge itself.[35]

Fundamentals of Cross-Bridge Kinetics

The general features of one of the most widely accepted chemomechanical models of muscle contraction are illustrated in Figure 5. The biochemical reaction scheme depicted is based largely on work carried out on vertebrate striated muscle.[31,39,40] The scheme is appropriate for striated muscles undergoing oscillatory work during steady state contractions, for which the rate-limiting (or slowest) step of the cross-bridge cycle (indicated by the bold

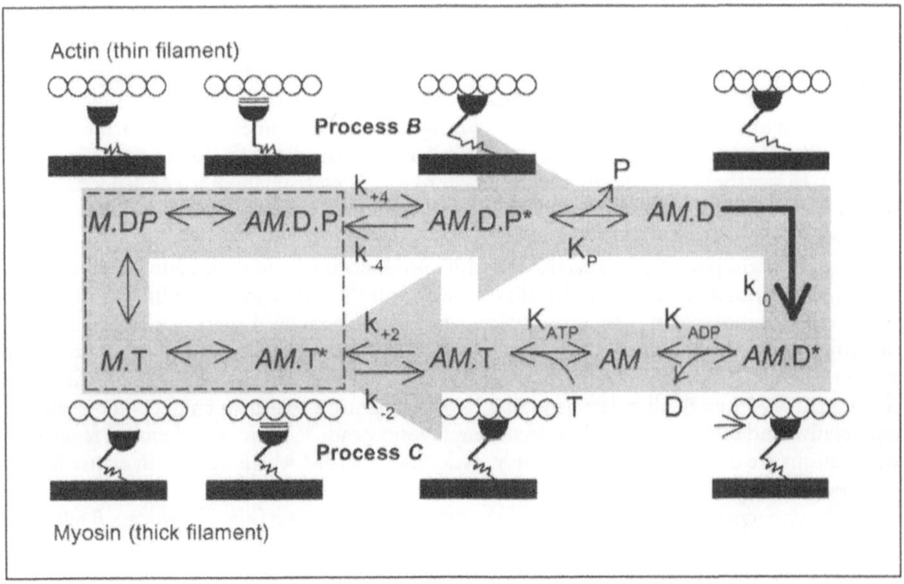

Figure 5. Cross-bridge kinetic scheme relating muscle work production (process *B*) and work absorption (process *C*) to actomyosin cross-bridge cycle. *A* is actin, *M* is myosin, T is ATP, D is ADP, and P is inorganic phosphate. The asterisk indicates a different conformational state. $k_{\pm 2}$ and $k_{\pm 4}$ are unidirectional rate constants, K_P is the myosin phosphate affinity constant, K_{ADP} is the MgADP affinity constant, and K_{ATP} is the myosin MgATP affinity constant.[31] The hatched box denotes weakly bound or detached states that cannot be distinguished on the basis of dynamic stiffness (complex modulus) measurements. The apparent rate constants of processes *B* and *C* are $2\pi b$ and $2\pi c$, respectively. For $b<<c$ (as in the example of Fig. 4), $2\pi b \sim [\eta\, k_2/(\eta\, k_2 + k_{-2})]\, k_4 + \xi\, k_{-4}$, and $2\pi c \sim \eta k_2 + k_{-2}$, where $\eta = K_{ATP}[T] / (1 + K_{ADP}[D] + K_{ATP}[T])$ and $\xi = K_P[P] / (1 + K_P[P])$.[31]

arrow) precedes ADP release, as appears to be the case for most vertebrate skeletal and cardiac muscles.[31,41,42] The individual chemical reactions that characterize the transitions between cross-bridges states have been investigated primarily in solution and, to a lesser extent, in fibers stripped of their membranes.

The fundamental mechanism, as developed primarily from the vertebrate studies, is as follows: Myosin (with the products of MgATP hydrolysis) binds to actin via weak ionic interactions at specific residues at the myosin-actin interface (*M*.DP to *AM*.D.P in Fig. 5). Calcium-regulated movement of thin filament proteins allows strong hydrophobic interactions to occur between myosin and actin, thereby creating a much tighter bond.[32,43] A major conformational change associated with strong binding (*AM*.D.P to *AM*.D.P*) and release of phosphate from the nucleotide binding site (*AM*.D.P* to *AM*.D) occurs within the myosin head.[31,44] The resulting strain is conveyed through the so-called 'converter' region to the end of a 20-kDa lever arm,[45,46] the long α-helix (cloaked by the two light chains) that extends from the motor domain to the rod portion of the myosin molecule. Concurrent with, or following this force-generating step, MgADP is released (*AM*.D. to *AM* in Fig. 5), followed by MgATP rebinding (*AM* to *AM*.T), myosin head detachment from actin (*AM*.T to *M*.T), and MgATP hydrolysis (*M*.T to *M*.D.P). Completing the cycle, the myosin head reattaches to actin (*M*.D.P. to *AM*.D.P) as soon as a binding site becomes available again. The process is stochastic, with billions of myosin heads within each muscle fiber operating more-or-less independently as they complete each cycle. The thin-filament based Ca^{2+}/troponin/tropomyosin system regulates the availability of binding sites, as may strain on attached crossbridges[47] (see chapter by Reedy, this volume).

Atomic models of the myosin head[45] docked to actin have been rebuilt using high-resolution tomographic electron micrographs of Ca^{2+} activated *Lethocerus* indirect flight muscle.[48] The reconstructions suggest that, in insect IFM, cross-bridges maintain a roughly perpendicular orientation to actin under isometric conditions. Thus, generation of force in insect flight muscle may occur with little or no lever-arm tilt during the pseudo-isometric phase at the peak of lengthening during the wing-beat, an action similar to the bending of a fishing pole (see Reedy chapter). In vertebrate striated muscle the extent to which a sarcomere shortens depends on the size of the load. In IFM, the in vivo change in sarcomere spacing during the wing beat is relatively constant and slight (2-5% of sarcomere length).

Figure 5 highlights the way MgATP and its products compete with actin for binding to myosin. The primary mechanism that drives the cycle is the difference in binding affinities and free energies of moieties of these three molecular species. The rugged energy landscape is one of potential wells whose barrier heights can be modulated by a variety of chemical and steric factors (chemical: availability of Ca^{2+} and $MgATP^{2-}$; steric: proximity of target sites and filament strain). Although all striated muscles have mechanisms that help confer some degree of cooperative and synchronous behavior among myosin heads,[49] steric and chemical regulatory mechanisms are of crucial importance in muscles like the IFM which operate in an oscillatory and temporally precise fashion.

In the conventional interpretation of the cross-bridge scheme (Fig. 5), process *B* reflects a transition from a weak electrostatic attachment of the myosin head to actin, to a strong hydrophobic attachment, followed by a conformational change in the myosin head that results in force, work, and power production.[31] Process *C* reflects a reverse sequence, that is, a conformational change in the myosin molecule and a transition from strong to weak binding that results in an exponential decay of force.[13,31] These assignments are still provisionary, however, as other interpretations merit attention. For example, there is evidence that during small amplitude sinusoidal length oscillations processes *B* and *C* derive primarily from, respectively, ambient thermal energy and externally-supplied mechanical work.[13] That is, during the lengthening (work absorbing) phase, the energy that drives the transition from strong to weak binding derives from external mechanical work, while during the shortening (work producing) phase the weak binding cross-bridges capture thermal energy to reform the strong binding state and generate force.[13] Process *C* has also been explained recently as an intermittent attachment and detachment of myosin heads to actin (modeled as a Hookian stiffness effective only during attachment) whose rates are governed stochastically by independent probability functions.[50]

Chemomechanics of *Lethocerus* Flight Muscle

Studies on skinned IFM of *Lethocerus*, like those on skinned fibers from vertebrate muscles, have generally supported the hypothesis that force generation is closely associated with phosphate release. Elevating inorganic phosphate (P_i) concentration strongly depresses maximal isometric force[51,52] while enhancing both the rate of force development,[51] the frequency of optimal oscillatory work production,[51] and the rate of force decay from rigor.[53] Phosphate-water oxygen exchange studies of *Lethocerus* flight muscle strongly suggests P_i release may be rate limiting during production of oscillatory work,[54] rather than a step associated with MgADP release as shown in Figure 5.

In *Lethocerus* flight muscle, as in vertebrate striated muscle, MgADP has an opposite effect to P_i. Isometric tension increases with added MgADP, while the frequency of optimal oscillatory work production decreases.[55] Through mass action, MgADP shifts the reaction to the tension-generating A.M.D state, but MgADP will also inhibit cross-bridge cycling by competing with MgATP for the A.M. state. Generally, reducing MgATP concentration has a similar effect of increasing MgADP concentration,[31] suggesting that the A.M.D and A.M states have similar levels of free energy.

Chemomechanics of *Drosophila* Muscle Myosin

Drosophila is currently the best system to study insect muscle chemomechanics, especially the role played by the myosin heavy chain. The availability of myosin nulls specific for flight and jump muscles, the ability to transgenically express different myosin isoforms and chimeras in these nulls (see chapter by Miller and Bernstein, this volume), and the advanced tools available to examine the transformed muscles and myosins (see Sparrow and Geeves chapter), all contribute to *Drosophila*'s excellence as a model organism.

For studying myosin chemomechanics, *Drosophila*'s most compelling advantage may be the manner by which the different isoforms of the myosin heavy chain are expressed in this species. *Drosophila* has a diverse array of MHC isoforms generated through alternative splicing of mRNA transcripts from the single copy *Mhc* gene.[56,57] There are 5 sets of alternatively spliced exons in *Mhc*, four in the head region alone. This contrasts with vertebrates where most MHC isoforms are each coded by a specific gene. Fifteen myosin isoforms have been identified to date in *Drosophila*, expressed in a wide variety of fiber types, including the slow supercontractile embryonic body wall muscles and the extremely fast indirect flight muscle.[58,59]

To directly test the role of myosin in setting mechanical and energetic properties of muscle, an embryonic body wall muscle myosin (EMB) was transgenically expressed in IFM fibers of an IFM myosin null[60] and compared to transgenic control fibers expressing wild type *Mhc* in IFM (IFI).[61] This substitution of EMB for the native fast myosin in *Drosophila* flight muscles transformed the IFM from high power-generating muscles that perform optimally at high oscillation frequencies, to ones that produce less power and function best at low frequencies (compare blue and red traces in Fig. 6). Instead of high power, EMB myosin appears adapted for longer more forceful contractions as EMB expressing fibers generated 3-fold higher calcium-activated isometric tension and two-fold higher maximum work.

Recent studies have attempted to correlate important functional properties of isolated myosin with the mechanical performance indices of skinned IFM to uncover details of the myosin cross-bridge mechanism versus attributes contributed by the presence of the organized myofilament lattice. Swank and colleagues[62] found that the in vitro actin sliding velocity on isolated myosin of the IFI isoform type is among the fastest reported for a type II myosin (6.4 $\mu m\ s^{-1}$ at 22°C) and is 9-fold greater than EMB velocity at the same temperature. The velocity difference is similar to the 8-fold difference in frequency of optimal power output in fibers,[61] suggesting both parameters are highly influenced by a similar underlying myosin cross-bridge property.

The unloaded velocity of actin sliding on myosin is equal to d_{uni}/t_{on}, where d_{uni} is the distance a single molecule moves actin per cross-bridge cycle, and t_{on} is the total time strongly attached to actin.[46] The results of optical trap experiments showed no difference in step size (d_{uni}).[62] Thus, the pronounced differences in actin sliding velocity between EMB and IFI must be due to differences in myosin cross-bridge kinetics (i.e., differences in t_{on} that are thought to be set primarily by rates of transitions between strongly bound myosin states).[46] In addition to cross-bridge kinetics, mechanical differences could also be influenced by differences in myosin stiffness as the fiber is subjected to a variable load during oscillatory contractions. Measurements of myosin stiffness are not yet accurate enough to determine if there is a cross-bridge stiffness difference between IFI and EMB.

Actin-activated ATPase assays revealed a 2-fold kinetic difference between EMB and IFI myosin.[27] The two-fold difference, however, is much less than the 9-fold difference in velocity. This suggests that the rate of the cross-bridge transition that limits velocity must be sped up more than a transition that limits ATPase activity in IFI relative to EMB. The limiting step for motility velocity is a transition between strongly bound states (states AM.D.P* to AMT in Fig. 5).[46] IFI ATPase rate (representing the average time a myosin takes to go through the entire cross-bridge cycle) cannot be limited by the same strongly bound state; instead, a slower transition involving weakly bound states must limit IFI ATPase rate (weakly bound states do not affect velocity, AMT* to AMD.P. in Fig 5).

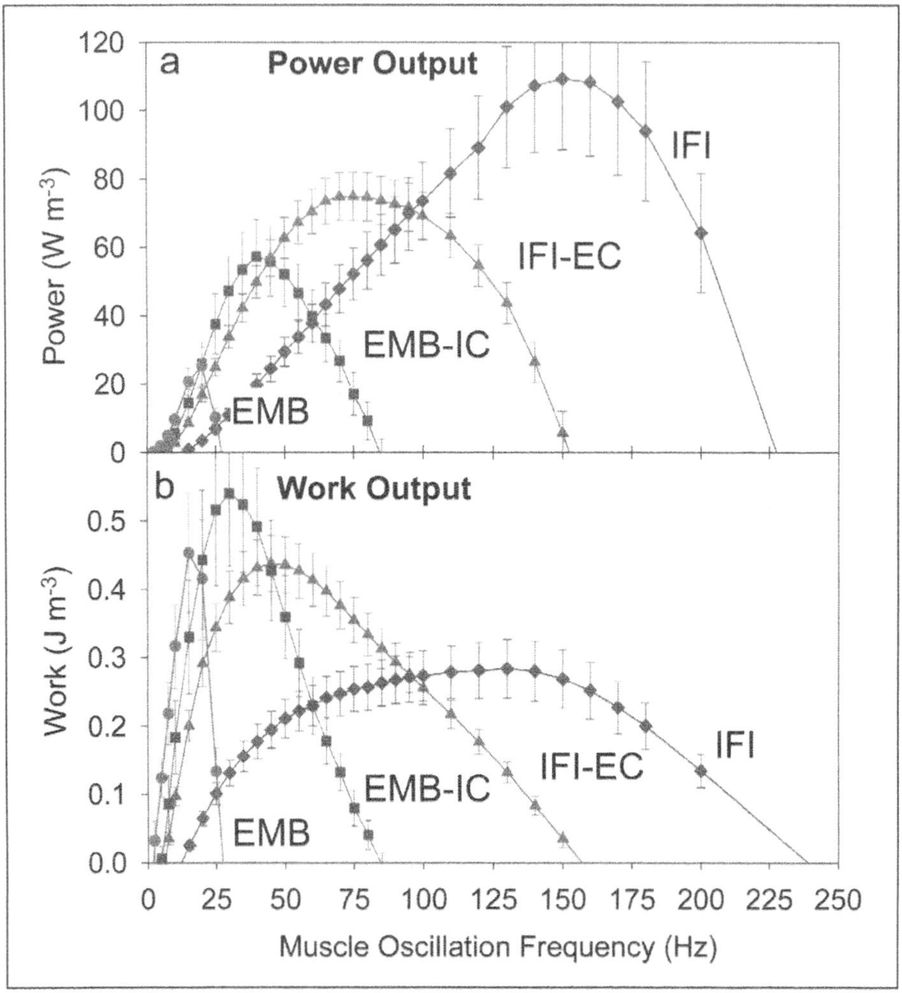

Figure 6. Substitution of EMB (red) for IFI (blue) muscle myosin is akin to converting the IFM from a fast to a slow fiber type. a) Maximum power output of EMB fibers is only 25% of IFI power at 15% of IFI fibers' optimal oscillation frequency of 150 Hz, thereby explaining the loss of flight ability in the fly lines expressing EMB. b) Maximum oscillatory work, however, is almost 2-fold higher from EMB fibers compared to IFI. EMB skinned fibers produce more work than IFI fibers at low frequencies, whereas IFI fibers generate more work at frequencies above 100 Hz. Fibers expressing chimeric myosins (green and purple traces), created by exchanging converter regions between IFI and EMB, had a large impact on power and work generation. EMB-IC is the embryonic myosin isoform with the exon 11 converter from IFI. IFI-IC is the IFM isoform with the exon 11 converter from the embryonic isoform. Adapted from Swank DM, Knowles AF, Suggs JA et al. Nat Cell Biol 2002; 4(4):312-316. A color version of this figure is available online at http://www.Eurekah.com.

The contrasting ATPase and motility differences also suggest that IFI myosin spends proportionally much less time strongly bound to actin of its total cross-bridge cycle time, referred to as a myosin's "duty ratio", than EMB. Littlefield and colleagues[27] calculated a 7-fold lower duty ratio for IFI compared to EMB myosin. A lower duty ratio for IFI is consistent with the lower isometric force and work levels observed in IFM fibers expressing IFI compared to IFM expressing EMB,[61] because isometric force is proportional to the number of cross-bridges × unitary stiffness of each cross-bridge × d_{uni} × duty ratio. The low IFI duty ratio may be an adaptation to not only reduce ATP use (by keeping IFI ATPase rate similar to the low EMB rate) but also to maintain high power generation by having a small number of heads at any given moment transition very rapidly through the strongly bound (power producing) cross-bridge states.

The Role of the Myosin Heavy Chain Alternative Exons in Setting Fiber Kinetics

Drosophila's method of alternative splicing to generate myosin isoforms has allowed researchers to locate and test regions of the myosin heavy chain (MHC) that determine the fundamental chemomechanical properties of the contractile system. The focus of recent mechanical studies has been the four variable regions in the head Figure 7.

The region with the greatest influence on *Drosophila* muscle mechanical properties is the converter region, encoded by exon 11. Exchanging converter regions between IFI and EMB and expressing these myosin chimeras in IFM had a large effect on isometric force, work, and power Figure 6. Notably, changes at the fiber level correspond to molecular level alterations in both actin-activated ATPase and actin sliding velocity, suggesting that both weak and strong binding rate-limiting steps can influence fiber kinetics and that both rate limiting steps are influenced by the converter region.[61] Surprisingly, the ATPase rate and actin velocity associated with myosin chimera IFI-EC (the embryonic converter exchanged into IFI) changed in directions opposite to that of the IFI parent isoform, demonstrating that the underlying rate limiting steps determining ATPase and velocity can be "uncoupled".

Preliminary results from fibers expressing a MHC chimera in which the embryonic version of exon 9 was exchanged into IFI (chimera IFI-ER) suggest that the exon 9 region has, after the converter region, the second greatest influence on the kinetics of power generation (Fig. 7). Exon 9 encodes a region of the molecule termed the relay loop. A large effect of the exon 9

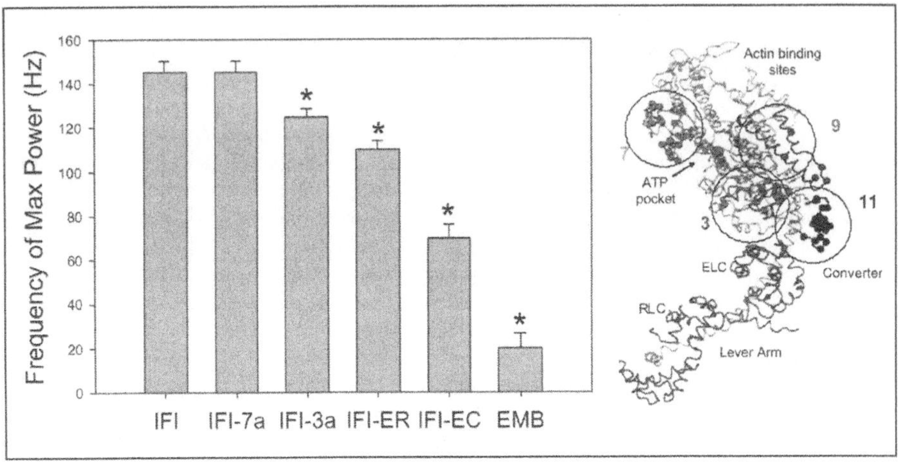

Figure 7. Maximum power output of IFM fibers expressing IFI, EMB or IFI myosin with one of the four alternative exon versions replaced by the equivalent EMB version. IFI-7a is the IFM myosin isoform with the exon 7 region from the embryonic isoform. IFI-3a is the IFM myosin isoform with the exon 3 region from the embryonic isoform. IFI-ER is the IFM myosin isoform with the exon 9 region from the embryonic isoform (the relay loop region). IFI-IC defined as in Figure 6. Asterisk denotes significant difference from IFI. The closely related 3D structure of the motor domain of chicken myosin is shown at right.[89] The circles denote the alternatively spliced regions encoded by exons 3, 7, 9 and 11 of the *Drosophila Mhc* gene. A color version of this figure is available online at http://www.Eurekah.com.

region on power generation and other mechanical parameters supports the hypothesis that interactions between the relay loop and converter are critical in transmitting structural changes at the nucleotide binding site to movement of the level arm.[63]

Exon 3 encodes a near N-terminal region that is located between the reactive sulfhydrils and the pivot point of the lever arm (Fig. 7). Experiments with IFM fibers expressing chimeras of the exon 3 region exchanged between the IFI and EMB isoforms (chimera IFI-3a) revealed a substantial influence on the frequencies at which maximum work and maximum power are generated.[64] Surprisingly, the molecular ATPase and actin velocity studies on these same chimeras[65] predict a substantial change in duty ratio (see chapter by Miller and Bernstein, this volume), however, there was no change in isometric force as seen with the exon 11 exchange when duty ratio was altered. This suggests exon 3 also influences myosin stiffness, which could offset the duty ratio's influence on isometric force.

Preliminary data from fibers expressing a myosin chimera with the EMB exon 7 exchanged into IFI (chimera IFI-7a) suggests this region may exert the least influence on muscle mechanical properties (Fig. 7). This is surprising because the exon 7 region is closest to the site of ATP hydrolysis.

Myosin Cross-Bridge Rate Constants

To investigate rates of specific steps of the cross-bridge cycle that are different between *Drosophila* myosin isoforms and those that are influenced by *Mhc* alternative exons, sinusoidal analysis has been conducted on skinned IFM under conditions in which Pi, ATP and ADP concentrations were varied.[66] Remarkably, the IFI fiber's response to Pi is *qualitatively* different than that of the typical vertebrate muscle fiber under conditions thought to exist in the intact fiber. In particular, tension *increases* and the optimal frequency of work production *decreases* with increasing Pi (Fig. 8), in contrast to all vertebrate studies reported to date in which tension *decreases* and the frequency of maximum oscillatory work output *increases* with increasing Pi.[31] With slow embryonic *Drosophila* myosin (EMB) expressed in IFM, the responses to altered Pi, ATP, and ADP qualitatively matched those of the typical vertebrate, suggesting that EMB conforms to the same standard biochemical scheme as its vertebrate counterpart. Thus, rate constants for some steps in the biochemical scheme for IFI must be very different, or some other substantial difference occurs in the cross-bridge cycle of IFI, compared to EMB and vertebrate myosins studied to date.

Further evidence that cross-bridge rate constants for at least some *Drosophila* myosins are dramatically different from the standard vertebrate model comes from recent transient kinetic solution studies.[67] The rate-limiting step for velocity of muscle shortening and work production is thought to be ADP release rate.[46] However, an investigation of S-1 fragments (catalytic and light chain binding region) of four *Drosophila* myosin isoforms[67] found that S-1 ADP off-rate does not correlate with previously measured actin motility velocity on the same set of myosins.[61] Miller and colleagues[67] further showed that the steps following ADP release (i.e., the 2nd order rate constant for ATP binding to actomyosin and myosin detachment from actin) were unlikely to be limiting, and thus hypothesized that ADP release and myosin detachment rate are not limiting actin filament velocity in *Drosophila* myosin isoforms.[62] Miller and colleagues[67] suggest that evolutionary pressure on a single muscle myosin gene may maintain a fast detachment rate in all isoforms, given the extremely fast nature of the IFM. As a result, the attachment rate and completion of the power stroke (steps associated with Pi release) or the equilibrium between A.M.D states (Fig. 5) may define actin filament velocity for these myosin isoforms. Thus, there is mounting evidence that some *Drosophila* myosins, at least, possess rate-limiting steps that do not conform to conventional cross-bridge kinetic schemes.

In this context it will be interesting to determine if other very fast muscle types respond to Pi in a manner similar to that of *Drosophila* IFM. Studies on toadfish swimbladder and rattlesnake shaker muscles (two of the fastest known in vertebrates) suggest very fast myosin detachment rates, that probably require a very fast rate of ADP release.[68,69] Preliminary studies using

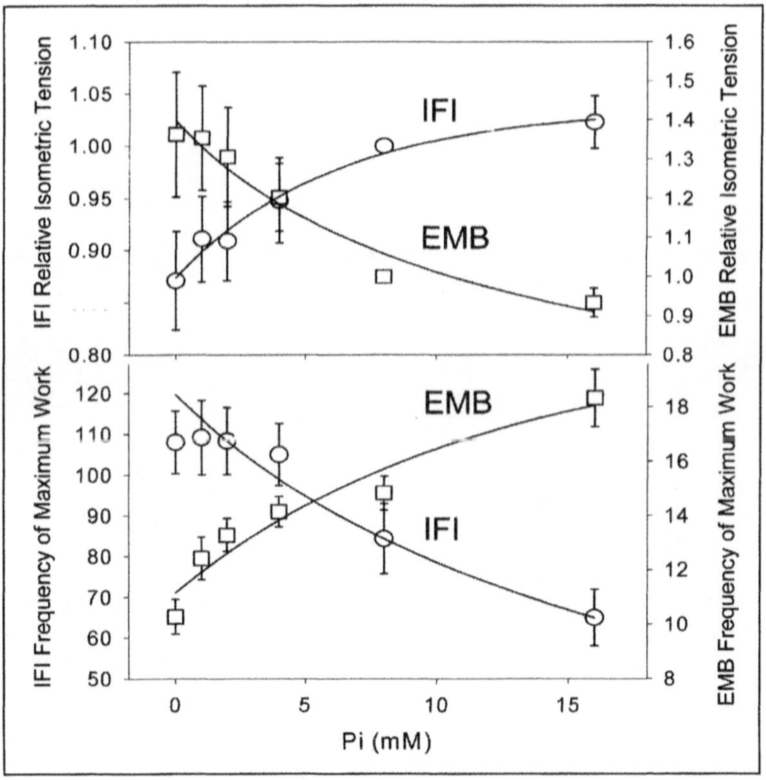

Figure 8. The effect of phosphate on isometric tension and frequency of maximum work production of *Drosophila* indirect flight muscle fibers expressing IFI and EMB myosins. Tension normalized to the level measured at 8 mM phosphate (note different y-axis scales). At pCa 4.5, 8 mM [Pi], and 5 mM [MgATP], isometric tension was 1.2 kN m^{-2} for IFI and 3.4 kN m^{-2} for EMB. Graphs reprinted with permission from: Swank DM, Maughan DW. Rates of force generation in *Drosophila* fast and slow muscle types have opposite responses to phosphate. In: Sugi H, ed. Molecular and Cellular Aspects of Muscle Contraction. New York: Plenum Publishing Co., 2003:459-468.

flight muscle of *Lethocerus* show similar responses of tension and frequency of maximum work output to phosphate as slow vertebrate muscle, suggesting the unusual Pi dependency seen in *Drosophila* IFM is not a generic property of all insect flight muscles. *Lethocerus* muscle oscillation frequency is one tenth that of *Drosophila*, similar to EMB fibers, suggesting a functional requirement for speed may be the critical evolutionary pressure.

Roles of Other Sarcomeric Proteins That Influence Chemomechanics

Figure 9 summarizes the structural relationship of myosin, actin, and other proteins of the myofilament lattice in *Drosophila* IFM. Previous sections have highlighted various regions of myosin that tune cross-bridge kinetics. In this section we briefly discuss other proteins of the IFM myofilament lattice that, in addition to their primary functions as structural and regulatory proteins, have certain generic features and, in some cases, evolved specializations, which help enhance oscillatory work and power output. The proteins illustrated in Figure 9 are discussed in detail in other chapters in this volume.

The myosin *essential* and *regulatory light chains* hug the α-helical backbone of subfragment 1, reinforcing and stiffening the lever arm that extends from the motor domain. A stiffer lever

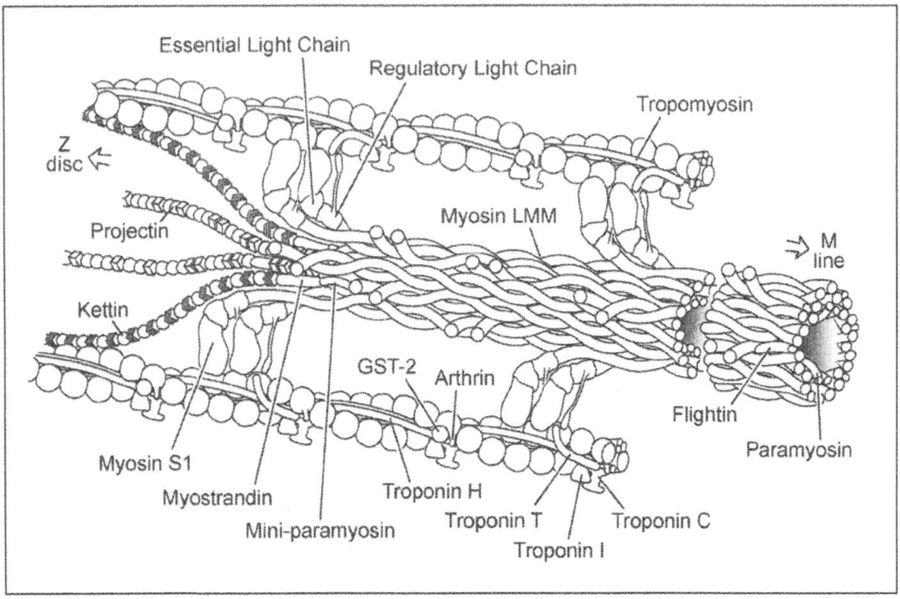

Figure 9. Schematic illustration of a portion of the *Drosophila* myofilament lattice at the A-I band junction. The ~200 kDa myosin molecules (the heavy chains) form dimers, comprised of two heads (S1) and a rod that forms the backbone of the thick filament. Two calmodulin-like molecules, the essential and regulatory light chains, stabilize the lever part of each motor domain. GST-2 (glutathione S-transferase) is a thin-filament associated enzyme believed to protect myofibrillar proteins from oxidative damage during flight.[90] Other proteins discussed in text. Reprinted with permission from: Maughan DW, Vigoreaux J. Nature's strategy for optimizing power generation in insect flight muscle. In: Sugi H, ed. Mysteries About the Sliding Filament Mechanism: Fifty Years after Its Proposal. New York: Kluwer/Plenum Publishing Co., 2004, in press.

enhances transmission of torque and power to the myofilaments. While little is known of the essential light chain's role in *Drosophila* outside this function, much is known about the modulatory features of the regulatory light chain (RLC). Disruption of the two conserved myosin light chain kinase-dependent phosphorylation sites on *Drosophila* RLC (at serine residues 66 and 67) markedly alters flight ability by reducing oscillatory power output of the IFM.[70] Sinusoidal length perturbation analysis[36] combined with in vivo X-ray diffraction studies[71] established that the reduced power output is due primarily to a reduced recruitment of cross-bridges into the work-producing pool, rather than to changes in the kinetics of actively-cycling cross-bridges.

IFM oscillatory power output in *Drosophila* is also enhanced by an N-terminal extension of the RLC that has an almost exact counterpart in vertebrate ELC.[72-74] In flies lacking nearly all 48 amino acids of the N-terminal extension, skinned IFM near in vivo lattice spacing exhibit significantly reduced net oscillatory work output at submaximal calcium activation,[75] a deficit that could account for an observed flight impairment.[76] It is possible that a N-terminus RLC-actin interaction exists, similar to that originally reported in vertebrate ELC,[77,78] shown later to be between a cluster of lysines near the ELC N-terminus and specific C-terminal residues of the actin monomer.[79,80] It is tempting to speculate that, like phosphorylation of RLC residues 66 and 67,[71] the N-terminal extension helps maintain a critical inter-filament spacing,[75] as proposed for vertebrate striated muscle ELCs,[73] and/or prepositions the head for optimum force production and oscillatory work output near the target site actins.

The primary structural element of the thick filament is *light meromyosin (*LMM), i.e., the long coiled-coil segment of the myosin molecule, with periodic charges that electrostatically bind one LMM alongside another. Laid alongside each other, the LMM reinforce and stiffen the thick filament. *Paramyosin,* which probably lies alongside LMM in the thick filament, also contributes to filament stiffness. Recent evidence demonstrates that an alanine-for-serine substitution in at least one of four phosphorylation sites (S9, S10, S13 and S18) reduces passive and dynamic stiffness,[81] leading to reduced stretch activation, power output, and flight ability. Additional stiffening of the thick filament probably results from paramyosin annealing adjacent LMMs along their length. *Flightin,* a unique 20 kDa protein associated with the thick-filament, may be responsible for cross-linking LMMs to one another by bridging myosin pairs through ionic bonds, possibly at the S2 hinge region of the myosin molecule.[82] A flightin null and a myosin rod mutant (R1559H) with low levels of flightin accumulation show marked reductions in passive and dynamic stiffness, commensurate with an almost complete loss of stretch activation and power output.[12]

Filamentous proteins of the titin family also contribute to passive stiffness of the sarcomere. *Projectin*[83] and *kettin*[84] link the thick filaments to Z-line proteins. Both contribute to the high passive stiffness of *Drosophila* flight muscle,[85] but their contributions to the enhancement of stretch-activation and effective power transmission are not well understood. Kettin at least appears to share the same close association that titin does with the thin filament in vertebrate muscle,[85] suggesting a role in stiffening the thin filament in the short I-band, thereby boosting oscillatory work and power output by reducing viscoelasticity of these elements. Passive tension borne by projectin and kettin may also be a prerequisite for the formation of both weak-binding and strong-binding cross-bridges in insect flight muscle.[11,86]

Various proteins also structurally reinforce thin filaments. A short, stiff connection from the ensemble of working cross-bridges to the Z-band, manifest as a very narrow I-band in the IFM, also provides a crucial link for effective power transmission. In addition to its calcium-dependent regulatory role, *tropomyosin* (TM) stiffens the thin filament by continuously running alongside the actin monomers.[87] *Troponin H* (TnH), an elongated TM fusion protein that is roughly two-thirds longer than the standard TM, either lies alongside or periodically replaces TM. Mutations in both standard tropomyosin and TnH establish their contribution to sarcomere stiffness and generation of power.[88] IFM from mutants with reduced expression of Tm and TnH exhibit significantly lower dynamic stiffness (complex modulus) and produce less power that wild type flies or flies rescued with the wild type standard tropomyosin gene, and IFM from both were compromised structurally, indicating both proteins are required for normal myofibril assembly. Surprisingly, the alterations were not nearly as severe in the TnH mutants as in the TM mutants at comparable levels of expression.[88] *Arthrin* is a ubiquitinated form of actin, whose role, if any, in reinforcing the thin filament and sarcomere has yet to be established (see chapters by Leonard and Bullard, and Sparrow, this volume). Arthrin, like flightin, is absent in vertebrate muscle, again highlighting the unique evolutionary track insects took to achieve enhanced power generation in their flight muscles.

Concluding Remarks

It is evident that we are now closer to understanding the fundamental mechanisms (if not all the details) of insect flight muscle contraction. There is considerable work ongoing, not only in uncovering details of the transition states of the cross-bridge kinetic scheme and the steps that set muscle fiber speed, but also in other areas involving proteins and protein complexes that assist in transmitting power to the ends of the muscle and, eventually, to the wings.

The tool that has proved most useful for studying the mechanical properties of insect flight muscle, even at the molecular level, has been the application of length perturbations and analysis of resultant force responses. This approach is based on the everyday principle that every child appreciates - that is, to begin to understand what is hidden inside a box, simply shake it. Step-wise and sinusoidal length perturbation analyses have been successful in producing a variety of physical and mathematical models, some simple (e.g., Fig. 2), some more complex (e.g., Fig. 5) involving multiple steps that are required to explain complex features of the transient and steady state responses.

The delayed response to stretch and shortening (stretch activation and shortening deactivation) is present in all striated muscle, but stretch activation is particularly pronounced in insect flight muscle. It is the enhanced nature of the stretch response that allows for the generation of positive work and power in muscles that do not relax in their normal operating mode (i.e., under conditions in which calcium is not removed between contractions).

Studies using the powerful genetic tools available with *Drosophila* have revealed insights into modifications of the myosin cross-bridge cycle for very fast muscle types and beckon further insight into how myosin has evolved for its wide variety of muscle functional roles. Alterations in chemomechanics of the cycle can be mapped to specific locations of the myosin S-1 region using myosin chimeras derived from the naturally occurring alternative splicing of one gene that generates all *Drosophila* muscle myosin isoforms.

A variety of accessory proteins contribute to structurally reinforcing the thick and thin filaments, thereby facilitating power transmission from the molecular motors to the ends of the muscle. Many of these proteins have been mutated and their structural and functional roles further defined. Mutating specific regions of myosin, actin, and other myofibrillar proteins in the flight muscles will continue to provide valuable information about the respective roles of each, not only in the basic steps of the chemomechanical mechanism of the motor protein, but throughout the entire power train.

Acknowledgements

NIH R01 HL68034 and R01 AR49425 supported the preparation of this paper. We thank Bill Barnes, Mark Miller, Brad Palmer, and Gary Nelson for assistance, and Masataka Kawai for his helpful comments.

References

1. Jewell BR, Ruegg JC. Oscillatory contraction of insect fibrillar muscle after glycerol extraction. Proc Roy Soc 1966; 164:428-459.
2. Machin KE, Pringle JWS. The physiology of insect fibrillar muscle II. Mechanical properties of a beetle flight muscle. Proc R Soc London Ser B 1959; 151:204-225.
3. Machin KE, Pringle JWS. The physiology of insect fibrillar muscle. III. The effect of sinusoidal changes of length on a beetle flight muscle. Proc R Soc 1960; (B)152:311-330.
4. Squire JM. The structural basis of muscular contraction. New York: Plenum, 1981.
5. Tregear RT. Physiology of insect flight muscle. Handbook Physiol 1983; 10:487-506.
6. Hill AV. The mechanics of active muscle. Proc R Soc Lond B Biol Sci 1953; 141(902):104-117.
7. Linari M, Reedy MK, Reedy MC et al. Ca-activation and stretch-activation in insect flight muscle. Biophys J 2004; 87:1101-1111.
8. Peckham M, Molloy JE, Sparrow JC et al. Physiological properties of the dorsal longitudinal flight muscle and tergal depressor of the trochanter muscle of Drosophila melanogaster. J Mus Res Cell Motil 1990; 11:203-215.
9. Machin KE. Feedback theory and its application to biological systems. Symp Soc Exp Biol 1964; 18:421-445.
10. White DCS. The elasticity of relaxed insect fibrillar flight muscle. J Physiol 1983; 343:31-57.
11. Granzier HLM, Wang K. Interplay between passive tension and strong and weak binding cross-bridges in insect indirect flight muscle: A functional dissection by gelsolin-mediated thin filament removal. J Gen Physiol 1993; 101:235-270.
12. Henkin JA, Maughan DW, Vigoreaux JO. Mutations that affect flightin expression in Drosophila alter the viscoelastic properties of flight muscle fibers. Am J Physiol Cell Physiol 2004; 286(1):C65-C72.
13. Maughan D, Moore J, Vigoreaux J et al. Work production and work absorption in muscle strips from vertebrate cardiac and insect flight muscle fibers. In: Sugi H, Pollack G, eds. Mechanism of Work Production and Work Absorption in Muscle. New York: Plenum Publishing Press, 1998:471-480.

14. Kawai M, Brandt PW. Sinusoidal analysis: A high resolution method for correlating biochemical reactions with physiological processes in activated skeletal muscle of rabbit, frog and crayfish. J Mus Res Cell Motil 1980; 1:279-303.
15. Jewell BR, Wilkie DR. The mechanical properties of relaxing muscle. J Physiol 1960; 152:30-47.
16. Edman KAP. Mechanical deactivation induced by active shortening in isolated muscle fibres of the frog. J Physiol 1975; 246:255-275.
17. Huxley AF, Simmons RM. Proposed mechanism of force generation in striated muscle. Nature London 1971; 233:533-538.
18. Ford LE, Huxley AF, Simmons RM. Proceedings: Mechanism of early tension recovery after a quick release in tetanized muscle fibres. J Physiol 1974; 240(2):42P-43P.
19. Piazzesi G, Linari M, Lombardi V. Kinetics of regeneration of cross-bridge power stroke in shortening muscle. Adv Exp Med Biol 1993; 332:691-700.
20. Abbott RH, Steiger GJ. Temperature and amplitude dependence of tension transients in glycerinated skeletal and insect fibrillar muscle. J Physiol 1977; 266:13-42.
21. Guth K, Kuhn HJ, Tsuchiya T et al. Length dependent state of activation - length change dependent kenetics of cross bridges in skinned insect flight muscle. Biophys Struct Mech 1981; 7:139-169.
22. Guth K, Poole KJV, Maughan D et al. The apparent rates of crossbridge attachment and detachment estimated from ATPase activity in insect flight muscle. Biophys J 1987; 52:1039-1045.
23. Thorson J, White DCS. Distributed representations for actin-myosin interaction in the oscillatory contraction of muscle. Biophys J 1969; 9:360-390.
24. Peterson JN, Nassar R, Anderson PA et al. Altered cross-bridge characteristics following haemodynamic overload in rabbit hearts expressing V3 myosin. J Physiol 2001; 536(Pt 2):569-582.
25. Huxley AF. Muscle structure and theories of contraction. Progress in Biophysics 1957; 7:255-318.
26. Podolsky RJ. Kinetics of muscular contraction: The approach to the steady state. Nature 1960; 188:666-668.
27. Littlefield KP, Swank DM, Sanchez BM et al. The converter domain modulates kinetic properties of Drosophila myosin. Am J Physiol Cell Physiol 2003; 284(4):C1031-C1038.
28. Chan WP, Dickinson MH. In vivo length oscillations of indirect flight muscles in the fruit fly Drosophila virilis. J Exp Biol 1996; 199(12):2767-2774.
29. Steiger GJ, Ruegg JC. Energetics and "efficiency" in the isolated contractile machinery of an insect fibrillar muscle at various frequencies of oscillation. Pflugers Arch 1969; 307(1):1-21.
30. White DCS, Thorson J. Phosphate starvation and the nonlinear dynamics of insect fibrillar flight muscle. J Gen Physiol 1972; 60:307-336.
31. Zhao Y, Kawai M. The effect of lattice spacing change on cross-bridge kinetics in chemically skinned rabbit psoas muscle fibers. II. Elementary steps affected by the spacing change. Biophys J 1993; 64:197-210.
32. Zhao Y, Kawai M. Kinetic and thermodynamic studies of the cross-bridge cycle in rabbit psoas muscle fibers. Biophys J 1994; 67:1655-1668.
33. Wang G, Kawai M. Effects of MgATP and MgADP on the cross-bridge kinetics of rabbit soleus slow-twitch muscle fibers. Biophys J 1996; 71:1450-1461.
34. Wang G, Kawai M. Force generation and phosphate release steps in skinned rabbit soleus slow-twitch muscle fibers. Biophys J 1997; 73(2):878-894.
35. Mulieri LA, Barnes W, Leavitt BJ et al. Alterations of myocardial dynamic stiffness implicating abnormal crossbridge function in human mitral regurgitation heart failure. Circ Res 2002; 90(1):66-72.
36. Dickinson MH, Hyatt CJ, Lehmann F-O et al. Phosphorylation-dependent power output of transgenic flies: An integrated study. Biophys J 1997; 73:3122-3134.
37. Blanchard E, Seidman C, Seidman J et al. Ca^{2+} sensitivity of isometric tension, oscillatory power, and complex stiffness parameters of cardiac muscle from an α-myosin heavy chain R403Q mouse model of human familial hypertrophic cardiomyopathy. Circ Res 1999; 84(4):475-483.
38. Heinl P, Kuhn HJ, Ruegg JC. Tension responses to quick length changes of glycerinated skeletal muscle fibres from the frog and tortoise. J Physiology 1974; 237(2):243-258.
39. Taylor EW. Mechanism of actomyosin ATPase and the problem of muscle contraction. CRC Crit Rev Biochem 1979; 6(2):103-164.
40. Cooke R. Actomyosin interaction in striated muscle. Physiol Rev 1997; 77(3):671-697.
41. Dantzig JA, Goldman YE, Millar NC et al. Reversal of the cross-bridge force-generating transition by photogeneration of phosphate in rabbit psoas muscle fibres. J Physiol 1992; 451:247-278.
42. Millar NC, Homsher E. Kinetics of force generation and phosphate release in skinned rabbit soleus muscle fibers. Am J Physiol 1992; 262(5 Pt 1):C1239-C1245.

43. Geeves MA, Lehrer SS. Dynamics of the muscle thin filament regulatory switch: The size of the cooperative unit. Biophys J 1994; 67:273-282.
44. Holmes KC, Angert I, Kull FJ et al. Electron cryo-microscopy shows how strong binding of myosin to actin releases nucleotide. Nature 2003; 425(6956):423-427.
45. Rayment I, Rypniewski WR, Schmidt-Base K et al. Three-dimensional structure of myosin subfragment-1: A molecular motor. Science 1993; 261:50-58.
46. Tyska MJ, Warshaw DM. The myosin power stroke. Cell Motil Cytoskeleton 2002; 51(1):1-15.
47. Agianian B, Krzic U, Qiu F et al. A troponin switch that regulates muscle contraction by stretch instead of calcium. EMBO J 2004; 23(4):772-779.
48. Tregear RT, Reedy MC, Goldman YE et al. Cross-bridge number, position, and angle in target zones of cryofixed isometrically active insect flight muscle. Biophys J 2004; 86(5):3009-3019.
49. Gordon AM, Homsher E, Regnier M. Regulation of contraction in striated muscle. Physiol Rev 2000; 80(2):853-924.
50. Palmer BM. Intermittant attachment and Hookian stiffness of myosin-actin interactions predict phases 1 and 2 of a step response and the C-process of sinusoidal analysis. Biophys J 2003; 84(2):248a.
51. Marcussen BL, Kawai M. Role of MgATP and inorganic phosphate ions in cross-bridge kinetics in insect (Lethocerus colossicus) flight muscle. Prog Clin Biol Res 1990; 327:805-813.
52. Yamakawa M, Goldman YE. Mechanical transients initiated by photolysis of caged ATP within fibers of insect fibrillar flight muscle. J Gen Physiol 1991; 98(4):657-679.
53. Goldman YE. Kinetics of the actomyosin ATPase in muscle fibers. Annu Rev Physiol 1987; 49:637-654.
54. Lund J, Webb MR, White DCS. Changes in the ATPase activity of insect fibrillar flight muscle during sinusoidal length oscillation probed by phosphate-water oxygen exchange. J Biol Chem 1987; 263:5505-5511.
55. Abbott RH, Mannherz GH. Activation by ADP and the correlation between tension and ATPase activity in insect fibrillar muscle. Pflugers Arch 1970; 321(3):223-232.
56. George EL, Ober MB, Emerson CP. Functional domains of the Drosophila melanogaster muscle myosin heavy-chain gene are encoded by alternatively spliced exons. Mol Cell Biol 1989; 9:2957-2974.
57. Rozek CE, Davidson N. Drosophila has one myosin heavy-chain gene with three developmentally regulated transcripts. Cell 1983; 32(1):23-34.
58. Hastings KE, Emerson Jr CP. cDNA clone analysis of six coregulated mRNAs encoding skeletal muscle contractile proteins. Proc Natl Acad Sci USA 1982; 79(5):1553-1557.
59. Zhang S, Bernstein SI. Spatially and temporally regulated expression of myosin heavy chain alternative exons during Drosophila embryogenesis. Mech Dev 2001; 101(1-2):35-45.
60. Wells L, Edwards KA, Bernstein SI. Myosin heavy chain isoforms regulate muscle function but not myofibril assembly. EMBO J 1996; 15(17):4454-4459.
61. Swank DM, Knowles AF, Suggs JA et al. The myosin converter domain modulates muscle performance. Nat Cell Biol 2002; 4(4):312-316.
62. Swank DM, Bartoo ML, Knowles AF et al. Alternative exon-encoded regions of Drosophila myosin heavy chain modulate ATPase rates and actin sliding velocity. J Biol Chem 2001; 276(18):15117-15124.
63. Holmes KC, Schroder RR. Switch 1 opens on strong binding to actin. Molecular and cellular aspects of muscle contraction. Adv Exp Med Biol 2003; 538:159-166.
64. Swank DM, Kronert WA, Maughan DW et al. Alternative versions of the myosin relay loop influence Drosophila muscle kinetics. Biophys J 2004; 86(1):656a.
65. Swank DM, Knowles AF, Kronert WA et al. Variable N-terminal regions of muscle myosin heavy chain modulate ATPase rate and actin sliding velocity. J Biol Chem 2003; 278(19):17475-17482.
66. Swank DM, Maughan DW. Rates of force generation in Drosophila fast and slow muscle types have opposite responses to phosphate. In: Sugi H, ed. Molecular and Cellular Aspects of Muscle Contraction. New York: Kluwer Academic /Plenum Publishers, 2003:459-468.
67. Miller BM, Nyitrai M, Bernstein SI et al. Kinetic analysis of Drosophila muscle myosin isoforms suggests a novel mode of mechanochemical coupling. J Biol Chem 2003; 278(50):50293-50300.
68. Rome LC, Cook C, Syme DA et al. Trading force for speed: Why superfast crossbridge kinetics leads to superlow forces. Proc Natl Acad Sci USA 1999; 96(10):5826-5831.
69. Rome LC, Lindstedt SL. The quest for speed: Muscles built for high-frequency contractions. News Physiol Sci 1998; 13:261-268.

70. Tohtong R, Yamashita M, Graham M et al. Impairment of flight ability and flight muscle function caused by mutations of phosphorylation sites of myosin regulatory light chain in Drosophila. Nature 1995; 374:650-653.
71. Irving TC, Maughan DW. In vivo X-ray diffraction of indirect flight muscle from Drosophila melanogaster. Biophys J 2000; 78(5):2511-2515.
72. Bhandari DG, Levine BA, Trayer IP et al. ^{1}H-NMR study of mobility and conformational constraints within the proline-rich N-terminal of the LC1 alkali ligth chain of skeletal myosin. Correlation with similar segments in other protein systems. Eur J Biochem 1986; 160(2):349-356.
73. Sweeney HL. Function of the N Terminus of the myosin essential light chain of vertebrate striated muscle. Biophys J 1995; 68:112s-119s.
74. Fewell JG, Hewett TE, Sanbe A et al. Functional significance of cardiac myosin essential light chain isoform switching in transgenic mice. J Clin Invest 1998; 101:2630-2639.
75. Irving T, Bhattacharya S, Tesic I et al. Changes in myofibrillar structure and function produced by N-terminal deletion of the regulatory light chain in Drosophila. J Muscle Res Cell Motil 2001; 22(8):675-683.
76. Moore JR, Dickinson MH, Vigoreaux JO et al. The effect of removing the N-terminal extension of the Drosophila myosin regulatory light chain upon flight ability and the contractile dynamics of indirect flight muscle. Biophys J 2000; 78(3):1431-1440.
77. Prince HP, Trayer HR, Henry GD et al. Proton nuclear-magnetic-resonance spectroscopy of myosin subfragment 1 isoenzymes. Eur J Biochem 1981; 121(1):213-219.
78. Trayer IP, Trayer HR, Levine BA. Evidence that the N-terminal region of A1-light chain of myosin interacts directly with the C-terminal region of actin: A proton magnetic resonance study. Eur J Biochem 1987; 164:259-266.
79. Sutoh K. Identification of myosin binding sites on the actin sequence. Am Chem Soc 1982; 21(15):3654-3661.
80. Andreev OA, Saraswat LD, Lowey S et al. Interaction of the N-terminus of chicken skeletal essential light chain 1 with F-actin. Biochemistry 1999; 38(8):2480-2485.
81. Hao Y, Miller MS, Swank DM et al. Mutation of paramyosin phosphorylation sites affects the passive stiffness of Drosophila indirect flight muscle. Biophys J 2004; 86(1):184a.
82. Reedy MC, Bullard B, Vigoreaux JO. Flightin is essential for thick filament assembly and sarcomere stability in Drosophila flight muscles. J Cell Biol 2000; 151(7):1483-1500.
83. Saide JD, Chin-Bow S, Hogan-Sheldon J et al. Characterization of components of Z-bands in the fibrillar flight muscle of Drosophila melanogaster. J Cell Biol 1989; 109(5):2157-2167.
84. Kulke M, Neagoe C, Kolmerer B et al. Kettin, a major source of myofibrillar stiffness in Drosophila indirect flight muscle. J Cell Biol 2001; 154(5):1045-1057.
85. Bullard B, Linke WA, Leonard K. Varieties of elastic protein in invertebrate muscles. J Muscle Res Cell Motil 2002; 23(5-6):435-447.
86. Granzier HLM, Wang K. Passive tension and stiffness of vertebrate skeletal and insect flight muscles: The contribution of weak cross-bridges and elastic filaments. Biophys J 1993; 65:2141-2159.
87. Cammarato A, Hatch V, Saide J et al. Drosophila muscle regulation characterized by electron microscopy and three-dimensional reconstruction of thin filament mutants. Biophys J 2004; 86(3):1618-1624.
88. Kreuz AJ, Simcox A, Maughan D. Alterations in flight muscle ultrastructure and function in Drosophila tropomyosin mutants. J Cell Biol 1996; 135(3):673-687.
99. Bernstein SI, Milligan RA. Fine tuning a molecular motor: The location of alternative domains in the Drosophila myosin head. J Mol Biol 1997; 271(1):1-6.
90. Clayton JD, Cripps RM, Sparrow JC et al. Interaction of troponin-H and glutathione S-transferase-2 in the indirect flight muscles of Drosophila melanogaster. J Muscle Res Cell Motil 1998; 19(2):117-127.

CHAPTER 21

Mapping Myofibrillar Protein Interactions by Mutational Proteomics

Joshua A. Henkin and Jim O. Vigoreaux

Abstract

The myofibril is a multiprotein complex that performs the contractile activity of the muscle cell. Biochemical experiments over the past decades have revealed protein interactions that are critical for regulated contractile activity and for maintaining myofibril integrity and structural stability. Most of this information has emerged from in vitro binding studies of only a handful of proteins, primarily from vertebrate systems. Here we present a large scale and unbiased approach, termed mutational proteomics, for identifying interacting proteins and mapping their binding sites. The approach entails a broad interrogation of mutant myofibrillar proteomes to identify proteins whose expressions are altered as a result of a particular mutation. We discuss the many advantages that are available from the study of *Drosophila* flight muscle to serve as a platform for a comprehensive survey of protein interactions, and the eventual goal of providing a complete map of all the protein-protein interactions in the myofibril.

Introduction

Movement is a fundamental property of living systems. Molecular motors, proteins that convert the chemical energy from ATP hydrolysis into mechanical energy, are largely responsible for most forms of movements, from the molecular to the organismal. The primary tissue responsible for animal locomotion is striated muscle that features prominent contractile organelles known as myofibrils. These intricate and highly organized networks of interconnected protein filaments are Nature's solution to the quest for fast, efficient, and immediate locomotory responses to varied external challenges. Within its confine, the myofibril possesses an ensemble of myosin II motors whose concerted nanometer advancements are manifested as macroscopic movements. The myofibril also contains all the protein elements required to regulate and propagate myosin's force and to respond to changes in cellular physiology. A detailed view of the functional and architectural arrangement of myofibrillar proteins will enhance our knowledge of the contractile mechanism of Nature's most sophisticated motility machine.

Our understanding of muscle contraction has benefited greatly from biochemical, biophysical, and structural biology approaches that have provided detailed information on the interactions among a handful of myofibrillar proteins.[1] Most studies, however, have focused on binary interactions among isolated proteins and no large scale approach has been undertaken to generate a higher order map of protein interactions. Much remains to be elucidated in terms of the complex network of interconnected myofibrillar proteins. As new sarcomeric and sarcomeric-associated proteins continue to be identified,[2] the need increases for a

Nature's Versatile Engine: Insect Flight Muscle Inside and Out, edited by Jim Vigoreaux.

wide-ranging and concerted effort in delineating how myofibrillar proteins are interconnected in the lattice. Information about protein interactions will provide a better mechanistic understanding of how the myofibril works, which in turn will help alleviate the gap that sometimes exists between molecular and physiological studies.[1] The ubiquitous nature of cellular protein complexes, and the realization that many important cellular functions are carried out by protein machines consisting of multiple components, provides additional impetus for the development of techniques to study the composition and organization of molecular machines.[3]

The advent of genomic science and mass spectrometry applications for protein analysis has given way to the discipline of proteomics, the study of the genome's complement of proteins or 'proteome'. While the goal of proteomics is first and foremost the identification of all the proteins encoded by the genome, a second noteworthy goal is the elucidation of protein function on a global scale. This entails the characterization of post-translational modifications, identification of protein-protein interactions, and elucidation of protein pathways. Given the importance of protein interactions in many cellular processes, techniques aimed at large scale identification of protein partners are of significant interest (Table 1).

The myofibril is the most stripped down version of the muscle machinery that still retains the organized array of filament characteristics of the intact muscle. As such, the myofibril possesses the proper spatial arrangement and native contacts among all protein components involved in regulated force generation and propagation. Many consider the myofibril the last frontier in muscle research, the study of which provides the bridge between physiological studies at the molecular and the organismal level. Therefore, much effort is being dedicated into the development of techniques to study the myofibril's structural and biomechanical properties.[4,5] Delineating the protein interactions that exist in the myofibril would be a huge step towards understanding the architectural design and engineering properties of this ensemble of proteins. A full catalogue of the myofibrillar proteome and of its protein interactions is of utmost importance.

A Glossary of Protein Interactions

The term protein-protein interaction generally implies the physical association between two proteins. This entails subcellular colocalization and atomic contact, long-lived or transient, that is amenable to biochemical and biophysical techniques. Protein interaction is also often used to imply functional association even in the absence of physical contact. Examples include two proteins acting on a single biochemical pathway, proteins on distinct pathways that functionally compensate, and proteins that form part of the same physical complex.

Protein interactions can be inferred from genetic experiments. Two genes are said to functionally interact if the double heterozygote progeny resulting from the mating of two mutants shows a phenotype distinct from either parent. A second strategy for uncovering functionally interacting proteins is to conduct genetic screens for second site noncomplementation. The mutants recovered from these screens represent modifiers (suppressors and enhancers) that may or may not physically associate with the product of the mutant gene.[6]

The study of protein interactions has been most prolific in the areas of cell signaling and cellular metabolism. Unlike the stably assembled myofibril, these networks are highly dynamic in time and space and the functional interactions are often mediated by post-translational modifications. Modular protein interaction domains that regulate the dynamic properties of signaling cascades have been identified and shown to be highly conserved across widely divergent species.[7] Several databases of protein interactions are available (e.g., DIP,[8] BIND[9]) as well as data mining tools (e.g., PreBIND: www.bind.ca) to help find information about protein interactions in the scientific literature. The Human Protein Reference Database (www.hprd.org) lists over 13,000 protein interactions despite its somewhat limited catalogue of proteins (~3300 proteins as of 11/26/03). Protein-protein interaction data is sometimes referred to as the interactome.

Table 1. Large scale approaches for identifying protein interactions

Method	Advantage	Disadvantage	Refs.
Protein microarrays	Sensitivity; chips can be screened for a variety of activities	Maintaining protein solubility, conformation and functional state	21,22
Surface plasmon resonance	Real time information on binding kinetics and affinity of proteins in native state	Attempts to increase capacity leads to increased nonspecific binding	58,59
Phage display	Partially defines binding site; millions of peptides can be screened; DNA clone available	Limited to peptides and small proteins	60
Yeast two-hybrid	Sensitivity (detects transient interactions); adaptability to meet specific needs; DNA clone available	Limited to binary interactions in nucleus; high rate of false positives (spurious transcription) and false negatives (incorrectly folded proteins that fail to bind)	13,14, 61
Tandem affinity purification	Isolates multiprotein complexes of different sizes; enriches low abundance proteins	Tagging may affect complex formation; Loosely bound proteins may be lost; high incidence of false positives and false negatives	16,61
Bimolecular fluorescence complementation	Allows detection in living cells	Difficult to implement for high throughput	62
Fluorescence resonance energy transfer	Allows detection in living cells; spatial and temporal resolution can monitor transient interactions	False negatives are common	63
Cellular co-localization	Use of epitope tags or GFP fusion allows detection in living cells	Epitope tagging (e.g., C-terminal GFP fusion) may cause mis-localization	23
Similar expression profile	Broad coverage under various conditions; Genes expressed coordinately are good candidates for interaction at the protein level	Predictive, indirect method	64
Gene co-expression	Uses evolution to filter out spurious gene expression links	Only maps those genes that have orthologs in other species	65
Computational methods	Exploit available genome information; describe structure of large scale networks	Predictions must be validated by direct experimentation; mostly limited to prokaryotes	25-27
Genetic screens with mutant arrays	Unbiased	Limited by mutant availability	33

Large Scale Approaches to Protein-Protein Interactions: An Overview

Several methods have been implemented for wide scale identification of protein-protein interactions (Table 1).[10] The methods can be categorized as experimental or computational (i.e., predictive in nature), or novel combinations of both.[11,12] While diverse, there are several underlying similarities among the experimental methods. Namely, the majority of these methods have been implemented in yeast and they represent scaled up versions of previously existing technologies adapted for high-throughput, automated processing. The two most common underlying technologies are the yeast two-hybrid assay and mass spectrometry of isolated protein complexes.

Many permutations of the original yeast two-hybrid have been developed to address specific needs or to overcome potential shortcomings of the parent technique.[13,14] Similarly, there are many methods for isolating protein complexes prior to mass spectrometry analysis. These include affinity chromatography,[15] tagging of specific proteins to act as baits for capturing the complex (e.g., tandem affinity purification),[16] and differential extraction/fractionation to obtain a subcellular structure or organelle.[17] Unlike yeast two-hybrid which identifies binary interactions, mass spectrometry of protein complexes allows identification of the proteins in the complex but does not yield information of their interactions or architectural organization.[17] A variation of the affinity purification method uses isotope-coded affinity tag (ICAT) reagents to obtain quantitative information about the protein components of the complex.[18]

Each method offers particular advantages and likewise, is handicapped by certain limitations (Table 1). Though high-throughput, no method as of yet can offer full coverage of the proteome, at least in the case of yeast.[19] Meta-analysis of high-throughput studies revealed that total coverage is poor, the rate of false positives remains high, and that agreement among studies is low. Some estimate that only between 5%-10% of protein interactions are identified using the yeast-two hybrid method.[20] The lack of reproducibility may indicate that only a limited subset of protein interactions is covered by any one method.[19] In fact, mostly 50% of all interactions detected in yeast two-hybrid genome wide screens are false and it is estimated that more than half of all current high-throughput data are spurious.[14,19] A footnote to this bleak assessment is that these technologies are still in their relative infancy and undergoing constant refinements and improvements. Additionally, higher accuracy can be obtained when the data sets from various methods are combined, however, this occurs at the expense of broader coverage.[19]

Recent developments in protein microarrays are promising but have yet to have a broad impact. Protein microarrays use technology similar to that of its predecessor, DNA microarrays, where known protein targets are printed on a glass slide (or synthesized on-chip), a sample of a labeled ligand is spread on the chip, and binding detected through fluorescence emission.[21] Other variations of protein microarrays include antibody chips used for quantifying a particular analyte in a biological sample.[22] Like protein microarrays, cell biological approaches, which identify interacting proteins based on their cellular colocalization, hold great promise but are still in their infancy.[23]

Computational methods that exploit genomic features are now available to predict protein function and protein interactions.[24] Genes that in some species are fused into a single transcription unit, and genes that are coinherited or evolved in a correlated fashion, are used as the basis of computational methods for predicting functional relationships among proteins.[25-27] In due course, all protein interactions predicted from computational methods must be validated by detailed biochemical/biophysical experimentation. New computational procedures that integrate data from experimental and predictive datasets are being used to generate more complete and accurate interactomes.[12]

Seldom do any of the methods outlined above offer information about the binding interface between two proteins. Methods that model interaction of proteins based on the known three-dimensional structure of the complex of homologous pairs have proven useful in evaluating interactions proposed from high-throughput methods such as yeast two-hybrids.[28,29] Detailed structural and thermodynamic analyses are nevertheless required to unequivocally map protein-protein recognition sites and to identify amino acids that participate in binding.[30]

Genetic experiments have been a constant source of information on protein interactions. From the classic experiments of Wood, Epstein, King, and others that allowed the elucidation of the stepwise assembly of bacteriophages,[31] to the more recent large scale synthetic lethal screens in yeast,[32] genetics has helped uncover significant protein interactions in a variety of biological processes. The bacteriophage studies elegantly illustrated how protein interactions in macromolecular assemblies are brought to bear by the symphony of genetic, biochemical, and ultrastructural (e.g., electron microscopy) approaches.

Mutational analysis of T4 bacteriophage assembly demonstrated one of the first applications of genetics for the dissection of protein interactions in biological structures. Analysis of distinct mutant phages gave evidence that the assembly of proteins is constrained to a specific spatial and temporal order dictated by the interactions among individual phage proteins. Thus, mutants were isolated that resulted in partial assembly of a phage substructure (head, tail, or tail fiber), while other mutants gave rise to a more or less completed assembly. Many of the mutations isolated disrupted the association between interacting proteins in such manner that if one protein was missing, none of the subsequent associations occurred and the process was aborted at the step where the missing protein was normally assembled.[31] Thus genetics can be used to dissect the sequential series of protein interaction steps involved in assembly of a complex molecular structure.

In *Drosophila*, the availability of the 'deficiency kit', a collection of mutant strains with overlapping large deletions that cover most of the euchromatic region of the chromosomes, permits a genome wide screen for second site noncomplementing mutations. Such an approach helps in the discovery of potentially novel interacting protein partners.[33] One clear benefit of a genetic approach is that, unlike a biochemical approach, it is not biased for high abundance proteins and it does not require any prior knowledge of a protein's suspected function.[6,19] Genetic strategies help confirm and expand the repertoire of protein interactions uncovered by traditional biochemical methods and modern high-throughput approaches.[6]

Mutational Proteomics in *Drosophila*: A Primer

Mutational proteomics is an attempt to combine the best attributes of proteomics (wide coverage, global survey of protein expression) and genetics (unbiased interrogation of protein function) to probe the myriad of protein interactions that constitute the myofibrillar lattice. There is currently a large collection of flight muscle mutants in *Drosophila melanogaster*, many of which have been characterized in terms of their molecular defect and developmental, ultrastructural, and physiological properties.[6,34] Mutational proteomics seeks to take advantage of new generation proteomic approaches to survey this collection of flight muscle mutants and gather information about their myofibrillar protein expression profiles. Hidden in these mutant proteomes is a treasure trove of information on the wide range effects of specific genetic perturbations, information that may reveal new and unsuspected functional and architectural relationships among myofibrillar proteins.

The roots of mutational proteomics can be traced back to the work of Hotta, Deak, and others beginning in the 1970's.[35,36] These investigators first reported that many single gene mutations affect the expression of multiple proteins as evidenced by one and two-dimensional gel electrophoresis (Fig. 1). One assumption implicit in those results was that the proteins whose expression was affected interacted, physically or functionally, with the product of the mutant gene. This assumption has never been formally tested but is supported by empirical evidence from several studies.[6] In their two-dimensional gel analysis of the myosin heavy chain (MHC) null mutant *Mhc*^{7} (formerly *Ifm(2)2*), Chun and Falkenthal showed that more than 10 protein spots, in addition to the myosin light chain subunits, failed to accumulate in the indirect flight muscles (IFM). All of the missing proteins were presumed to be thick filament

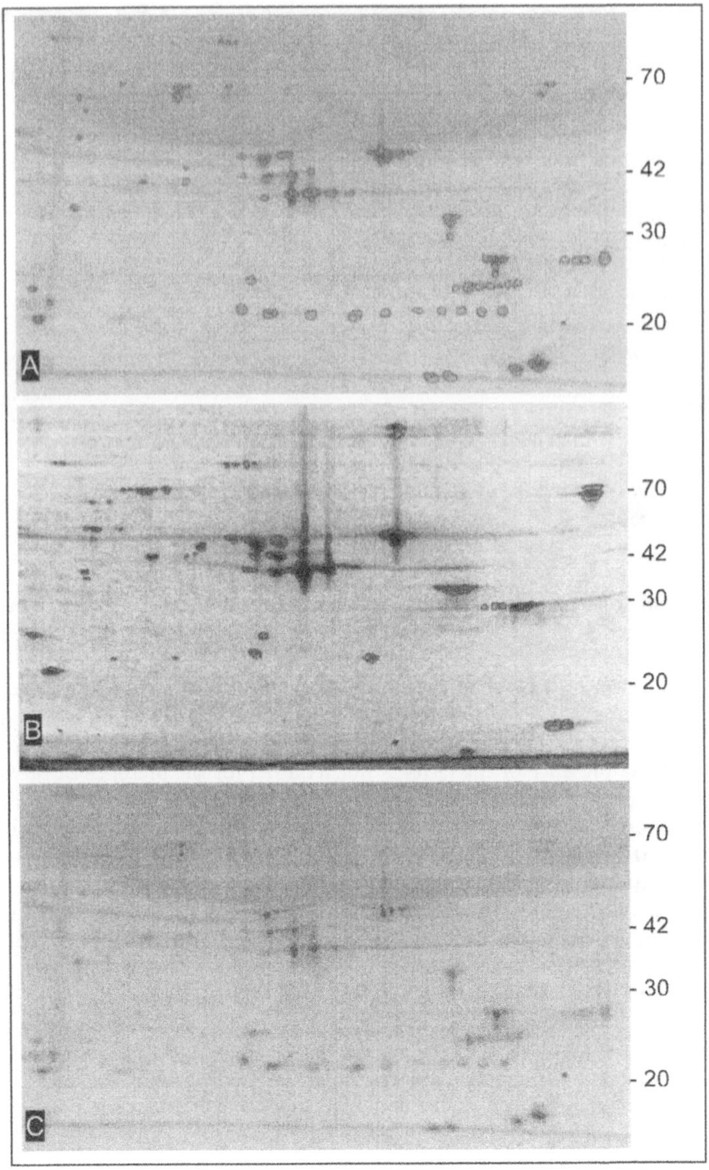

Figure 1. Two-dimensional gel analysis of IFM skinned fibers from wild-type and *Mhc*[7] flies. A) Red open circles indicate spots detected in the wild-type proteome; B) blue open circles indicate spots detected in the *Mhc*[7] proteome. C) The image on the bottom superimposes the two gels, A and B. Green open circles represent spots in the wild-type proteome that are also found in the *Mhc*[7] proteome, while orange open circles indicate spots detected in the wild-type proteome that are not found in the *Mhc*[7] proteome. More than 30 spots found in the wild-type gel are missing in the *Mhc*[7] gel. All gels are 12% SDS-PAGE, IEF pH 4-7. Basic end is to the left. Approximate molecular weights are indicated on the right side of each gel. A color version of this figure is available online at http://www.Eurekah.com.

Table 2. Mutational proteomics

Advantages	
	Unbiased, broad coverage in *vivo* technique
	No experimental manipulations
	Exploits available resources
	May reveal binding residues
	Provides qualitative & quantitative information
	Insight into role of post-translational modifications in binding
	Cooperative binding, more than one protein
	Broad applicability (stages, states)
	Results evaluated within context of functional data
Disadvantages	
	Untested
	Requires full proteome coverage
	Requires multiple mutants for each protein
	Unlinked effects may be difficult to filter out
	Requires confirmation by follow-up biochemical studies

components since electron microscopy revealed that thin filaments and Z bands appeared marginally affected in this mutant.[37] To date, only flightin, among the missing proteins has been identified and shown to be a myosin binding protein.[38,39]

Mogami and Hotta described two general classes of protein relationships inherent in their analysis of mutant IFM protein expression.[35] The first revealed that the presence of certain proteins (e.g., actin) is necessary for the presence of other proteins (e.g., tropomyosin), but that the presence of the latter is not required for the presence of the former. The second revealed pairs of proteins whose expression is interdependent (e.g., flightin isovariants). Subsequent characterization of a handful of these mutants showed that several of the relationships implied by the mutant proteome analysis are consistent with known interactions of myofibrillar proteins.[6]

The analysis of mutant proteomes also has the potential of revealing interactions among proteins other than the one encoded by the gene mutated in the strain being analyzed. For example, a strict interrelationship between one isoform of TnH (spot 34)[40] and an unknown protein (spot 184, possibly TnC) has been uncovered through the analysis of 15 actin *Act88F* mutant strains.[6] Examination of multiple mutant alleles for many different proteins should permit the assembly of an interaction matrix that can serve as a template for the design of targeted, hypothesis-driven biochemical experiments to probe physical contacts.

Table 2 summarizes some of the key advantages and disadvantages of mutational proteomics. The analysis of mutant proteomes is complementary to the increasingly popular "organellar proteomics" where the protein composition of isolated organelles (or differentially extracted cellular fractions) is obtained after several purification steps.[41] Because no preparation is expected to be fully devoid of contaminants, one disadvantage of these methods is that cross-contaminating proteins may be difficult to identify as such. Proteins that had not been previously identified (i.e., gene products) likewise face the omen of verification as bona fide components of the organelle. Mutational proteomics offers an independent means for validating the information from organellar proteomics by assessing the expression of a questionable protein in several mutants of known organellar components. One example in which this approach would be immediately useful is in the characterization of a putative 'myofibrillar' ADP/ATP translocase (ANT/*sesB*) in *Drosophila.* This integral membrane protein is known to be abundant in the inner mitochondrial membrane but a proteome analysis of purified IFM myofibrils also revealed the presence of this protein.[42] Given the preponderance of mitochondria in IFM, it would not be unexpected if proteins from this organelle contaminate the myofibrillar preparation. In conjunction with immunolocalization studies that establish the intracellular distribution of ANT, one could establish if the expression of ANT is consistently affected among certain mutations of known myofibrillar components, evidence that can be construed as a possible physical or functional interaction between ANT and the myofibril (see chapter by Vishnudas and Vigoreaux for further discussion of ANT).

Recently, CuraGen Corporation (New Haven, Conn) completed the world's first comprehensive protein interaction map for a multicellular organism. Using the yeast two-hybrid method to screen more than 10,500 predicted transcripts in *Drosophila*, a high confidence map of 4780 interactions among 4679 proteins was generated representing a network with two levels of organization: local connections (interactions among proteins in small complexes) and global connections (higher order communication between complexes).[43] None of these interactions corresponded to known interactions among myofibrillar proteins except for an interaction between TnC and TnI. For example, the screen uncovered five proteins that interact with MHC (two known proteins, sns and Or49a, and three potential gene products) and eleven proteins that interact with the myosin regulatory light chain (RLC; 10 potential gene products and CKIIb), but not the well documented interaction between these two subunits of myosin. Similarly, only one protein (tsr) and one potential gene product (CG6873) were found to interact with IFM actin encoded by the *act88F* gene, while many of actin's well-characterized binding proteins (e.g., itself, TnI) were not revealed. These results demonstrate the inadequacy of the yeast two-hybrid system for studying myofibrillar protein interactions and the need for developing other large scale approaches that are appropriate for this organelle. Other investigators have successfully used yeast two hybrid to identify binding interactions among vertebrate myofibrillar associated proteins.[44]

A successful, large scale mutational proteomics endeavor rests upon three pillars: (i) the availability of multiple mutants for each myofibrillar protein component and feasibility of generating new ones, (ii) the means to identify myofibrillar proteins and to ascertain changes in the proteome comprehensively and quantitatively, and (iii) computational methods to assemble an interaction database from proteomic results. *Drosophila* IFM is conveniently suited for mutational proteomics as mutations affecting IFM exclusively are easily obtained and maintained.[6,34] Over 100 mutations that affect flight muscle function have been reported in the literature and many more can be generated given the accessibility to DNA clones for most, if not all, *Drosophila* genes, and the means for generating transgenic lines that express mutant genes exclusively in the IFM. Of particular interest is the availability of distinct classes of mutant strains: from null mutations that disrupt myofibril assembly and severely impact muscle function, to single amino acid changes that minimally affect muscle properties.[6] Furthermore, the genome wide *Drosophila* P element disruption project, charged with generating at least one mutant for every gene, is another source of potential IFM mutants.[45] The ability to systematically interrogate the *Drosophila* genome for mutations that affect flight muscle function affords a tremendous opportunity to implement the mutational proteomics approach.

The second pillar involves a complete listing of all the proteins that make up the myofibril, i.e., the myofibrillar proteome, and the changes that take place in mutant IFM. Major steps have been taken towards the accomplishment of the first goal including the completion of the sequencing of the *Drosophila* genome, the development of techniques for the subfractionation of the muscle cell proteome, and techniques to purify myofibrils to homogeneity.[40,46,47] One method entails the separation of proteins based on their differential solubility properties. Distinct cytosolic, organelle, and cytomatrix fractions are obtained within microliter volumes from a small number of fibers (Fig. 2). The microdrop fractionation method offers several advantages over the myofibril purification method for the application of mutational proteomics. First, the fractionation entails a minimal number of manipulations and does not require chromatography, centrifugation or other steps that may result in potential loss of protein. Second,

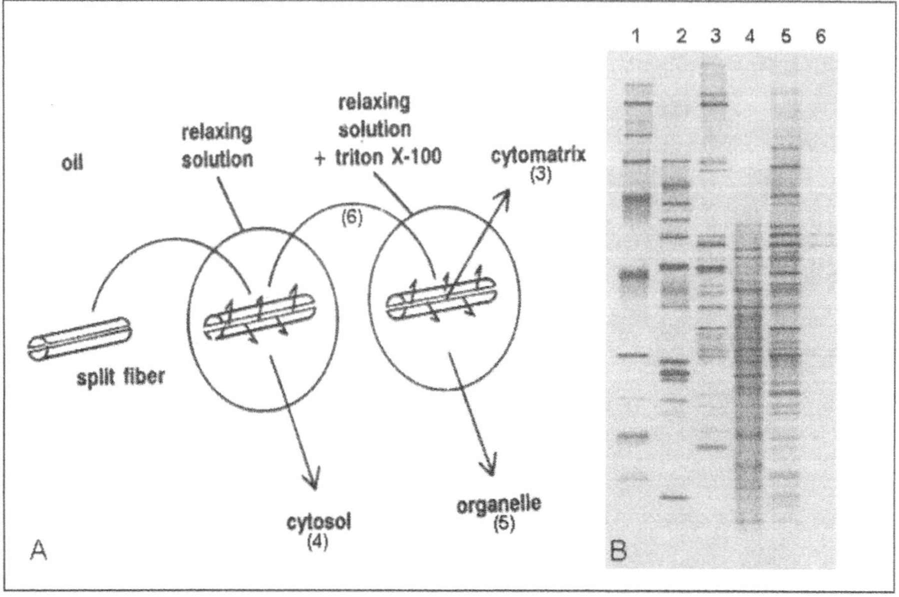

Figure 2. SDS-PAGE of IFM subcellular fractions. A) Cartoon representation of the microdrop fractionation method.[6] Numbers in parenthesis represent lanes in the corresponding gel. B) Silver stained SDS-PAGE showing the distinct composition of the three subcellular fractions. Lanes 1 and 2, protein standards; lane 3, cytomatrix; lane 4, cytosol; lane 5, organelle; lane 6, wash. Note that each fraction has a distinct protein profile. Modified from: Maughan DW, Godt RE. Parvalbumin concentration and diffusion coefficient in frog myoplasm. J Muscle Res Cell Motil 1999; 20(2):199-209.

the method is suited for small scale separation of proteins from one or a few fibers. Third, because the fractionation preserves the cytosolic and organelle fractions, it allows examination of these subproteomes to reveal possible changes in protein subcellular distribution.

Current efforts are underway to decipher the complete proteome of *Drosophila* (http://www.ebi.ac.uk/proteome). While the study of the myofibrillar proteome stands to gain from this endeavor, a focused approach on an enriched myofibrillar preparation is likely to reveal details that may not be evident on a whole fly proteome. Two such efforts have been undertaken, both with important but limited success. In one study, two-dimensional gels were used to identify proteins that are enriched in IFM (over other thoracic tissues) and to identify proteins that are present in myofibril preparations.[40] Based on these results, twenty proteins were judged to be myofibrillar components.

Ashman and colleagues used a different approach to prepare myofibrils and a combination of matrix assisted laser desorption ionization time-of-flight mass spectrometry (MALDI-TOF MS) and nanospray MS/MS to identify proteins separated by one-dimensional SDS-PAGE.[42] Of a total of 23 bands, they were able to identify 16 proteins, ten of which correspond to proteins also identified by the two-dimensional PAGE approach. Missing from both of these studies are several proteins that have been identified by conventional approaches. Clearly the gel approach (whether one-dimensional or two-dimensional) has its limitations, namely the exclusion of high molecular weight proteins, and other complementary approaches are needed to identify the full proteome. More complete coverage of the proteome can be achieved with powerful new approaches (shotgun proteomics) that rely on orthogonal liquid chromatography separation of complex peptide mixtures that are then directly analyzed by tandem mass spectrometry.[48,49]

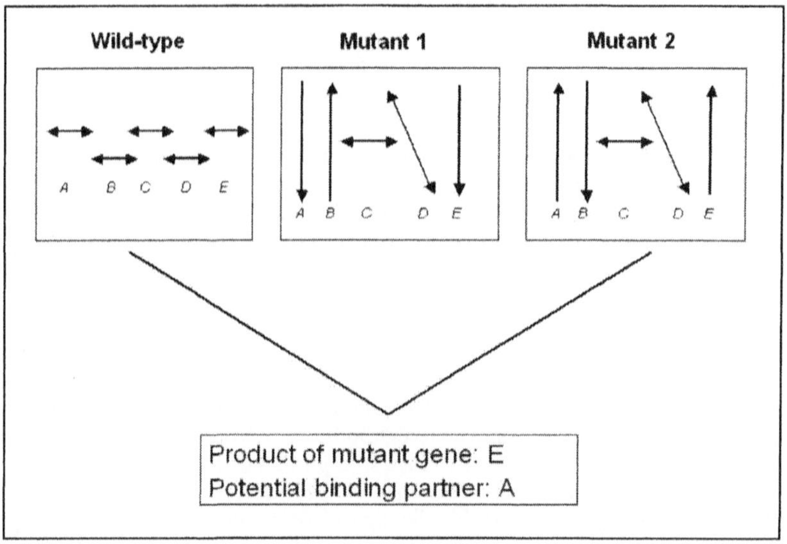

Figure 3. Annotation of myofibrillar protein interactions from the analysis of mutant proteomes. Each letter represents one protein spot on a 2DE. Two headed arrows are spots whose accumulation does not change in the mutants with respect to the wild-type control. Arrows pointing down are spots whose accumulation is decreased, while arrows pointing up are those whose accumulation increases in the mutant. Diagonal arrows represent qualitative changes (e.g., phosphorylation). In the example shown, spot *E* is the product of the mutant gene and spot *A* is a candidate binding protein because its expression parallels that of spot *E* in the two mutants.

The third pillar entails the implementation of informatics tools to collect, analyze, manage, and model the experimental results. Proteomes from mutant IFM are compared to the wild-type proteome to determine if the expression of each individual protein is increased, decreased, or not changed as a consequence of the mutations (Fig. 3). In the case of 2DE, several companies offer gel image analysis software to identify protein spots that are differentially expressed among gels.[50] From this data we can learn the proteome wide effects of each individual mutation. The application of computational tools (e.g., association analysis) to datasets of mutant proteomes permits the identification of protein changes that recur in concert. This information will form the basis for the formulation of models that will predict meaningful associations among the proteins of the myofibril. Conclusions derived from these analyses should be taken as hypotheses to be validated by other experimental means.

The following example of our ongoing work, while limited in scope, illustrates the path of a mutational proteomics approach.

Flightin-Myosin Interaction: A Case Study of Mutational Proteomics

Flightin is a novel, 20 kD *Drosophila* protein that is expressed exclusively in the IFM. Immunolocalization and biochemical fractionation studies showed that flightin is a component of the myofibril.[51] The localization of flightin to the sarcomeric A band was the first indication that this protein may be associated with the thick filament. Early studies of the MHC null mutant *Ifm(2)2* by two-dimensional gel electrophoresis had shown that protein spots 158 and 159, later identified as flightin,[51] were missing in this mutant.[35,37] In contrast, flightin was found associated with the thick filament-enriched cytomatrix of the *Act88F* null mutant *KM88*.[38,52] The association of flightin with thick filaments was corroborated by its presence in the myosin fraction after high salt extraction of skinned IFM fibers.[52] Together these results provided strong evidence that flightin is a thick filament component, but did not reveal the nature of the flightin-thick filament interaction or the identity of the flightin binding partner(s).

The first clue that flightin is a myosin rod binding protein came from the analysis of the myofibrillar proteome of *Mhc*^{13} by one and two-dimensional gel electrophoresis.[53] This MHC mutant strain carries a single amino acid substitution in the light meromyosin (LMM) region of the myosin rod. Aside from the appearance of a new peptide corresponding to a proteolytic product of MHC, flightin was the only protein whose expression appeared to be affected by this mutation. A second myosin rod single amino acid mutant, *Mhc*^{6}, mapping only five residues downstream from *Mhc*^{13}, also interfered with the expression of flightin.[53] Both of these mutations map to an MHC constitutive exon that is expressed in every *Drosophila* muscle. Nevertheless, the mutations were shown to have a pronounced effect on the structure and function of the IFM but little to no effect on other muscles.[53] In addition, the severity of the IFM defect worsened in direct proportion to the reduction in flightin abundance: *Mhc*^{13} in which little or no flightin is present, has a more severe muscle defect than *Mhc*^{6}, in which some unphosphorylated flightin is present. That flightin is associated with the rod region of myosin is also consistent with the observation that flightin copurifies with thick filaments from *Mhc*^{Y57}, a transgenic strain that expresses a headless myosin.[52] The discovery of *fln*^{0}, a null mutation in flightin whose phenotype is remarkably like that of *Mhc*^{13} despite having a normal MHC, suggest that the phenotypic manifestations of *Mhc* rod alleles result from their effect on flightin abundance rather than to some intrinsic effect on the myosin molecule.[52,54] These results lead to the hypothesis that *Mhc*^{6} and *Mhc*^{13} define residues involved in flightin binding to the myosin rod.

Binding of flightin to the myosin rod in vitro was demonstrated using two different solid state binding assays.[39] These studies further showed that flightin binding to myosin is negated by the point mutation in *Mhc*^{13} (Glu 1554 to Lys), suggesting that Glu 1554 is part of the flightin binding site in myosin. These in vitro studies provide evidence that the absence of flightin in *Mhc*^{13} IFM results from its inability to bind the myosin rod. In summary, myofibrillar proteome analysis of myosin mutants have led to the identification of flightin as a thick filament component and as a myosin rod binding protein, and have potentially identified one myosin residue (Glu 1554) that is critical for flightin binding.

Concluding Remarks

Mutational proteomics is a valuable addition to the arsenal of experimental approaches aimed at elucidating the molecular basis of muscle function. A fully comprehensive understanding of myofibrillar protein interactions will no doubt necessitate the confluence of methodologies that address different aspects of protein function, as no method alone can possibly provide unambiguous and complete data. Mutational proteomics adds to the mix of well-established methodologies by providing information at the interface of genetics and proteomics. Mutational proteomics complements and validates genetic and biochemical studies and can contribute towards the interrelation and integration of datasets derived from these two approaches, thereby improving the confidence of interactions detected by those methods. Such an approach would be highly relevant to muscle research where biochemical approaches have tended to dominate vertebrate studies and genetic approaches have been the norm in model organisms, such as flies and worms. By providing concurrently a global view of the proteome and a snapshot of potential protein interaction domains, mutational proteomics bridges the macromolecular to the atomic and fills an important niche in the interplay of data-driven and hypothesis-driven approaches. As conceptualized, mutational proteomics fills an unmet need for in vivo approaches that permit identification of biologically significant protein interactions and contribute towards the elucidation of global protein function.

The broad view afforded by mutational proteomics allows one to look beyond the primary site of the genetic defect to more completely ascertain the molecular events that may give rise to the mutant phenotype. This approach is particularly useful for the study of human myopathies that arise from mutations in contractile protein genes. Knowledge of the genetic defect has not translated explicitly into an understanding of the pathogenesis or clinical manifestations of the disease.[55] Information such as that derived from the analysis of the mutant proteome, together with functional studies of isolated myofibrils, would contribute towards a mechanistic understanding of the disease, particularly in those cases where the relationship between gene defect and clinical profile is weak.

Many challenges notwithstanding, the application of graph-theoretic approaches to mutational proteomic databases can provide the impetus for generating a top-down view of the myofibril. At first sight, the myofibril exhibits features of a scale-free network[56] with major proteins such as actin and myosin heavy chain acting as large hubs and the majority of accessory proteins having few connections. The structure of the network is likely to reflect the evolutionary history of contractile machines, with the more recently evolved proteins showing fewer connections than the more ancestral proteins. A major strength of *Drosophila* is that the network of protein interactions uncovered by mutational proteomics is evaluated in a biologically relevant context by exploiting the variety of experimental approaches that address function from the molecular to the organismal level. A multidisciplinary approach employing genetic, cell biological, biochemical, physiological, and cellular and locomotory mechanics, together with bioinformatics approaches should provide an integrated, system-wide view of muscle function by which whole-organism physiological performance can be explained in molecular terms.[6,57]

Acknowledgements

We would like to thank Eileen Brown for technical support and David Maughan for helpful discussions and for providing Figure 2. This work was supported by NSF grants 00910768 and 0315865 to JOV.

References

1. Gordon AM, Homsher E, Regnier M. Regulation of contraction in striated muscle. Physiol Rev 2000; 80(2):853-924.
2. Clark KA, McElhinny AS, Beckerle MC et al. Striated muscle cytoarchitecture: An intricate web of form and function. Annu Rev Cell Dev Biol 2002; 18:637-706.
3. Alberts B. The cell as a collection of molecular machines. Cell 1998; 92(3), entire issue.
4. Nyland LR, Maughan DW. Morphology and transverse stiffness of Drosophila myofibrils measured by atomic force microscopy. Biophys J 2000; 78(3):1490-1497.
5. Kulke M, Neagoe C, Kolmerer B et al. Kettin, a major source of myofibrillar stiffness in Drosophila indirect flight muscle. J Cell Biol 2001; 154(5):1045-1057.
6. Vigoreaux JO. Genetics of the Drosophila flight muscle myofibril: A window into the biology of complex systems. Bioessays 2001; 23:1047-1063.
7. Pawson T, Nash P. Assembly of cell regulatory systems through protein interaction domains. Science 2003; 300(5618):445-452.
8. Xenarios I, Salwinski L, Duan XJ et al. DIP, the Database of Interacting Proteins: A research tool for studying cellular networks of protein interactions. Nucleic Acids Res 2002; 30(1):303-305.
9. Bader GD, Betel D, Hogue CW. BIND: The biomolecular interaction network database. Nucleic Acids Res 2003; 31(1):248-250.
10. Zhu H, Bilgin M, Snyder M. Proteomics. Annu Rev Biochem 2003; 72:783-812.
11. Tong AH, Drees B, Nardelli G et al. A combined experimental and computational strategy to define protein interaction networks for peptide recognition modules. Science 2002; 295(5553):321-324.
12. Jansen R, Yu H, Greenbaum D et al. A bayesian networks approach for predicting protein-protein interactions from genomic data. Science 2003; 302(5644):449-453.
13. Uetz P, Giot L, Cagney G et al. A comprehensive analysis of protein-protein interactions in Saccharomyces cerevisiae [see comments]. Nature 2000; 403(6770):623-627.
14. Phizicky E, Bastiaens PI, Zhu H et al. Protein analysis on a proteomic scale. Nature 2003; 422(6928):208-215.
15. Blagoev B, Kratchmarova I, Ong SE et al. A proteomics strategy to elucidate functional protein-protein interactions applied to EGF signaling. Nat Biotechnol 2003; 21(3):315-318.

16. Gavin AC, Bosche M, Krause R et al. Functional organization of the yeast proteome by systematic analysis of protein complexes. Nature 2002; 415(6868):141-147.
17. Andersen JS, Lyon CE, Fox AH et al. Directed proteomic analysis of the human nucleolus. Curr Biol 2002; 12(1):1-11.
18. Ranish JA, Yi EC, Leslie DM et al. The study of macromolecular complexes by quantitative proteomics. Nat Genet 2003; 33(3):349-355.
19. von Mering C, Krause R, Snel B et al. Comparative assessment of large-scale data sets of protein-protein interactions. Nature 2002; 417(6887):399-403.
20. DeFrancesco L. Probing protein interactions. The Scientist 2002; 16:28.
21. Zhu H, Bilgin M, Bangham R et al. Global analysis of protein activities using proteome chips. Science 2001; 293(5537):2101-2105.
22. MacBeath G. Protein microarrays and proteomics. Nat Genet 2002; 32(Suppl):526-532.
23. Huh WK, Falvo JV, Gerke LC et al. Global analysis of protein localization in budding yeast. Nature 2003; 425(6959):686-691.
24. Schachter V. Protein-interaction networks: From experiments to analysis. Drug Discov Today 2002; 7(11 Suppl):S48-54.
25. Marcotte EM, Pellegrini M, Thompson MJ et al. A combined algorithm for genome-wide prediction of protein function. Nature 1999; 402(6757):83-86.
26. Enright AJ, Iliopoulos I, Kyrpides NC et al. Protein interaction maps for complete genomes based on gene fusion events. Nature 1999; 402(6757):86-90.
27. Date SV, Marcotte EM. Discovery of uncharacterized cellular systems by genome-wide analysis of functional linkages. Nat Biotechnol 2003; 21(9):1055-1062.
28. Aloy P, Russell RB. The third dimension for protein interactions and complexes. Trends Biochem Sci 2002; 27(12):633-638.
29. Aloy P, Russell RB. Interrogating protein interaction networks through structural biology. Proc Natl Acad Sci USA 2002; 99(9):5896-5901.
30. Lockless SW, Ranganathan R. Evolutionarily conserved pathways of energetic connectivity in protein families. Science 1999; 286(5438):295-299.
31. Wood WB. Bacteriophage T4 morphogenesis as a model for assembly of subcellular structure. Q Rev Biol 1980; 55(4):353-367.
32. Tong AH, Evangelista M, Parsons AB et al. Systematic genetic analysis with ordered arrays of yeast deletion mutants. Science 2001; 294(5550):2364-2368.
33. Halsell SR, Kiehart DP. Second-site noncomplementation identifies genomic regions required for Drosophila nonmuscle myosin function during morphogenesis. Genetics 1998; 148(4):1845-1863.
34. Bernstein SI, O'Donnell PT, Cripps RM. Molecular genetic analysis of muscle development, structure and function in Drosophila. In Rev Cytol 1993; 143:63-152.
35. Mogami K, Hotta Y. Isolation of Drosophila flightless mutants which affect myofibrillar proteins of indirect flight muscle. Mo Gen Genet 1981; 183:409-417.
36. Deak, II, Bellamy PR, Bienz M et al. Mutations affecting the indirect flight muscles of Drosophila melanogaster. J Embryol Exp Morphol 1982; 69:61-81.
37. Chun M, Falkenthal S. Ifm(2)2 is a myosin heavy chain allele that disrupts myofibrillar assembly only in the indirect flight muscle of Drosophila melanogaster. J Cell Biol 1988; 107:2613-2621.
38. Vigoreaux JO. Alterations in flightin phosphorylation in Drosophila flight muscles are associated with myofibrillar defects engendered by actin and myosin heavy chain mutant alleles. Biochem Genet 1994; 32:301-314.
39. Ayer G, Vigoreaux JO. Flightin is a myosin rod binding protein. Cell Biochem Biophys 2003; 38:41-54.
40. Mogami K, Fujita SC, Hotta Y. Identification of Drosophila indirect flight muscle myofibrillar proteins by means of two-dimensional electrophoresis. J Biochem (Tokyo) 1982; 91:643-650.
41. Taylor SW, Fahy E, Ghosh SS. Global organellar proteomics. Trends Biotechnol 2003; 21(2):82-88.
42. Ashman K, Houthaeve T, Clayton J et al. The application of robotics and mass spectrometry to the characterisation of the Drosophila melanogaster indirect flight muscle proteome. Letters Peptide Science 1997; 4:57-65.
43. Giot L, Bader JS, Brouwer C et al. A protein interaction map of Drosophila melanogaster. Science 2003; 302(5651)1727-1736.
44. Lu S, Carroll SL, Herrera AH et al. New N-RAP-binding partners alpha-actinin, filamin and Krp1 detected by yeast two-hybrid screening: Implications for myofibril assembly. J Cell Sci 2003; 116(Pt 11):2169-2178.
45. Spradling AC, Stern D, Beaton A et al. The Berkeley Drosophila genome project gene disruption project: Single P-element insertions mutating 25% of vital Drosophila genes. Genetics 1999; 153(1):135-177.

46. Saide JD, Chin-Bow S, Hogan-Sheldon J et al. Characterization of components of Z-bands in the fibrillar flight muscle of Drosophila melanogaster. J Cell Biol 1989; 109:2157-2167.
47. Warmke J, Yamakawa M, Molloy J et al. Myosin light chain-2 mutation affects flight, wing beat frequency and indirect flight muscle contraction kinetics in Drosophila. J Cell Biol 1992; 119:1523-1539.
48. Link AJ, Eng J, Schieltz DM et al. Direct analysis of protein complexes using mass spectrometry. Nat Biotechnol 1999; 17(7):676-682.
49. McDonald WH, Yates III JR. Shotgun proteomics: Integrating technologies to answer biological questions. Curr Opin Mol Ther 2003; 5(3):302-309.
50. Chakravarti DN, Chakravarti B, Moutsatsos I. Informatic tools for proteome profiling. Biotechniques 2002; (Suppl)4-10:12-15.
51. Vigoreaux JO, Saide JD, Valgeirsdottir K et al. Flightin, a novel myofibrillar protein of Drosophila stretch-activated muscles. J Cell Biol 1993; 121(3):587-598.
52. Reedy MC, Bullard B, Vigoreaux JO. Flightin is essential for thick filament assembly and sarcomere stability in Drosophila flight muscles. J Cell Biol 2000; 151:1483-1499.
53. Kronert WA, O'Donnell PT, Fieck A et al. Defects in the Drosophila myosin rod permit sarcomere assembly but cause flight muscle degeneration. J Mol Biol 1995; 249:111-125.
54. Henkin JA, Maughan DW, Vigoreaux JO. Mutations that affect flightin expression in Drosophila alter the viscoelastic properties of flight muscle fibers. Am J Physiol Cell Physiol 2004; 286:C65-C72.
55. Taylor MR, Carniel E, Mestroni L. Familial hypertrophic cardiomyopathy: Clinical features, molecular genetics and molecular genetic testing. Expert Rev Mol Diagn 2004; 4(1):99-113.
56. Bray D. Molecular networks: The top-down view. Science 2003; 301(5641):1864-1865.
57. Maughan DW, Vigoreaux JO. An integrated view of insect flight muscle: Genes, motor molecules, and motion. News Physiol Sci 1999; 14:87-92.
58. Merchant M, Weinberger SR. Recent advancements in surface-enhanced laser desorption/ionization-time of flight-mass spectrometry. Electrophoresis 2000; 21:1164-1177.
59. Nedelkov D, Nelson RW. Surface plasmon resonance mass spectrometry: Recent progress and outlooks. Trends Biotechnol 2003; 21(7):301-305.
60. Mullaney BP, Pallavicini MG. Protein-protein interactions in hematology and phage display. Exp Hematol 2001; 29(10):1136-1146.
61. Ho Y, Gruhler A, Heilbut A et al. Systematic identification of protein complexes in Saccharomyces cerevisiae by mass spectrometry. Nature 2002; 415(6868):180-183.
62. Hu CD, Kerppola TK. Simultaneous visualization of multiple protein interactions in living cells using multicolor fluorescence complementation analysis. Nat Biotechnol 2003; 21(5):539-545.
63. Sorkin A, McClure M, Huang F et al. Interaction of EGF receptor and grb2 in living cells visualized by fluorescence resonance energy transfer (FRET) microscopy. Curr Biol 2000; 10(21):1395-1398.
64. Ge H, Liu Z, Church GM et al. Correlation between transcriptome and interactome mapping data from Saccharomyces cerevisiae. Nat Genet 2001; 29(4):482-486.
65. Stuart JM, Segal E, Koller D et al. A gene-coexpression network for global discovery of conserved genetic modules. Science 2003; 302(5643):249-255.

Index

F

G

H

I

K

L

M

N

O

P

R

S

T

U

X

Y

Z

GPSR Compliance

The European Union's (EU) General Product Safety Regulation (GPSR) is a set of rules that requires consumer products to be safe and our obligations to ensure this.

If you have any concerns about our products, you can contact us on ProductSafety@springernature.com

In case Publisher is established outside the EU, the EU authorized representative is:

Springer Nature Customer Service Center GmbH
Europaplatz 3
69115 Heidelberg, Germany

Batch number: 10755466

Printed by Printforce, the Netherlands